INTRODUCTION TO THE
PHYSICS and CHEMISTRY of MATERIALS

INTRODUCTION TO THE
PHYSICS and CHEMISTRY of MATERIALS

Robert J. Naumann

CRC Press
Taylor & Francis Group
Boca Raton   London   New York

CRC Press is an imprint of the
Taylor & Francis Group, an **informa** business

CRC Press
Taylor & Francis Group
6000 Broken Sound Parkway NW, Suite 300
Boca Raton, FL 33487-2742

© 2009 by Taylor & Francis Group, LLC
CRC Press is an imprint of Taylor & Francis Group, an Informa business

No claim to original U.S. Government works
Printed in the United States of America on acid-free paper
10 9 8 7 6 5 4 3 2 1

International Standard Book Number-13: 978-1-4200-6133-8 (Hardcover)

Library of Congress Cataloging-in-Publication Data

Naumann, Robert J.
    Introduction to the physics and chemistry of materials / Robert J. Naumann.
        p. cm.
    "A CRC title."
    Includes bibliographical references and index.
    ISBN 978-1-4200-6133-8 (alk. paper)
    1. Materials science. I. Title.

TA403.N3325 2008
620.1'1--dc22   2008044088

Visit the Taylor & Francis Web site at
http://www.taylorandfrancis.com

and the CRC Press Web site at
http://www.crcpress.com

Contents

Contents

Contents

Preface

The purpose of this book is to prepare students from diverse engineering and science backgrounds for advanced study in materials science. Since the Huntsville campus did not offer an undergraduate program in materials science when the Tricampus Materials Science PhD program was instituted between the Huntsville, Birmingham, and Tuscaloosa campuses of the University of Alabama system, it was customary to require students with physics and chemistry degrees to audit an undergraduate engineering course in structures and properties of materials in order to acquaint them with the basic concepts of materials science and engineering. Similarly, engineering students, with little or no background in quantum mechanics found taking a graduate course in solid-state physics from texts such as Kittel or Ashcroft and Mermin somewhat daunting.

I felt the students would be better served by offering a course at the first year graduate level that would build on their undergraduate backgrounds in mathematics and whatever science and engineering courses they may have had, and give them the background they need for more advanced work in materials science at a level commensurate with their education. The result was a two-semester core course in our program: the first semester emphasized chemical bonding, crystal structure, mechanical properties, phase transformations, and a brief introduction to the processing of materials, while the second semester dealt with the thermal, electronic, photonic, optical, and magnetic properties of materials. The topics discussed in this book are arranged accordingly.

For students with little or no background in modern physics or quantum mechanics, a chapter has been added at the beginning to introduce quantum concepts and wave mechanics through a simple derivation of the Schroedinger equation, the electron-in-a-box problem, and the wavefunctions of the hydrogen atom. Clearly, the electron-in-a-box problem is essential to understand the Fermi energy in a metal or a semiconductor. Knowledge of hydrogen-like wave functions is needed to understand chemical bonding and the periodic table. I especially wanted students to appreciate the mathematical origin of the quantum numbers that specify the various quantum states so they do not seem to be simply pulled out of the air. I also tried to give an historical prospective to the development of this knowledge so the student will have at least heard the names of some of the giants on whose shoulders we stand.

In treating the chemical bond, hydrogen-like wavefunctions are also used to derive the sp^2 and sp^3 hybrid bonds essential to understanding the structure and properties of graphite, nanotubes, and semiconductors. Techniques for evaluating the Madelung sum are discussed and the Coulomb potential is used to compute the lattice energies of ionically bonded systems. At this level, the metallic bond has to be treated somewhat empirically, but it is shown that the cohesive energy per volume for simple metals is roughly proportional to the electron density and, like the ionic bond, is inversely proportional to approximately the fourth power of the atomic radius. The energy per volume of the transition metals is shown to peak when the d-band is about half filled, and the covalent bonds become saturated as would be expected.

The treatment of structure emphasizes the role of size and bond type in determining the structure of materials. Point and space groups are introduced as well as Pearson and Strukturbericht notations so the student would at least recognize these symbols in the literature. Drawings to help visualize stacking in diamond and wurtzite structures are included. Unusual structures found in electron-deficient metals are briefly mentioned along with those of selected polymers.

Concepts of the reciprocal lattice and Brillouin zones are introduced in the chapter on X-rays along with a more formal treatment of the scattering of X-rays or wave-like particles. This formal treatment for obtaining the Laue conditions for constructive interference leads to a deeper understanding of the scattering process and gives students the ability to calculate the structure factors for various materials. The applications of X-ray diffraction to material identification, structure determination, and crystal characterization are briefly discussed.

Elastic behavior is introduced with the stress and strain tensors and the elastic coefficient matrix and it is shown that for a material with cubic symmetry the coefficient matrix reduces to only three elastic coefficients. Relationships between the various elastic moduli are derived. The concept of a simple central force potential such as the Coulomb potential is taken to the limits of its validity and is used to obtain the bulk modulus and to estimate the thermal expansion of ionically bonded materials. The Coulomb potential is then modified to represent a unidirectional potential and used to estimate the elastic coefficients of these materials. The Morse potential is used to relate the elastic constants of metals to their theoretical strengths.

The role of defects in solids is introduced with a derivation of vacancy defects followed by a discussion of other point defects. Dislocations are introduced as an explanation of why materials fall short of their theoretical mechanical properties, setting the stage for various strengthening mechanisms introduced in later chapters that are designed to inhibit the motions of dislocations. The concept of surface energy is introduced by using the Madelung sum for a (100) plane of an NaCl structure with the Madelung sum for the solid to obtain the surface energy excess.

Methods for mechanical testing of materials are briefly introduced along with various strengthening mechanisms. The number and surface area of the slip systems in metals and in ceramics are shown to be responsible for the ductility (or the lack of it) and for ductile-to-brittle transitions. Griffith's theory of brittle fracture is used to introduce fracture mechanics and to develop the concept of fracture toughness. The viscoelastic behavior of polymers is briefly discussed.

The concept of free energy is used to describe single and multiple component phase transitions. The classical theory of nucleation is presented along with various undercooling methods and their role in the development of bulk glassy metals. The role of excess heat of mixing is demonstrated by calculating simple binary phase diagrams for solid solution systems, intermetallic systems, and eutectic systems from their free energies. Peritectic and monotectic reactions along with their solid-state counterparts are also described in terms of their free energies. Actual phase diagrams of similar systems are presented as examples.

The problem of macrosegregation during alloy solidification from the melt is presented along with methods for dealing with it. The constitutional supercooling criterion for maintaining a planar solidification interface is developed along with a description of the microsegregation that results when the interface breaks down into dendritic growth.

A chapter introducing the Bose–Einstein, Maxwell–Boltzman, Planck, and Fermi–Dirac distribution functions follows before discussing the thermal, electronic, magnetic, and optical properties for the benefit of students who have not been exposed to quantum statistical mechanics. This chapter is a logical beginning for the second half of this book since these concepts are essential to an understanding of these properties. Similarly, the Maxwell equations are used to derive the equations for absorption and normal reflection of electromagnetic waves in the chapter on optical properties. The band structure of metals and semiconductors is examined and the relation between absorption spectra and inter-band transitions is shown.

Recent developments in organic electronics and in polymer light-emitting diodes are discussed along with the use of superlattices to construct quantum confinement devices such as resonance tunneling diodes and quantum well cascade lasers. Recent developments in memory storage devices such as flash memories and magnetic storage based on giant magnetoresistance phenomena are also taken into account.

Examples are provided to evaluate how well-simplified theories, such as assuming only nearest neighbor interactions in lattice vibrations, predict the actual performance of various materials. Problems are assigned wherever possible to help students apply what they have learned.

Although the text is organized as a two semester introductory course in materials science, it could equally serve as a one semester advanced undergraduate course in structures and properties of materials by choosing the first 14 chapters. It could also serve as a one semester advanced undergraduate course or first-year graduate course in solid-state physics or chemistry by omitting chapters 9 through 14.

I would like to express my appreciation to my colleagues, Professor Michael Banish, Professor James Baird, and Professor William Kaukler at the University of Alabama in Huntsville and to Dr. Martin Volz, Dr. Frank Szofran, and Dr. Donald Gillies at the NASA Marshall Space Flight Center for their suggestions and help in preparing this book. I also thank Chantell Marsh for proofreading the manuscript. Most of all, I appreciate the patience and understanding of my wonderful wife, Sally, for putting up with me during the time it took to complete this book.

Author

Robert J. "Bob" Naumann received his BS, MS, and PhD in physics and mathematics from the University of Alabama. After a long career as a scientist and a division chief in the Space Sciences Laboratory at the NASA Marshall Space Flight Center, he joined the University of Alabama in Huntsville as a professor of materials science. In addition to his teaching duties, he also served as an interim director of the Center for Materials Research and as the director of the Tricampus Materials Science PhD program between the University of Alabama campuses at Tuscaloosa, Birmingham, and Huntsville.

1

Introduction to Materials Science

1.1 What Is Materials Science?

Materials science is a multidisciplinary science that has evolved from the combination of a number of specialized fields of the metallurgist, the ceramist, the solid-state physicist, the electronic engineer, the polymer chemist, and the biotechnologist in which the common threads that weave these diverse fields into a generalized field of study have been recognized. Modern science has transformed what used to be largely an empirical art into a multidisciplinary exact science involving physics, chemistry, various fields of engineering, and, to some extent, even biology that allows us to begin to understand the behavior of materials and to manipulate matter at the most fundamental or atomic level. This capability allows materials scientists to understand a material's properties in terms of its structure, to design material structures to achieve certain properties, and to develop processes to achieve the desired structures.

The difference between engineering and science has become increasingly blurred as technology has become more complex. Clearly, today's engineers must have a broad knowledge of science to understand the source of the knowledge they apply and today's scientists must be aware of the practical applications of their scientific quest for knowledge if they wish to guide their research in the most fruitful directions. Although there are still "pure" scientists who seek knowledge for knowledge's sake and could care less if it has any application, in today's competition for research funding, it pays to direct one's interest to questions of practical nature. Materials science by its nature is an applied science. Many institutions combine the scientific and engineering aspects and offer degrees in materials science and engineering. If one has to make a distinction between the two fields, it could be said that materials science is more interested in the basic understanding of the relationship between the structure and properties of materials in order to design new materials with specific properties and materials engineers are more interested in how to produce the materials economically, assure that the materials produced have the desired properties, and that the materials selected for specific applications do their intended jobs.

1.2 Role of Materials in History

Materials have played such a critical role in the evolution of technology throughout the history of man that historical epochs, such as the Stone Age, the Bronze Age, the Iron Age, and the Silicon Age have been named for the materials available to mankind at the time.

During the Neolithic Period (the New Stone Age) humans gradually undertook domestication of animals, cultivation of crops, production of pottery, and building of towns, such as Jericho, by around 7000 BC. Copper tools and carved ivory found in Palestine are believed to date between 5000–4000 BC. Bronze was used by the Canaanites as early as 3000 BC and walls of their towns were fortified with a plaster-like material.

The ancient Romans made extensive use of concrete. The oldest surviving concrete structure is the Temple of Vesta, built in Tivoli during the first century BC. Other examples are the brick-faced concrete walls of the Camp of the Praetorian Guard, built by Sejanus in AD 21–23, the octagonal domed fountain hall of Nero's Golden House (AD 64–68), and Hadrian's Pantheon of about AD 118–128, whose solid concrete dome is 43.2 m in diameter and 1.5 m thick and is supported by 6 m thick walls of brick-faced concrete.

Battles have been won and lost because of subtle differences in arms and armor. The Greeks defeated the Persians at the battle of Marathon in 490 BC because their bronze armor had been annealed to a soft condition, which maximized its work of fracture. As a result, the Persian arrows only dented the Greek armor and did not penetrate far enough to cause a fatal wound. Similarly the English won the battle of Crècy in 1346 and the battle of Agincourt in 1415 partly because they used flax for their longbow strings. Flax has a higher Young's modulus than the animal sinew used by the French, which gave their arrows greater kinetic energy. Also the expensive and prestigious iron armor of the French knights contained a high carbon content, making it brittle and easily penetrated by the hard-hitting arrows from the English longbows.

Although the French naval architecture and shipbuilding was usually superior to the British during the eighteenth century, their cast iron naval cannons were generally less reliable than those used by the British and the bursting of the French cannons was influential in the 1805 battle of Trafalgar.

Copper was the first metal used by man because it can be found naturally in its uncombined metallic state. Robert Peary, while exploring Greenland in 1894, discovered that the local Eskimos had been breaking off chunks of a copper meteorite they had found and were using the metal for tools and weapons. Although rare, other copper meteorites have been found throughout the world and had apparently provided ancient civilizations with this metal. Its use was known in eastern Anatolia (the Asian portion of Turkey, a crossroad of civilization in its day) as early as 6500 BC. Copper was being extracted from a quarry near Varna, Bulgaria around 4400 BC. The Egyptians were casting copper in molds as early as 4000 BC. Much of the early Roman supply of copper came from Cyprus (Cyprium is Latin for copper meaning metal from Cyprus) where, somewhere around 3500 BC, the Cypriots had learned to extract copper from copper pyrites ($CuSFe_2$) by reducing the ore in charcoal fires. Since the melting point of pure copper is fairly high (1083°C), they also discovered, probably by accident, that the addition of small amounts of tin would lower the melting temperature, make the material easier to cast, and improve the strength of the material. Thus the Chalcolithic (Copper-Stone) Age gave way to the Bronze Age.

Although the use of iron as a precious metal was known as far back as 3000 BC, its superiority over bronze was not recognized until around 1200 BC in Europe and the Middle East. Since charcoal or coke was heated with iron ore to reduce it to iron, the resulting cast iron had excess carbon, which caused it to be brittle. Even so, the widespread production of iron for tools and weapons changed the face of Europe as Asia for the next 2000 years.

Early ironworkers removed this excess carbon by heating and hammering. Japanese and Syrian sword makers repeatedly folded layers of iron over each other and continued to hammer to break up the carbon so it could combine with oxygen to form carbon monoxide, although they probably did not know what they were actually doing. The result was a fairly soft but very tough sword that would bend before it would break. Then to harden the edge, they packed the sword in charcoal and reheated it to diffuse the carbon back into the cutting region to harden it. The "heat and beat" procedure was effective, but very expensive and such handcrafted, high-quality steel implements were limited to the affluent.

Despite the long history of the use of metals in ancient civilizations, the art of metallurgy was passed on from country to country along trade routes. It was not until 1530 when Georgius Agricola wrote *de re Metallica* (published posthumously in 1556), which described the practice of mining and smelting of ores, that a comprehensive survey of the art of metallurgy existed.

In 1784, Henry Cort discovered that it was possible to stir small "pigs" or puddles of iron to mix the oxide forming on the surface so that it could combine with the excess carbon, a process called puddling. This process eliminated the expensive process of hammering the iron to remove the excess carbon, which made good iron more affordable and, with the invention of the blast furnace by Abraham Darby, became the forerunner of modern steel making.

The Bessemer converter, invented by Sir Henry Bessemer in 1856, removed the excess carbon from the iron by blowing air through the molten pig iron and made the economical production of steel in tonnage quantities possible. The availability of large quantities of relatively inexpensive iron and steel made our present industrial society possible.

In the beginning of the twentieth century, the structures and machines generally were massive enough that relatively little thought was given to analyzing the actual strength of the structures. Even so, catastrophic failures did occur. Perhaps the most noteworthy was the sinking of the *Titanic* in which it is now believed that the steel used for her plates may have undergone a ductile-to-brittle transition at the temperature she was operating in. Such a transition could cause a relatively small crack in her hull to propagate and cause a catastrophic failure.

With the advent of aircraft and rockets, a great deal more attention had to be paid to carefully choosing materials and designing structures to provide just the needed strength with a minimum of weight. The famous Japanese Zero fighter plane that proved so effective in the early days of World War I owed its agility to its very light weight which was achieved using a very high strength-to-weight precipitation-hardened aluminum alloy. This alloy loses it strength over time due to overaging or ripening of the precipitates so that they are no longer a strengthening mechanism. As a result, such aircraft (of the few that survived) are no longer considered flight-worthy. Of course long service life was not a major design consideration for a combat aircraft.

Even in more recent times, catastrophic materials failures still occur. Three DeHavilland Comets (the first jet-powered commercial airliner) disintegrated in mid-air because their designers had forgotten (or failed to learn) the lessons their colleague, A.A. Griffin, at the Royal Aircraft Establishment at Farnborough, had taught concerning fracture mechanics in 1920.

1.3 How Materials Are Classified

Materials can be classified into four broad categories: metals, ceramics, semiconductors, and polymers. Metals, because of their detached electrons, are good reflectors of light, good conductors of heat and electricity, and tend to be ductile. Ceramics generally are poor conductors of electricity because their valence electrons are tied up in chemical bonds, although the recent discovery of ceramic superconductors is a notable exception. Their strong chemical bonds make them refractory, hard, and brittle. Semiconductors are a special class of ceramics whose valence electrons can be readily promoted to conduction electrons which makes them the basis for modern electronic and photonic devices. Polymers are perhaps the most versatile class of materials because of the virtually infinite combinations with which polymer chemists can assemble the molecules that make up polymers. Like ceramics, their valence electrons tend to be tied up in forming chemical bonds and some of the best electrical insulators are polymers. However, the recent discovery of conductive polymers has opened a new field of molecular electronics with the promise of large flexible TV screens or display panels.

Glasses and composites are important forms of materials but are not considered as specific classes of materials. Composites are simply a mixture of two or more of the other types of materials combined in such a way as to reinforce the favorable qualities and reduce the unfavorable qualities of the components for a specific application. Most metals and ceramics form a polycrystalline structure when they solidify. However, it is possible to cause both metals and ceramics to solidify in an amorphous state without the order associated with the crystalline structure. Such an amorphous solid is called a glass.

1.4   Overview of the Classes of Materials and Their Properties

1.4.1   Metals

Metals are mostly comprised of the elements on the left-hand side of the periodic table. The metallic bond forms when the outer, or valence, electrons become detached from their parent atoms when these atoms are brought in close vicinity, leaving positively charged ion cores swimming in a sea of electrons. When a metal solidifies, these ion cores arrange themselves in regular arrays called unit cells that repeat to form a crystalline structure. The sea of electrons can respond to electrical fields as well as carry heat, which makes metals good thermal and electrical conductors. They also respond to rapidly oscillating electrical fields associated with light, which makes them shiny and good reflectors.

The metallic bond is not as strong as the other primary bonds and, since the atoms are positively charged and all the same size in an elemental metal, it is relatively easy for them to slide over one another in the presence of a dislocation (a missing line of atoms in the array) which makes pure metals soft and easily deformable. For this reason, it is necessary to mix different metals together to form alloys or to use other techniques to strengthen them for most applications. Finding ways to toughen metals (strengthening them while still maintaining some degree of ductility so that they will bend before they break) has been a major quest for metallurgists throughout the centuries.

1.4.2   Ceramics

Ceramics are generally compounds formed from two or more elements that are held together by ionic or covalent bonds, or a mixture of both. Ionic bonds generally form between elements on the opposite sides of the periodic table where there is a large difference in electronegativity. Electronegativity difference is the tendency of an atom, near the right-hand side of the periodic table, to grab one or more electrons from a metal atom with loose electrons to become a negatively charged anion. The resulting positively charged metal ion, or cation, is electrically attracted to the anion. Again, the resulting ionically bonded molecules form crystalline structure comprised of repeating unit cells whose structure depends on the ratio of the cation to anion diameter.

Since all of the electrons are tied up in forming chemical bonds, there are no free electrons available to carry electricity and heat. Consequently, ceramics are generally poor conductors of heat and are good electrical insulators (an important exception is the new class of high-temperature superconductors). They are generally transparent to light as single crystals or glasses but will be translucent or opaque if the structure is polycrystalline (many small crystalline grains with random orientation) because of scattering of light by the grain boundaries.

The ionic bond is very strong which explains why ceramics generally have high melting temperatures. Since the anions are generally larger than the cations and are oppositely charged, it is difficult for the ions to slide over one another, which explains why ceramics are hard and brittle. Trying to find ways to make ceramics less brittle, so that that their high strength at elevated temperatures can be utilized, is one of the goals of modern ceramists.

1.4.3 Semiconductors

Semiconductors are a special case of ceramic materials so designated because the energy required to free a valence electron and make it a conduction electron is on the order of the energy of a photon of visible light. This small energy gap between the valence and conduction bands allows some electrons to be thermally excited into the conduction band at room temperature, making this class of materials poor conductors or semiconductors. However, it is possible to dope intrinsic (pure) semiconductors with impurity atoms to form donor states in order to create a large number of conduction electrons to form *n*-type material, or to form acceptor states in order to create holes in the valence band to form *p*-type material. Junctions between *n*- and *p*-type materials can be formed which are the basis for diodes, transistors, and other devices essential to the modern electronics revolution. Since the bandgap is on the order of infrared or visible light, photoelectric detectors can be made from these materials. Conversely, light emitted when electrons fall from the conduction back to the valence band is the basis for light-emitting diodes (LEDs) and solid-state lasers.

Elemental semiconductors are the lower atomic number members of the group IV family, i.e., carbon, silicon, and germanium that form sp^3 hybrid covalent bonds. A covalent bond forms when two atoms share their electrons to fill an electron shell. Group IV elements have two *s*-electrons that sit at a lower energy level than their two *p*-electrons (*s*- and *p*-designate the subshells that the valence electrons occupy). In the heavier group IV elements, e.g., tin and lead, these outer two *p*-electrons detach themselves to form a metallic bond. But in the lighter elements, one of the *s*-electrons is promoted to the *p*-level where it can interact with other *s*- and *p*-electrons to form the very strong directional sp^3 bond. The extra energy required to promote the *s*-electron to *p*-level is regained in the overall energy of the sp^3 bond.

Covalent bonds are directional and in the case of the sp^3 hybrid bond, they form an open, diamond-like crystal structure. The diamond form of carbon is also formed from the sp^3 bond that is extremely strong, which is why diamond is so hard. (Diamond is not generally considered to be a semiconductor because its bandgap is much larger than visible light.) The sp^3 bond gets progressively weaker as the atoms in the group IV family get larger and almost disappears with tin, the next member above germanium. Compound semiconductors are formed by combining elements from group III (aluminum, gallium, indium) with those from group V (phosphorous, arsenic, antimony) to form gallium arsenide, indium phosphide, etc., or from group II (zinc, cadmium, mercury) with group VI (sulfur, selenium, tellurium) to form mercury telluride, cadmium selenide, etc. These compounds also form open structures with sp^3-directed covalent bonds, although these may be mixed with some ionic bonding.

Modern technology has made it possible to combine these materials into structures that can be controlled at the atomic scale to form exotic new photonic and electronic devices such as the new highly efficient white LEDs, highly efficient double heterojunction lasers, single electron transistors, quantum wires, quantum dots, etc.

1.4.4 Polymers

Polymers can be thought of as a bowl of spaghetti, the noodles representing long chains of repeating groups of molecules called "mers," hence the name "polymer." These mers in the chains are covalently bonded together so that the chains themselves are quite strong. However, generally the chains are only loosely bonded to each other through secondary van der Waals-type bonds, which allows some degree slipping past each other, especially at elevated temperatures, hence the plastic nature of polymers. The mechanical properties of polymers can be controlled by the length of the chains (average molecular weight) and by the side groups of the mer units that can make it more difficult for them to slide past one another. Polymers are generally amorphous, meaning they have no regular structure, although in some circumstances the chains can fold back and forth on themselves to form regions of crystallinity in the form of ordered repeating units in a three-dimensional array. Side chains can be attached to lower the interaction between chains by keeping the farther apart if a noncrystalline, low density, fairly soft material is required. For applications requiring the material to resist deformation, the chains can be cross-linked by opening bond along the length of the chains (this occurs naturally in some plastics upon exposure to ultraviolet light) or by adding sulfur atoms to form disulfide bridges between the chains (vulcanization of rubber). The wide variety of molecules available to the polymer chemist and nearly infinite possible ways in which they can be arranged has produced a remarkable array of produces from soft fabrics to bulletproof vests, from plastic wrap to high-strength polycarbonate windows, and from rubber bands to truck tires.

Since all of the valence electrons are involved in forming chemical bonds, for the most part, polymers are extremely good insulators. However, things can be arranged so that charge transfer can take place along the chains so that it is possible to make conductive polymers. Recently, materials scientists have been able to form molecular diodes and transistors, opening up an exciting new field of molecular electronics. In the future, you may unroll a large flat screen TV and roll it back up when you are no longer watching it.

1.4.5 Glasses

Glass is not a type of material in the sense of metals, ceramics, or polymers, but refers to the amorphous state of a material. Generally, metals and ceramics in the solid phase prefer to arrange themselves in a crystalline structure, meaning their atoms are arranged in regular unit cells that repeat over and over in each direction. This repetition of the unit cells produces what is called long-range order, meaning the atoms in one unit cell will be arranged just as they are in any other cell many cell spacings away. Such an arrangement allows all of the bonds to be satisfied and the solid material is said to be its lowest energy state.

At elevated temperatures, the thermal energy is sufficient to break some of these bonds allowing the atoms (or molecules) to move about in a random fashion. Long-range order is destroyed and the system is said to be amorphous. In one sense, a glass can be thought of as a liquid that has been frozen in place.

The lack of long-range order in a glass has its advantages. There are no grain boundaries to scatter light so glasses tend to be transparent, making them an inexpensive substitute for single crystals for applications such as windows. Since grain boundaries are attacked preferentially by corrosive chemicals, their absence makes glasses less vulnerable to chemical attack. The random network makes it more difficult for dislocations to form and move, making glasses strong but brittle.

In order to form a crystalline solid, one or more nucleation events must take place in which an embryonic crystal is formed. This embryo forms a sort of template on which the other atoms can arrange themselves to produce the long-range crystalline order. As a melt is cooled, its viscosity increases rapidly so that the atoms or molecules cannot move about as readily. To form a glass from the melt, it is necessary to delay nucleation until the temperature falls to the point where the viscosity becomes high enough to prevent the atoms or molecules from arranging themselves in an orderly manner.

Ceramic melts tend to be more viscous than metals, hence they are generally good glass formers, especially eutectic mixtures of ceramics, which have a lower melting point than their individual components. The glasses we are most familiar with are ceramics, usually oxides of silicon with various additives to tailor the properties of the glass for a specific application. Such glasses are transparent because the grain boundaries that would scatter the light in a polycrystalline solid are absent in an amorphous solid. However, it is also possible to form metallic glasses which, because of their free electrons, are not transparent.

1.5 Contemporary Materials Science

1.5.1 Bioinspired Materials

Until recently, man had thought of metals such as bronze and later iron and steel as the ultimate in structural materials and had neglected the elegant manner in which nature tended to optimize her structures using only organic materials. Her handiwork is now beginning to be taken seriously in the science of biomimetrics or bioinspired materials, the study of plant and animal materials and the attempt to synthesize materials with the desirable properties that nature has evolved. One of the most remarkable feats of nature is the mechanical properties of soft tissues such as worm cuticle, the walls of arteries, veins, and the intestines as well as the skins of various animals. We are used to thinking of materials such as metals that have high Young's moduli, but typically operate in less than 0.1% elongation. Most soft tissues in nature can operate with strains of 100%–150% elongation and the cuticle of a pregnant locust can extend by 1200% when she lays her eggs and recover elastically. If metals could be strained by 20% without rupturing, they would have a stored energy content approximately equal to their bond energy and, when ruptured, the energy release would be equivalent to that of a high explosive. Elastomers such as rubber can be extended by as much as 800%, but the stress increases with strain, causing a tendency to form aneurysms, as can be demonstrated by squeezing a long toy balloon. Furthermore, there is a large stored energy release when rupture occurs as when a balloon is pricked with a pin. Nature's soft tissues behave more like the surface of a liquid in which the force required to stretch the surface remains constant over a large portion of the strain, just the behavior of surface tension. This property resists the formation of aneurysms and surfaces, when pricked, are self-healing.

Tendons, which transmit the force from muscles to the bones, are a composite of elastin and collagen. They can store up to 20 times the energy of spring steel. They are also an example of what we would now call a "smart material." They let you know when you are in danger of overstressing them through the pain you feel. There is a considerable effort these days to develop materials that can monitor and configure themselves in response to their environment.

Wood is another of nature's finest products and is certainly the oldest construction material used by man. It is a remarkable composite composed of soft cellulose and hard lignin. It has a tensile strength per unit mass equal to that of mild steel and is a tougher per unit mass than the toughest steels. Its unusually high work of fracture is due to the soft cellulose which causes it to split easily along the grain. These soft layers tend to blunt crack propagating across the grain.

Muscles provide nature's motive power by converting chemical energy directly to mechanical energy. Since they are not heat engines, they are not limited by the Carnot efficiency. (If they were, we could not get out of bed.) However, they are only ~30% efficient, making them about as efficient as most internal combustion engines. Since their efficiency is not limited by something as fundamental as the Carnot efficiency, there could be substantial payoff if higher efficiency artificial muscles could be produced.

1.5.2 Polymers

Perhaps the greatest advances in materials made during the twentieth century have been in polymer science. The invention of nylon in 1938 by Wallace Carothers at Dupont laid the foundation for the synthetic fiber industry and the miracle fabrics we enjoy today. Leo Baekeland developed a method for producing bakelite (phenol formaldehyde), the first thermosetting plastic, in 1909. This material was the forerunner of the enormous plastics industry that developed during and after World War II. Other polymer products from synthetic rubber to adhesives and coatings have found their way into virtually every aspect of our present day lifestyle.

In 1977, Alan Heeger, Alan MacDiarmid, and Hideki Shirakawa were awarded the Nobel Prize for their work on the production of high electrical conductivity in poylacetylene, paving the way for the exciting new field of molecular electronics. Shortly thereafter, molecular diodes were fabricated, followed by molecular transistors. In 1987, Ching Tang and Steven Van Slyke at Eastman Kodak developed small-molecule organic LEDs (OLEDs). In 1990, Jeremy Burroughes at the University of Cambridge developed large molecule or polymer LEDs (PLEDs) that can emit brighter colors over the entire spectrum, making flat screen video and other displays possible. Some of the OLED displays are already on the market in cell phone and other small displays. Large size flexible displays are still several years from market. One difficulty with such devices is that they are relatively inefficient, as were the early semiconductor LEDs. In 1999, Steven Forrest at Princeton University and Mark Thompson at University of Southern California found a way to control the spin of the electrons so that all of them could emit light instead of heat.

More recently, electroactive polymers have been developed that change shape in response to electrical stimulus. A spin-off company, Artificial Muscles, Inc. (Sunnyvale, California) has been created to commercialize the pioneering work by the Stanford Research Institute in this area.

1.5.3 Superconductors

When the first superconductor with a transition temperature of 95 K was discovered by Jim Ashburn and M.K. Wu at the University of Alabama in Huntsville in 1986, there was great excitement in the field since the highest previous transition temperature was only 23 K. The implications of this discovery was that now superconductors could operate with liquid nitrogen that boils at 77 K instead of the much more expensive and difficult to store liquid helium that boils at 4 K. Also, the fact that this was accomplished with a ceramic material ($YBa_2Cu_3O_{7-}$) rather than the traditional metallic systems (Nb_3Ge and

Nb46.5 wt% Ti) suggested a possible new mechanism for superconductivity that was stronger than the electron–phonon coupling described by the Bardeen–Schrieffer–Cooper theory and held out hope for the possibility of a room temperature superconductor. Despite extensive work in the field of high-temperature superconductivity (HTSC), the mechanism involved is still not completely understood and the transition temperature has been stalled at 134 K. Unfortunately, even with their high transition temperatures, the new HTSC materials have found only limited use because their brittle nature makes it difficult to fabricate them into wires for making large superconducting magnets for high-field magnetic resonance imaging (MRI) and other applications that are presently fulfilled by conventional Nb46.5 wt% Ti superconducting magnets cooled by liquid helium.

1.5.4 Computational Process Modeling

With the availability of large computational capability much attention is now being given to modeling the solidification process in order to design the heat flow in a casting or crystal growth process to produce the desired final microstructure. For such an endeavor to be successful, it is necessary to not only know the fundamental laws governing the evolution of the microstructure but also the thermophysical properties of the melt. Unfortunately, very little data is available on the properties of liquid metals and recent space experiments suggest that the existing diffusion and thermal conductivity data are contaminated by convection. Obtaining accurate values for mass and heat transport coefficients for molten metals in normal gravity is difficult and may require experiments in the reduced gravity of low Earth orbit.

1.5.5 Metallic Glasses

Glass is an amorphous phase lacking crystalline order and can be thought of as a liquid frozen in place. Ceramics that are very viscous in the melt can form glasses with modest cooling rates because the high viscosity keeps the atoms from moving into an ordered crystalline lattice. Most molten metals have low viscosities and do not readily form glasses; those that do require very rapid cooling rates obtained by splat cooling or melt spinning (directing a stream of molten metal onto a rapidly spinning drum). Such processes result in a thin sheet or a thin ribbon. Metallic glasses are not transparent (since metals tend to be good reflectors), but the absence of grain boundaries gives them some interesting and useful properties. Since corrosive chemicals attack grain boundaries where the bonding is weakest, both metallic and ceramic glasses tend to be more resistant to chemical attacks. Since grain boundaries tend to make magnetic materials more difficult to demagnetize, amorphous iron–boron–silicon alloys are being used to reduce power loss due to hysteresis in power transformers. The amorphous structure of a metal makes it difficult for dislocations to form, which makes the metals very hard and highly resilient.

It is possible to add various components to an alloy which increases the viscosity of the melt and makes it more difficult for a crystal to form. William Johnson at the California Institute of Technology was able to measure the thermophysical properties of various candidate compositions for metallic glass formers on one of the Spacelab flights and was able to use these data to design a new class of materials called bulk metallic glasses (BMGs). Since then a new company, Liquidmetals Technology (Rancho Santa Margarita, California) has been formed to manufacture BMGs, which are now finding use in everything from sporting goods and medical implants to industrial coatings that can withstand harsh environments (see an impressive demonstration of the resiliency of a BMG on www. liquidmetal.com).

1.5.6 Advanced Structural Materials

The search goes on for lighter stronger materials, materials that maintain their strength at high temperatures, tougher materials (strength combined with ductility), better ways of combining the ductility and thermal conductivity of metals with the hardness of ceramics to make cermets for cutting tools, methods for toughening ceramics either by making them machinable or casting them to final form, and methods for strengthening metal or ceramic matrix composites (metals or ceramics) with a dispersed second phase material that is harder or stronger than the metal, to name a few examples.

Candidates for the dispersed phase in metal or ceramic matrix composites include metallic whiskers (very thin single crystals that have near theoretical strength), silicon carbide or nitride fibers, and carbon fibers. In 1965, William Watt at the Royal Aircraft Establishment in Farnborough, England discovered a method for pyrolyzing polyacrylonitrile fibers to form a chain of graphite rings, which are extremely strong. Since then methods for producing cheaper carbon fibers were developed in the United States which led to the development of carbon–carbon (C–C) composites now used for the thermal protection system on the nose and the leading edge of the wings of the space shuttle.

1.5.7 Fullerenes

In 1985, Richard Smalley at Rice University discovered the first fullerene, a hollow sphere that comprised of 60 carbon atoms, so named because its structure resembles the geodesic structure designed by Buckminster Fuller. Since then nanotubes or buckey tubes have been discovered that look like rolled-up chicken wire. Single-wall nanotubes (SWNTs) are extremely strong and have, close to the theoretical strength of the sp^2 carbon bond, one of the strongest chemical bonds. At present, these nanotubes are only a few nanometers long, but attempts are being made to extend their lengths so they can be used as reinforcing fibers or perhaps woven into cables.

SWNTs have other interesting properties. They can be semiconductors or insulators depending on their chirality (whether their structure forms left- or right-handed helices). They are superconductors with modest transition temperatures. Properly doped, they have possible application for hydrogen storage.

1.5.8 Semiconductors

The invention of the transistor, in 1947, by John Bardeen, Walter Brattain, and William Shockley of the Bell Laboratories, as a replacement for the bulky energy consuming vacuum tube, promised to revolutionize the world of electronics. But the early transistors were individually made and were unsuitable for high-frequency applications. The real break-throughs in technology that enabled the mass production of the large variety of inexpensive electronic components that range from watches to computer motherboards had to come about in the production of large diameter, dislocation-free single crystal silicon wafers with purities of less than 1 part in 10 trillion. Also needed was the development of lithographic technology that could produce patterns with submicron details. Advances in this later technology, brought about by collaborations between polymer chemists and electronic and optical engineers, has made possible the ever-increasing scale of device integration predicted by Moore's law (the number of transistors on a chip will double every 18 months).

1.5.9 Microelectromechanical Systems

This technology has also paved the way for microelectromechanical systems (MEMS). MEMS technology can produce new photonic switching devices and routers for the telecommunications industry, miniaturized accelerometers, and optical gyroscopes for guidance systems, chemical laboratories on a chip for analyzing toxic atmospheres as well as biohazards, and other advanced chemical and biosensors.

1.5.10 Photonics

Similarly, the development of extremely high purity silica fibers and new photonic materials that provide laser transmitters and receivers have revolutionized landline communication systems. Other developments in photonic materials have made possible highly efficient LEDs in all colors that are being used in traffic lights and other displays. Very small efficient short wavelength solid-state lasers have also made possible massive optical storage which enabled the development of CDs and DVDs.

1.5.11 Magnetic Storage

Advances in magnetic materials, exploiting the giant magnetoresistance (GMR) effect, and thin film fabrication technology have made computer hard drives with capacities of 40–80 GB available for only a few hundred dollars. Fundamental ongoing work on spintronics, the storage and transport of information by controlling the spin of electrons, offers even greater advances in information storage and computational capability.

1.5.12 Quantum Electronics

Even newer technologies that allow device fabrication on an atomic scale have made possible exotic devices such as highly efficient heterojunction lasers, single electron transistors, and various quantum confinement devices that act as artificial atoms in which the energy levels can be controlled by design.

1.5.13 Structure of Biological Macromolecular Crystals

Some of the most exciting breakthroughs in biotechnology are sure to come from advances in structural biology in which the three-dimensional structure of very complex biological macromolecules (proteins, membranes, DNA fragments, etc.) is determined, allowing us to understand the mechanism by which these substances act at the molecular level. Many of these macromolecules are associated with the pathway of a disease state such as HIV reverse transcriptase, an enzyme needed by the HIV retrovirus to insert its genetic code into a healthy cell. Finding the active site for this molecule and determining its three-dimensional structure provides the information needed to design a molecule that will fit into that site and block the action of the enzyme. Drug discovery is turning from its previous trial-and-error approach to rational drug design based on molecular structure. The rate-limiting step in this process is obtaining the crystals of the target macromolecule needed by the structural biologist to obtain the x-ray diffraction information needed to solve the structure. Here are excellent opportunities for material scientists who are knowledgeable in crystal growth to collaborate with structural biologists and medicinal chemists in order to find cures for some of mankind's most pernicious diseases.

1.6 What Is the Future of Materials Science?

The future of materials science is open, limited only by the imagination of the materials scientist and the 92 stable elements nature has given us to work with.

Bibliography

Ashley, S., Artificial muscles, *Sci. Am.*, Oct. 2003, 53–59.
Encyclopedia Britannica, Deluxe CD-ROM, 2004.
Gordon, J.E., *The Science of Structures and Materials*, Scientific American Library, New York, 1988.
Howard, W.E., Better displays with organic films, *Sci. Am.*, Feb. 2004, 76–81.

2

Fundamental Principles

Most students will have had the material in this chapter somewhere in their undergraduate curriculum; it will not hurt for them to see it again and perhaps see it in a different light. For those readers who may be unfamiliar with this material, we will not make heavy use of it in the rest of the book, but it is the basis of much of what we will be doing, so it will help to have some understanding of the basic principles that govern the behavior of matter. The material is presented with a historical prospective so that the reader might appreciate how this knowledge evolved.

2.1 Review of Atomic Structure

2.1.1 History of Spectra

In 1666, Sir Isaac Newton, using a prism to refract a narrow beam of sunlight in a dark room, demonstrated that white light was a mixture of many colors. Much later William Wollaston, using a slit to collimate the light, found dark bands in the continuous spectrum of colors. In 1814 Joseph von Fraunhofer carefully repeated Wollaston's earlier experiment passing the light through a narrower slit and showed that the spectrum of the Sun's radiation has many dark lines, sometimes called Fraunhofer lines. These lines were later interpreted by Gustav Kirchoff as absorption spectra of Ca, Na, and other elements that exist in relative small quantities on the Sun. By 1885 the hydrogen spectrum, which had been observed in many stars, had been extended to 14 lines and Johann Balmer had developed a numerical relationship relating the wavelengths of the visible portion of the spectrum. Other series were discovered in the ultraviolet by Theodore Lyman, and in the infrared by A.H. Pfund, F.S. Brackett, and Friedrich Paschen. Johannes Rydberg extended Balmer's work in 1890 and found a general rule applicable to many elements,

$$\frac{1}{\lambda} = R\left(\frac{1}{n_1^2} - \frac{1}{n_2^2}\right), \tag{2.1}$$

where
  R is the empirical Rydberg constant $= 109{,}677.58 \text{ cm}^{-1}$ for hydrogen
  n_1 and n_2 can take on only integer values (see Figure 2.1)

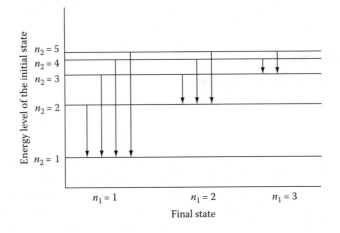

FIGURE 2.1
Schematic of the transitions between the energy levels in the hydrogen atom. The transition from $n_2 = 2$ to $n_1 = 1$ emits a photon with a wavelength of 121.5 nm which is called the Lyman-α for short because it is the principal line in the series. The transition from $n_2 = 3$ to $n_1 = 2$ emits a photon with a wavelength of 656.3 nm and is usually referred to as the H-α line instead of the Balmer-α line.

For
$n_1 = 1$; $n_2 = 2, 3, 4$, Lyman series (ultraviolet, 121.5 nm)
$n_1 = 2$; $n_2 = 3, 4, 5$, Balmer series (visible, 656.3 nm)
$n_1 = 3$; $n_2 = 4, 5, 6$, Paschen series (infrared 1.875 μm)
$n_1 = 4$; $n_2 = 5, 6, 7$, Bracket series (infrared 4.050 μm)
$n_1 = 5$; $n_2 = 6, 7, 8$, Pfund series (infrared 7.400 μm)

Before the twentieth century, there was no theoretical explanation for these observed spectral lines or why other atoms should produce their own characteristic spectra. Finally, in 1913, Niels Bohr developed the first theoretical model that offered a basis for understanding these observations.

2.1.2 Bohr's Theory

For an electron orbiting a nucleus with charge, Z, Bohr equated the

$$\text{Coulomb force} = \frac{-Ze^2}{4\pi\varepsilon_0 r^2} \text{ with the centrifugal force} = \frac{m_e v^2}{r}$$

$$\text{to obtain the velocity } v^2 = \frac{Ze^2}{4\pi\varepsilon_0 m_e r} \text{ and the momentum } p^2 = \frac{Ze^2 m_e}{4\pi\varepsilon_0 r}.$$

From this he was able to write

$$\text{KE} = \frac{m_e v^2}{2} = \frac{Ze^2}{8\pi\varepsilon_0 r}, \quad \text{PE} = -\int_{\infty}^{r} \frac{Ze^2}{4\pi\varepsilon_0 r^2} = -\frac{Ze^2}{4\pi\varepsilon_0 r}, \tag{2.2}$$

and the total energy is

$$\text{KE} + \text{PE} = -\frac{Ze^2}{8\pi\varepsilon_0 r}. \tag{2.3}$$

But classical theory allows any energy whereas the observed spectra suggested transitions between discrete energy states. Also a charge moving around an orbit is constantly being accelerated. Electromagnetic theory requires an accelerated charge to radiate energy. This would cause the electron to quickly spiral into the nucleus! What is wrong here?

Here is where Bohr made the bold assumption that angular momentum was somehow quantized in units of $h/2\pi$, where h was Planck's constant that had been postulated in 1900 to explain black body radiation and was invoked by Einstein in 1905 to explain the photoelectric effect. Orbits in which the angular momentum was an integral number of $h/2\pi$ were assumed to be stable against radiating energy and spectral emission (or absorption) involved a photon with energy equal to the difference between such orbits. Therefore, the angular momentum could be written as

$$m_e vr = pr = \frac{nh}{2\pi} = n\hbar. \tag{2.4}$$

Squaring this and using the previous result for p^2,

$$p^2 r^2 = \frac{Ze^2 m_e}{4\pi\varepsilon_0 r} r^2 = n^2 \hbar^2, \tag{2.5}$$

from which he obtained

$$\frac{1}{r} = \frac{Ze^2 m_e}{4\pi\varepsilon_0 n^2 \hbar^2}, \tag{2.6}$$

and the energy becomes

$$E = -\frac{Ze^2}{8\pi\varepsilon_0}\frac{1}{r} = -\frac{Ze^2}{8\pi\varepsilon_0}\left(\frac{Ze^2 m_e}{4\pi\varepsilon_0 n^2 \hbar^2}\right) = -\frac{Z^2 e^4 m_e}{2(4\pi\varepsilon_0)^2 \hbar^2}\left(\frac{1}{n^2}\right). \tag{2.7}$$

Note the $1/n^2$ dependence in energy, just as in Rydberg's empirical model.

Bohr went on to assume that spectra were caused by electrons jumping from one value of n to another, so the difference in energy would be given by

$$h\nu = \frac{hc}{\lambda} = \Delta E = \frac{Z^2 e^4 m_e}{2(4\pi\varepsilon_0)^2 \hbar^2}\left(\frac{1}{n_1^2} - \frac{1}{n_2^2}\right). \tag{2.8}$$

Also, if you put the numbers into Equation 2.8 with $Z=1$, the

$$\frac{1}{\lambda} = \frac{2\pi^2 e^4 m_e}{(4\pi\varepsilon_0)^2 h^3 c}\left(\frac{1}{n_1^2} - \frac{1}{n_2^2}\right) = 1.097 \times 10^5 \left(\frac{1}{n_1^2} - \frac{1}{n_2^2}\right) = R\left(\frac{1}{n_1^2} - \frac{1}{n_2^2}\right), \tag{2.9}$$

where $R = 1.097 \times 10^5$ is the empirical Rydberg constant. What a major triumph! (Or not, since the theory only works for hydrogen and is not in agreement with our present understanding of the electron.)

The reader may be wondering why the observed spectra from the Sun were in the form of dark lines instead of emission lines described in Figure 2.1. The answer is that the observers were viewing the bright surface of the photosphere of the Sun through the cooler solar atmosphere. The surface of the Sun emits a continuum of wavelengths referred to as black body radiation, similar to an incandescent lamp. Photons with wavelengths corresponding to the transitions shown in Figure 2.1 are absorbed in the cooler solar atmosphere

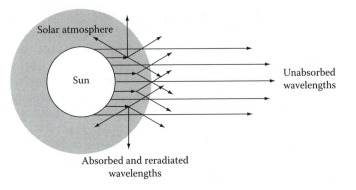

FIGURE 2.2
Schematic showing how light that is spectroscopically absorbed and reradiated shows up as a dark band against a bright background.

by kicking electrons from lower states $n_1 = 1$, $n_2 = 2$, to various excited states and the solar atmosphere reradiates this energy by emitting photons corresponding to the transitions shown in Figure 2.1. The re-emitted radiation is omnidirectional, so the radiation at these frequencies seen by the observer is small compared to the radiation at other wavelengths that does not get absorbed. Thus the solar atmosphere appears to absorb those wavelengths that correspond to allowed electronic transitions. Figure 2.2 attempts to show this effect. Since the most plentiful elements on the Sun are H and He, their emission spectra in the solar atmosphere exceed the black body radiation from the photosphere, although this was not realized until 1925. (In fact, He was discovered on the Sun in this manner before it was known on Earth.)

Actually, Bohr's theory was an inspired (lucky) guess and, since it is valid only for the hydrogen atom, it is not particularly useful as a general theory. Attempts were made by Sommerfeld to extend the theory to two-electron systems, but without much success. To move on, we must understand more about the electron.

2.2 The Electron

Ever since its discovery by J.J. Thompson in 1897 while investigating cathode rays, the electron was found to have some very unusual properties.

Around the turn of the nineteenth century, there were two distinct theories concerning the properties of light. Earlier physicists, such as Newton, believed that light consisted of little particles called corpuscles. In the early nineteenth century, Huygens, Fresnel, Fraunhofer, and others clearly demonstrated interference and the diffraction of light, which could only be explained by a wave-like behavior and in 1865 James Clerk Maxwell showed that light was an electromagnetic wave that could propagate through a vacuum. However, in 1905, Einstein explained the photoelectric effect by assuming that light consisted of packets of energy called photons whose energy was given by $h\nu$, where ν is the frequency and h is Planck's constant that was derived by Planck to explain the spectral distribution of black body radiation. The momentum of a photon is given by $h\nu/c$, where c is the velocity of light. Since the frequency of light $\nu = c/\lambda$, its momentum can be written as h/λ.

2.2.1 de Broglie Wavelength

In 1924 Louis de Broglie, struck by the dualism of the particle and wave nature of light, argued that if a photon of radiation had momentum given by $p = h\nu/c = h/\lambda$, then maybe a mass particle had a wave nature where $\lambda = h/p = h/mv$. So one can see that Bohr's assumption $m_e vr = pr = hr/\lambda = nh/2\pi$ implies that $2\pi r = n\lambda$, or that the allowed orbits correspond to integral number of matter waves.

In 1927, Davisson and Germer demonstrated that electrons did indeed have wave-like properties by showing that a beam of electrons directed toward a metallic single crystal exhibited diffraction patterns, much like x-rays.

2.2.2 Heisenberg Uncertainty Principle

In the same year, Werner Heisenberg formulated the uncertainty principle, which asserts that it is impossible to simultaneously determine the position and momentum of a mass particle to a precision better than $\Delta x \Delta p_x \geq h$, where h is Planck's constant. There are several ways to understand this. One argument stems from the fact that that if a particle is represented by a single frequency (or wavelength), its momentum ($p = h/\lambda$) is known precisely, but a plane wave extending from $-\infty$ to $+\infty$ provides no information about the position of the particle. To localize the particle, waves with slightly different frequencies must be added so their amplitudes bunch up to form a wave packet as shown in Figure 2.3. The relative width of the wave packet can be shown to be inversely proportional to the spread in frequencies or $\Delta x/\lambda \approx \lambda/\Delta\lambda$ or $\Delta x \Delta\lambda \approx \lambda^2$. Since $p = h/\lambda$, $\Delta p = -\Delta\lambda h/\lambda^2$ and $\Delta x|\Delta p| \approx h$.

Since the electron's mass turns out to be 1/2000 that of the proton or neutron, its wavelength is on the order of atomic dimensions and its wave-like nature becomes dominant; whereas protons and neutrons with heavier masses and much shorter wavelengths exhibit more particle-like behavior. (Neutron diffraction, a powerful tool for analyzing the structure of crystals, takes advantage of the short wave-like properties of the neutron which allows the positions of the lighter elements such as hydrogen, which are not seen by x-rays, to be determined.)

If we divide the Δx by r and multiply Δp_x by the same r, the Heisenberg uncertainty relationship can be written as $(\Delta x/r) \Delta(mv_x r) = \Delta\theta\Delta L \geq h$, where L is the angular momentum. Recall that the angular momentum of a Bohr orbit is $nh/2\pi$, so if we specify the ΔL to within one Bohr orbit, the $\Delta\theta \geq 2\pi$ and we cannot specify where the electron is located within its orbit. Therefore, when dealing with electrons, we must abandon classical mechanics and use a new method of wave mechanics or quantum mechanics to describe their behavior. Instead of trying to locate the electron's position, we must be content to finding its wavefunction, usually designated as $\psi(x, y, z)$, which when multiplied by its complex conjugate $\psi(x, y, z)^*$ gives the probability of the finding the electron in that location. As it turns out, this is quite useful, because we can multiply the charge of the electron $-q$ times $\psi\psi^*$ to get the charge distribution and use the result to calculate the energy of chemical bonds and other properties.

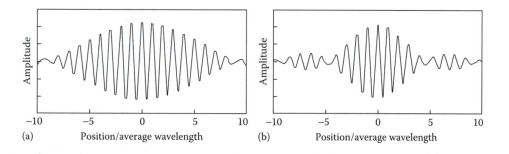

FIGURE 2.3
(a) Particle is localized to $\sim 10\lambda$ when $\Delta\lambda/\lambda = 0.1$. (b) Particle is localized to $\sim 5\lambda$ when $\Delta\lambda/\lambda = 0.2$.

A corollary to the uncertainty principle involves energy and time. Since $E = p^2/2m$, $\Delta E = p\Delta p/m$ and $\Delta x \Delta p = \Delta x \Delta E m/p = \Delta E \Delta x/v = \Delta E \Delta t \geq h$. This relationship is useful for estimating the rest energies of particles with short lifetimes.

A formal method of dealing with the wave nature of particles was developed by Irwin Schrödinger in 1926. A greatly simplified derivation of the time-independent Schrödinger wave equation is presented in Section 2.3.

2.3   Schrödinger Wave Equation

A general wave equation can be written for the propagation of any wave (elastic, sound, electromagnetic) as

$$\nabla^2 \Psi = \frac{1}{c^2} \frac{\partial^2 \Psi}{\partial t^2}, \tag{2.10}$$

where
   c is the velocity of propagation
   Ψ is the amplitude

The solution can be written in the form

$$\Psi(x, y, z, t) = \psi(x, y, z)e^{i\omega t}, \tag{2.11}$$

where
   ω is the angular frequency
   $\psi(x, y, z)$ is the time-independent wavefunction

When Equation 2.10 is substituted back into the partial differential equation (PDE),

$$\nabla^2 \psi + \frac{\omega^2}{c^2} \psi = 0. \tag{2.12}$$

Since $\omega = 2\pi\nu = 2\pi c/\lambda$, this can be written as

$$\nabla^2 \psi + \frac{4\pi^2}{\lambda^2} \psi = 0. \tag{2.13}$$

Introduce the de Broglie wavelength $\lambda = h/mv$,

$$\nabla^2 \psi + \frac{m^2 v^2}{\hbar^2} \psi = 0$$

where

$$\hbar = \frac{h}{2\pi}. \tag{2.14}$$

Now substitute $mv^2 = 2(E - V)$ where E is the total energy and V is the potential energy and multiply by $-\hbar/2m$ to get the time-independent Schrödinger wave equation that describes the behavior of particle waves in terms of their wavefunction ψ,

$$-\frac{\hbar^2}{2m}\nabla^2\psi + V\psi = E\psi. \tag{2.15}$$

The wave equation is often written as an eigenvalue equation in the form

$$H\psi = E\psi, \tag{2.16}$$

where
   H is the Hamiltonian operator defined as $H = ((-\hbar^2/2m)\nabla^2 + V)$
   E represents the energy eigenvalues, i.e., those values of E for which a solution to
   Equation 2.15 exists

2.3.1 Electrons in a Box

Now consider N electrons in a box with volume $V = L^3$. The potential inside the box
($0 \leq x \leq L$) is 0 and infinite outside (i.e., the electrons are confined to the box or crystal).
The Schrödinger equation can be written as

$$-\frac{\hbar^2}{2m}\nabla^2\psi = E\psi, \quad 0 \leq x, y, z \leq L. \tag{2.17}$$

The general solution can be written as

$$\psi(x, y, z) = Ae^{i(k_x x + k_y y + k_z z)}, \tag{2.18}$$

where
   $\mathbf{k} = \vec{i}k_x + \vec{j}k_y + \vec{k}k_z$ is the propagation or wave vector
   A is a normalization constant chosen so that $\iiint \psi\psi^* dV = 1$

(We will make frequent use of the propagation vector $\mathbf{k}$ to represent the momentum of a
wave-like particle in the rest of this chapter. The scalar momentum $p = h/\lambda = 2\pi\hbar/\lambda = \hbar k$
where $k = 2\pi/\lambda$.)
   Putting this solution back into the differential equation,

$$-\frac{\hbar^2}{2m}\nabla^2\psi = \frac{\hbar^2}{2m}\left(k_x^2 + k_y^2 + k_z^2\right)\psi = \frac{\hbar^2}{2m}k^2 = E\psi, \tag{2.19}$$

therefore, having ignored the boundary conditions, we get

$$E = \frac{\hbar^2 k^2}{2m} = \frac{p^2}{2m}$$

where the momentum

$$p = \hbar k. \tag{2.20}$$

This is the energy of a free electron (not confined to a crystal). Note that this energy is
continuous (not quantized) and that it goes as k^2.
   Now we apply the boundary conditions that require the wavefunction to vanish at the
walls of the box. From Euler's formula, $e^{ix} = \cos x + i \sin x$, the solution may be written as

a product of sines or cosines depending on the choice of the leading coefficient. Since $\cos(\theta) = 1$, this choice does not match the prescribed boundary conditions. Therefore, the appropriate solution is

$$\psi = A \sin\left(\frac{n_x \pi x}{L}\right) \sin\left(\frac{n_y \pi y}{L}\right) \sin\left(\frac{n_z \pi z}{L}\right), \tag{2.21}$$

where n_x, n_y, n_z are integers. When this is put back into the Schrödinger equation, the energy has discrete values given by

$$E_n = \frac{\hbar^2 \pi^2}{2mL^2}\left(n_x^2 + n_y^2 + n_z^2\right) = \frac{\hbar^2 \pi^2}{2mL^2} n^2, \tag{2.22}$$

where $n^2 = \mathbf{n} \cdot \mathbf{n} = \left(\vec{i}n_x + \vec{j}n_y + \vec{k}n_z\right)^2$.

The uncertainty principle tell us that if we confine an electron to dimension L, the momentum uncertainty will be $\Delta p = h/L$, which could be satisfied if $p = \pm h/2L$ or if $p^2 = h^2/4L^2$. This would correspond to an energy $E = p^2/2m = h^2/8mL^2 = \hbar^2\pi^2/2mL^2$, which is the same as the first energy level given by Equation 2.18. Thus, in some cases, the uncertainty principle can be used to estimate the lowest energy state of a system.

Now the problem becomes, what value of n^2 corresponds to the lowest energy states of N valence electrons we put into the crystal? We could add up the individual states. Let us start with the lowest state with $n_x = 1$, $n_y = 1$, $n_z = 1$ or E_{111} which hold two electrons, E_{112}, E_{121}, and E_{211} each hold two electrons: that accounts for the first 8 electrons. It should be apparent we will never get to 10^{23} electrons this way, so we must find a better way.

Let us represent n_x as points on the x-axis, n_y as points along the y-axis, and n_z as points along the z-axis. The lowest energy states then become the values of n_x, n_y, n_z that are included inside the sector of a sphere with radius $n = (n_x^2 + n_y^2 + n_x^2)^{1/2}$. The number of states is just the volume of this segment of a sphere with radius n or

$$\text{Number of states} = \frac{1}{8}\frac{4\pi}{3}n^3, \tag{2.23}$$

and since two electrons can occupy each state, (see Section 2.5.1)

$$\text{Number of electrons } (N_e) = \frac{2}{8}\frac{4\pi}{3}n^3 = \frac{\pi}{3}n^3. \tag{2.24}$$

Putting this back into the expression for energy, we find that the energy of the last electron added to the crystal is

$$E_F = \frac{\hbar^2\pi^2}{2mL^2}\left(\frac{3N_e}{\pi}\right)^{2/3} = \frac{\hbar^2\pi^2}{2m}\left(\frac{3N_e}{\pi V}\right)^{2/3} = \frac{\hbar^2\pi^2}{2m}\left(\frac{3n_e}{\pi}\right)^{2/3}, \tag{2.25}$$

where n_e is the number of valence electrons per unit volume. This E_F is the ground state energy of the system and is called the Fermi energy of the electrons. Since the electron density in metals is on the order of $10^{29}/m^3$, the Fermi energy is ~ 5 eV, which is the same order as the metallic bonding energy. Remember that 1 eV is the energy an electron has when accelerated by a potential of 1 V or 1.6×10^{-19} J. The thermal energy of an atom is given by kT where k is Boltzmann's constant $= 1.38 \times 10^{-23}$ J/atom-deg. Equating the two energies and solving for T, we find that 1 eV is equivalent to $\sim 11,000$ K, so the ground state energy of the electrons in a metal corresponds to $\sim 55,000$ K.

Notice that the energy E_n increases as n^2. The density of states $N(E)$ (the number of states with energies E and $E + dE$) can be obtained by differentiating Equation 2.24 to obtain $dN = \pi n^2\, dn$. Differentiating Equation 2.22, to obtain $dE = (\hbar^2 \pi^2/mL^2)n\, dn$,

$$N(E) = \frac{dN}{dE} = \frac{\pi}{2}\left(\frac{2mL^2}{\hbar^2 \pi^2}\right)n = \frac{V}{2\pi^2}\left(\frac{2m}{\hbar^2}\right)^{3/2} E^{1/2}. \tag{2.26}$$

The average energy can be found by integrating

$$\langle E \rangle = \frac{\displaystyle\int_0^{E_\mathrm{F}} EN(E)dE}{\displaystyle\int_0^{E_\mathrm{F}} N(E)dE} = \frac{\displaystyle\int_0^{E_\mathrm{F}} E^{3/2}\, dE}{\displaystyle\int_0^{E_\mathrm{F}} E^{1/2}\, dE} = \frac{3}{5}E_\mathrm{F}. \tag{2.27}$$

2.3.2 Wavefunctions for the Hydrogen Atom

Now let us apply the Schrödinger equation to the hydrogen atom. The hydrogen atom with its single electron lends itself to a closed form solution of the Schrödinger equation. The purpose of this exercise is to show how the quantum numbers that will be used to characterize the wavefunctions of more complicated atoms arise from pure mathematical considerations.

If we write ∇^2 in polar form

$$-\frac{\hbar^2}{2\mu}\left[\frac{1}{r^2}\frac{\partial}{\partial r}\left(r^2\frac{\partial}{\partial r}\right) + \frac{1}{r^2 \sin^2 \theta}\frac{\partial^2}{\partial \phi^2} + \frac{1}{r^2 \sin \theta}\frac{\partial}{\partial \theta}\left(\sin \theta \frac{\partial}{\partial \theta}\right)\right]\psi = (E - V)\psi, \tag{2.28}$$

where the reduced mass $\mu = M/M + m$ has been substituted for m. (This allows us to work in the center-of-mass coordinate system.)

One method for attempting to find an analytical solution to a PDE is to see if the variables can be separated. Let $\psi(r, \phi, \theta) = R(r)\Theta(\theta)\Phi(\phi)$ where $R(r)$ is a function of r only, etc. Put this back into the PDE, divide by $R\Theta\Phi$, and multiply by $-2\mu r^2 \sin^2 \theta/\hbar^2$ to obtain

$$\frac{1}{\Phi}\frac{\partial^2 \Phi}{\partial \phi^2} + \frac{\sin^2 \theta}{R}\frac{\partial}{\partial r}\left(r^2 \frac{\partial R}{\partial r}\right) + \frac{\sin \theta}{\Theta}\frac{\partial}{\partial \theta}\left(\sin \theta \frac{\partial \Theta}{\partial \theta}\right) + \frac{2\mu}{\hbar^2}r^2 \sin^2 \theta(E - V(r)) = 0. \tag{2.29}$$

Note that the first term is a function of Φ only; therefore, it must equal be to some constant, say $-m^2$. We now have an ordinary differential equation (ODE) for Φ, i.e.,

$$\frac{d^2 \Phi}{d\phi^2} = -m^2 \Phi. \tag{2.30}$$

The solution is just

$$\Phi = Ae^{im\phi}, \tag{2.31}$$

which can be verified by direct substitution. In order for Φ to be single valued,

$$\Phi(m\phi) = \Phi(\phi + 2\pi) \tag{2.32}$$

or

$$e^{im\phi} = e^{im(\phi+2\pi)} = e^{im\phi} \, e^{im2\pi} \tag{2.33}$$

Therefore,

$$e^{im2\pi} = 1. \tag{2.34}$$

Since $e^{im2\pi} = \cos 2\pi m + i \sin 2\pi m = 1$, we see that m is restricted to values of $0, \pm 1, \pm 2, \ldots$. The m is the azimuthal quantum number which, for reasons that become clear later, is also called the magnetic or projection quantum number. The normalization coefficient is $A = 1/\sqrt{2\pi}$.

Now if we put the $-m^2\Phi$ back into Equation 2.26 $\partial^2\Phi/\partial\phi^2$ and divide by $\sin^2\theta$, we get

$$\frac{m^2}{\sin^2\theta} - \frac{1}{\Theta\sin\theta}\frac{\partial}{\partial\theta}\left(\sin\theta\frac{\partial\Theta}{\partial\theta}\right) = \frac{1}{R}\frac{\partial}{\partial r}\left(r^2\frac{\partial R}{\partial r}\right) + \frac{2\mu}{\hbar^2}r^2(E - V(r)) = 0. \tag{2.35}$$

The left side is now a function of θ only. We set it equal to a separation constant $\ell(\ell+1)$. Now the Θ term can be obtained by solving the ODE

$$\frac{1}{\sin\theta}\frac{\partial}{\partial\theta}\left(\sin\theta\frac{\partial\Theta}{\partial\theta}\right) - \frac{m^2\Theta}{\sin^2\theta} = -\ell(\ell+1)\Theta. \tag{2.36}$$

This is an associated Legendre equation whose solution may be written

$$\Theta = B\sin^{|m|}\theta P_\ell^{|m|}(\cos\theta), \tag{2.37}$$

where $P_\ell^{|m|}(\cos\theta)$ are associated Legendre polynomials whose series converge only for $\ell = 0, 1, 2, 3, \ldots$ and $m = \pm 0, \pm 1, \pm 2, \ldots, \pm \ell$. The normalization coefficient is given by

$$B = \sqrt{\frac{(2\ell+1)(\ell-|m|)!}{2(\ell+|m|)!}}.$$

Values for the associated Legendre polynomials and the normalized polar wavefunction $\Theta_{m\ell}$ are listed in Table 2.1.

TABLE 2.1

Associated Legendre Polynomials and Normalized Polar Wavefunction

ℓ	m	$P_\ell^m(\cos\theta)$	$\Theta_{m\ell}$
0	0	1	$\sqrt{1/2}$
1	± 1	1	$\sqrt{3/4}\,\sin\theta$
1	0	$\cos\theta$	$\sqrt{3/2}\,\cos\theta$
2	± 2	3	$\sqrt{15/16}\,\sin^2\theta$
2	± 1	$3\cos\theta$	$\sqrt{15/4}\,\sin\theta\cos\theta$
2	0	$1/2\,(3\cos^2\theta - 1)$	$\sqrt{10/16}\,(3\cos^2\theta - 1)$

Source: From White, H.E., *Introduction to Atomic Spectra*, McGraw-Hill, New York, 1934.

TABLE 2.2

Derivatives of the First Several Leguerre Polynomials

n	ℓ	$L_{n+\ell}^{2\ell+1}(x)$
1	0	$-1!$
2	0	$2x - 4$
3	0	$-3x^2 + 18x - 18$

Source: From White, H.E., *Introduction to Atomic Spectra*, McGraw-Hill, New York, 1934.

Putting $-\ell(\ell+1)$ back into Equation 2.36, the PDE for the Θ term, we have the radial equation,

$$\frac{1}{r^2}\frac{\partial}{\partial r}\left(r^2\frac{\partial R}{\partial r}\right) - \frac{\ell(\ell+1)}{r^2}R + \frac{2\mu}{\hbar^2}(E - V(r))R = 0 \tag{2.38}$$

to be solved for R. For a potential function $V(r) - Ze^2/4\pi\varepsilon_0 r$, the solution becomes

$$R_{n,l}(x) = Cx^\ell e^{-x/2}L_{n+l}^{2\ell+1}(x), \tag{2.39}$$

where $n = 1, 2, 3, \ldots$; $\ell = 0, 1, 2, \ldots, (n-1)$; $x = 2rZ/na_1$; a_1 is the first Bohr radius

$$a_1 = \frac{4\pi\varepsilon_0\hbar^2}{\mu e^2} = 0.531 \times 10^{-8} \text{ cm} \tag{2.40}$$

and $L_{n+l}^{2\ell+1}(x)$ are derivatives of the Laguerre polynomials. The first several values of $L_{n+\ell}^{2\ell+1}(x)$ are given in the Table 2.2.

The designation, s, p, d, f, for electrons with $\ell = 0, 1, 2, 3$, respectively, is a carryover from an old spectral notation which stood for sharp, principle, diffuse, and fine.

The normalization coefficient is

$$C = \sqrt{\frac{4(n-\ell-1)!Z^3}{[(n+\ell)!]^3n^4a_1^3}}. \tag{2.41}$$

For $E > 0$ (unbound states), solutions exist for all values of E. For $E < 0$ (bound states), solutions exist only for

$$E = \frac{\mu Z^2 e^4}{2(4\pi\varepsilon_0)^2\hbar^2}\left(\frac{1}{n^2}\right). \tag{2.42}$$

The eigenfunctions $\psi_{n,\ell,m}$ can now be constructed

$$\psi_{n,\ell,m}(r, \phi, \theta) = R_{n,\ell}(r)\Theta_{\ell,|m|}(\theta)\Phi_m(\phi). \tag{2.43}$$

The first several normalized wavefunctions are given by

$$\psi_{1s} = \frac{1}{\sqrt{\pi}}\left(\frac{1}{a_1}\right)^{3/2}e^{-\rho}, \tag{2.44}$$

$$\psi_{2s} = \frac{1}{4\sqrt{2\pi}} \left(\frac{1}{a_1}\right)^{3/2} (2 - \rho)e^{-\rho/2},$$ (2.45)

$$\psi_{2p0} = \frac{1}{4\sqrt{2\pi}} \left(\frac{1}{a_1}\right)^{3/2} \rho e^{-\rho/2} \cos\theta,$$ (2.46)

$$\psi_{2p\pm1} = \frac{1}{4\sqrt{2\pi}} \left(\frac{1}{a_1}\right)^{3/2} \rho e^{-\rho/2} \sin\theta e^{\pm i\phi},$$ (2.47)

where $\rho = r/a_1$.

The product $\psi\psi^*$ may be interpreted as the probability density function for the electron. The normalization coefficients A, B, C were chosen so that

$$\iiint\limits_{\text{all space}} \psi\psi^* dV = 2\pi \int_0^\pi \int_0^\infty \psi\psi^* r^2 \sin\theta d\theta dr = 1,$$ (2.48)

which says that the probability of finding the electron somewhere is unity.

Note that the wavefunction ψ for the 1s electron peaks at $r = 0$ and has no angular dependence as shown in Figure 2.4a. But when the product $2\pi r^2 \psi\psi^*$ is taken, the probability of finding the electron is maximized at $r = a$, the first Bohr radius, as shown in Figure 2.4b.

The radial probability distribution functions for 2s and 2p electrons are shown in Figure 2.5a and b, respectively. Note that there is a small probability of finding the 2s electron at the first Bohr radius, but the most probable location is around 5 Bohr radii (recall the Bohr model predicts the electron shells to be spaced according to n^2). On the other hand the 2p electron has the highest probability density at 4 Bohr radii.

Another way to visualize the probability distributions is by the use of density plots in Figures 2.6 and 2.7. The highest probability density for both the 1s and the 2s states is spherically distributed as shown in Figure 2.5a and b. From these plots, one is tempted to think of the electrons in circular orbits around the nuclei. But remember, the electron in an

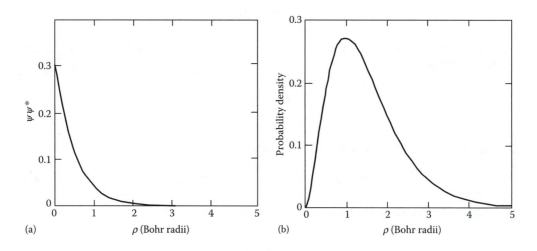

(a) $\psi\psi^*$ ρ (Bohr radii)

(b) Probability density ρ (Bohr radii)

FIGURE 2.4
(a) Wavefunction product $\psi\psi^*$ for 1s electrons. (b) Radial probability distribution for 1s electrons.

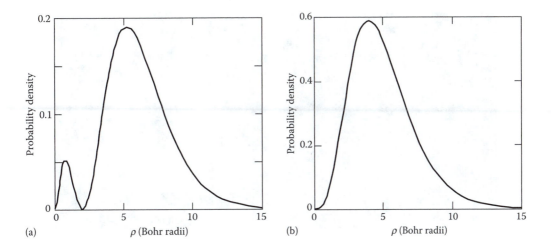

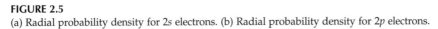

(a) ρ (Bohr radii)   (b) ρ (Bohr radii)

FIGURE 2.5
(a) Radial probability density for 2s electrons. (b) Radial probability density for 2p electrons.

s-state has no angular momentum. Thus, one must think of the electron as simply smeared out over region indicated by the figure.

Because of the $\cos\theta$ term in the ψ_{2p0} wavefunction (Equation 2.45), the probability density has the figure eight-like lobe about the z-axis as shown in Figure 2.7a, but the $\psi_{2p\pm1}$ wavefunction is complex and it loses its ϕ-dependence when $\psi\psi^*$ is taken, giving the probability density in the form of a toroidal distribution about the z-axis as shown in Figure 2.7b. What is special about the z-axis? Why should the electron be distributed in a lobe about this axis and not the other two axes?

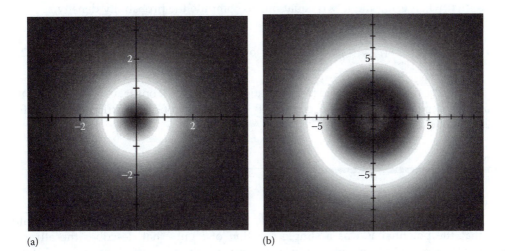

(a)   (b)

FIGURE 2.6
(a) Probability density of an electron in a 1s state. The most likely location is at 1 Bohr radius. (b) Radial probability density of an electron in a 2s state. There is a small probability of the electron at 1 Bohr radius, but the most likely location is at ~5 Bohr radii.

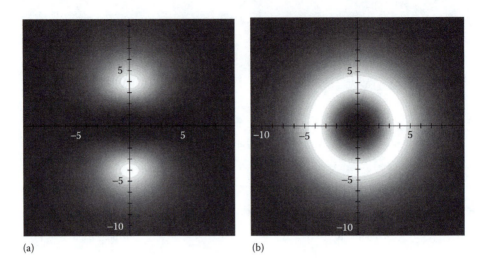

(a)   (b)

FIGURE 2.7
(a) Probability density for an electron in the $2p_0$ or $2p_z$ state. (b) Probability density for an electron in the $2p\pm1$ state in the $\theta = \pi/2$ plane.

Recall that any linear combination of solutions to a differential equation is also a solution. Using this linear combination of atomic orbitals (LCAO), we can construct two new real wavefunctions that have lobes about the x- and y-axes, respectively, i.e.,

$$p_x = \frac{\psi_{2p+1} + \psi_{2p-1}}{2} = \frac{1}{4\sqrt{2\pi}} \left(\frac{1}{a_1}\right)^{3/2} \rho e^{-\rho/2} \sin\theta \cos\phi$$
$$p_y = \frac{\psi_{2p+1} - \psi_{2p-1}}{2i} = \frac{1}{4\sqrt{2\pi}} \left(\frac{1}{a_1}\right)^{3/2} \rho e^{-\rho/2} \sin\theta \sin\phi \tag{2.49}$$

Notice that when using these two equations to compute the probability density,

$$p_x p_x^* + p_y p_y^* = \frac{1}{32\pi} \left(\frac{1}{a_1}\right)^3 \rho^2 e^{-\rho} \sin^2\theta = \psi_{2p\pm1} \psi_{2p\pm1}^*, \tag{2.50}$$

we get the same result as we would from Equation 2.49. In fact, if one takes $p_x p_x^* + p_y p_y^* + p_z p_z^*$, the angular dependence disappears completely and the electrons are distributed in a uniform shell about the nucleus.

2.4 One Electron Approximation

The reason the Schrödinger equation could be solved exactly for the hydrogen atom was that the potential V was a simple function of r only that describes the attraction between the single electron and the nucleus. In a system with many electrons, the potential function would have to include the repulsive interaction between an electron and all the other electrons. This is an extremely difficult calculation which can only be carried out numerically using an iterative scheme. A guess is made for the wavefunction of each electron from the hydrogen wavefunction. Then the trial wavefunctions are adjusted until the Schrödinger equation is satisfied. There are several procedures for doing this known as the Hartree method and the Hartree–Fock method. However, it is possible to gain some understanding of many electron systems by making some simplifying assumptions.

Even though the electrons react strongly with one another, it is possible to obtain a satisfactory description of the electronic structure by assuming that each electron moves more or less independently in an effective field. This effective field must take into account the screening field of the other electrons which reduces attraction of the electron in question to the nucleus. If one time-averages the motion of the other electrons, it is a reasonable approximation to consider their distribution as spherically symmetrical. If the screening field is taken to be spherically symmetric, the potential function is a function of r only, just as in the case of the hydrogen atom. This will affect the radial wavefunction and give different energy states, but the angular portions of the wavefunction will be preserved along with the three quantum numbers n, ℓ, and m that describe the electronic states.

2.5 Periodic Table

The early nineteenth century saw a rapid advance in analytical chemistry in which many new elements were identified and attempts were made to classify them. J.W. Döbereiner found triads of elements such as Cl, Br, I and Li, Na, K that had similar chemical properties and noticed that the atomic weight of the middle element was close to the average of the other two. In 1869 Mendeleyev proposed the periodic law in which "the elements arranged according to the magnitude of atomic weights show a periodic change of properties."

2.5.1 Pauli Exclusion Principle

In 1925, Samuel Goudsmit and George Uhlenbeck, while studying the doublet lines of the sodium spectra, realized that the electron must have a magnetic moment and assigned a spin quantum number of $\pm 1/2$, giving it an angular momentum of $1/2\,\hbar$. In the same year Wolfgang Pauli, in order to reconcile the periodic table with the new findings in quantum mechanics, proposed his exclusion principle which states than no two electrons can exist in the same quantum state. (This is interesting considering that this was supposedly 2 years before Schrödinger published his wave mechanics method that led to the other three quantum numbers. Apparently, the concept of the n, ℓ, and m quantum numbers and their relations was deduced empirically before the quantum mechanical solution of the hydrogen atom.)

Later it was also found that there were other elementary particles that had half-integral spins such as particles in the class of leptons (e.g., electrons, muons), baryons (e.g., neutrons, protons, lambda particles), and nuclei of odd mass number (e.g., tritium, helium-3, uranium-233). All particles with half-integral units angular momenta are classified as fermions and obey Fermi–Dirac statistics; whereas particles with integral units of angular momenta, e.g., photons and nuclei of even mass number, are classified as bosons and obey Bose–Einstein statistics. (See Chapter 15 for further detail.)

2.5.2 Theoretical Basis for the Periodic Table

The fact that the one electron approximation preserves the angular part of the hydrogen wavefunction along with the m, ℓ, and n quantum numbers allows us to use these quantum numbers with the Pauli exclusion principle to construct the periodic table and to use the

TABLE 2.3

Possible Quantum States through $n = 4$

n	ℓ	m	s	Maximum Occupancy
1	0	0	$\pm 1/2$	$1s^2$
2	0	0	$\pm 1/2$	$2s^2$
2	1	$\pm 1, 0$	$\pm 1/2$	$2p^6$
3	0	0	$\pm 1/2$	$3s^2$
3	1	$\pm 1, 0$	$\pm 1/2$	$3p^6$
3	2	$\pm 2, \pm 1, 0$	$\pm 1/2$	$3d^{10}$
4	0	0	$\pm 1/2$	$4s^2$
4	1	$\pm 1, 0$	$\pm 1/2$	$4p^6$
4	2	$\pm 2, \pm 1, 0$	$\pm 1/2$	$4d^{10}$
4	3	$\pm 3, \pm 2, \pm 1, 0$	$\pm 1/2$	$4f^{14}$

Notes: Principal quantum number, $n = 1, 2, 3, \ldots$; angular momentum quantum number, $\ell = 0, 1, 2, \ldots, (n-1)$; magnetic or projection quantum number, $m = +\ell, \ell - 1, \ldots, -\ell$.

angular part of the hydrogen wavefunction to describe properties of many electron systems with hydrogen-like wavefunctions. This will be seen later when we investigate the s–p hybridization of the wavefunctions of the carbon atom to obtain the diamond and graphite structures.

Now the periodic table has a solid theoretical basis. Since the Pauli exclusion principle allows only 1 fermion (electron) per quantum state, and since fermions have spin $s = \pm 1/2$, the possible states are shown in Table 2.3.

The electronic configuration is specified by the principal or n quantum number followed by the angular momentum or ℓ quantum number, using the old spectrographic notation, and the number of electrons having those quantum numbers is indicated by a superscript. From the above table, one can see that the first shell ($n = 1$), sometimes called the K-shell, can have only two electrons. The next or L-shell ($n = 2$) can have eight electrons. The third, or M-shell, can have 18 electrons, etc. One would expect that the electrons with the lowest principal quantum number are closest to the nucleus, hence would be deeper in the energy well and will fill first, completing one shell before starting the next. Things do start off that way. Helium with its $2s$ electrons completes the K-shell. The L-shell is completed with neon having filled the s and p subshells with 10 electrons. However, a departure is seen in filling the M-shell ($n = 3$) in that the $4s$ electrons start to fill before the $3d$ electrons (Table 2.4). Recall that the $4s$ electrons, with no orbital angular momentum, have penetrating orbits which allows them to spend a fraction of their time inside of the circulating orbits of the $3d$ electrons. When inside these lower n orbits, these $4s$ electrons are no longer screened by the other electrons and therefore are more tightly bound than the orbiting electrons. Thus the M-shell ends with argon, which has the configuration of Ne $+ 3s^2 3p^6$ and has a total of 18 electrons.

As shown in Table 2.4, the N-shell ($n = 4$) starts with potassium, which has the configuration of Ar $+ 4s^1$. The 10 $3d$ electrons start filling with scandium and end with krypton, which has the configuration Ar $+ 3d^{10} 4s^2 4p^6$.

As may be seen in Table 2.4, the s-electrons of the next highest n number fall below the d-electrons from $n = 4$ to $n = 6$, the higher n rows are omitted in the table. This causes the $n = 5$ or O-shell to terminate with $4d^{10}$ plus $5s^2 5p^6$ and accounts for the first two rows of 18 elements which contain the transition metals. The $n = 6$ or P-shell starts off like the N- and O-shells with cesium, which is Xe $+ 6s^1$, but after lanthanum, which is Xe $+ 5d^1 6s^2$, the $4f$ electrons start to fill before the $5d$ electrons. To preserve the similarity of properties of the $n = 6$ series with their $n = 4$ and 5 counterparts, the next 14 elements are split off into a separate group called the lanthanide series so that hafnium, which is configured

TABLE 2.4

Atomic Configuration of the Elements (Lanthanides and Actinides Omitted)

	IA	IIA	IIIB	IVB	VB	VIB	VIIB	VIIIB	VIIIB	VIIIB	IB	IIB	IIIA	IVA	VA	VIA	VIIA	VIIIA
K	H $1s$																	He $1s^2$
L	Li $2s$	Be $2s^2$											B $2s^2$ $2p$	C $2s^2$ $2p^2$	N $2s^2$ $2p^3$	O $2p^4$ $2s^2$	F $2p^5$ $2s^2$	Ne $2p^6$ $2s^2$
M	Na $3s$	Mg $3s^2$											Al $3s^2$ $3p$	Si $3s^2$ $3p^2$	P $3s^2$ $3p^3$	S $3p^4$ $3s^2$	Cl $3p^5$ $3s^2$	Ar $3p^6$ $3s^2$
N	K $4s$	Ca $4s^2$	Sc $3d$ $4s^2$	Ti $3d^2$ $4s^2$	V $3d^3$ $4s^2$	Cr $3d^5$ $4s$	Mn $3d^5$ $4s^2$	Fe $3d^6$ $4s^2$	Co $3d^7$ $4s^2$	Ni $3d^8$ $4s^2$	Cu $3d^{10}$ $4s$	Zn $3d^{10}$ $4s^2$	Ga $3d^{10}$ $4p$ $4s^2$	Ge $3d^{10}$ $4p^2$ $4s^2$	As $3d^{10}$ $4p^3$ $4s^2$	Se $3d^{10}$ $4p^4$ $4s^2$	Br $3d^{10}$ $4p^5$ $4s^2$	Kr $3d^{10}$ $4p^6$ $4s^2$
O	Rb $5s$	Sr $5s^2$	Y $4d$ $5s^2$	Zr $4d^2$ $5s^2$	Nb $4d^4$ $5s$	Mo $4d^4$ $5s$	Tc $4d^6$ $5s$	Ru $4d^7$ $5s$	Rh $4d^8$ $5s$	Pd $4d^{10}$	Ag $4d^{10}$ $5s$	Cd $4d^{10}$ $5s^2$	In $4d^{10}$ $5p$ $5s^2$	Sn $4d^{10}$ $5p^2$ $5s^2$	Sb $4d^{10}$ $5p^3$ $5s^2$	Te $4d^{10}$ $5p^4$ $5s^2$	I $4d^{10}$ $5p^5$ $5s^2$	Xe $4d^{10}$ $5p^6$ $5s^2$
P	Cs $6s$	Ba $6s^2$	La $5d$ $6s^2$	Hf $4f^{14}$ $5d^2$ $6s^2$	Ta $4f^{14}$ $5d^3$ $6s^2$	W $4f^{14}$ $5d^4$ $6s^2$	Re $4f^{14}$ $5d^5$ $6s^2$	Os $4f^{14}$ $5d^6$ $6s^2$	Ir $4f^{14}$ $5d^9$	Pt $4f^{14}$ $5d^9$ $6s$	Au $4f^{14}$ $5d^{10}$ $6s$	Hg $4f^{14}$ $5d^{10}$ $6s^2$	Tl $4f^{14}$ $5d^{10}$ $6p$ $6s^2$	Pb $4f^{14}$ $5d^{10}$ $6p^2$ $6s^2$	Bi $4f^{14}$ $5d^{10}$ $6p^3$ $6s^2$	Po $4f^{14}$ $5d^{10}$ $6p^4$ $6s^2$	At $4f^{14}$ $5d^{10}$ $6p^5$ $6s^2$	Rn $4f^{14}$ $5d^{10}$ $6p^6$ $6s^2$

TABLE 2.5

Selected Ionization Energies
of the Elements in Electron Volts

H	13.60
Li	5.39
Na	5.14
K	4.34
Rb	4.18
Cs	3.89
He	24.59
Ne	21.56
Ar	15.76
Kr	14.00
Xe	12.13
Ra	10.74

as $Xe + 4f^{14}5d^26s^2$, has the same outer electron configuration as zirconium and titanium in the lower series. A similar extraction of the actinide series is made in the $n = 7$ series.

Thus all of the elements at the end of the period (He, Ne, Ar, Kr, Xe, and Rn) have closed shells, s^2 for He and $ns^2 \, np^6$ for the others, which causes these electrons to be tightly bound and therefore do not readily interact with other elements. Compare the ionization energy of these so-called noble gases with the alkali metals where the single outer electron is shielded from the nuclei by the inner core electrons in Table 2.5.

2.6   Summary

A review of the foundations of modern physics that underlies all of our understanding of matter is given with a historical perspective. Beginning with the early attempts of the Bohr model of the atom to explain observed spectra, a review is given of the particle and wave nature of material and the resulting uncertainty principle. A simplified derivation of the time-independent Schrödinger equation is presented along with its application to electrons confined to a crystal and to the hydrogen atom. The quantum mechanical behavior of confined electrons (electrons in a box) is key to understanding the physics and chemistry of metals and will be frequently drawn on in later chapters. Particular attention is given to the hydrogen orbital wavefunctions because their angular dependence are similar to the wavefunctions that combine to form the molecular orbitals responsible for the chemical bonding of all materials. Also the quantum numbers associated with these wavefunctions are shown to come about naturally from the mathematical solutions to the Schrödinger equation. These quantum numbers, together with the Pauli exclusion principle, form the basis of the periodic table and the electronic configurations of the elements.

Bibliography

Encyclopedia Britannica, Deluxe CD-ROM, 2004.

Hoddeson, L., Braun, E., Teichmann, J., and Weart, S., *Out of the Crystal Maze: Chapters from the History of Solid State Physics*, Oxford University Press, New York, 1992.

Krane, K., *Modern Physics*, 2nd edn., John Wiley & Sons, New York, 1996.

Kittel, C., *Introduction to Solid State Physics*, 7th edn., John Wiley & Sons, New York, 1966.

Schiff, L.I., *Quantum Mechanics*, McGraw-Hill, New York, 1955.

White, H.E., *Introduction to Atomic Spectra*, McGraw-Hill, New York, 1934.

Problems

1. Fe–heme electronic structure in hemoglobin can be modeled as a 1 by 1 nm square in a plane with 26 delocalized electrons. Why is blood red? (Hint: Treat the molecule as electrons in a two-dimensional box. Find the ground state energy for the 26th electron. This is the highest occupied molecular orbital (HOMO). Now find the energy of the next available state. This would be the lowest unoccupied molecular orbital (LUMO). As a result of a HOMO–LUMO transition, what wavelength of light does this energy difference correspond to? Also note that E_{12}, corresponding to $n_x = 1$, $n_y = 2$, has the same energy as E_{21}, but is a different state and each state can hold two electrons.)

2. Before the discovery of the neutron by James Chadwick in 1932, it was thought that the nucleus contained electrons and neutrons. Use the uncertainty principle to show that an electron cannot exist inside the nucleus. (Compare the attractive Coulomb potential to the energy uncertainty resulting from the confinement within the nucleus. Take the radius of the nucleus to be 10^{-15} m.)

3. Al has 18×10^{28} electrons/m^3. (a) Find its Fermi energy in electron volts. (b) Find the next available energy level in a 1 cm^3 crystal of Al. (c) Do the same for a 10 nm cube quantum dot.

3

Chemical Bonding

The ability of atoms and molecules to form chemical bonds is the defining feature of the structure and properties of solids. The types of bonds that are formed determine if the material will be a metal, a ceramic, or a polymer, and whether the material will conduct electricity, transmit light, or be magnetic.

3.1 What Holds Stuff Together?

All matter that we deal with on an everyday basis is held together by electrical forces that form chemical bonds. These forces are manifested in different ways, depending upon which elements are involved. There are three type of primary bonds: (1) the metallic bond in which electrons become detached from atoms when they come together so the ion cores become mutually attracted to the sea of electrons surrounding them; (2) the covalent bond in which atoms become mutually attracted by sharing electrons in order to form closed electron shells; and (3) the ionic bond in which a mutual attraction occurs when one or more electrons leaves a metal atom to complete an atomic shell of a nonmetallic atom forming an oppositely charged ion pair. Much weaker bonds, such as the hydrogen bond, which arise from dipolar attractions between molecules when a hydrogen atom becomes covalently bonded to an O, N, or F atom, or to the van der Waals bond, which arises from induced dipole–dipole interactions, play a secondary role in the structure of materials. Understanding these basic forces that hold materials together is crucial to understanding the structure and properties of materials. We shall start with the ionic bond since conceptually it is the easiest to visualize and it lends itself to a simple analytical model.

3.2 Ionic Bonding

The ionic bond is the strongest chemical bond, ranging from 10.5 eV for LiF to 5.8 eV for CsI, but it can only act between two (or more) dissimilar atoms. Closing the electron shell of an atom lowers its energy whether this is accomplished by sharing electrons to form covalent bonds or by electron transfer to form ionic bonds. The ionic bond arises from the transfer of electrons to form ions that are then attracted to one another electrostatically. However, many ionically bonded compounds also have some degree of covalent bonding in which electrons are partially shared instead of being completely transferred.

3.2.1 Electronegativity and Electron Affinity

The tendency to form an ionic bond between two components depends on the ability of one of the components to attract an electron to fill its outer cell and the ability of the other component to give up an outer electron. The ability of an atom to attract or give up an electron is defined as its electronegativity, thus the tendency to form ionic bonds is determined by the difference in electronegativity between the two components. In 1932, Linus Pauling developed a method for quantifying the electronegativity of the elements in which he arbitrarily assigned an electronegativity of 4 to fluorine, the most electronegative atom. Cesium has the lowest ionization potential, making it the most electropositive. Therefore, the Cs–F bond was considered to be purely ionic and Pauling then used the measured heat of formation of various compounds as well as that of their components to estimate the degree of covalent bonding in order to obtain the electronegativity scale for the other elements. Mulliken later showed that the sum of the electron affinity (the energy released when an electron is acquired to form a negative ion) and the ionization potential (the work required to remove an electron) were almost directly proportional to Pauling's electronegativity. In fact, the sum of the electron affinity and ionization energy in kilocalorie per mole divided by 125 gives the Pauling electronegativity to within a few percent.

The electronegativity increases as one moves to the right-hand side of the periodic table, reaching a maximum at group VII, the halogens, and decreases with increasing atomic number. As a result, ionic bonding always dominates in compounds formed between group I and group VII elements and plays a major role in group II and group VI compounds.

3.2.2 Coulomb Potential

The attractive part of the bonding force is the Coulomb potential, which results from the electrostatic force between two charged particles,

$$U_C = \frac{q_i q_j}{4\pi\varepsilon_0 r_{ij}}, \tag{3.1}$$

where
  r_{ij} is the distance between the ith and jth ion
  q_i is the charge on the ith ion

If the ith and jth ions have the same charge, the potential is positive and the force repulsive; if the charges are opposite, the force is attractive. Note the $4\pi\varepsilon_0$ in the denominator. We will use rationalized meter, kilogram, second (MKS) units throughout.

If the ions get close enough that their core electronic wavefunctions begin to overlap, the Pauli principle—which prevents electrons from occupying the same quantum state—forces these electrons into higher energy states, causing a strong short-range repulsive force to act. This force is represented empirically by a positive potential that increases rapidly with decreasing r. Born suggested this repulsive force be represented by an inverse power law

$$U_R = \frac{B}{r_{ij}^n}, \tag{3.2}$$

where B and n are constants to be determined empirically (typically, $n = 6$–10). An alternative form for describing the repulsive potential is the Born–Mayer potential given by $U_R = B\exp(-\beta r)$ where B and β are constants to be determined empirically. The formalism is similar using either the inverse power law or the Born–Mayer equation to represent the repulsive potential as shown in Section 3.7.1.

3.2.3 Madelung Constant

The total energy of the system may be found by taking the sum of these two potential functions over all of the ions in the system,

$$U = \frac{1}{2} \sum_{i \neq j} \left(\frac{q_i q_j}{4\pi\varepsilon_0 r_{ij}} + \frac{B}{r_{ij}^n} \right), \tag{3.3}$$

where the $1/2$ is to avoid double counting when summing over j for each i and vice versa.

It is convenient to write r_{ij} as $r p_{ij}$, where r is the nearest neighbor distance and p_{ij} is a dimensionless parameter relating the distance between the ith and jth charges in units of r. Now instead of summing over both i and j, we arbitrarily chose some i as the 0th charge, sum over j and multiply by N, the number of ion pairs.

$$U(r) = N \left(-\frac{e^2}{4\pi\varepsilon_0 r} \sum_{j>0} \frac{\pm 1}{p_{0j}} + \frac{B}{r^n} \sum_{j>0} \frac{1}{p_{0j}^n} \right) = N \left(-\frac{e^2}{4\pi\varepsilon_0 r} A + \frac{B}{r^n} \sum_{j>0} \frac{1}{p_{0j}^n} \right), \tag{3.4}$$

where $A = \sum_{j>0} \pm 1/p_{0j}$ is called the Madelung sum for the lattice. The Madelung sum can be obtained by directly summing over the lattice taking the plus sign for unlike charge pairs and minus sign for like charges, but convergence is very slow because of the $1/r$ dependence of the Coulomb potential. This sign convention is chosen to make $A > 0$ and the total potential energy negative for a stable lattice. Several methods (the Evjen solution and the method of Ewald) have been developed for computing the Madelung constant for different lattices (see Appendix for discussions on computing the Madelung sum).

The value of A depends on the structure, which is determined by the relative sizes of the cations and the anions, which are generally larger than the cations. Since each ion must be surrounded by counterions, a close-packed structure is not possible. It is necessary for the smaller cations to be able to hold off the surrounding anions so that they do not come in contact with each other, otherwise the strong repulsive forces would make the lattice unstable. If the cations are nearly the size of the anions, it is possible to arrange eight counterions around each ion without them contacting each other. If the cations are smaller than the anions, it is possible to arrange only six counterions without them touching, and if the cation is very much smaller than the anion, only a coordination number (number of nearest neighbors) of 4 is possible. This issue will be discussed in more detail in Chapter 5.

A structure with a coordination number of 8 is known as the CsCl structure. It can be described as two interpenetrating simple cubic lattices with the anions on the corners of one lattice and the cations located on the corners of the second interpenetrating lattice arranged such that they sit in the centers of the cubes of the anion lattice so that each ion is surrounded by 8 counterions. The Madelung constant for this structure is equal to 1.763.

The NaCl or rock salt structure has a coordination number of 6. It consists of two interpenetrating face-centered cubic (fcc) lattices with the anions on the lattice sites of one lattice and the cations on the lattice sites of the interpenetrating lattice arranged such that the cations sit on the octahedral interstitial sites (points on the edges of the face-centered cubes half-way between the corners. Despite the terminology, the coordination number of an octahedral site is 6 instead of 8 as will be explained in Chapter 4. For this structure, $A = 1.748$.

The zinc blende structure has a coordination number of 4 and is a diamond-like structure with opposite charges on every other occupied lattice point. The A for this structure is equal to 1.638. (These and other lattices will be discussed in more detail in Chapter 5.)

3.2.4 Lattice Energy

Equation 3.4 for $U(r)$ is valid for any r. However, at equilibrium when the lattice is under no strain, the potential is at a minimum and the attractive Coulomb forces are balanced by the repulsive forces. To enforce this condition, we differentiate the potential function with respect to r and set the derivative to 0 for $r = r_0$, the equilibrium distance. This gives

$$\frac{e^2 A}{4\pi\varepsilon_0 r_0} = \frac{nB}{r_0^n} \sum_{j>0} \frac{1}{p_{0j}^n}. \tag{3.5}$$

Using this result to eliminate $B \sum_{j>0} 1/p_{0j}^n$ in Equation 3.4, the potential function becomes

$$U(r) = -\frac{NAe^2}{4\pi\varepsilon_0 r_0} \left(\frac{r_0}{r} - \frac{1}{n} \left(\frac{r_0}{r} \right)^n \right). \tag{3.6}$$

At equilibrium, $r = r_0$, and the lattice energy is given by

$$U_{\text{Lat}} = U(r_0) = -NA \frac{e^2}{4\pi\varepsilon_0 r_0} \left(\frac{n-1}{n} \right). \tag{3.7}$$

For the NaCl structure, there are 4 ion pairs per unit cell and the unit cell has a volume of $(2r_0)^3$, so $N = 1/2r_0^3$. Therefore, Equation 3.7 can be written as

$$\frac{U}{\text{Vol}} = -A \frac{e^2}{8\pi\varepsilon_0 r_0^4} \left(\frac{n-1}{n} \right). \tag{3.8}$$

Thus in Equation 3.6 we have a simple relationship between the Madelung constant A, the equilibrium nearest neighbor distance r_0 (which can be measured using x-ray diffraction), the lattice energy U_{Lat}, and an arbitrary constant n. If the lattice energy is known, n can be computed from Equation 3.7. The bond-energy function $U(r)$ determines mechanical properties such as the compressibility, bulk modulus, Young's modulus, etc. Also, a detailed knowledge of the bond energy as a function of r can be used to determine properties such as thermal expansion and theoretical strength of the material. These details will be discussed in Chapter 7 and subsequent chapters. Here we are able to see how we can begin to make predictions of macroscopic material properties from atomistic considerations! The bond-energy function from Equation 3.6 is shown in Figure 3.1.

3.2.5 Born–Haber Cycle

The lattice energy described in Equation 3.6 is defined as the energy required to separate the individual ions in the compound. This is distinguished from the cohesive energy which is the energy required to separate the components of the solid into free neutral particles,

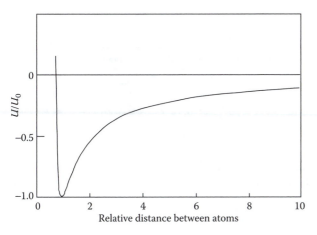

FIGURE 3.1
Potential function for the ionic bond with $n=9$. Note the long range of the attractive Coulomb potential, which is the reason why so many terms must be considered in computing the lattice energy.

although many texts do not make this distinction and use cohesive energy in place of lattice energy. (Actually, this distinction is important only in ionically bonded compounds.) The problem arises because the lattice energy is difficult to measure directly. What is generally measured is the heat of formation, which is roughly equivalent to the cohesive energy as defined here. To convert cohesive energy to lattice energy, the energies required to vaporize and disassociate the individual elemental components to form monatomic gases must first be computed and then the ionization energies are added. This process is sometimes called the Born–Haber cycle and is illustrated in Figure 3.2.

The measured lattice energies/volume for several systems with the NaCl structure are shown in Table 3.1. Also shown are the values of the parameter n obtained from measured lattice energies using Equation 3.8.

Notice that the energy/volume drops rapidly with increasing distance between nearest neighbors, as would be expected from Equation 3.8. Also note that the derived value of n increases with increasing nearest neighbor distance. Figure 3.3 shows a log–log plot of the lattice energy/volume as a function of r_0. The measured slope is -3.87 instead of -4 expected from Equation 3.8. The difference is due to the change of n with nearest neighbor distance.

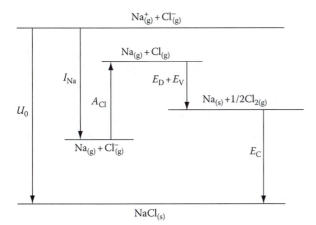

FIGURE 3.2
Born–Haber cycle for NaCl. The lattice energy U_0 is equal to the ionization energy of Na (I_{Na}), less the electron affinity of Cl (A_{Cl}), plus the heat of disassociation of Cl (E_D) and the heat of vaporization of Na (E_V), less the cohesive energy of NaCl (E_C). (Energy levels are drawn to scale.)

TABLE 3.1

Derived Values for n

	U_0/V_0 (GJ/m^3) Measured	r_0 (nm) Measured	n Derived
LiF	102.9	0.2014	6.23
LiCl	40.66	0.2570	8.30
NaCl	28.34	0.282	8.98
KF	34.45	0.2674	8.21
KCl	18.39	0.3147	9.46
KBr	15.33	0.3298	9.92
KI	11.74	0.3533	10.74

Source: From Kittel, C., *Introduction to Solid State Physics*, John Wiley & Sons, New York, 1966.

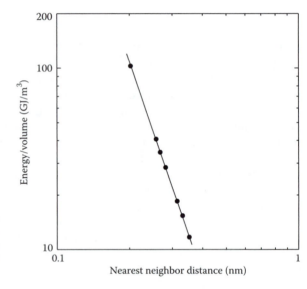

FIGURE 3.3
Lattice energy vs. nearest neighbor distance for ionic crystals with the NaCl structure. (Note the approximate 4th power dependence on nearest neighbor distance as predicted by Equation 3.6.)

3.3   Covalent Bond

3.3.1   Heitler–London Theory

In 1927, Walter Heitler and Fritz London successfully applied the quantum theory to the problem of the covalent bond between hydrogen atoms using what is known as the variational method. This was one of the great achievements of the quantum theory in that it provided a "first principles" explanation of this bond which hitherto had been developed phenomenologically by the chemists. Essentially they used a linear combination of the unperturbed wavefunctions for the hydrogen atom as a trial function to compute the energy from

$$E = \frac{\int d\tau \psi \hat{H} \psi^*}{\int d\tau \psi \psi^*},$$

(3.9)

where

$\psi = C_1\psi_1 + C_2\psi_2$ and ψ_1 and ψ_2 are the ground state hydrogen radial wavefunctions centered about their relative positions

$\hat{H}$ is the Hamiltonian, which is an operator that includes the potential energy of attraction between the unlike charges as well as the repulsive energy between particles with like charges.

Putting these wavefunctions into Equation 3.9, the energy becomes

$$E = \frac{C_1^2 \int d\tau\psi_1\hat{H}\psi_1 + 2C_1C_2 \int d\tau\psi_1\hat{H}\psi_2 + C_2^2 \int d\tau\psi_2\hat{H}\psi_2}{C_1^2 \int d\tau\psi_1^2 + 2C_1C_2 \int d\tau\psi_1\psi_2 + C_2^2 \int d\tau\psi_2^2}. \tag{3.10}$$

Now we find the coefficients that minimize the energy in Equation 3.10 by taking the partial derivatives of E with respect to C_1 and C_2 and setting them to zero,

$$\begin{aligned}\frac{\partial E}{\partial C_1} &= C_1(H_{11} - ES_{11}) + C_2(H_{12} - ES_{12}) = 0 \\ \frac{\partial E}{\partial C_2} &= C_1(H_{12} - ES_{12}) + C_2(H_{22} - ES_{22}) = 0\end{aligned} \tag{3.11}$$

where

$\int d\tau\psi_i\hat{H}\psi_j = H_{ij}$
$\int d\tau\psi_i\psi_j = S_{ij}$

It may be seen that the only nontrivial solution for C_1 and C_2 exists when the determinant

$$\begin{vmatrix} H_{11} - ES_{11} & H_{12} - ES_{12} \\ H_{12} - ES_{12} & H_{22} - ES_{22} \end{vmatrix} = 0. \tag{3.12}$$

Normalizing the wavefunctions, $S_{11} = S_{22} = 1$. By symmetry, $H_{11} = H_{22}$. Solving the determinant for E, we get

$$E_s = \frac{H_{11} + H_{12}}{1 + S_{12}} \quad \text{and} \quad E_a = \frac{H_{11} - H_{12}}{1 - S_{12}}, \tag{3.13}$$

where

s is the symmetric solution
a is the antisymmetric solution
H_{12} represents the lowering of the energy between hydrogen ions, hence is negative
S_{12}, which represents the overlap of the two electron wavefunctions, is positive, but less than 1 since they are not in the same position.

Hence, it may be seen that $E_s < E_a$ and that $E_a > 0$. Putting these values back into Equation 3.10, we find that $C_1 = C_2$ for E_s and that $C_1 = -C_2$ for E_a. We can now construct the symmetric and antisymmetric wavefunctions,

$$\psi_s = C_1(\psi_1 + \psi_2) \quad \text{and} \quad \psi_a = C_1(\psi_1 - \psi_2). \tag{3.14}$$

Normalizing the two wavefunctions and constructing the Hamiltonian from the electrostatic attraction and repulsion between the electrons and the nuclei, the calculated value for $E_s = -1.77$ eV, which represents the bonding state and $E_a > 0$ represents the antibonding state. The actual disassociation energy for the hydrogen molecule is 2.78 eV. So the theory is not exact, which is not surprising considering the approximations that were made. The important feature of this exercise is the fact that a bonding state resulted from the symmetric combination of wavefunctions; whereas an antibonding state resulted from the antisymmetric combination.

3.3.2 LCAO Approach

These concepts are used in the simpler linear combination of atomic orbitals (LCAO) approach. We construct a molecular orbital by starting with the wavefunctions of isolated atoms and take a linear combination of these wavefunctions to describe the state of the electron in the molecule. Consider, for example, a diatomic molecule consisting of atoms A and B. For sufficiently large distance between them, the molecular wavefunction can be written as $\psi_+ = \psi_A + \psi_B$ and $\psi_- = \psi_A - \psi_B$ where ψ_A and ψ_B are the electron wavefunctions of the electrons in A and B when they are isolated and noninteracting. Recall that radial wavefunction of the s-electron has the form, $\psi = \exp(-r/a)$, which peaks at $r = 0$. The ψ_+ and ψ_- are shown schematically below. The symmetric wavefunction ψ_+ is called the bonding orbital because $\psi_+^2 = \psi_A^2 + \psi_B^2 + 2\psi_A\psi_B$ and results in an increase of charge between the two atoms. Similarly, the antisymmetric wavefunction ψ_- is called an antibonding orbital because the electron density is excluded between the two atoms as shown in Figure 3.4.

The two positively charged ions are attracted to the increased electronic charge between them, which lowers the energy below E_A or E_B, the energies of the electron in the isolated state. Reducing the charge in this region increases the repulsion and raises the energy of the antibonding level, as shown in Figure 3.5. The closer the two ions are brought together, the greater the difference between the bonding and antibonding energy, until their inner electron shells begin to overlap and produce a strong repulsive force.

3.3.3 Sigma and Pi Bonds

The p-orbitals are directional in nature and can form two types of bonds, the so called σ and π bonds. In the π bond, the overlap is perpendicular to the line between the two

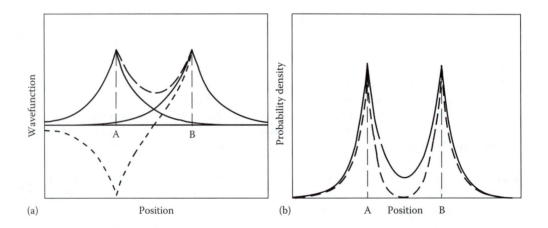

FIGURE 3.4

(a) Schematic of 1s overlapping wavefunctions of two atoms as they are brought together (solid lines). The long dashed line is the sum of the two wavefunctions and the short dashed line is their difference. (b) The probability density of the two wavefunctions in Figure 3.4b. The solid line represents the ψ_+^2 and the dashed line represent the ψ_-^2. The negative charge between the two atoms in the case of ψ_+^2 produces the bonding state.

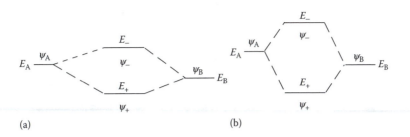

(a)   (b)

FIGURE 3.5
Energies of the bonding and antibonding molecular orbitals in a diatomic molecule. Atoms A and B have energies E_A and E_B, respectively, when separated. As they are brought together and their wavefunctions begin to overlap, they form a bond and an antibond energy level as shown in (a). As they are brought closer, the more the separation between the two levels as shown in (b).

atoms; whereas, in the σ bond, the overlap of the p-wavefunction is along the line between the two atoms as shown in Figure 3.6.

The $+$ and $-$ signs used in this type of diagram indicate the phases of the wavefunctions, not the electrical charge. Overlap of two $+$ or two $-$ regions indicates that the two wavefunctions add constructively; whereas, overlap of a $+$ and $-$ region indicates destructive interference. The σ bond is generally stronger because of the greater overlap of the wavefunction. There is no restriction to the rotation about the axis of the σ-bond, whereas rotation is not allowed about the π-bond.

To illustrate how this works with a multielectron atom, consider the N_2 molecule. Here we must accommodate 14 electrons, 7 from each atom. As the two N atoms are brought together, each of the atomic levels split into a bonding and an antibonding state as shown in Figure 3.7.

Note that the splitting between the $2p_x$ bonding and antibonding levels is greater than between the splitting between the $2p_y$, $2p_z$ levels. This is because it is assumed that a σ bond exists along the x-axis. π bonds exist between the y and z components. Both the bonding and antibonding orbitals are occupied for the inner electrons, but only the bonding orbitals are occupied in the $2p$ states. The six $2p$ electrons form a very strong triple bond between the two atoms, which is the reason why the N_2 molecule is so stable.

The O_2 molecule introduces two additional electrons. Since there are no more bonding states available for them to occupy, they must go to the next available $2p_y$ or $2p_z$ antibonding states which causes that lobe to become saturated and no longer able to form a bond. Thus the disassociation energy of the double bond in the O_2 molecule is considerably less that the N_2 molecule. This accounts for the fact that the primary atmospheric constituents at low-earth orbit altitudes are atomic oxygen and molecular nitrogen.

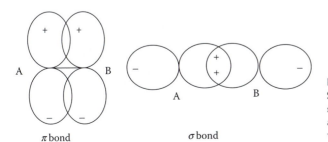

π bond σ bond

FIGURE 3.6
Schematic of π and σ bonds. The σ bond is stronger because it has more overlap, but allows rotation about the bond line. The weaker π bond does not allow rotation.

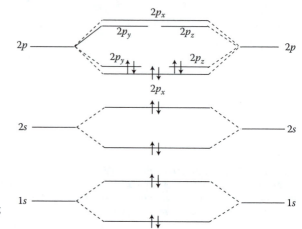

FIGURE 3.7
Schematic of the bonding and antibonding
levels in the N_2 molecule.

3.3.4　s–p Bonds

Another type of bond is the s–p bond shown in Figure 3.8. In this case the s-wavefunction
has no negative component and the greatest overlap occurs when the positive portion of
the p-wavefunction is directed toward the center of the s-electron. The H_2O molecule is
formed by hydrogen making s–p bonds with the p_x and p_y lobes of the oxygen atom as
shown in Figure 3.9. Not shown is the bond bending from the mutual repulsion of the two
naked H^+ ions, which produces the well-known 104.5° bond angle between the H^+ ions in
the H_2O molecule.

　　A similar situation occurs in the NH_3 molecule. Instead of the three hydrogen ions
making 90° angles with one another as would be expected from the p-electron lobes, the
actual bond angle is 106.7° because of mutual repulsion between the H^+ ions.

3.3.5　Hybridization

The carbon atom, with its $1s^2 2s^2 2p^2$ configuration has a valence of 4. However, in many of
the systems involving C, these valence electrons behave as if they were identical and form
directed bonds toward the vertices of a tetrahedron with a bond angle of 109.5° surrounding the atom (e.g., diamond, CH_4, etc.). This can be explained by assuming that the $2s$ and
$2p$ orbitals can form a linear combination such that the four orbitals are equivalent and
point to the opposite corners of a cube with the C atom at the center. The $n=2$ hydrogen-
like wavefunctions from Chapter 2 are given in Cartesian space by

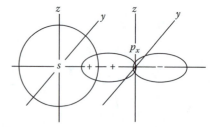

FIGURE 3.8
Schematic of the overlapping wavefunctions in the s–p bond.

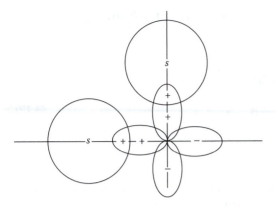

FIGURE 3.9
s–p bonds formed between the p_x and p_y lobes of oxygen with hydrogen to form the H_2O molecule. The mutual repulsion between the two naked H^+ ions distorts the 90° bond angle to 104.5°.

$$\psi_{2s} = \frac{1}{4\sqrt{2\pi}} \left(\frac{1}{a_0}\right)^{3/2} (2-\rho)e^{-\rho/2}$$

$$\psi_{2p_x} = \frac{1}{4\sqrt{2\pi}} \left(\frac{1}{a_0}\right)^{3/2} \rho e^{-\rho/2} \sin\theta \cos\phi$$

$$\psi_{2p_y} = \frac{1}{4\sqrt{2\pi}} \left(\frac{1}{a_0}\right)^{3/2} \rho e^{-\rho/2} \sin\theta \sin\phi \qquad (3.15)$$

$$\psi_{2p_z} = \frac{1}{4\sqrt{2\pi}} \left(\frac{1}{a_0}\right)^{3/2} \rho e^{-\rho/2} \cos\theta$$

where $\rho = r/a_0$.

It is left as a problem to show that the combination

$$\psi_{\mathrm{I}} = \frac{1}{2}\left(\psi_{2s} + \psi_{2p_x} + \psi_{2p_y} + \psi_{2p_z}\right)$$

$$\psi_{\mathrm{II}} = \frac{1}{2}\left(\psi_{2s} + \psi_{2p_x} - \psi_{2p_y} - \psi_{2p_z}\right)$$

$$\psi_{\mathrm{III}} = \frac{1}{2}\left(\psi_{2s} - \psi_{2p_x} + \psi_{2p_y} - \psi_{2p_z}\right) \qquad (3.16)$$

$$\psi_{\mathrm{IV}} = \frac{1}{2}\left(\psi_{2s} - \psi_{2p_x} - \psi_{2p_y} + \psi_{2p_z}\right)$$

has this property. Essentially one of the $2s$ electrons is promoted to the $2p$ state which can then hybridize with the remaining $2s$ electron to form four overlapping bonds with neighboring atoms to provide an energy reduction that more than compensates for the additional energy needed to promote the $2s$ electron. This is known as the sp^3 hybrid bond. Similarly, other group IV atoms can form similar hybrids (Si, Ge, and α-Sn) as well as III–V compounds such as GaAs, GaN, InP, BN, etc. The tendency to form this hybrid bond diminishes as the interatomic distance increases. This accounts for the reason that Sn can exist either as a metal or a semimetal and reason why Pb is always a metal as will be discussed further in Chapter 20. Correspondingly, the bond strength diminishes from 7.30 eV for diamond, 4.64 eV for Si, and 3.87 eV for Ge.

Carbon can also hybridize in a threefold structure by taking the following linear combinations:

$$\psi_I = \frac{1}{\sqrt{3}}\psi_{2s} + \sqrt{\frac{2}{3}}\psi_{2p_x}$$

$$\psi_{II} = \frac{1}{\sqrt{3}}\psi_{2s} - \frac{1}{\sqrt{6}}\psi_{2p_x} + \frac{1}{\sqrt{2}}\psi_{2p_y} \cdot \qquad (3.17)$$

$$\psi_{IIi} = \frac{1}{\sqrt{3}}\psi_{2s} - \frac{1}{\sqrt{6}}\psi_{2p_x} - \frac{1}{\sqrt{2}}\psi_{2p_y}$$

This is known as the sp^2 hybrid and it is left to the student to show that his forms a planar structure with bond angles 120° apart. Graphite is structured with sp^2 bonds which gives it a very strong planar hexagonal network. The remaining electron delocalizes and is free to meander through the structure as a conduction electron. The layers are bonded only by van der Waals forces, which accounts for the slippery nature of the material. This sp^2 bond is also responsible for the formation of buckyballs and carbon nanotubes, which are in the forefront of research in nanotechnology (see Chapter 5).

3.4 Metallic Bond

A vast majority of the elements are metals or metal-like with cohesive energies ranging from ~1 to 9 eV. Figure 3.10 shows how these energies depend on their position in the periodic chart.

The first two columns are the alkali metals with only s electrons forming a weak metallic bond. The bond energies rise rapidly in the transition metals, peak, and then fall to a minimum at column 12. But something funny seems to be going on in columns 4–9 in the atoms with $3d$ electrons. Then another rise and fall in the cohesive energies is seen with a peak occurring at column 14. How do we account for this behavior?

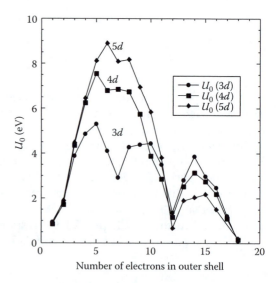

FIGURE 3.10
Cohesive energies of the $3d$, $4d$, and $5d$ metallic elements as a function of their outer shell electronic configuration.

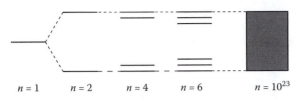
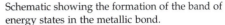

$n = 1$ $n = 2$ $n = 4$ $n = 6$ $n = 10^{23}$

FIGURE 3.11
Schematic showing the formation of the band of energy states in the metallic bond.

3.4.1 Simple Metals

Recall from the discussion of the covalent bond that bringing two atoms close together formed a pair of electron orbitals of different energies (a bonding and an antibonding orbital) and that the difference in energy levels increases as the atoms are brought close together. Bringing two more atoms adds two more orbitals with energies between the bonding and antibonding states of the two closest atoms. Two more atoms add two more orbitals and so on for n atoms. Thus in a typical metal in which there are $\sim 10^{23}$ atoms, there will be an almost continuous band of $n = \sim 10^{23}$ very closely spaced states that can be considered a continuous band (Figure 3.11).

Each of these states can accommodate 2 electrons (one spin up and one spin down), so for the alkali metals, these states will be only half full. This leaves plenty of available states for the electrons to move into in response to an applied electric field, which accounts for the high conductivity and other electronic properties of metals. It would appear at first glance that the group II metals would completely fill the available bands and would have no available states left to move into. This would be the case if it were not for the fact that the 2s bands overlap into the p bands which provide the additional closely spaced states for the electrons to move into (Figure 3.12).

Since there are so many closely spaced states from so many atoms, the valence electrons delocalize and can move freely throughout the solid. Thus, one can think of a metal as an electron gas which moves through a lattice of positive ion cores. The binding comes from the presence of the electron density between the ion cores whose attractive force overcomes the repulsive force between the ions, very much like the presence of the anions between the cations in the case of ionic bonding. Since the electrons are delocalized, the metallic bond

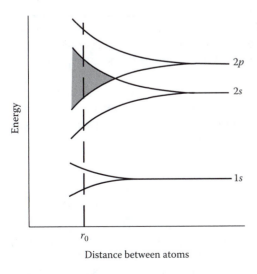

FIGURE 3.12
Showing overlap of 2s and 2p bands that makes electrical conductivity in divalent metals possible.

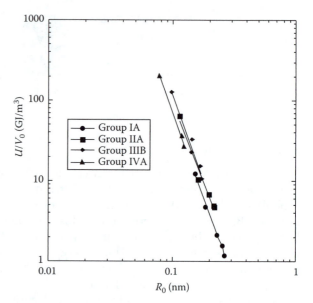

FIGURE 3.13

Energy per volume vs. radius for the simple metals and group IVA elements. The slopes for each of the groups are approximately −4.

loses the directionality of the covalent bond and the ion cores tend to form close-packed structures.

Qualitatively, we can see that bond energies increase in going from column 1 (IA) to 2 (IIA) and from column 13 (IIIA) to 14 (IVA) in Figure 3.10. This can be understood simply by the fact that more electrons are involved in the forming the bond. Also we see that the bonds tend to be weaker with increasing atomic numbers in the same column. As the atoms get larger, they tend to be farther apart and the bond energy falls off with distance.

Unfortunately there are no simple theories to predict the cohesive energies of the metals like the coulomb attraction in ionic crystals. More sophisticated quantum mechanical theories using pseudopotential or other modeling techniques are generally required. There are some interesting correlations, however.

Figure 3.13 plots U/V_0 as a function of the radius for group I (Li, Na, K, Rb, and Cs), group II (Be, Mg, Ca, Sr, and Ba), group III (B, Al, Ga, In, and Tl), and the diamond structured group IV (C, Si, and Ge) elements. The group IV elements are not metals, but were included because they fit the pattern of the simple metals. The curves almost fall on top of each other and the slope of each of the groups is close to −4 (the average slope is −4.11). The R_0^{-4} dependence of the energy/volume is quite similar to the energy dependence of the ionic-bonded systems.

The cohesive energy/volume is plotted against valence electron density in Figure 3.14. Again the curves for the different groups fall close to one another, but there are some differences in the slopes. The slopes of groups I and II average 1.34, which is consistent with $U/V_0 \sim N_{val}/R_0^4$. The slopes of groups III and IV were 0.75 and 0.50, respectively.

3.4.2 Why No Metallic Hydrogen?

Hydrogen and lithium belong to the same group in the periodic table, but at ambient temperature hydrogen forms a H_2 molecule and exists as a gas while Li forms a metal. Why?

The difference lies in the availability of nearby states to the *s* states of the metal-forming systems. In H the next available state (2*s*) is more than 9 eV above the ground state. Chemical bonds typically run 4–5 eV. In the case of H, there are no other states available

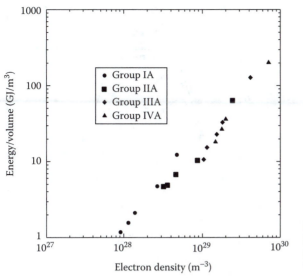

FIGURE 3.14
Energy per volume vs. valence electron density for simple metals and group IVA elements. The average slope is 0.85.

so the 1s bonding orbital is filled by forming H_2, and the bond is saturated. In the case of the alkali metals such as Li, there are the $2p$, $3s$, $3p$, etc., states that are only 1–2 eV away from the $2s$ state. The presence of these available states allows the Li atom to share its $2s$ electron with other atoms in order to form a metallic bond.

It is believed that metallic hydrogen can exist at pressures on the order of 100 GPa and that the core of large planets, such as Jupiter, may consist of metallic hydrogen. Such pressures can be created for an instant of time in the laboratory using shock compression techniques and have produced evidence of metallic hydrogen, but the hydrogen does not remain in the metallic form after the pressure is relieved (see Weir et al., 1996).

3.4.3 Transition Metals

In addition to the metallic bond formed by the s-electrons, the transition metals have some degree of covalent bonding through their p- and d-electrons, which accounts for their greater strength. The pure metallic bond is quite weak in the absence of p and d electrons and ranges from 3.32 eV for Be to 0.67 eV for Hg when these covalent bonds are saturated. In Figure 3.10 we saw the cohesive energies begin to increase rapidly in column 3 when the d-shells start to fill (Sc, Y, and La) and, except for the anomalous behavior of the 3d transition metals, reach a peak, and start to decrease when the d-shells are approximately half full when the added electrons are forced to fill antibonding states. The cohesive energies reach a minimum when the d-shell is completely full in column 12 (Zn, Cd, and Hg).

One would expect the peak in the cohesive energies to peak in column 7 when the d-shells are half full. This does not quite happen in Figure 3.10. However, if the cohesive energy/volume instead of energy/atom of the transition metals is plotted (Figure 3.15), the curves are seen to be more symmetrical about column 7 than in Figure 3.10. The anomalous behavior of the 3d metals (Cr, Mn, Fe, Co, and Ni) can be explained by the fact that some of their d-electrons are tied up in magnetic interactions and do not overlap to make covalent bonds (see Chapter 25).

As with the simple metals, a theoretical treatment of the cohesive energies of the transition metals requires a more sophisticated quantum mechanical treatment. A few trends can be observed, however. Unlike the simple metals, the strongest bonds occur

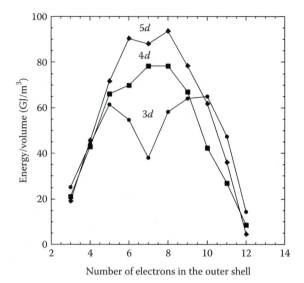

FIGURE 3.15

Cohesive energy per volume vs. column number for the $3d$, $4d$, and $5d$ transition metals. In this plot, the curves tend to be more symmetrical about column 7 where the number of bonding states should be maximized.

between the atoms with the highest atomic numbers when their d-orbitals are approximately half full. Since the atomic radii of the $5d$ metals remain approximately the same as their $4d$ counterparts (see Section 3.5), there will be greater overlap of the larger d-shells of the $5d$ elements, hence stronger bonds. This trend is reversed as the d-orbitals are just starting to fill or are nearly saturated and the strongest bonds are formed among the lower atomic weight elements as in the case of the simple metals.

3.5   Atomic and Ionic Radii

Because the bonding force between atoms falls rapidly with distance, the effective size of the atoms or ions plays a crucial role in the binding energies. X-ray diffraction offers an extremely precise (1 part in 10^5) method for determining the spacing between atomic planes in a crystal as well as how the atoms are configured within the planes (see Chapter 6). Assuming the atoms (or ions) touch each other, one can infer nearest neighbor distances and, if the atoms are all the same, one can assign an atomic radius to that particular atom. Figure 3.16 is a plot of the atomic radii of the metallic elements in rows 4–6 in the periodic table.

Comparing this plot with Figure 3.10 or with Figure 3.15, it may be seen that the smaller the radius, the stronger the bond for the simple metals. Note how the size decreases with increasing electron density for a given row and how much larger the atoms of the alkali metals are than the transition metals. It is also interesting that row 5 and row 6 transition metals have nearly the same radii even though the row 6 elements have many more core electrons. The combination of their high atomic weights and small atomic diameters is responsible for the high densities of transition metals.

Determining the atomic radii from x-ray measurements of the spacing between planes of atoms gets a little more complicated in binary systems, especially if the atoms are ionically bonded. One can certainly measure the distance between atoms A and B, but how much does one assign to A and to B? Strictly speaking, the nearest neighbor spacing does in fact depend somewhat on the species, the bonding type, and the coordination number. For example, the nearest neighbor distance in sp^3-bonded carbon (diamond) is 0.154 nm, so we

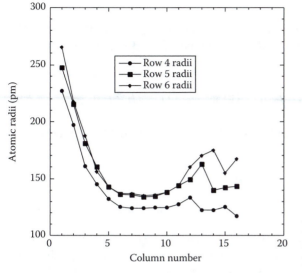

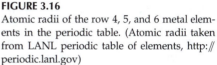

FIGURE 3.16
Atomic radii of the row 4, 5, and 6 metal elements in the periodic table. (Atomic radii taken from LANL periodic table of elements, http://periodic.lanl.gov)

assign an atomic radius of 0.077 nm to carbon. Similarly, 1/2 the nearest neighbor distance in sp^3-bonded silicon is 0.117 nm. Silicon carbide has the same crystal structure except that each Si atom is surrounded by 4 C atoms and vice versa. Can we add 0.077 and 0.117 to get 0.194 nm for the nearest neighbor distance in SiC? Well almost. The observed nearest neighbor distance in SiC is 0.189 nm. Though not absolutely precise, taking the atomic radius from nearest neighbor distances of like atoms is good enough for the most part to predict nearest neighbor distances and crystal structures for metallic and covalent bonding systems.

Things are decidedly different for ionically bonded systems. Generally when an electron leaves a metallic atom to form a cation, the cation will be effectively smaller than the atom in a metallic bond even though the valence electron in the metal is delocalized. Similarly, the size of the anion that has gained the electron is generally larger than the neutral atom. A self-consistent set of ionic radii have been worked out for systems with a coordination number of 6 (rock salt structure). A correction of +0.008 nm must be added to the sum of the standard ionic radii for coordination number of 8 and a correction of −0.011 must be subtracted for tetrahedrally coordinated structures.

Because the electrons in a metal become delocalized, it may seem strange that the hard sphere size of the ions cores consistent with the measured density of the solid is larger than the ionic diameter used to calculate the size of ionic compounds. The explanation is that a delocalized electron is not completely lost from the ion core as it is when transferred to an electronegative atom. The delocalized electrons are shared among all of the ion cores so the effective size of the ion core is less diminished when its valence electrons delocalized.

3.6 Secondary Bonding

The weaker secondary bonding mechanisms generally play a lesser role in the presence of primary bonds such as providing the attractive forces between covalently bonded chains in a polymer. However, they provide the cohesive forces that hold materials together when no primary bonds are present. All of these secondary bonds involve dipole–dipole interactions.

3.6.1 Electric Dipole

An electric dipole consists of a pair of opposite charges separated by a distance **d**. The electric field produced by an electric dipole along a line including **d** may be written as

$$E = \left(\frac{q}{4\pi\varepsilon_0 r^2} - \frac{q}{4\pi\varepsilon_0 (r+d)^2} \right) \frac{\mathbf{d}}{|\mathbf{d}|} = \frac{q}{4\pi\varepsilon_0 r^2} \left(1 - \frac{1}{(1+d/r)^2} \right) \frac{\mathbf{d}}{|\mathbf{d}|}. \tag{3.18}$$

If $d \ll r$, the bracket term can be expanded to give $1 - (1+d/r)^{-2}$ so that the field becomes

$$E = \frac{2dq}{4\pi\varepsilon_0 r^3} = \frac{2\mathbf{p}}{4\pi\varepsilon_0 r^3}, \tag{3.19}$$

where **p** is the dipole moment which is taken to be positive in the direction of the positive charge.

The potential energy of a dipole in an electric field is given by $U = -\mathbf{p} \cdot \mathbf{E}$. Therefore, an attractive potential exists between two electric dipoles that are aligned nose to tail that goes as $-p_1 p_2 / r^3$ and the force between them goes as $-p_1 p_2 / r^4$. Thus a dipole–dipole bond is directional and short ranged.

The electric field from one dipole can induce a dipole moment in an adjacent atom that does not have a permanent dipole moment according to

$$\mathbf{p_2} = \alpha \mathbf{E} = \frac{2\alpha \mathbf{p_1}}{4\pi\varepsilon_0 r^3}, \tag{3.20}$$

where α is the polarizability of the atom. Therefore, the interaction energy between a dipole and an atom or molecule with an induced dipole goes as $-\alpha p_1 p_2 / r^6$ and the force between them is maximized along the axis of the permanent dipole.

3.6.2 Hydrogen Bond

When a hydrogen atom combines with an electronegative atom such as O, N, or a halogen, its s-electron is essentially lost to the other atom, leaving a virtually bare proton sitting on the molecule. The strong dipole field that results from this bare positive charge can attract a negative ion, another polar molecule, or even induce a dipole in a nonpolar atom. All of these possibilities are forms of hydrogen bonding. Several examples are shown in Figure 3.17.

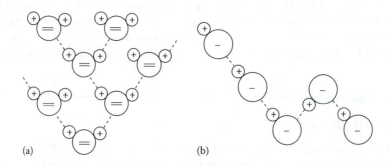

FIGURE 3.17
Schematic of hydrogen bonding in polar substances (a) H_2O and (b) HF. The two H atoms in H_2O allow water ice to form open tetrahedral-like structures whereas as a single functional molecule such as HF can only form chains.

Because of the small size of the hydrogen ion, only two counterions can get close to it without touching each other. Therefore the hydrogen atom is always doubly coordinated in a hydrogen bond. The strength of the hydrogen bond is typically 0.1–0.5 eV, which is less than the primary bonding energies, but large enough to hold substances together at ambient temperatures in the absence of any of the primary bonding mechanisms. Thus the hydrogen bond is very important in polymers and other organic solids, in biological systems, and in polar molecular systems such as water. The tendency for such bonds to form tetrahedral open structures in the case of water is responsible for many of the unique properties of water such as ice having a lower density than liquid water.

3.6.3 van der Waals Bond

As shown in Section 3.6.1, a permanent dipole can induce a dipole in a nonpolar atom or molecule. It is also possible for the induced dipole to induce a dipole in another nonpolar atom or molecule. Thus any two neutral nonpolar atoms or molecules can be attracted to each other by inducing dipoles in each other according to

$$U = -\frac{2p_1}{4\pi\varepsilon_0 r^3} \cdot \frac{\alpha 2p_1}{4\pi\varepsilon_0 r^3} = -\frac{4\alpha p_1^2}{(4\pi\varepsilon_0)^2 r^6}. \tag{3.21}$$

But how does this process get started between two neutral nonpolar particles? Fluctuations of the charge distribution are continuously occurring because of the zero-point energy in the system in the absence of any other mechanism. A fluctuation in the charge distribution in the first particle produces a temporary dipole moment which induces a dipole moment in an adjacent particle. This induced dipole in the second particle induces a dipole moment in the first particle and so the process continues.

This resulting force between the two particles is known as the van der Waals force (also called the London dispersion force). Like the metallic and ionic bonds, the force acts along the line between any two particles, so it is best described as a central force. Because there is no charge on the particles, close-packed structures are favored in van der Waals solids. Note that the attractive (negative) part of the potential goes as r^{-6} (or V^{-2}, which is the correction for the molecular interaction in the equation of state for a van der Waals gas).

As one might imagine, the van der Waals bond is extremely weak (on the order of ambient thermal energy), but it is always there and works when nothing else does. It provides the mechanism for condensation and solidification of noble and molecular gases. Its strength increases with increasing number of electrons, which explains why F_2 and Cl_2 are gases at ambient temperature while Br_2 is liquid and I_2 is solid. The van der Waals bonding becomes even more important in macromolecular systems such as polymers and biomolecules.

3.6.4 Lennard–Jones 6–12 Potential

One widely used expression for the short-range interaction energy between two atoms or molecules is the Lennard–Jones 6–12 potential between the *i*th and *j*th particle, which may be written as

$$U(r_{ij}) = A\left[\left(\frac{\sigma}{r_{ij}}\right)^{12} - \left(\frac{\sigma}{r_{ij}}\right)^6\right], \tag{3.22}$$

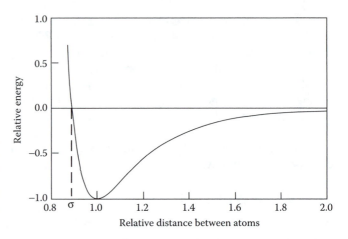

FIGURE 3.18
Lennard–Jones potential function. The quantity, σ, is the closest distance one particle falling from infinity can approach another. Note the short range of the Lennard–Jones potential by comparing this figure with Coulomb potential in Figure 3.2.

where A and σ are constants to be determined for a particular system. The 6th power attractive term is chosen to model the van der Waals interaction while the 12th power repulsive term is purely arbitrary. As shown in Figure 3.18, the potential is zero when r_{ij} equals σ, which is the closest distance one particle falling from infinity can come to the other particle. Thus σ can be interpreted as the diameter of the particles in a hard sphere model.

Differentiating to find r_0 that minimizes the potential, we find $r_0 = 2^{1/6}\,\sigma$. Putting this back into Equation 3.2, we can express the coefficient A in terms of the binding energy U_0 (taken as a positive number) and write the potential as a function of r/r_0.

$$U(r_{ij}) = 4U_0\left[\left(\frac{\sigma}{r_{ij}}\right)^{12} - \left(\frac{\sigma}{r_{ij}}\right)^{6}\right] = U_0\left[\left(\frac{r_0}{r_{ij}}\right)^{12} - 2\left(\frac{r_0}{r_{ij}}\right)^{6}\right]. \tag{3.23}$$

This form is useful in modeling molecular vibrations and for relating the pressure of a van der Waals gas to the attractive potential between the molecules.

We can apply the Lennard–Jones potential to a crystalline solid such as a condensed noble gas in which van der Waals forces are the only bonding mechanism. We replace r_{ij} with $p_{ij}r$ as we did for the case of ionic bonding and write the binding energy for a particular configuration,

$$U(r) = \frac{1}{2}\sum_{i\neq j}U(r_{ij}) = NA\sum_{j>0}\left[\left(\frac{\sigma}{p_{0j}r}\right)^{12} - \left(\frac{\sigma}{p_{0j}r}\right)^{6}\right] = NA\left[\left(\frac{\sigma}{r}\right)^{12}\Sigma_{12} - \left(\frac{\sigma}{r}\right)^{6}\Sigma_{6}\right], \tag{3.24}$$

where Σ_{12} and Σ_6 are short for the lattice sums $\sum_{j=1}p_{0j}^{-12}$ and $\sum_{j=1}p_{0j}^{-6}$, respectively. Differentiating Equation 3.24 with respect to r to find the r_0 that minimizes the U_{lat},

$$\left(\frac{\sigma}{r_0}\right)^{6} = \frac{\Sigma_6}{2\Sigma_{12}}. \tag{3.25}$$

Putting this back into Equation 3.23 gives

$$U_0 = -NA\left[\left(\frac{\Sigma_6}{2\Sigma_{12}}\right)^{2}\Sigma_{12} - \left(\frac{\Sigma_6}{2\Sigma_{12}}\right)\Sigma_6\right] = \frac{-NA}{4}\frac{\Sigma_6^2}{\Sigma_{12}}. \tag{3.26}$$

Putting Equations 3.26 and 3.25 back into Equation 3.24, the Lennard–Jones potential may be written as a function of r in the form,

$$U(r) = U_0 \left[\left(\frac{r_0}{r} \right)^{12} - 2 \left(\frac{r_0}{r} \right)^6 \right]. \tag{3.27}$$

The van der Waals bond is universal in that it applies to all atoms. But since the polarizability $\alpha \sim 10^{-40}$, the force is very weak ($U_{lat} \sim 0.01 - 0.1$ eV) and is usually overshadowed by other stronger bonds. However, in the absence of other bonds (e.g., for the noble gases), it becomes the primary bonding mechanism. It is short ranged and nondirectional, hence when acting as a primary bond, it favors a high coordination number found in either fcc of hcp structures.

3.6.5 Lattice Sums

The lattice sums for the van der Waals bond for an fcc structure are

$$\sum_{j>0} p_{0j}^{-6} = \Sigma_6 = 14.45392; \quad \sum_{j>0} p_{0j}^{-12} = \Sigma_{12} = 12.13188; \quad \frac{\Sigma_6^2}{\Sigma_{12}} = 17.220398. \tag{3.28}$$

Similarly for a hcp structure,

$$\sum_{j>0} p_{0j}^{-6} = \Sigma_6 = 14.45489; \quad \sum_{j>0} p_{0j}^{-12} = \Sigma_{12} = 12.13229; \quad \frac{\Sigma_6^2}{\Sigma_{12}} = 17.222127. \tag{3.29}$$

It looks as if the hcp lattice has slightly lower energy and would be favored. Interestingly, most van der Waals-bonded substances, primarily the noble gases actually form fcc structures. The only explanation is that it is not a perfect theory. Remember, the assumed form for the potential repulsive potential was arbitrary and has no theoretical basis.

3.7 Other Potential Functions

A number of empirical potential functions have been suggested that can be fitted to various molecular interactions as well as to various solids. These will be reviewed in this section.

3.7.1 Born–Mayer Potential

As mentioned previously, some texts prefer to consider only nearest interactions for the repulsive term and write it in the form $U_R = B \exp(-\beta r)$.

$$U(r) = -\frac{NAe^2}{4\pi\varepsilon_0 r} + B \exp(-\beta r). \tag{3.30}$$

Taking derivatives to obtain the equilibrium energy and configuration as before, we find that

$$U(r) = -\frac{NAe^2}{4\pi\varepsilon_0 r_0} \left(\frac{r_0}{r} - \frac{e^{\beta(r_0 - r)}}{r_0 \beta} \right) \tag{3.31}$$

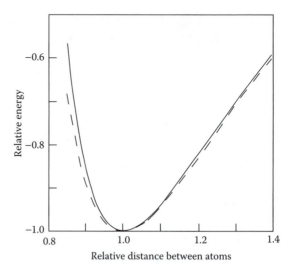

FIGURE 3.19
Comparison of Born–Mayer potential (dashed line) with the Born potential $(1/r^n)$ for the repulsive term (Equation 3.6) (solid line) with $r_0\beta = n$. Only the region around $r = r_0$ is shown because this is where the difference between the two is the largest.

and the energy at r_0 is

$$U(r_0) = -\frac{NAe^2}{4\pi\varepsilon_0 r_0}\left(1 - \frac{1}{r_0\beta}\right). \tag{3.32}$$

Comparing this result with Equation 3.7, we see that $r_0\beta$ plays the same role as n in the power-law formulation. A comparison of the two potentials is shown in Figure 3.19.

3.7.2 Mie Potential

A generalized empirical potential function proposed by Mie can be written as the sum of an attractive and repulsive term in the form

$$U(r) = -Ar^{-m} + Br^{-n}, \tag{3.33}$$

where A, B, m, and n are constants to be determined empirically. Taking a derivative with respect to r and setting it to zero at r_0 as before allows the constant B to be eliminated by setting

$$B = A\frac{m}{n}r_0^{n-m}. \tag{3.34}$$

Using this result to relate the equilibrium spacing r_0 to the lattice energy at equilibrium,

$$U(r_0) = -\frac{A}{r_0^m}\left(\frac{n-m}{n}\right). \tag{3.35}$$

Putting this back into the general equation (Equation 3.33), the potential becomes

$$U(r) = \frac{U(r_0)}{n-m}\left[n\left(\frac{r_0}{r}\right)^m - m\left(\frac{r_0}{r}\right)^n\right]. \tag{3.36}$$

One can see that by setting $m = 1$, Equation 3.36 reduces to Equation 3.6. Also, the Lennard–Jones potential is obtained by setting $m = 6$ and $n = 12$ which is the same as Equation 3.27.

3.7.3 Buckingham Potential

The Buckingham potential can be thought of as a generalized Born–Mayer potential with an exponent other than 1 assigned to the attractive potential. One form is

$$U(r) = \frac{U_0 mn}{m - n} \left[\frac{1}{m} \left(\frac{r_0}{r} \right)^m - \frac{1}{n} e^{n(1 - r/r_0)} \right]. \tag{3.37}$$

If we set $m = 1$ and $n = \beta r_0$, we obtain

$$U(r) = \frac{U(r_0) \beta r_0}{1 - \beta r_0} \left[\left(\frac{r_0}{r} \right) - \frac{1}{\beta r_0} e^{\beta(r_0 - r)} \right], \tag{3.38}$$

which is the Born–Mayer potential given by substituting Equation 3.32 into Equation 3.31.

3.7.4 Morse Potential

We define a displacement from equilibrium x as $x = (r - r_0)/r_0$, from which $r_0/r = (1 + x)^{-1}$. Using Equation 3.36, we can write

$$U(r) = \frac{U(r_0)}{n - m} \left[n \left(\frac{r_0}{r} \right)^m - m \left(\frac{r_0}{r} \right)^n \right] = \frac{U(r_0)}{n - m} [n(1 + x)^{-m} - m(1 + x)^{-n}]. \tag{3.39}$$

For small x, $e^x = 1 + x + \cdots$. Substituting e^{-mx} for $(1 + x)^{-m}$,

$$U(x) = \frac{U(r_0)}{n - m} [n e^{-mx} - m e^{-nx}]. \tag{3.40}$$

We now set $n = 2m$

$$U(x) = U(r_0) \left[2e^{-mx} - e^{-2mx} \right], \tag{3.41}$$

which is the Morse potential. The Morse potential is often used for molecular modeling as well as for metals. Changing the sign in the Morse potential gives an antibonding potential with 4 times the energy of the bonding potential at the equilibrium point.

The Morse potential is sometimes written in the form $\Delta U(x) = U(r_0) - U(x)$ or

$$\Delta U(x) = U(r_0) - U(r_0) \left[2e^{-mx} - e^{-2mx} \right] = U(r_0) \left[1 - 2e^{-mx} + e^{-2mx} \right]$$
$$= U(r_0)[1 - e^{-mx}]^2. \tag{3.42}$$

3.8 Summary

The three primary chemical bonds are the ionic bond, the covalent bond, and the metallic bond. The ionic bond requires an electron transfer one atom to another atom with a higher electron affinity and forms between elements that have a large difference in electronegativity. The attractive forces between the resulting anions and cations are purely electrostatic and can be described by Coulomb's law. The ionic bond is the strongest of the chemical bonds, has a long range since the energy falls off as $1/r$, and is nondirectional in nature. Since it is necessary for each ion to be surrounded by counterions, the coordination number (number of nearest neighbors), hence the crystal structures that are possible, is determined by the relative sizes of the cations and anions.

The covalent bond forms by atoms sharing electrons with one another to form filled electron shells which lowers the energy of the system. As two atoms are brought together, their wavefunctions began to overlap, forming a bonding state and an antibonding state. The strength of the covalent bond can be quite large and depends along with other things on how many electrons are allowed go into the bonding state. Since the p, d, and f electrons have angular dependence, bonds formed by these overlapping wavefunctions are directional which leads to open diamond-type structures. Covalent bonds may form between like atoms as well as between dissimilar atoms, although the bonds between dissimilar atoms may be partially ionic, depending on the difference in their electronegativity.

Metallic bonds form by valence electrons becoming delocalized from atoms on the left hand side of the periodic table that have only a few loosely attached outer shell electrons. These electrons form a sea of negative charge that surrounds and attracts the positively charged ion cores with the electron sea filling the role of the anions in the ionic bond. The metallic bond is generally weaker than either the ionic or the covalent bond and is nondirectional. Since there is no restriction on the number of nearest neighbors other than geometry, metals tend to form close-packed structures with a coordination number of 12 although many metals prefer to form body-centered cubic structure with a coordination number of 8. The covalent bond is also formed among the d electrons in the transition metals which accounts for their strength, especially when the d shells are nearly half full so that none of the electrons have to fill antibonding states.

Secondary bonds are dipolar in nature. A polar molecule has a neutral charge but an inhomogeneous charge distribution such that one end of the molecule is positive and the other end negative. Such a molecule forms an electric dipole and is surrounded by an electric field that attracts other dipoles. An H atom that is covalently bonded to a halogen or to an O atom effectively loses its electron to the host atom which forms a strong electric dipole. Dipole–dipole bonds mediated by an attached H atom are said to be hydrogen bonded. The hydrogen bond, though considerably weaker than the primary bonds, is very important in the bonding of organic solids and as a secondary bond in polymer systems. The bond is short-ranged since the dipole field falls as $1/r^3$ and is directional along the dipole axis. Single functional molecules such as HF and HCl tend to form chains with nose to tail structures. Bifunctional molecules such as H_2O form open structures that cause ice to expand when it freezes.

An electric field can induce a dipole moment in a nonpolar molecule by pushing the positive and negative charges in opposite directions. A fluctuation in charge distribution in one nonpolar molecule can induce a dipole moment in an adjacent nonpolar molecule causing an attractive force to occur. London dispersion forces (also called van der Waals forces) are extremely weak and short ranged, falling as $1/r^6$, but they work when no other bonding forces are present. Consequently, they provide the mechanism by which noble and molecular gasses condense at very low temperatures. They also hold neutral layered solids such as graphite together and provide interactions between the covalently bonded chains in polymers.

A number of semitheoretical models have been proposed to describe the bond-energy curve in ionic-bonded and van der Waals-bonded systems in which the attractive force can be described accurately by an inverse power law. The repulsive potential is described by an exponential or a high order inverse power law. Such potentials are useful for relating physical properties such as the bulk modulus and elastic modulus to the lattice energy. Unfortunately, there are no simple lattice models that can be applied to metallic or covalently bonded systems.

Appendix: Madelung Summation

To illustrate what is involved in computing a Madelung sum, we will apply the method of Evjen to a two-dimensional (2D) lattice. Consider the lattice sum for a plane of ions on the (100) face of a rock salt structure shown in Figure A.3.1.

Pick an arbitrary ion near the center of the array and draw squares around it. (In a three-dimensional sum, we draw cubes around it.) We simply add the contributions from the ions in each square. For the first square, $R_1 = -4/1 + 4/2^{1/2}$. The contribution from the second square would be $R_2 = 4/2 - 8/(2^2 + 1^2)^{1/2} + 4/(2^2 + 2^2)^{1/2}$. The third term would be $R_3 = -4/3 + 8/(3^2 + 1^2)^{1/2} - (3^2 + 2^2)^{1/2} + 4/(3^2 + 3^2)^{1/2}$, etc. A general term can be written as

$$R_n = \frac{4(-1)^n}{n} + \sum_{i=1}^{n-1} \frac{8(-1)^{n+i}}{\sqrt{n^2 + i^2}} + \frac{4}{\sqrt{n^2 + n^2}} \qquad (A.3.1)$$

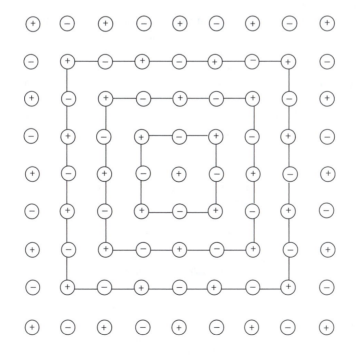

FIGURE A.3.1
Charges on the (100) face of an NaCl crystal.

TABLE A.3.1

Slow Convergence of the Lattice
Using the Direct Method

N	R_n	A_n
1	1.172	1.172
2	0.163	1.335
3	0.080	1.415
4	0.044	1.459
5	0.028	1.487
6	0.020	1.507
7	0.014	1.521
8	0.011	1.532
9	0.087	1.541
10	0.070	1.548
100	1.786×10^{-5}	1.609
1000	1.786×10^{-7}	1.6151

and the lattice sum may be written as

$$A_n = \sum_n R_n = -4 \left[\sum_{i=1}^{N} \frac{(-1)^i}{i} + \sum_{i=1}^{n} \sum_{j=1}^{n} \frac{(-1)^i}{\sqrt{i^2 + j^2}} \right], \tag{A.3.2}$$

where the sign has been reversed to make the Madelung constant positive in accordance with the convention.

The problem with this straightforward approach is that the convergence is extremely slow. For example, let us examine the first 10 terms and their sum shown in Table A.3.1.

Notice how slowly the sum is converging. Even after 1000 squares, the sum has only converged to the third decimal.

Convergence may be hastened by summing the contributions of the charges contained within each of the squares instead of the ions on the squares. For example, since only 1/2 of the 4 face atoms and 1/4 of the 4 corner atoms lie within the first square, the contribution from the charges within the first square is $S_1 = -2/1 + 1$. The contributions from the charges between the first and second square include the remainder of the ions outside the first square, $-2/1 + 3/(1^2 + 1^2)^{1/2}$, as well the portion of the charges that lie inside the second square, i.e., $+2/2 - 4/(2^2 + 1^2)^{1/2} + 1/(2^2 + 2^2)^{1/2}$. The contributions when summed in this manner are shown in Table A.3.2.

TABLE A.3.2

Faster Convergence by Grouping
Contributions from Squares

N	$-S_n$	A_n
1	1.292893	1.292893
2	0.313981	1.606874
3	0.003648	1.610522
4	0.002987	1.613510
5	0.000981	1.614491
6	0.000441	1.614933
7	0.000225	1.615158
8	0.000127	1.615285
9	0.0000767	1.615361
10	0.0000491	1.615410
100	4.058×10^{-9}	1.615542498
1000	3.985×10^{-13}	1.615542627

Taking the contributions from the ions on each square provides the same number of positive and negative charges, but if only the contributions from the portions of the ions within each square are considered, there are more unlike charges than like charges and the attractive contributions are maximized, which helps the sum to converge more rapidly. With this latter scheme, convergence to 3 decimal places is achieved after only 7 squares and to 6 decimal places after 100 squares. If the sum of total ion contributions is taken as in Equation A.3.2, convergence to the first decimal place is not reached until the 44th square and convergence to the third decimal place is not achieved after the 1000th square.

Since summing over the ions between each square gives back the contribution from the outer portion of the previous square, one can simply compute the lattice sum by using Equation A.3.2 to sum over all of the ions up to the nth square and then simply add the contribution from the ions with the next square, which is given by

$$A(n) = -4\left[\sum_{i=1}^{N} \frac{(-1)^i}{i} + \sum_{i=1}^{n}\sum_{j=1}^{n} \frac{(-1)^i}{\sqrt{i^2+j^2}} + \frac{(-1)^{n+1}}{2(n+1)} \right.$$

$$\left. + \sum_{i=1}^{n} \frac{(-1)^{n+1+i}}{\sqrt{(n+1)^2+i^2}} + \frac{1}{4\sqrt{2(n+1)^2}} \right]. \tag{A.3.3}$$

Bibliography

Borg, R.J. and Dienes, G.J., *The Physical Chemistry of Solids*, Academic Press and Harcourt Brace Jovanovich Publishers, Boston, MA, 1992.

Gilman, J.J., *The Electronic Basis of the Strength of Materials*, Cambridge University Press, UK, 2003.

Ibach, H. and Lüth, H., *Solid State Physics*, 3rd edn., Springer-Verlag, New York, 1990.

Kittel, C., *Introduction to Solid State Physics*, 7th edn., John Wiley & Sons, New York, 1966.

Ladbury, R., Livermore's big guns produce liquid metallic hydrogen. *Phys. Today*, 49, May 1996, 17–18.

Srivasava, C.M. and Srinivasan, C., *Science of Engineering Materials*, Wiley Eastern, New Delhi, 1987.

Weir, S.T., Mitchell, A.C., and Nellis, W.J., *Phys. Rev. Lett.*, 76, 1996, 1860.

Problems

1. Show that the coordination number for a close-packed structure is 12. Now show why ionically bonded systems cannot have a close-packed structure.

2. Use the angular dependence of the wavefunctions in Equation 3.16 to construct the sp^3 wavefunctions, which should produce four lobes of charge distribution. Find the directions (θ and ϕ) of these four lobes. (Ignore the radial part of the wavefunctions.)

3. Show that the sp^2 hybrid wavefunctions given by Equation 3.17 produce a three-lobed charge distribution 120° apart.

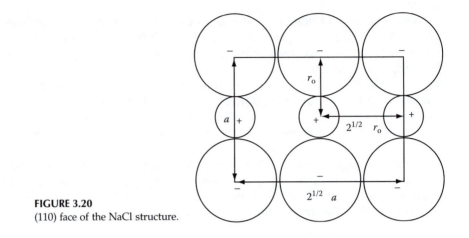

FIGURE 3.20
(110) face of the NaCl structure.

4. Arrange two electric dipoles with dipole moments $q\mathbf{d}$ nose to tail at distance $\mathbf{r}$ ($r \gg d$) along a line including their axis. Use Coulomb's law to show that the force between them varies as r^4.

5. The (110) surface of the NaCl structure is shown in the Figure 3.20. Find the 2D Madelung constant for this surface.

4

Crystals and Crystallography

Fine glassware is mistakenly referred to as crystal because of it beauty and clarity—it is actually glass and not crystalline. This confusion came about because Baroque period artisans cut natural quartz crystals to make crystal chandeliers. Later, glassmakers were able to produce cut glass that rivaled the quartz crystals in beauty, but the term "crystal" remained. So what is a crystal?

4.1 What Are Crystals?

In materials science, the word crystal implies three-dimensional (3D) long-range order in the structure, meaning that the atomic arrangement is repeated over and over again in all directions. Glasses, on the other hand are amorphous and their atoms have no long-range order. It is this macroscopic symmetry of crystals that first attracted the attention of philosophers as well as scientists. Most metals and ceramics we deal with have a crystalline structure, but instead of being single crystals, they are made up from many small crystal-lites which form a polycrystalline structure. Soils also contain microscopic crystals of various minerals. Even though these crystals are microscopic in size, their size is still very large compared to atomic dimensions and the requirement for long-range order is met. (In the new science of nanostructured materials, the size of the clusters of atoms is on the order of atomic dimensions and such long-range order is no longer possible, which is what gives such structures unique properties.)

In polycrystalline materials, the individual grains are randomly oriented and their regular periodicity is interrupted in the regions where they intersect. These regions are called grain boundaries. Since not all of the chemical bonds are satisfied in these regions, grain boundaries are points of weakness and are generally more vulnerable to chemical attack than the grains themselves. Metallurgists etch their samples with corrosive chemicals in order to reveal the grain boundaries. Even though grain boundaries may be points of weakness, polycrystalline materials are generally stronger than single crystals because the grain boundaries tend to block the motion of dislocations and act as a strengthening mechanism. (This topic is discussed in greater detail in Chapter 9.)

Grain boundaries can scatter photons of light as well as electrons and holes moving across them. Also the mobility of electrons and holes as well as the velocity of photons depends upon their direction of motion within a crystal. Since designers of electronic and photonic devices want to have a material with uniform electrical and optical properties, they generally require the semiconductors they use to be in the form of single crystals. The ability to grow large single crystal silicon wafers and to lay down the patterns for hundreds of individual devices simultaneously on them was the enabling technology that permitted the boom in consumer electronic devices.

4.1.1 Unit Cell

To understand the nature of materials, it is necessary to begin with an understanding of the nature of crystals. The basic building block of crystalline structure is a repeating volume of atoms (or molecules) called the unit cell, which is described by the lattice vectors $\mathbf{a}$, $\mathbf{b}$, and $\mathbf{c}$ and the angles α, β, and γ between them (see Figure 4.1). The volume of a unit cell is given by $V = \mathbf{a} \times \mathbf{b} \cdot \mathbf{c}$.

The magnitudes of these lattice vectors are called the lattice parameters. They and the angles between them are unique to each element and are used for material identification. The smallest volume of repeating volume of atoms is called a primitive unit cell; however, for reasons that will become apparent later, it is often convenient to define a larger unit cell that has a higher symmetry than the primitive unit cell.

4.1.2 Crystal Lattice and the Translation Group

A crystal lattice is a 3D array of points generated by translating the unit cell an integral number of times along each of its lattice vectors and forms the framework of the structure. Atoms or molecules may or may not be located on lattice points. The operation of displacing a unit cell parallel to itself is called a translation operation denoted by $\mathbf{T} = n_1\mathbf{a} + n_2\mathbf{b} + n_3\mathbf{c}$. The totality of all such operations, i.e., for all integral values on n_1, n_2, n_3, is the translation group.

The position of the jth atom within the unit cell is specified by $\mathbf{r}_j = x_j\mathbf{a} + y_j\mathbf{b} + z_j\mathbf{c}$ where $j = 1, 2, \ldots$, number of atoms in the unit cell and x_j, y_j, and z_j are the projections of the jth atom's locations onto the lattice vectors defined such that $0 < x_j, y_j, z_j < 1$, as indicated in Figure 4.2. The set of atomic coordinates within the unit cell is called the basis of the structure. Adding the basis to the lattice defines the complete crystalline structure. All atom positions may be generated by translating the unit cell integral distances along each axis, i.e., $\mathbf{r}'_j = \mathbf{r}_j + n_1\mathbf{a} + n_2\mathbf{b} + n_3\mathbf{c}$ (Figure 4.2). The fact that every atom in the crystal has regularly spaced counterparts in all three dimensions allows a large number of atoms to

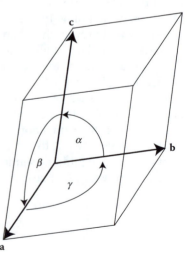

FIGURE 4.1
Unit cell vectors and the angles between them.

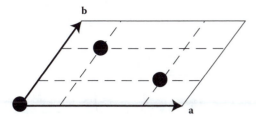

FIGURE 4.2
Oblique unit cell in two dimensions with a basis of (0,0), (1/4,3/4), (3/4,1/4).

coherently diffract x-rays and makes it possible to make precise measurements of the lattice parameters as well as to determine the position and types of atoms making up the unit cell.

The use of the word "group" has a special mathematical meaning. A group is defined as a set of operations that obey the following set of rules: (1) the result of any two operations must also be an element of the set; i.e., $A \otimes B = C$; (2) the operations must be associative; i.e., $A \otimes (B \otimes C) = (A \otimes B) \otimes C$; (3) there must exist an identity operation; $I \otimes A = A$ (achieved by setting $n_1, n_2, n_3 = 0$); and (4) there must be an inverse operation that when applied to a previous operation element returns the system to its original state; i.e., $A^{-1} \otimes (A \otimes B) = B$, which implies $A^{-1} \otimes A = I$. In general, group elements are not required to be commutative; i.e., $A \otimes B = B \otimes A$ is not required. If the elements do happen to commute, the group is called an abelian group. (We will not make further use of the mathematical concept of a group in this book. Readers interested in the application of group theory to crystallography are referred to Harrison's *Solid State Theory*.)

4.1.3 Crystallographic Directions

Directions within a crystalline lattice are defined in terms of the projections of the vector in question relative to the three lattice vectors. Directions are specified as the smallest set of whole number projections with no common denominator. All distances are expressed in unit cell vector lengths, hence are dimensionless. For example, the direction along the a-axis in Figure 4.3 has a projection on **a** of 1 and 0 on **b** and on **c**, hence it is the [100] direction. Similarly, the vector in the a–b plane with equal projections on both **a** and **b** is the [110] direction and a vector with equal projections on **a**, **b**, and **c** is the [111] direction. Notice that this is not a Cartesian coordinate system; $\mathbf{a} \neq \mathbf{b} \neq \mathbf{c}$ and $\alpha \neq \beta \neq \gamma \neq 90°$. The procedure for determining lattice direction is the same for any coordinate system.

Since the direction vectors are specified in the smallest set of whole numbers, they specify only direction, not actual lengths. The vector [222] reduces to [111] and both specify the same direction.

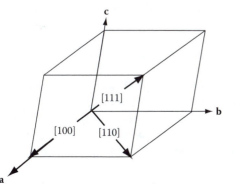

FIGURE 4.3
Basic crystallographic direction vectors.

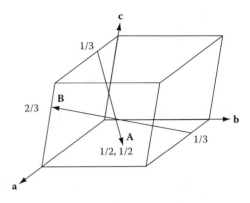

FIGURE 4.4
Generalized vector directions. Using vector subtraction, vector **A** would be designated [13$\bar{6}$] and vector **B** would be [2$\bar{3}$2].

However, the projections of a vector onto the axes need not be whole numbers. The location of the central atom in a body-centered cubic (bcc) unit cell has coordinates relative to the unit cell (1/2,1/2,1/2), so the direction of a vector from the origin to that central atom is written as [111]. Similarly, the location of the middle atom of a face-centered unit cell in the a–b plane is (1/2,1/2,0), so the direction from the origin to it would be written as [110]. Nor do the vectors have to pass through the origin. Consider a more general case shown in Figure 4.4. Vector **A** goes from (1/3,0,1) to (1/2,1/2,0). Vector subtraction yields (1/6,1/2,−1). Multiplying by 6 to clear the fractions, vector **A** will have direction given by [13$\bar{6}$]. Similarly, vector **B** goes from (1/3,1,0) to (1,0,2/3). Vector subtraction yields (2/3,−1,2/3). Multiplying by 3, we obtain the direction [2$\bar{3}$2].

Notice that the lattice directions are written within square brackets with no commas. Negative directions are indicated with a bar over the number as [$\bar{1}\bar{1}$0]. Families of similar directions are denoted by an angle bracket ⟨100⟩, which consists of the six vectors, [100], [010], [001], [$\bar{1}$00], [0$\bar{1}$0], and [00$\bar{1}$]. Note that [100] is antiparallel to [$\bar{1}$00], etc.

The above rules apply for any geometry of the unit cell. For the case of cubic unit cells, it is easy to obtain the angles the various direction vectors make with each other by using the definition of the dot product $\mathbf{A} \cdot \mathbf{B} = |\mathbf{A}||\mathbf{B}| \cos(\mathbf{AB})$. For example, the direction from the central atom in a bcc unit cell to one of the corner atoms can be written as [111], and the direction from the central atom to the atom in the opposite corner on the same face is [$\bar{1}\bar{1}$1]. Taking the dot product, the angle between these two corner atoms may be found:

$$\cos(\mathbf{AB}) = \frac{(1)(-1) + (1)(-1) + (1)(1)}{\left(\sqrt{(-1)^2+(-1)^2+(1)^2}\right)\left(\sqrt{(1)^2+(1)^2+(1)^2}\right)} = -\frac{1}{3}. \tag{4.1}$$

Taking the arc cosine of −1/3 gives the angle between **A** and **B,** which is found to be 109.47°.

4.1.4 Miller Indices

The orientation of a plane is defined by its Miller indices, h, k, ℓ. They are generated by taking the reciprocals of the intercepts of the plane with the three lattice vectors measured in lattice vector units, multiplying them by a factor that produces the smallest set of whole numbers with no common denominator, and writing them in parentheses as three numbers with no commas between them. For example, a plane intercepting **a** at 1/2, **b** at 1/3, and **c** at 2/3 in Figure 4.5 would have Miller indices given by (2/1,3/1,3/2) × 2 = (463) and would be referred to as the "four–six–three" plane. The Miller indices do not refer to a

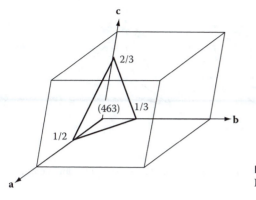

FIGURE 4.5
Illustration of the (463) plane.

specific plane, but to any plane that has the same orientation, so a plane that intersected at 1,2/3,4/3 would also be the (463) plane.

If a plane does not intersect an axis as in Figure 4.6a, the intercept is taken to be ∞ and its Miller index would be zero. The plane intersects the **a**-axis at 1/2 and the **b**-axis at ∞. We have to add another unit cell below the **a**–**b** plane in Figure 4.6b to see that it intersects the **c**-axis at −1. So the Miller indices are (20$\bar{1}$). Negative indices are expressed with a bar over the index, just as in the case of lattice directions.

What if the plane contains an axis or goes through the origin and does not intersect any axis? For example, the plane containing the **b** and **c** lattice vectors in Figure 4.7 can be moved out one unit along the **a**-direction and would be referred to as (100). Similarly, a plane that contains the **c**-axis and bisects the angle between the −**a** and +**b**-axes could be moved forward one unit along the **a**-axis (or **b**-axis) so that it intercepts **a** at 1, **b** at 1, and **c** at ∞ and would be written as (110). A plane that contains the origin and intersects no axes can be brought forward until it intercepts the three axes as in the case of the (111) plane in Figure 4.7.

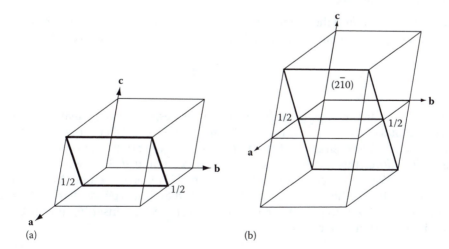

(a)   (b)

FIGURE 4.6
(a) Plane to be specified does not intersect one of the axes and (b) unit cell added below to allow plane (20$\bar{1}$) to intersect the **c**-axis at −1.

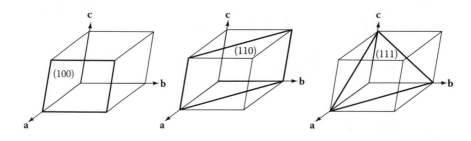

FIGURE 4.7
Planes that contain one or more axes or go through the origin can be shifted by unit cells so that their intercepts can be located.

Just as in the case of lattice directions, families of planes, such as the six face planes (100), (010), (001), ($\bar{1}$00), (0$\bar{1}$0), (00$\bar{1}$) are designated with a special symbol, in this case, {100}. However, note that (100) and ($\bar{1}$00), (010) and (0$\bar{1}$0), (001) and (00$\bar{1}$) are equivalent.

As stated previously, the Miller indices refer to a family of parallel planes. However, sometimes it is necessary to refer to a specific plane. In this case, the plane is specified without reducing it to the lowest set of whole numbers. For example, a plane intercepting **a** at 1/2, **b** at ∞, and **c** at ∞ would be written as (200) instead of (100). The reason is that the atoms in that particular plane are of interest and the (200) plane may very well have a different set of atoms than the (100) plane. This notation is sometimes referred to as Laue notation and much use is made of this notation in x-ray crystallography as seen in Chapter 6.

So far, no assumptions have been made concerning the properties of the lattice vectors **a**, **b**, and **c**, so all of the notations we have developed thus far apply to any set of lattice vectors. For cubic systems only, the Miller indices of a plane are the same as the direction indices of a vector that is perpendicular to it; i.e., the (321) plane is perpendicular to the [321] vector as demonstrated in Section 4.1.5.

4.1.5 Interplanar Spacing

Often it is necessary to find the distance from the origin to a specified plane, or the distance between a set of parallel planes. The equation for a plane in Cartesian coordinates is

$$\frac{x}{x'} + \frac{y}{y'} + \frac{z}{z'} = 1, \tag{4.2}$$

where x', y', z' are the intercepts of the plane on the respective coordinate axes. For the orthorhombic lattice, $\mathbf{a} \neq \mathbf{b} \neq \mathbf{c}$, $\alpha = \beta = \gamma = 90°$, the plane $hk\ell$ intercepts the coordinate axes at a/h, b/k, and c/ℓ. We wish to construct a vector **d** from the origin that is perpendicular to the plane. To enforce this condition, we require $\mathbf{d} \cdot \mathbf{A} = 0$ and $\mathbf{d} \cdot \mathbf{B} = 0$ where **A** and **B** lie in the plane hkl. We construct vectors **A** and **B** in the hkl plane by setting $\mathbf{A} = -\hat{\mathbf{i}}a/h + \hat{\mathbf{j}}b/k$ and $\mathbf{B} = -\hat{\mathbf{i}}a/h + \hat{\mathbf{k}}c/\ell$. The vector from the origin that is perpendicular to this plane is found by taking $\mathbf{A} \times \mathbf{B}$ to give $\mathbf{d} = \text{const}(\hat{\mathbf{i}}h/a + \hat{\mathbf{j}}k/b + \hat{\mathbf{k}}\ell/c)$. Inserting this in Equation 4.2, to find the constant, we obtain

$$\mathbf{d}_{hk\ell} = \begin{pmatrix} h/a \\ k/b \\ \ell/c \end{pmatrix} \left(\frac{1}{(h/a)^2 + (k/b)^2 + (l/c)^2} \right). \tag{4.3}$$

The magnitude of **d** is

$$d_{hk\ell} = \frac{1}{\sqrt{h^2/a^2 + k^2/b^2 + \ell^2/c^2}}. \tag{4.4}$$

For cubic lattices, $\mathbf{a} = \mathbf{b} = \mathbf{c}$ and Equations 4.3 and 4.4 reduce to

$$\mathbf{d}_{hk\ell} = \begin{pmatrix} h \\ k \\ \ell \end{pmatrix} \left(\frac{a}{h^2 + k^2 + \ell^2} \right). \tag{4.5}$$

$$d_{hk\ell} = \frac{a}{\sqrt{h^2 + k^2 + \ell^2}}. \tag{4.6}$$

Notice that the direction of the vector perpendicular to the $(hk\ell)$ plane is $[hk\ell]$ for the case of cubic symmetry.

If we multiply the intercepts a/h, b/k, and c/ℓ by an integer n, Equation 4.2 becomes

$$\frac{xh}{a} + \frac{yk}{b} + \frac{z\ell}{c} = n, \tag{4.7}$$

and the distance from the origin to the nth plane is given by

$$d_{nhk\ell} = \frac{n}{\sqrt{h^2/a^2 + k^2/b^2 + \ell^2/c^2}}, \tag{4.8}$$

so that Equation 4.7 represents a family of planes whose spacing between them is given by Equation 4.4 (or Equation 4.5 in the case of cubic symmetry). Since the $(hk\ell)$ plane cuts through the unit cell, it may not contain any lattice points or atoms. To construct a plane containing the lattice points, chose n to be the common factor for h, k, and ℓ, i.e., set $n = 6$ for the (321) plane.

Equation 4.7 may be used as a test to see if a particular atom lies in the plane. For example, consider the atom in the center of the bcc system $(a = b = c)$ whose x, y, z coordinates are $a/2$, $a/2$, $a/2$. Setting $h = 1$, $k = 1$, $\ell = 0$, and $n = 1$, we see that Equation 4.7 is satisfied and the central atom does indeed lie in the (110) plane.

A common misconception among students is that the central atom in the bcc system lies in the (111) plane. Equation 4.8 with $h = 1$, $k = 1$, $\ell = 1$ is not satisfied by $x = y = z = a/2$. This misconception is caused by the assumption that a cube is divided in half by the (111) plane. Actually two different (111) planes can pass through a cube; one intercepting at (1,0,0). (0,1,0), (0,0,1) and the other at (1,0,1), (0,1,1), and (1,1,0). These two planes are parallel but distinct as may be seen by viewing the system along the $[\bar{1}10]$ direction in Figure 4.8. The central atom lies in the region between the two planes.

4.1.6 Miller–Bravais Notation

In dealing with hexagonal systems, it is sometimes convenient to introduce a four-number notation for lattice directions and the Miller indices. Three axes, **a**, **b**, **c**, lie 120° apart in the basal plane and the fourth is normal to this plane as shown in Figure 4.9. Since one of the three indices in the basal plane is redundant (specifying two indices automatically fixes the third) the convention is to require the sum of the three indices to be zero. There is a formal method for converting from the three index to the four index system, i.e., $h = n/3$

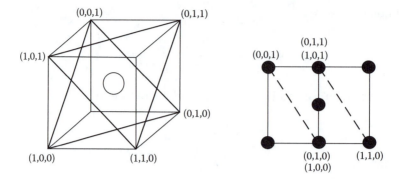

FIGURE 4.8
A bcc structure showing two distinct (111) planes with their intercepts (left) and as seen from along the [$\bar{1}$10] direction (right). The center atom lies between the two (111) planes indicated by heavy dashed lines.

$(2h' - k'); j = n/3(2k' - h'); k = -(h + k); \ell = n\ell'$; where the primes represent the indices in the three number system. Vector **A** in Figure 4.9 would be [$\bar{1}$00] in the three coordinate system. Using the equations above, it transforms into [$\bar{2}$110] when n is set to 3 in order to clear the fractions. If vector **A** had intercepted the vertical axis at 1, making it [$\bar{1}$01], it would transform into [$\bar{2}$113]. If vector **B** is displaced to the right where it intersects **a** and **b** at 1, it goes from 0, 1, 0 to 1, 0, 0. Vector subtraction yields [1$\bar{1}$0]. Transforming to the four-index system gives [1$\bar{1}$00]. Notice that in this case, n is set to 1 so if this vector also had a vertical component of 1, [1$\bar{1}$1], it would transform into [1$\bar{1}$01].

The designation of planes is more straightforward and proceeds as before. Plane A intersects the **a**-axis at $-1/2$ and axes **b** and **c** at 1. If it is perpendicular to the basal plane, it would be designated as ($\bar{2}$110). If it also intersected the **z**-axis at one unit, it would be ($\bar{2}$111). Plane **B** intercepts the **b**-axis at 1, the **a**-axis at -1, and the **c**-axis not at all. It would thus be denoted (1$\bar{1}$00). Again note that the sum of the first three Miller–Bravais indices is always zero in this notation.

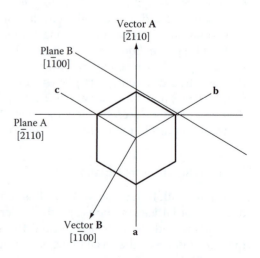

FIGURE 4.9
Miller–Bravais scheme for hexagonal crystals.

4.2 Crystal Systems and Symmetry

Crystal structures are classified according their symmetry. Why is symmetry important? Many physical properties of materials such as their stiffness, electrical and thermal conductivity, dielectric function, and magnetic properties are tensor quantities, which means they are dependent on the direction relative to the crystalline axes. Such properties reflect the symmetry of the lattice. A fundamental postulate relating the symmetry of the structure and the symmetry of its properties is known as Neumann's principle: "The magnitude of a particular physical property, when measured along a specific direction, is unchanged when the material is rotated, reflected, or inverted into a new orientation corresponding to one of the symmetry elements of its point group."

4.2.1 Point Symmetry Operations

One type of point symmetry is rotational symmetry. If there is an axis about which the lattice can be rotated by $1/n$ of a revolution and all lattice points can be brought back into congruence, this is referred to as an axis of n-fold symmetry. Another type of point symmetry is mirror symmetry. If a plane can be chosen such that every point above this plane corresponds to a point below it, the system is said to have mirror symmetry about that plane. In addition to the symmetry operations described above, there are three other symmetry operations that can be performed about a point: the operation of inversion in which the coordinates of a point are inverted along a line through a point called the center of inversion; a rotation-reflection, in which a rotation is followed by a reflection in the plane perpendicular to the axis of rotation; and a rotation-inversion, which is an n-fold rotation followed by an inversion. These operations are included because they may bring a system a crystal system into congruence with itself when a simple rotation may not. There are two systems for describing the point symmetry of crystal classes: the older Schoenflies symbols and the Hermann–Manguin (H–M) or international symbols. The latter are generally preferred, although both are seen in the literature.

4.2.2 Basic Crystal Systems

Now we may ask, what types of lattices are possible? Think of the unit cells as a bunch of identical building blocks that must fill all space when assembled. There are only seven basic shapes that meet this requirement and these shapes constitute the seven basic crystal systems.

4.2.2.1 Triclinic System

The triclinic system, shown in Figure 4.10a, has the properties $a \neq b \neq c$ and $\alpha \neq \beta \neq \gamma \neq 90°$ or $120°$ (where α is the angle between b and c, etc.), and has the least symmetry. It has no rotational symmetry other than the trivial onefold axis of symmetry, meaning that it is necessary to rotate a full $360°$ about an axis in order to bring all points back into congruence. It does, however, posses a center of inversion. The Schoenflies symbol for this class is C_1, the C for "cyclic" and the subscript 1 indicates only an n-fold axis of rotational symmetry; in this case $n = 1$. The international symbol is $\bar{1}$, the bar over the 1 indicating the center of inversion.

4.2.2.2 Monoclinic System

The monoclinic system, shown in Figure 4.10b, is characterized, $a \neq b \neq c$ and $\alpha = \gamma = 90°$, $\beta \neq 90°$, has a twofold axis of symmetry along the [010] direction, and a plane of mirror symmetry (the (020) plane). The Schoenflies symbol for this class is C_{2h}. h indicates a horizontal mirror plane perpendicular to the axis of rotational symmetry. The international symbol is $2/m$, 2 for the axis of twofold rotational symmetry, and m indicates a mirror plane perpendicular to the axis of rotational symmetry.

Why is there no biclinic system in which $a \neq b \neq c$ and $\alpha = 90°$, $\beta \neq \gamma \neq 90°$? This is because it produces no different symmetry than the triclinic system.

4.2.2.3 Trigonal System

The trigonal (sometimes called rhombohedral) system has $a = b = c$ and $\alpha = \beta = \gamma \neq 90°$ or 120° as shown in Figure 4.10c. It has 3 threefold axes of symmetry, 3 twofold axes, and three mirror planes perpendicular to the twofold axes. It is easier to think of the trigonal lattice as a cube that has been distorted by pulling on opposite corners. The Schoenflies symbol for this class is D_{3d}, where D stands for dihedral to indicate the twofold rotational axes. The

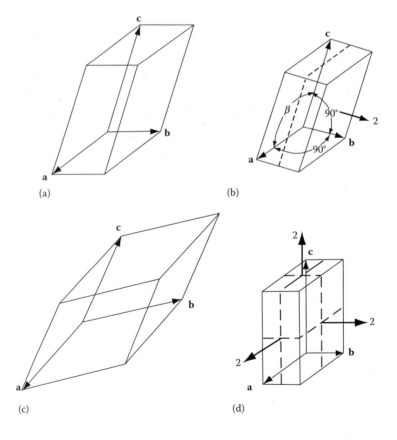

FIGURE 4.10
(a) Triclinic system, $a \neq b \neq c$ and $\alpha \neq \beta \neq \gamma \neq 90°$ or 120°; (b) monoclinic system, $a \neq b \neq c$ and $\alpha = \gamma = 90°$, $\beta \neq 90°$. "2" indicated the twofold rotation axis and the dashed lines indicate mirror planes; (c) trigonal system, $a = b = c$ and $\alpha = \beta = \gamma \neq 90°$ or 120°; (d) orthorhombic system, $a \neq b \neq c$ and $\alpha = \beta = \gamma = 90°$.

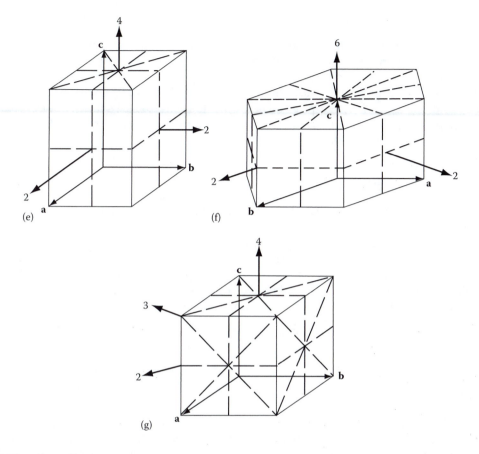

FIGURE 4.10 (continued)
(e) Tetragonal system, $\mathbf{a}=\mathbf{b}\neq\mathbf{c}$ and $\alpha=\beta=\gamma=90°$; (f) hexagonal system, $\mathbf{a}=\mathbf{b}\neq\mathbf{c}$ and $\alpha=\beta=120°$, $\gamma=90°$; and (g) cubic system, $\mathbf{a}=\mathbf{b}=\mathbf{c}$ and $\alpha=\beta=\gamma=90°$.

subscript d stands for diagonal and indicates the mirror planes that contain the threefold axis and bisect the angles between the twofold axes. The complete international symbol is $\bar{3}2m$, usually abbreviated as $\bar{3}m$, the bar above the 3 indicating a rotation-inversion axis (a rotation followed by inverting the coordinates of each point along a line through a point called the center of inversion).

4.2.2.4 Orthorhombic System

The orthorhombic system, which has $\mathbf{a}\neq\mathbf{b}\neq\mathbf{c}$ and $\alpha=\beta=\gamma=90°$, has three mutually perpendicular axes of twofold symmetry (along the $\langle100\rangle$ directions) and three mutually perpendicular mirror planes as may be seen in Figure 4.10d. The Schoenflies symbol for this class is D_{2h}. The h indicates a mirror plane perpendicular to the n-fold axis as before. The existence of the other mirror planes is implied by the rotational symmetry. The complete H–M symbol is $2/m2/m2/m$ indicating three mutually orthogonal mirror planes perpendicular to the rotational axes. The twofold rotational axes are implied by the existence of the three mirror planes. Sometimes this is abbreviated as $2/mmm$ or to just mmm since the mirror planes also imply the rotational axes.

4.2.2.5 Tetragonal System

The tetragonal system, $\mathbf{a} = \mathbf{b} \neq \mathbf{c}$ and $\alpha = \beta = \gamma = 90°$, has one axis of fourfold symmetry in the [001] direction, two sets of orthogonal axes of twofold symmetry in the $\langle 100 \rangle$ and $\langle 110 \rangle$ directions, a set of three orthogonal mirror planes as well as two other mirror planes, (110) and ($\bar{1}$10) as shown in Figure 4.10e. The Schoenflies symbol for this class is D_{4h}, using the same reasoning as for D_{2h}, except in this case $n = 4$. The full H–M symbol is $4/m2/m2/m$ abbreviated to $4/mmm$, where the $/mmm$ indicates a mirror plane normal to the fourfold axis and mirror planes parallel to it.

4.2.2.6 Hexagonal System

The hexagonal system has a sixfold axis of symmetry along [0001] and two sets of three twofold rotational axes, one set along the $\langle 2\bar{1}\bar{1}0 \rangle$ family (only one shown in Figure 4.10f) and the other along the $\langle 1100 \rangle$ family (only one shown). There are two sets of three mirror planes, $\{\bar{2}110\}$ (only one shown) and $\{1100\}$ (only one shown), as well as the (0002) mirror plane. The Schoenflies symbol for this class is D_{6h}, implying the horizontal mirror plane in addition to the mirror symmetry implied in D_{6d}. The full international symbol is $6/m2/m2/m$ abbreviated to $6/mmm$, indicating a mirror plane normal to the sixfold axis and mirror planes parallel to the sixfold axis.

4.2.2.7 Cubic System

The cubic system has the most symmetry of all as can be seen in Figure 4.10g. It has three mutually orthogonal axes of fourfold symmetry along the $\langle 100 \rangle$ directions, four axes of threefold symmetry along the $\langle 111 \rangle$ directions, and six axes of twofold symmetry along the $\langle 110 \rangle$ directions. Mirror planes consist of the $\{200\}$ and $\{110\}$ families. The Schoenflies symbol for the full symmetry of this class is O_h, the O standing for octahedral. The full international symbol is, $4/m\bar{3}2/m$ indicating mirror planes perpendicular to the fourfold axes as well as to the twofold axes and a center of inversion on the threefold axes. (The $\{111\}$ family of planes are not mirror planes.) However, this symbol is usually abbreviated as $m\bar{3}m$ (in some text, this is written as $m3m$).

4.2.3 Restricted Symmetry

The symmetries for the seven basic crystal systems described above assume the full symmetry or holohedry of each of the lattices. When the basis is added, some of these symmetries may be restricted. For example, a face-centered cubic (fcc) crystal such as Al has the full cubic symmetry. However, diamond also has the fcc structure with atoms occupying the lattice points as well as every other tetrahedral interstitial point. Its point group is $\bar{4}3m$, which implies a rotation-inversion on the fourfold axes. The threefold symmetry is preserved without the threefold rotation-inversion. The twofold symmetry is no longer preserved and the only mirror symmetry is along the $\{110\}$ planes.

In another example, at temperatures $>393\,K$, barium titanate has the perovskite structure, which is simple cubic with all of the symmetry elements of the cubic lattice, so its point group is O_h or $m\bar{3}m$. As the temperature is reduced to its Curie temperature, the lattice contracts and the oxygen ions on the faces of the cube squeeze the titanium ion in the center of the cube so that it is displaced in one direction while the oxygen ions are displaced in the opposite direction, destroying the inversion symmetry as well as the mirror symmetry about the central plane and the rotational symmetry about several of the axes. The resulting crystal must then be described by C_{4v} in the Schoenflies notation or by 4*mm* in the international notation indicating a tetragonal lattice with fourfold symmetry about the **c**-axis and two sets of vertical mirror planes.

Each of the other crystal systems has similar restricted symmetries and it can be shown that there is a total of 32 unique sets of point symmetry operations or point groups. The symmetry of every crystalline structure may be described by one of these 32 point groups. Such classification of point symmetries is useful in the search for materials with certain properties. For example, if one is looking for materials with permanent dipole moments, one would look only at systems that are noncentrosymmetric, i.e., systems that do not possess a center of inversion symmetry. The 10 noncentrosymmetric point groups are 1, 2, 3, 4, 6, *m*, 2*mm*, 3*m*, 4*mm*, and 6*mm*.

Notice that rotation symmetry only exists for $n = 1$, 2, 3, 4, and 6; five- and sevenfold rotations are not allowed because bodies with these symmetries cannot fill all space for the same reason that you cannot tile a floor with pentagons or with septagons. Icosahedral quasicrystals with fivefold symmetry can form, but cannot grow into crystalline solids in the strict sense of the word. Even so, such quasicrystals are extremely interesting both theoretically as well as from applications that utilize their unusual properties, as discussed later.

4.2.4   Bravais Lattices

Now consider the set of points that was shown in Figure 4.11 representing a two-dimensional (2D) lattice. One could connect any group of three nearest points to form a set of lattice vectors that would generate the entire structure by translations along these two vectors. However, instead of choosing two oblique primitive vectors such as **a** and **b**, which form an oblique unit cell, this particular lattice geometry makes it is possible to choose two vectors **a**′ and **b**′ that make a right angle with one another but whose unit cell has a lattice point in the center of the rectangle they describe. Such a choice of a unit cell is considered nonprimitive (since it contains a lattice point and is not the smallest possible unit cell), but one can see that it has more symmetry than the primitive unit cell. In order to recognize this special symmetry in 2D lattices, this particular lattice would be referred to as a face-centered rectangle. Thus there are five 2D crystal lattices: oblique, rectangular, face-centered rectangular, hexagonal, and square. We had a face-centered rectangle, why not a face-centered square? A face-centered square is simply a smaller square, so no new symmetry is being introduced. Similarly, a base-centered hexagon is just a smaller hexagon.

Similar special symmetries arise in the 3D structures. For example, if a lattice point is placed in the center of a cube, the primitive lattice would be trigonal. But because of the

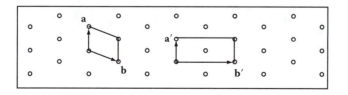

FIGURE 4.11
Face-centered rectangular lattice in 2D. All the lattice points could be generated by translating along the oblique primitive lattice vectors **a** and **b**, or by the nonprimitive vectors **a**′ and **b**′ with a basis of (0,0) and (1/2,1/2).

special symmetry associated with the cubic lattice, it is called the body-centered cubic or bcc lattice and this nomenclature is generally used instead of the primitive trigonal lattice as seen in Figure 4.12a. A similar situation arises when lattice points fall on the centers of the faces of the cube. This lattice is also given the special name of face-centered cubic or fcc lattice shown in Figure 4.12b.

Other members of the seven basic crystal systems can be constructed from simpler primitive vectors if lattice points are placed in their centers or on their faces. The reader may wonder why some of these seven basic crystal systems have separate body-centered, face-centered, or base-centered lattices and others do not. The answer is, if the additional lattice points produce a system of higher symmetry than the primitive lattice, the non-primitive lattice is given a special name. A body-centered tetragonal (bct) shown in Figure 4.12c has the symmetry of the tetragonal system, which is higher than the trigonal primitive lattice. But then, why not a face-centered tetragonal lattice? Because it

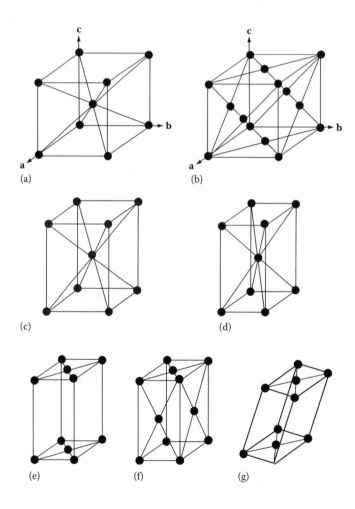

(a) (b) (c) (d) (e) (f) (g)

FIGURE 4.12
(a) bcc lattice, (b) fcc lattice. Both of these lattices could be represented by a primitive trigonal lattice by taking the lattice vectors from the origin to the three nearest body-centered points in the case of the bcc or the three nearest face-centered points in the case of the fcc. (c) bct lattice, (d) body-centered orthorhombic lattice. (e) base-centered orthorhombic lattice, (f) face-centered orthorhombic lattice (not all lattice points shown), and (g) face-centered monoclinic lattice (not all lattice points shown).

would simply result in another bct. The same argument would apply to a base-centered tetragonal lattice.

On the other hand, the orthorhombic lattice can have body-centered, base-centered, and face-centered lattices, shown in Figure 4.12d through f, which are not redundant because the three nonprimitive lattice vectors all have different lengths.

It is difficult to get more primitive than the tetragonal or triclinic lattice, and the base-centered monoclinic lattice (Figure 4.12g) rounds out the 14 Bravais lattices.

M.L. Frankheim in 1842 was the first to classify the possible crystal lattices including the special body-centered and face-centered nonprimitive lattices. However, he had mistakenly added a 15th structure that turned out to be redundant. Auguste Bravais was the first in 1845 to correctly characterize the 14 unique lattices that now bear his name.

4.2.5 Hexagonal Close-Packed Lattice

It was mentioned that the base-centered hexagon is not considered as one of the Bravais lattices because it adds no new symmetry. There is an important hexagonal structure similar in some respects to the fcc lattice called the hexagonal close-packed (hcp) lattice. Like the base-centered hexagon, the hcp lattice is not considered to be one of the 14 Bravais lattices either. The structure consists of a sandwich of two vertically aligned hexagonal planar arrays with three atoms between them that sit on three of the six vertices as shown in Figure 4.13.

4.2.6 Space Groups

All crystal systems can be classified into one of the 14 Bravais lattices which can be subdivided into 32 crystal classes or point groups. If certain other translation operations that do not have point symmetry are considered, such as a translation combined with a mirror reflection (glide plane operation) or a translation combined with an *n*-fold rotation (screw axis), the 32 point groups can be subdivided into 230 possible space groups that completely describe the symmetry of all possible crystal systems. These are enumerated in the *International Tables for Crystallography*, vol A (Ed. Th. Hahn, 2006).

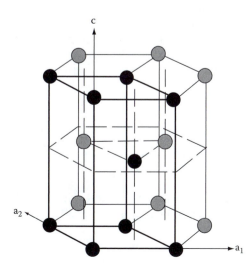

FIGURE 4.13
hcp lattice. The unit cell capable of generating this structure is outlined in heavier lines and the lattice points are black instead of gray.

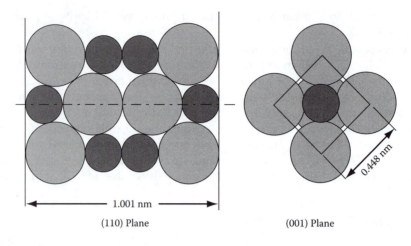

(110) Plane   (001) Plane

FIGURE 4.14

Hg_2Cl_2 molecule with space group I4*mmm*. The smaller Hg atoms are dark gray and the Cl atoms are light gray. Note that no atoms sit on any of the lattice points, but the environment (in this case back-to-back Cl) at each lattice point is the same and the space group uniquely describes the configuration of the unit cell.

Space group notation begins with a capital letter, *P* for primitive, *C* for base centered, *F* for face centered, *I* for body-centered (from the German *Interzentrum*), or *R* for rhombohedral. The point group is designated next with subscripts indicating the various screw axis and glide plane operations that leave the structure invariant. Care is taken to eliminate any duplications where two different sets of operations produce the same result such as the equivalence of $\bar{2} = m$ and $\bar{6} = 3/m$. For example, an fcc crystal would have a space group designated by $Fm\bar{3}m$, a bcc crystal's space group is $Im\bar{3}m$, a bct crystal is I4*mmm*, etc.

For a further example of how the point and space group describes a crystal system, consider the mercurous chloride (Hg_2Cl_2) system. The crystalline structure consists of parallel chains of covalently bonded Cl–Hg–Hg–Cl molecules aligned along the crystallographic c-axis as shown in Figure 4.14. (The Cl^- anions are shown in light gray and the Hg^+ cations in darker gray. Their sizes are to scale.)

The point group can easily be seen to be 4/*mmm* from the symmetry. The space group is I4/*mmm*. This additional piece of information tells us that the unit cell is body centered. Note there are no ions at the center of the unit cell or at any of the other lattice points, but the environment at the center, which is a point between two adjacent Cl^- ions, is the same as at every other lattice point. (Note: We could have just as easily chosen the lattice points to be between the two adjacent Hg^+ ions.)

4.3 Structural Relationships

The basis for an fcc structure is (0,0,0) (1/2,0,0) (0,1/2,0) (0,0,1/2). The fcc unit cell contains four atoms; eight atoms on the corners, 1/8 of which are contained in the unit cell, and six atoms on the sides, 1/2 of which are in the unit cell.

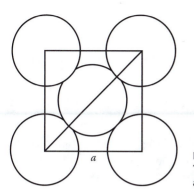

FIGURE 4.15
The (100) face of an fcc structure showing the relationship between the atomic radii and the length of the unit cell.

In the diagram of the (100) face of an fcc structure in Figure 4.15, note that the diagonal is four times atomic radii R. Thus the relation between the unit cell length a and the atomic radius for the fcc structure is

$$4R = \sqrt{2a^2} \quad \text{or} \quad a = 2\sqrt{2}R. \tag{4.9}$$

This same diagonal would be along the [111] direction in the bcc structure. Therefore, for the bcc structure,

$$4R = \sqrt{3a^2} \quad \text{or} \quad a = \frac{4}{\sqrt{3}}R. \tag{4.10}$$

The basis for the bcc system is $(0,0,0)$ $(1/2,1/2,1/2)$; so there are eight corner atoms and one interior atom for a total of two atoms per unit cell.

4.3.1 Density and Packing Calculations

4.3.1.1 Atomic Density

The fcc unit cell contains four atoms and the volume of the unit cell is a^3 or $8^{3/2}R^3$. Therefore, the atomic density or number of atoms per unit volume is given by

$$n_{\text{at}} = \frac{4}{8^{3/2}R^3} = \frac{\sqrt{2}}{8R^3}. \tag{4.11}$$

Similarly, the atomic density of a bcc structure is given by

$$n_{\text{at}} = \frac{2}{R^3}\left(\frac{\sqrt{3}}{4}\right)^3. \tag{4.12}$$

4.3.1.2 Mass Density

Since the mass of the individual atoms in the unit cell is A_W/N_A where A_W is the atomic weight and N_A is Avogadro's number, the mass density of an fcc metal is given by

$$\rho = \frac{\sqrt{2}A_W}{8R^3N_A}. \tag{4.13}$$

Similarly, the density of a bcc structure is given by

$$\rho = \frac{2A_w}{R^3 N_A}\left(\frac{\sqrt{3}}{4}\right)^3. \tag{4.14}$$

4.3.1.3 Atomic Packing Factor

The atomic packing factor (APF) is the ratio of the hard sphere volume inside the unit cell to the volume of the unit cell. For an fcc system with four atoms inside the unit cell, the hard sphere volume is $4 \times 4R^3/3$ and the unit cell volume is $8^{3/2}R^3$. Therefore, the APF is

$$\text{APF} = \frac{16\pi}{3 \times 8^{3/2}} = 0.740. \tag{4.15}$$

For the bcc structure,

$$\text{APF} = \frac{8\pi}{3}\left(\frac{\sqrt{3}}{4}\right)^3 = \frac{\pi\sqrt{3}}{8} = 0.68. \tag{4.16}$$

4.3.1.4 Planar Density

The planar density (number of atoms per unit area) of the (111) face of an fcc structure can be found from Figure 4.16. The altitude of the equilateral triangle inscribed by the six circles is $\sqrt{12}R$. Its area is therefore $4\sqrt{3}R^2$. The triangle contains three 1/2 circles and three 1/6 circles, so the number of atoms included is 4. The planar density becomes

$$\text{Planar density} = \frac{2}{4\sqrt{3}R^2} = \frac{1}{2\sqrt{3}R^2}, \tag{4.17}$$

for the (111) plane in the fcc lattice.

The bcc structure has no close-packed planes like the (111) plane in the fcc structure. The planes with the highest planar densities are the {101}, {211}, and {321} families. The (101) face shown in Figure 4.17 contains two atoms and has an area given by $\sqrt{2}a^2$ or $16\sqrt{2}/3R^2$. The planar density is given by

$$\text{Planar density} = \frac{3}{8\sqrt{2}R^2}. \tag{4.18}$$

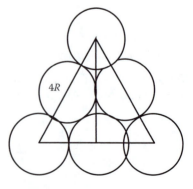

FIGURE 4.16
The (111) high planar density face of the fcc structure showing how the atomic planar density is calculated.

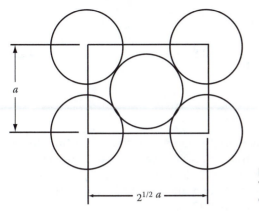

FIGURE 4.17
The (110) face of a bcc structure has the highest planar density.

The atoms in the (211) and (321) bcc planes are shown in Figure 4.18. The area of a triangle, whose sides are A, B, C, can be found from

$$\text{Area} = \frac{\sqrt{(A+B+C)(A+B-C)(A-B+C)(-A+B+C)}}{4}. \tag{4.19}$$

4.3.1.5 Planar Fraction

The planar fraction is the area covered by the atoms whose centers lie in the plane divided by the area of the plane. Unlike the planar density, which contains the radius of the atoms, the planar fraction is a pure number and is perhaps more meaningful. Planes with the highest planar fraction are the slip planes along which metals deform plastically. The importance of this parameter is discussed in more detail in Chapter 9.

The planar fraction is just the planar density multiplied by πR^2. From Equation 4.17, the planar fraction of the (111) plane in the fcc system is

$$\text{Planar fraction} = \frac{\pi}{2\sqrt{3}} = 0.907, \tag{4.20}$$

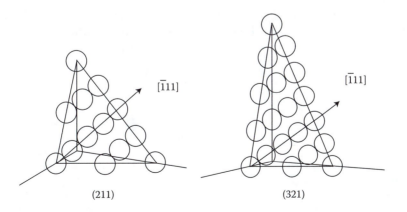

FIGURE 4.18
Atoms sitting in the (211) and (321) planes of the bcc structure. Atoms in the (211) plane have the coordinates (1,0,0), (0,2,0), (0,0,2), (0,1,1) and (1/2,1/2,1/2) and satisfy the equation $2x+y+z=2$. Atoms in the (321) plane satisfy $3x+2y+z=6$.

and from Equation 4.18, the planar fraction for the bcc system is

$$\text{Planar fraction} = \frac{3\pi}{8\sqrt{2}} = 0.833. \tag{4.21}$$

4.3.1.6 Linear Densities

Linear density refers to the number of atoms per length along a specified direction. In the (111) plane in the fcc system, the directions with the highest linear densities are the [$\bar{1}$01], [0$\bar{1}$1], and [$\bar{1}$10]. (Note that these vectors lie in the (111) plane, so their dot products with the plane must be zero.) Each of these vectors is $4R$ long and contains two atoms, so the linear density of the $\langle \bar{1}01 \rangle$ family in the fcc system is $1/(2R)$.

The directions of highest linear density in the (110) plane in the bcc system are [$\bar{1}$11] and [1$\bar{1}$1] or the $\langle \bar{1}11 \rangle$ family. Again it may be seen that the linear density in these directions is $1/(2R)$.

4.3.1.7 Linear Fraction

The linear fraction is the fraction of a line in a specified direction that is covered by atoms whose centers lie on the line. Since the width of an atom is $2R$, the linear atomic fraction of both the $\langle \bar{1}01 \rangle$ family in fcc and the $\langle \bar{1}11 \rangle$ fraction in bcc will be 1.

4.4 Interstices

The positions between the atoms in a unit cell are called interstices or interstitial sites and are important because they provide paths for diffusion in solids and are sites for small impurity atoms that are often added to harden the material. They also affect the solubility of small atoms such as carbon in iron, which plays a major role in steel making.

4.4.1 Interstitial Sites in the Simple Cubic Lattice

In the simple cubic structure, the only interstitial site (shown in gray in Figure 4.19) is located at $(1/2,1/2,1/2)$ and is coordinated by the eight atoms on the corners of the cube.

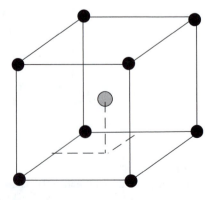

FIGURE 4.19
Interstitial site for a simple cubic lattice is located at $(1/2,1/2,1/2)$.

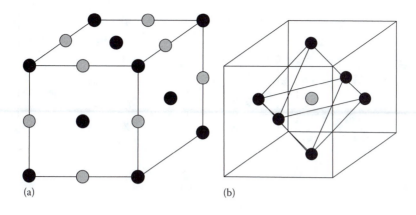

FIGURE 4.20
(a) Edge octahedral interstitial sites in the fcc lattice shown in gray (not all points shown) and (b) central octahedral site showing the six nearest neighbors forming the octahedron about the site.

4.4.2 Interstitial Sites in the Face-Centered Cubic Lattice

In the fcc structure, the interstitial site at the center of the cube (1/2,1/2,1/2) and those that are between the atoms that lie on the coordinate axes at (1/2,0,0), etc., have six nearest neighbors and are said to be octahedral sites because the six atoms surrounding them form an eight-sided double pyramid called an octahedron as shown in Figure 4.20a. (This is an unfortunate nomenclature since most students tend to think of an octahedral site as having a coordination number of 8. It is not 8—it is 6.) Perhaps this octahedron is easier to visualize if we only look at the (1/2,1/2,1/2) octahedral site and omit the corner atoms, as shown in Figure 4.20b. There are four such sites per unit cell.

The size of the octahedral site in the fcc lattice can be shown to be

$$r = \left(1 - \sqrt{2}\right)R = 0.414R. \tag{4.22}$$

The interstitial sites located at (1/4,1/4,1/4), etc., have four nearest neighbors that form a tetrahedron and are called tetrahedral sites. There are eight such sites per unit cell as may be seen in Figure 4.21a. The tetrahedral coordination is best seen for the (3/4,3/4,3/4) in Figure 4.21b.

The size of the tetrahedral interstitial site in the fcc structure can be shown to be

$$r = \left(\sqrt{3/2} - 1\right)R = 0.225R. \tag{4.23}$$

4.4.3 Interstitial Sites in the Body-Centered Cubic Lattice

In the bcc lattice, the three octahedral sites are located at the center of the faces of the unit cell (0,1/2,1/2), etc., as shown on Figure 4.22a and the 12 tetrahedral sites are located on the faces at (1/2,1/4,0), etc., as shown on Figure 4.22b. The sizes of these interstitials are given in Table 4.1.

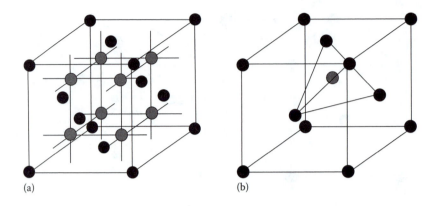

(a)   (b)

FIGURE 4.21
(a) Eight tetrahedral sites in the fcc lattice and (b) the site at $(3/4,3/4,3/4)$ is easiest to visualize because it is coordinated with an outside corner atom and the three exposed face atoms.

4.4.4 Interstitial Sites in the Hexagonal Close-Packed Lattice

The unit cell for the hcp lattice consists of two **a**-vectors $120°$ apart and a **c**-vector normal to the two **a**-vectors. The ideal c/a ratio is $\sqrt{8/3}$.

In an hcp unit cell, the two octahedral sites are located at $(1/3,2/3,1/4)$ and $(1/3,2/3,3/4)$, etc., and the internal tetrahedral sites are at $(2/3,1/3,1/8)$ and $(2/3,1/3,7/8)$ as may be seen in Figure 4.23a and b. There are additional tetrahedral sites on the edges at $(0,0,3/8)$, $(0,1,3/8)$, $(1,1,3/8)$, $(1,0,3/8)$ and another set of four at $c=5/8$. Since these holes sit on edges of the unit cell, they only count as one interior hole, giving the total number of tetragonal interstitial sites to 3 per unit cell.

4.4.5 Comparison of Interstitial Sites in the Metallic Lattices

The number of the interstitial sites and their radius r relative to the atomic radius R for the three lattice systems are given in Table 4.1. Even though the bcc lattice seems more open

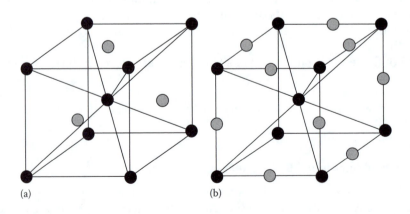

(a)   (b)

FIGURE 4.22
(a) Octahedral interstitial sites (gray) in the bcc lattice lie on the faces and are coordinated with the four corner atoms plus two center atoms (not all shown) and (b) tetrahedral interstitial sites (gray) lie on the edges of the bcc lattice and are coordinated with two corner atoms and the two middle atoms (not all shown).

TABLE 4.1

Tetrahedral Sites

Structure	Atoms per Unit Cell	Number of Tetrahedral Sites	Tetrahedral Sites r/R	Number of Octahedral Sites	Octahedral Sites r/R
bcc	2	3	0.291	3	0.155
fcc	4	8	0.225	4	0.414
hcp	2	3	0.225	2	0.414

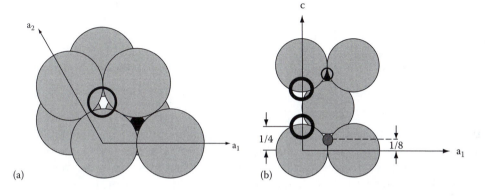

FIGURE 4.23
(a) Octahedral interstitial (open circle) and tetrahedral interstitial (black dot) in the (0001) planes of the hcp unit cell and (b) octahedral interstitial (larger open circle) and tetrahedral interstitial (black dot) sites in the ($1\bar{1}00$) face of an hcp unit cell showing their positions relative to the **c**-axis.

than the fcc or hcp lattice, notice that their octahedral sites are actually smaller than those for the fcc and hcp because they are constrained by the short distance between the central atoms in the bcc lattice system (Figure 4.23). This difference between the size of the interstitial sites in the fcc and bcc plays a significant role in the solidification of the iron–carbon system as discussed in Chapter 14.

4.5 Quasicrystals

Having gone to great lengths to stress the importance of long-range order in crystals brought about by repeated translations of unit cells that fill all space, we now have to inform the reader that nature makes exceptions to these rules of crystal growth. There is another close-packed structure with a coordination number of 12 besides the fcc and the hcp configurations and that is the icosahedron. The icosahedron, shown in Figure 4.24, is a polyhedron with 20 triangular sides. It is possible for crystals to nucleate and grow in this configuration, but since an icosahedron has six axes with fivefold symmetry, it cannot fill all space, hence form a crystal with a periodic structure. It can, however, form an aggregate of small icosahedral grains now called quasicrystals. The structure of the grains can be quite complex, but there are several ways in which they can arrange themselves in a long-range repeating structure so that x-ray diffraction (XRD) can reveal their fivefold (and sometimes higher fold) symmetry.

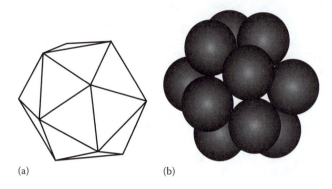

(a)   (b)

FIGURE 4.24
Vertices of an icosahedron also represent a close-packed structure with a coordination number of 12 (see Figure 5.19). Crystals can nucleate in this configuration, but since icosahedra cannot fill all space; they themselves do not form periodic structures.

While working with rapidly solidified Al–Fe–Mn alloys at the NBS (now NIST) in 1981, Dan Shechtman observed an unmistakable XRD pattern with tenfold symmetry. (Crystals with n-fold symmetry, where n is odd, can diffract with $2n$ symmetry for reasons that will be discussed in Chapter 6.) It was several years before the concept of quasicrystals became generally accepted, but then intensive research was focused on searching for new quasicrystals and understanding their properties. Most of the early work focused on systems produced by melt spinning (a technique for rapid solidification) because it was thought that quasicrystals were metastable since they were not in a configuration in which all the bonds were satisfied. Most of these systems were Al-transition metals or Al–Mg-transition metals. However, it was later found that some quasicrystals such as $Al_{65}Cu_{20}Fe_{15}$, $Al_{65}Cu_{20}Ru_{15}$, $Al_{65}Cu_{20}Os_{15}$, and $Al_{70}Pd_{20}V_5Co_3$ were in fact stable.

How can icosahedra, which we know cannot fill all space, form solids with enough long-range order to diffract x-rays? One of the simplest ways would be to place each icosahedron on a Bravais lattice site and fill in the space between them with other atoms. Actually, most quasicrystal structures are more complicated than this and will not be dealt with here.

Quasicrystals have some very unusual properties that are discussed in subsequent chapters along with the properties of ordinary crystals. Suffice it to say at this point that, although they consist of metals, their properties are very unmetal-like. One of their most interesting properties is their extremely low surface energy, which means nothing wants to stick to them. Taking advantage of this property, Lynntech, Inc. (College Station, Texas) has developed an electrodeposition process for coating Al-3004 alloy with $Al_{65}Cu_{23}Fe_{12}$ quasicrystals to form nonstick wear-resistant cookware.

4.6 Summary

A crystal is an array of atoms (or molecules) that repeat in three dimensions over distances that are long compared to the interatomic spacing. Since this configuration represents the lowest possible internal energy, most solids are in the form of crystals, although these crystals may be microscopic in size. A polycrystalline solid consists of many randomly oriented crystallites bonded together through their grain boundaries. Some applications require macroscopic single crystals in which the long-range order is extended over millimeters or even centimeters. Noncrystalline or amorphous solids can be formed in which the atoms themselves are randomly oriented resembling liquids that have been frozen in place. Such solids are called glasses and have no grain boundaries. It is also possible to form nanocrystalline materials in which the order extends only over a few atomic spacings. Such materials possess some of the properties of ordered solids as well as amorphous solids.

A unit cell of a crystal consists of a volume which contains the array of atoms that are repeated. The smallest possible unit cell that can form the structure is a primitive unit cell although is it often convenient to describe to solid using nonprimitive unit cells that have higher symmetries. A unit cell is described by three lattice vectors that define the directions the cell can be propagated to form the crystal lattice or framework of the structure. The basis of the crystal describes the positions of the atoms within the unit cell. The basis plus the lattice defines the structure. The set of all possible translations along the lattice vectors, forms the translation group.

Directions relative to the lattice vectors are defined by the projections onto the lattice vectors on these axes converted to whole numbers with no common denominator and set in square brackets with no commas. Negative numbers are denoted by bars above the number. Equivalent families of vectors, such as directions along the lattice vectors, are denoted by pointed brackets.

Planes are denoted by their Miller indices h, k, and ℓ generated by taking the reciprocals of the intersections of the plane with the lattice vectors and converting them to whole numbers with no common denominator. The indices are set in parentheses with no commas and negative numbers are denoted by bars above the number. The indices represent not only a single plane, but all parallel planes that have the same spacing. For rectangular lattices, the distance from the origin to the $(hk\ell)$ plane (as well as between planes with the same indices) is given by

$$d_{hk\ell} = \frac{1}{\sqrt{(h^2/a^2) + (k^2/b^2) + (\ell^2/c^2)}}.$$

Families of equivalent planes, such as planes perpendicular to the lattice vectors, are denoted by curly brackets. For cubic lattices only, the vector perpendicular to a plane has the same indices as the Miller indices of the plane.

Sometimes, particularly when describing higher order reflecting planes in XRD, it is necessary to define a specific plane. In this case, the Miller indices are not reduced to the nearest common denominator.

There are seven different polyhedra with different symmetries that can fill all 3-D space. These polyhedra form the primitive cells of seven basic crystal systems. It is also possible to form an additional seven nonprimitive systems with higher symmetry by adding lattice points in the center or on the faces of the basic systems, thus forming the 14 Bravais lattices that describe all crystals.

Many of the physical properties of a crystal are determined by its symmetry, which can be described by its point group and space group. There are 32 possible point group operations (operations than involve rotations or reflections about a point of a plane of symmetry). By including certain other spatial operations, the 32 point groups can be subdivided into 230 possible space groups that completely describe the symmetry of all possible crystal systems.

Relations between the atomic radius R and the lattice dimension a in the cubic systems are important for determining properties such as mass density, APF, planar densities, and line densities. These latter two parameters determine the slip systems that are important in the deformation of metals. For the fcc lattice

$$a = 2\sqrt{2}R \quad \text{or} \quad 4R = \sqrt{2a^2}$$

and for the bcc lattice

$$a = \frac{4}{\sqrt{3}}R \quad \text{or} \quad 4R = \sqrt{3a^2}.$$

The interstitial spaces between the atoms in the lattice offer pathways for diffusion to occur and places for impurity atoms to reside. There are two types of interstitial sites characterized by their coordination numbers: tetrahedral interstitial sites whose four nearest neighbors form a tetrahedron and octahedral sites whose six nearest neighbors form an octahedron. The small octahedral sites in bcc material result from the short distance between the central atoms in the bcc structure and are responsible for the low solid solubility of interstitial impurities in bcc material.

Just as we think we know everything about how matter is constructed, nature always has a way of surprising us. Quasicrystals, crystals with fivefold symmetry that were not supposed to exist, actually do. Their discovery has opened exciting new areas of research into their structure and properties, especially when it was found that, even though they are metallic systems, in many respects they behave more like ceramics and semiconductors than metals.

Bibliography

Allen, S.M. and Thomas, E.L., *The Structure of Materials*, Wiley MIT Series in Materials Science and Engineering, John Wiley & Sons, New York, 1999.

Callister, W.D., *Materials Science and Engineering: An Introduction*, 7th edn., John Wiley & Sons, New York, 2007.

Dunois, J.-M., *Useful Quasicrystals*, World Scientific Publishing, Hackensack, NJ, 2005.

Hahn, T., Ed., Space-group symmetry in *International Tables for Crystallography: Volume A*, 5th edn., Institut für Kristallographie, Technische Hochschule Aachen, Germany, 2005.

Harrison, W., *Solid State Theory*, McGraw Hill, New York, 1970.

Liboff, R.L., *Primer for Point and Space Groups*, Springer, New York, 2004.

International Tables for Crystallography, Vol. A, Th. Hahn, Ed., Springer, 2006, available online through www.springerlink.com

For an online source of point and space groups posted by the University of Oklahoma Crystallographic Laboratory, see http://xrayweb.chem.ou.edu/notes/symmetry.html.

Problems

1. Al forms an fcc lattice and has a density of 2.7 g/cm^3 and a molecular weight of 26.98. Calculate the lattice parameter, the atomic radius, and the atomic density.

2. Rank the planar fractions in order of the (100), (101), (111), (211), and (321) planes in the bcc system.

3. What are the directions of highest linear density in the planes mentioned in Problem 2?

4. Verify the relative sizes of the interstitial sites presented in Table 4.1.

5. Find the APF for the hcp structure.

6. Find the planar density fraction in the $(1\bar{1}00)$ plane of an hcp lattice? What is the direction of highest planar density?

7. What happens in a trigonal lattice when the angle between axes becomes 120°?

5

The Structure of Matter

The properties of matter are intimately tied to its structure, as we shall see in subsequent chapters, and the structure of matter is primarily determined by the sizes of the atoms and ions of its constituents and by the types of bonding between them. In this chapter, we will explore some of the myriads of ways nature has found to assemble materials, starting with simple one-component systems and then moving to compounds and finally to polymeric molecular structures.

5.1 Structure of Metals

Because the metallic bond is nondirectional and because the positive ion cores attract each other through the intervening sea of electrons that flows between them, metals tend to form close-packed structures, meaning they try to get as close together as they can in order to maximize their coordination number. Johannes Kepler (a mathematician as well as the famous astronomer) conjectured that the maximum number of identical spheres that could surround and touch another sphere was 12. Now, let us see what kinds of structures are possible with a coordination number of 12.

5.1.1 Face-Centered Cubic Versus Hexagonal Close-Packed Structures

Imagine packing oranges into a crate and you want to put as many oranges as possible into a given volume. There are two possible ways of doing this in such a way as to fill all space with repeating structures. You would start by arranging the first layer in a hexagonal array so that each orange had six nearest neighbors (label this layer A). You would put the next layer in three of the six vertices formed by the oranges in the first layer (label this layer B). Now when you start the third layer, you have two choices: (1) You could place the oranges in the third layer in the vertices directly over the ones in the first layer, forming an ABABAB repeating structure (Figure 5.1a), or (2) you could make a C layer by placing the third layer in the other three vertices formed by the second layer to form an ABCAB-CABC repeating structure (Figure 5.1b). Some metals prefer to solidify in the ABABAB structure, which is called hexagonal close-packed (hcp), while others prefer to solidify in the ABCABC or face-centered cubic (fcc) structure.

The relation between the ABC stacking and the fcc structure can be better understood from the diagram in Figure 5.2. The atoms that lie in the (111) A, B, and C planes are shown in Figure 5.2a. These planes are projected in the (110) face in Figure 5.2b to show the orientation and spacing of these planes. These are the planes with the highest planar

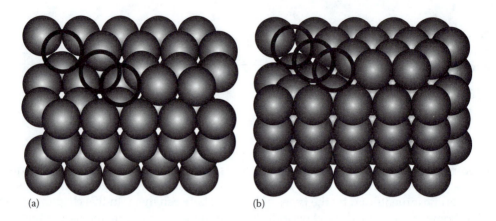

(a)   (b)

FIGURE 5.1
(a) ABA stacking to form the hcp structure. Heavy black circles indicate where the next ball would be placed in rows 2, 3, and 4. Note that every other vertex is uncovered. (b) ABC stacking to form the fcc structure. Heavy black circles indicate where the next ball would be placed in rows 2, 3, and 4.

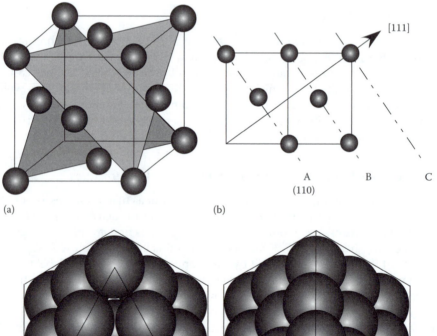

(a)   (b)

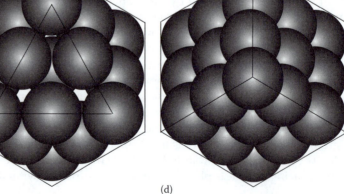

(c)   (d)

FIGURE 5.2
(a) (111) planes in the fcc lattice. (b) Projection of the (111) ABC planes on the (110) face. (c) High planar density atomic configuration of the (111) A plane looking at the fcc cube along the [111] direction. (d) Addition of atoms in the B and C planes forms the [111] corner of the fcc cube. The outline of the full cube is shown for reference.

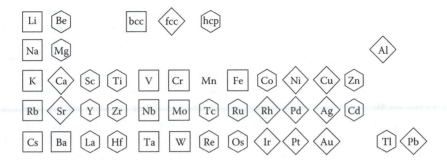

FIGURE 5.3

Periodic table of the metals that form fcc, hcp, and bcc structures. The transition metals tend to go from hcp → bcc → hcp → fcc except for Mn, Fe, and Co whose magnetic properties apparently influence their structures. Mn has a complex cubic structure with a space group *I34m*.

density as may be seen in Figure 5.2c. Adding the B- and C-layer atoms, we see the [111] corner of the cube in Figure 5.2d.

Both fcc and hcp structures have the same coordination number (number of nearest neighbors) of 12 and the same atomic packing fraction of 0.74 (ratio of the volume of oranges to the container volume), but metals that form the fcc structure are generally more ductile than hcp metals because there are more planes with high area fraction which provide slip systems in the fcc structure. This will be discussed in more detail in Chapter 9.

Metals solidifying with the fcc structure (space group $Fm\bar{3}m$) include Al, Ag, Au, Ca, Cu, Ir, Ni, Pb, Pd, and Yb. Metals solidifying with the hcp structure (space group $P6_3/mmc$) include Be, Co, Gd, Ho, Mg, Ti, Tl, and Zn (see Figure 5.3). None of the hcp structures have the ideal c/a ratio of 1.633. Mg and Co come the closest with c/a values equal to 1.624 and 1.623, respectively. Be has the smallest with 1.568 and Zn has the largest with 1.856.

Note that the space group for the hcp structure is $P6_3/mmc$. The point group is $6/mmm$, but the subscript 3 next to the 6 signifies additional symmetry in the space group in the form of a threefold screw axis parallel to the sixfold rotational axis. Similarly, the c in place of the last m signifies a glide plane parallel to the c-axis.

5.1.2 Double Close-Pack Structures

α-La and some of the rare earth metals, e.g., Pr, Nd, and Am form a double close-pack structure, ABACABAC, etc., essentially doubling the length of the c-axis. Its space group is also $P6_3/mmc$.

5.1.3 Body-Centered Cubic Structures

Other metals tend to solidify into a body-centered cubic (bcc) structure with a coordination number of 8. These include Ba, Cr, Cs, Eu, Fe, K, Li, Mo, Nb, Rh, Ta, V, and W. The space group is $Im\bar{3}m$.

5.1.4 What Determines Which Structure a Metal Will Have?

We have just shown that the fcc and hcp structures have the same number of neighbors, so there could not be much difference in the energies of the two systems. So what determines which structure a metal will chose to solidify in? And why do some metals solidify in bcc structures instead of close-packed structures such as fcc or hcp? The number of nearest neighbors for the bcc structure is 8 compared to 12 for the close-packed structures, but the number of second nearest neighbors in the bcc structure is 6, and these are closer than the second nearest neighbors in the close-packed structures. Therefore, with 14 nearest and not-to-far-away neighbors, one would expect the difference in binding energy between bcc and the close-packed systems may not be very large. We know in fact that changes in temperature or pressure can cause a metal to change its structure. Fe, for example, changes from bcc to fcc at 912°C and back to bcc at 1394°C before it melts at 1538°C. These different forms of the same material are called allotropes.

So then how does a metal decide what structure it should solidify into? Examining the periodic table of the metals that form fcc, hcp, and bcc structures in Figure 5.3 reveals some semblance of a pattern. All of the monovalent alkali metals are bcc. Except for the magnetic elements (Mn, Fe, and Co), the transition metals go from hcp $\rightarrow$ bcc $\rightarrow$ hcp $\rightarrow$ fcc. The structures of the other elements seem to have no particular order.

There is no simple explanation other than to say that the free energies of each of the elements are determined by the temperature, pressure, and the distance between the ion cores and their electronic environment. Subjecting elements to high pressure alters their internuclear distance and makes a $P-V$ contribution to the free energy which can produce solid-state phase changes. The only simple explanation seems to be that those phases that happen to have the lowest free energies under ambient conditions are the ones that appear in Figure 5.3.

The ability to compute the free energies of various possible structures has evolved over the years beginning with early attempts to obtain self-consistent solutions to the Schrödinger equation for a many electron system. This computational capability was greatly enhanced in the 1970s by the development of the density functional theory (DFT) that earned Walter Kohn the 1998 Nobel Prize in chemistry. The essence of the DFT is the replacement of the many-bodied wavefunction, which has $3N$ variables, with a density function containing only three variables.

5.2 Intermetallic Compounds

If there is sufficient negative heat of mixing between two metals, or a metal and a semimetal, meaning that two unlike atoms are more attracted to each other than to like atoms, a compound may form. Such unlike attractions can arise from differences in electron negativity or from their ability to form covalent bonds. Unlike solid solution systems where one component can substitute randomly for the other, intermetallic compounds have a definite composition and tend to have ordered structures. They are generally hard and brittle and can have a higher melting temperature than either of their two components because of their bond energy. Case in point, NiAl has a melting point that is 283°C higher than pure Ni.

Strictly speaking an intermetallic compound is a compound formed by two metals, such as Ni_3Al, Al_2C, etc., although the definition has been broadened to include metal–metalloid (Si, Ge, and As) and even nonmetals (Fe_3C). There are over 11,000 known intermetallic compounds, discovered primarily from their phase diagrams. It is estimated that there could be as many as 500,000 possible ternary systems and 10^7 quaternary systems with only a small fraction actually known (see Villars, 1994). Intermetallic compounds possess many unique mechanical, chemical, electronic, optical, and magnetic properties. Their uses range from strengthening or even replacing superalloys (Ni–Al), to catalysts (Raney Ni), protective surface coatings (Al–Cu–Fe icosohedral phase), semiconductors (GaAs and AlN), thermoelectric materials (Bi_2Te_3), super-hard magnets ($SmCo_5$ and $Fe_{14}Nd_2B$), and even dental amalgams (Ag_2Hg_3 and Sn_8Hg) that solidify at 37°C and have the crush strength of cast iron. The following few examples are intended to introduce the reader to the enormous scope of these fascinating materials and to emphasize that there is still much to be learned in the field of metallurgy.

5.2.1 Strukturbericht and Pearson Notation

The structures of intermetallic phases are often designated in the literature by Strukturbericht and Pearson symbols instead of by the space group. The Strukturbericht method for classifying metallic structures consists of a capital letter followed by a number and a subscript. The A-series were supposed to be elemental structures (with an exception for A-15); the B-series, AB compounds; the C-series, AB_2 compounds, etc. The number and subscript following the letter designated the lattice. Examples are A1 for fcc, A2 for bcc, A_a for bct, A3 for hcp, A4 for diamond, etc., up to A20. For the AB binary systems, B1 is the NaCl structure, B2 is the CsCl structure, B3 is zinc blende, B4 is wurtzite, etc. up to B37. C designates the AB_2 compounds, D the A_nB_m compounds, etc.

Pearson symbols are also used to indicate the crystal symmetry and the number of atoms in the unit cell. The Bravais lattice is designated by a lower case letter indicating the basic crystal class (c-cubic, h-hexagonal or trigonal, t-tetragonal, o-orthorhombic, m-monoclinic, and a-asymmetric [triclinic]), followed by B, F, I, or P to indicate symmetry of the lattice, and then the number of atoms in the unit cell. For example, NaCl would be designated cF8, Fe would be cI2, perovskite would be cP5, Hg_2Cl_2 would be tI8, etc.

5.2.2 NiAl Intermetallic Phases

The NiAl intermetallic phases are of particular importance because of their use in strengthening Ni-based alloys, especially the superalloys used in high-temperature gas turbine blades (the Ni–Al phase diagram is shown in Figure 12.12). Nickel aluminide (Ni–Al) is the simplest of these compounds. It is described by the Strukturbericht symbol B2, which tells us it has the CsCl structure, and the Pearson symbol cP2 tells us we have a cubic lattice with two atoms per unit cell. Therefore NiAl has Ni atoms on the corners of a cube and an Al atom in the center (or vice versa). (Note: this is not a bcc structure because the atom in the center is not the same as those on the corners.) The width of this phase in the phase diagram tells us that some solid solution is possible within the phase.

Next consider Ni_3Al designated as $L1_2$ or cP4. The L designates an alloy, but unless one were familiar with Strukturbericht notation, it would be necessary to look up the structure in a catalog of Strukturbericht crystal structures such as the Naval Research Laboratory online catalog of crystal structures (www.nrl.navy.mil/lattice/). The Pearson symbol is perhaps more informative, telling us we have a primitive cubic crystal with four atoms per unit cell, and with a little imagination, one could guess that Al atoms must occupy the corners while Ni atoms sit on the six faces.

Now, what about Al_3Ni? Would not it be the reverse of Ni_3Al? One might think so, but the notation DO_{11} or oP16 tells us that it has a much more complicated structure. Perhaps it might help if we knew the space group was *Pmna*, or not. We can see immediately that the unit cell is orthorhombic and contains 16 atoms, but one would have to consult a catalog to get the actual structure from the Strukturbericht notation.

5.2.3 Laves Phases

Laves phases have what is known as a topological close-packed (TCP) structure which is often undesirable because they can act as stress risers that initiate cracks because of their hardness and morphology and because they tie up atoms needed for the more desirable coherent γ' phases that are involved in precipitation hardening of alloys.

A number of AB_2 intermetallic compounds with the A/B diameter ratio approximately 1.225 form what are called Laves phases. In the cubic Laves phase (C15), the A-atoms sit on the cubic diamond sites and the B-atoms form tetrahedra about the A-atoms. Examples include $MgCu_2$, $CsBi_2$, and $RbBi_2$. The hexagonal Laves structure (C14) is similar to the cubic structure except the A-atoms sit on hexagonal diamond sites. Examples include $MgZn_2$, $CaMg_2$, $ZrRe_2$, KNa_2, $TaFe_2$, $NbMn_2$, and UNi_2. Another class of hexagonal Laves phase, the dihexagonal (C36) class, stacks according to ABACABAC. This class includes $NbZn_2$, $ScFe_2$, $ThMg_2$, $HfCr_2$, and UPt_2.

5.2.4 MAX Phases

Recently, a new class of materials has been introduced, the so-called MAX phases, which actually stands for $M_{n+1}AX_n$, where M is an early transition metal, A is from either IIIA or IVA group, and X is C or N. Examples are Ti_3SiC_2, Ti_2AlN, and $Ti_2AlC_{0.5}N_{0.5}$. This new class of materials has both metallic and ceramic properties. Like metals, they are electrically and thermally conductive, easily machinable, and deform plastically at high temperatures. Like ceramics, they have low densities, are elastically rigid, maintain their strength at high temperatures, and are resistant to oxidation. They show remarkable promise for high temperature applications such as gas turbine blades since they are lighter and can withstand higher temperatures than the Ni-based superalloys presently used for that application.

The MAX phases are classified according to their n-number; for example, Ti_2AlN ($n = 1$) is referred to as 211, Ti_3SiC_2 ($n = 2$) as 312, and Ti_4AlN_3 ($n = 4$) as 413. The unit cells are hexagonal (space group $P6_3/mmc$) and contain near close-packed transition metal carbides and nitrides, interleaved with planes of pure A-atoms, every third layer for 211, every fourth layer for 312, and every fifth layer for 413.

5.2.5 Interstitial Compounds

Small atoms such as C and N have radii less than 2/3 of some of the transitional metals and can form interstitial compounds. The C and N atoms form covalent bonds with transition metals, such as Ti, V, Zr, Nd, Hf, and Ta, to form carbides and nitrides that are extremely hard and have very high melting points. The best-known example of a transitional metal-carbide is Fe_3C, also known as cementite, which is the primary hardening component in steel. It has a complex DO_{11} structure similar to Al_3Ni. There is another form of Fe_3C known as bainite (Pearson symbol hP8—apparently no Strukturbericht symbol) which forms in steels at a lower temperature than cementite.

Hydrogen can also readily enter the interstities of metals to form stoichiometric hydrides such as TiH_2 and ZrH_2 or nonstoichiometric compounds. Palladium has the unique capability of absorbing up to 900 times its own volume of hydrogen under standard conditions. Hydride formation generally destroys the ductility of metals and leads to hydrogen embrittlement. This can be a serious problem in metals used in nuclear reactors where there is a high neutron flux.

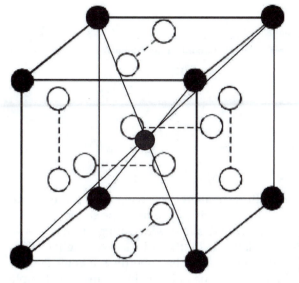

FIGURE 5.4
A-15 structure of A_3B compounds. The A-atoms form aligned chains on the faces of the bcc lattice.

5.2.6 A15 Superconductors

A number of intermetallics with the composition A_3B, where A is a transition metal and B is either another transition metal of a group IV semiconductor, form what is known as the A15 phase. (Why this deviation from the Strukturbericht designation is not clear.) Many of the A15 compounds, e.g., Nb_3Al, Nb_3Ge, and Nb_3Sn, are superconductors with transition temperatures $\sim 20\,K$ (the highest known transition temperatures until the discovery of the high-temperature ceramic superconductors). The structure shown in Figure 5.4 is bcc with the A-atoms on the bcc lattice sites and the B-atoms, two per side, aligned in orthogonal chains along the sides of the unit cell as shown. The space group is $Pm\bar{3}n$, the n signifying a diagonal glide plane.

5.3 Ionic Compounds

The ionic bond, like the metallic bond, is nondirectional and favors larger coordination numbers than directionally bonded covalent compounds. The coordination number, however, is limited by Pauling's first rule of ionic crystal formation which requires every ion to be surrounded by a polyhedron of counterions and the stability of the system requires the ion to be in direct contact with each counterion. These requirements establish the allowable coordination number (number of nearest neighbors) based on the ratio of cation/anion size. Since the negatively charged anions have received electrons from the cations, they tend to be the larger of the two.

We saw in the case of metals that it was possible to have close-packed structures with a coordination number of 12. Even if the cations and the anions were the same size, it is not apparent how one could surround each ion with a polyhedron of 12 counterions in a repeating structure that would fill all space. If the cation is slightly smaller than the anion, it will not be able to touch each of the 12 surrounding anions and according to Pauling's first rule, the structure will be unstable. Another way of looking at this is, since there is strong electrostatic repulsion between like ions, the cation must be large enough to prevent

TABLE 5.1

Coordination Numbers

Ratio of Cation/Anion Diameter	Possible Coordination Number
1.00	12
0.732–1.0	8
0.414–0.732	6
0.225–0414	4
0.155–0.225	3
<0.155	2

the surrounding anions from direct contact with each other. Otherwise, the repulsive force between the counterions will be greater than the attractive force from the cation.

The next largest coordination number that would allow a repeating structure is 8, which would be in the form of a simple cubic lattice with the anions at the eight corners and the cation at the interstitial position at the center of the cube (or the equivalent in which the cations sit on the corners of the cube and the anion is in the middle). Simple geometry shows that the cation/anion diameter must be greater than 0.732 to prevent the surrounding ions from touching each other. Following this line of reasoning, the coordination number for various ranges of cation to anion diameter ratios is summarized in Table 5.1.

The following are examples of the simplest and most common of the many possible structures for ionic compounds.

5.3.1 Cesium Chloride Structure (B2)

If the cation to anion ratio is between 1 and 0.732, the coordination number can be 8 and the compound will form an interpenetrating simple cubic structure with the anions (or cations) at the corners of the cube and the counterion in the center of the cube as in Figure 5.5. This is known as the cesium chloride structure and it is not a bcc structure because the ion in the middle is not the same as those on the corners. The space group is $Pm\bar{3}m$.

It is necessary that the net charge in the unit cell be neutral. The cation in the middle is completely inside the unit cell and contributes a charge of +1. The eight anions on the corners have only 1/8 of their charge inside the unit cell, so they collectively contribute −1 to balance the cation charge.

Compounds with this structure include CsBr, CsI, RbCl, AlCo, AgZn, BeCu, MgCe, RuAl, and SrTl.

5.3.2 Rock Salt or Sodium Chloride Structure (B1)

If the cation to anion ratio is between 0.732 and 0.414, the cation is not large enough to keep eight nearest neighbor anions from touching one another, so the coordination number must be reduced to six. Now the anions will form an interpenetrating fcc structure (Figure 5.6) with the smaller cations occupying the octahedral interstitial positions between the anions (recall that the term octahedral refers to the fact that the six nearest neighbors form an octahedron about the interstitial ion). This arrangement is known as the sodium chloride or rock salt structure. Its space group is $Fm\bar{3}m$.

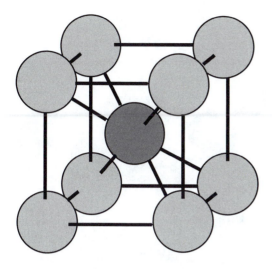

FIGURE 5.5
CsCl structure with coordination number 8. This is a simple cubic lattice, not a bcc lattice, with the counterion in the middle.

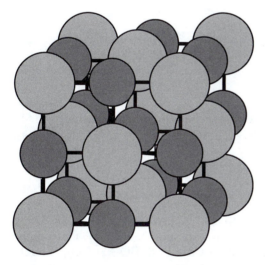

FIGURE 5.6
Rock salt structure is fcc with the counterions occupying all of the octahedral sites. The ionic bonds are shown as heavy lines.

Again the charge of the cations (1 in the middle plus 12 on the edges, each contributing 1/4 to the unit cell for a total of +4) is balanced by the charge of the anions (eight on the corners times 1/8 plus the six on the faces, each contributing 1/2 for a total of −4).

Compounds with this structure include AgCl, BaS, CaO, CeSe, DyAs, GdN, KBr, Lap, LiCl, LiF, MgO, NaBr, NaF, NiO, PrBi, PuC, RbF, ScN, SrO, TbTe, UC, YN, YbO, and ZrO.

Now that we have examined a few of the simpler AX ionic structures, we will look at a few of the more complicated ionically bonded structures.

5.3.3 Fluorite Structure (C1)

The fluorite structure (CaF_2) has the formula AX_2 (A is the cation and X the anion). We now must have twice the number of anions than cations. The structure works out to be fcc with the cations on the fcc sites and the anions occupying all of the tetrahedral positions (space

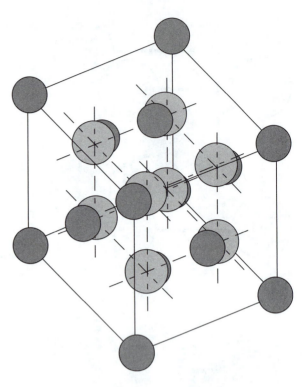

FIGURE 5.7
Fluorite structure, typical for AX_2 compounds, places the divalent A cations on the fcc sites and the monovalent X anions on all of the tetrahedral sites in order to balance the charge. Each anion is coordinated with one corner and three face cations. Each cation is coordinated by eight anions.

group $Fm\overline{3}m$) in Figure 5.7. Other compounds with the fluorite structure include AmO_2, $AuAl_2$, $AuIn_2$, BaF_2, Be_2B, CO_2, CdF_2, CeO_2, $CoSi_2$, EuF_2, HgF_2, Ir_2P, Li_2O, Na_2O, $NiSi_2$, $PtAl_2$, Rb_2O, $SrCl_2$, SrF_2, ThO_2, and ZrO_2.

5.3.4 Perovskite Structure (E2₁)

The perovskite structure (ABX_3) is simple cubic with the quadravalent A cation located in the center, the divalent B cations located on the eight corners, and the three divalent anions on the faces as seen in Figure 5.8. Examples include $CaTiO_3$ $BaTiO_3$, $SrZrO_3$, $SrSnO_3$, $PbTiO_3$, and $PbZrO_3$. The space group is $Pm\overline{3}m$. The titanates and zirconates only have this symmetry above a certain temperature called the Curie temperature. At temperatures below the Curie temperature, the lattice shrinks and there is insufficient room for the B cation to remain in the center. The anions on the faces will force the B ion in one direction while being displaced in the opposite direction, distorting the lattice from simple cubic to tetragonal, as mentioned in Chapter 3. This distortion results in a permanent charge displacement, making the crystal not only a piezoelectric (like quartz), but also a ferroelectric, meaning it has a permanent electric dipole moment. There will be more discussion about this in Chapter 23 on ferroelectrics.

5.3.5 Spinel and Inverse Spinel Structure (H1₁)

Another important ionic structure is the spinel structure, the mineral name for $MgAl_2O_4$, but also the generic name for AB_2X_4 compounds where A is a divalent metal, B a trivalent metal, and X is a group VI element, usually oxygen. Normal spinel is fcc with space group $Fd\overline{3}m$. The X anions occupy the fcc sites, the A^{2+} ion occupies one of the eight tetrahedral

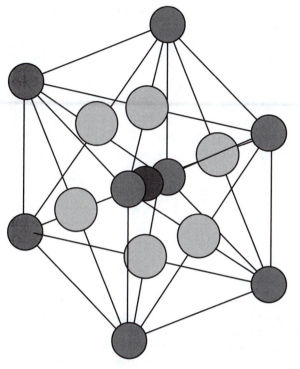

FIGURE 5.8
Perovskite structure for ABX_3 compounds. The quadravalent A cation is located in the center (darker sphere), octahedrally coordinated by the six divalent anions on the faces. The divalent cations sit on the corners, octahedrally coordinated by the six anions on the faces.

sites, and the B^{3+} ions sit on two of the four octahedral sites. Other compounds with this structure include $ZnAl_2Se_4$, $CrAl_2S_4$, $CaIn_2S_4$, $CdAl_2S_4$, $ZrCr_2Se_4$, and $ZnMn_2Te_4$.

The inverse spinel structure has one of the B^{3+} cations on the tetrahedral site and the A^{2+} cation and the other B^{3+} cation on the octahedral sites. The tetrahedrally coordinated cation shares an O^{2-} ligand with one of the two octahedrally coordinated cations. This feature is responsible for the interesting magnetic properties of the ferrites, a class of ceramic magnets that will be discussed in more detail in Chapter 25.

5.3.6 Structures with Small Cation-to-Anion Ratios

As the cation to anion ratio is further reduced to between 0.414 and 0.225, it is no longer possible for six anions to surround an anion without touching each other, so the coordination number is reduced to four and the anions form a tetrahedron about the cation. We will encounter this type of structure again in the case of the zinc blende structure and in the orthosilicate ion (Figure 5.18). The bonding in these cases may be a mixture of ionic and covalent bonds. In the case of ZnS or ZnSe, the cation/anion diameter is less than 0.414 so that in their cases the zinc blende or wurtzite structures could be the result of pure ionic bonding rather than covalent bonding. However, the cations of the remaining group II elements are large enough so that the other II–VI systems (CdS, HgTe, etc.) must form the tetrahedral configuration through directed covalent bonds.

A further reduction of cation to anion ratio between 0.225 and 0.150 allows a coordination number of only three forming an equilateral triangle such as the BO_3^{3-} ion. The units can form sheet structures by sharing a bridging O atom at the corners, as in the case of the SiO_4^{4-} ion. A regular crystalline structure is not possible, but because there is no barrier to rotation about the bridging O atoms, it is possible to form a three-dimensional (3D) random network such as B_2O_3 glass illustrated in Figure 5.21.

For $r_c/r_a < 0.15$, the maximum coordination number is 2. Because of the small size of the H^+ ion, substances that are hydrogen bonded have a coordination number of 2 and can only form linear chains unless the molecules involved are multifunctional such as in the case of H_2O, which can form a variety of crystalline ices by hydrogen bonding.

The reader should be cautioned that the radius rules above are by no means absolute and in fact are only obeyed ∼50% of the time. More ionic compounds have the rock salt (NaCl) structure and fewer have the CsCl structure than would be predicted from the cation/anion radius. Still, the use of the radius ratio is a good place to start in trying to understand the crystalline structure of binary ionically bonded compounds.

5.4 Covalent Structures

5.4.1 Diamond Structure (A4)

The sp^3 covalently bonded carbon, silicon, germanium, and α-Sn (gray tin which is stable below 291 K) form the cubic diamond structure as shown in Figure 5.9. Recall from Chapter 4 that the point group for the diamond structure is $4\bar{3}m$, but additional symmetry causes the space group to be given as $Fd\bar{3}m$, the d indicating a diagonal glide plane. Compared to metallic structures, the diamond structure is a very open structure with an atomic packing factor of 0.34. As pointed out previously, the sp^3 bond involves a cube with an atom in its center surrounded by atoms on opposite corners. This cube, however, does not constitute a suitable unit cell because the entire structure cannot be generated by translating such a cube along its axes. Instead, a proper unit cell for the diamond structure is an fcc lattice with atoms on each lattice site and on every other tetrahedral site as seen in Figure 5.9a.

Since only every other interstitial is occupied, the stacking sequence of the (111) plane of the diamond structure is AA′BB′CC′AA′BB′CC′, or double the ABCABC stacking of the fcc structure with the prime planes containing the interstitial atoms as shown in Figure 5.9b, which also shows projections of the diamond structure in the [001], [$\bar{1}$10], and [111] directions. The planes containing the various atoms are indicated by their respective letters in the [$\bar{1}$10] projections. Note that in this projection the intestinal atoms are hidden by the atoms on the fcc sites.

Four [110] projections of the diamond structure are combined in Figure 5.9c so that the distorted hexagonal channels in the ⟨110⟩ directions may be seen.

5.4.2 Sphalerite or Zinc Blende (B3)

Sphalerite or zinc blende is the chief ore of ZnS. ZnS as well as many of the III–V and II–VI compound semiconductors such as GaAs form a diamond-like structure with one type of atom on the fcc lattice sites and the other type of atom on every other tetrahedral site. The space group is $F\bar{4}3m$ (the d-glide plane symmetry in diamond is lost by the fact that different atoms occupy the interstitial sites). Since there are four atoms on the lattice points of the fcc unit cell, the stoichiometry is maintained if half of the eight tetrahedral sites are occupied by the second atom. For these systems, the double stacking described above would have different atoms in doubled layers, i.e., A(Zn) A′(S) B(Zn) B′(S) C(Zn) C′(S), etc. This type of structure is the same as shown in Figure 5.9 if the black spots are

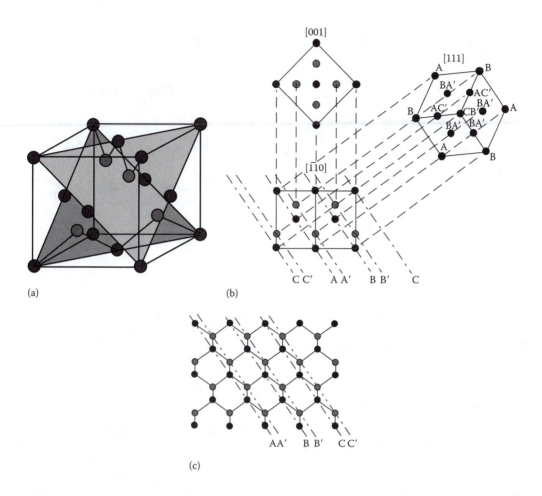

FIGURE 5.9

(a) Diamond structure is fcc (black) with every other tetrahedral interstitial sites occupied (gray). Some bonds are omitted for clarity. A (dark gray) and C (light gray) planes are also shown. (b) Views of the diamond structure are in three projections. The AA'BB'CC' stacking sequence can be seen in the [110] direction. The planes containing atoms in the [111] projection are indicated by the letters in the [110] projection. Notice that the interstitial atoms are hidden by the fcc atoms in [111] projection. (c) View of four of the unit cells shown in Figure 5.9b. Note the distorted hexagonal channels that run through the crystal.

assigned to one type of atom and gray to the other. Crystals with this structure include AlAs, AlP, BN, BP, CdS, CdTe, GaAs, GaP, GaSb, HgTe, HgSe, InP, InSb, β-SiC, ZnS, ZnSe, and ZnTe.

5.4.3 Hexagonal Diamond

All of the above diamond-like structures were based on the basic fcc lattice. It is also possible for sp^3 bonded compounds to form hcp-like structures. Consider the stacking sequence AABBAABB. Carbon stacked in this sequence is known as lonsdaliete or hexagonal diamond and is found in meteorites. The structure (space group $P6_3/mmc$) is similar to the wurtzite structure shown in Figure 5.10 except all of the atoms are the same.

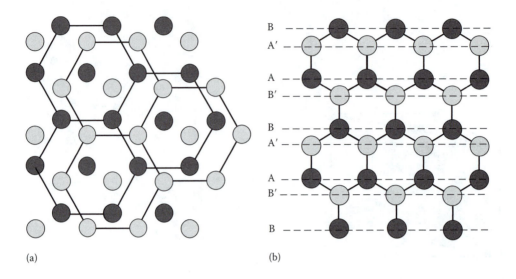

(a)   (b)

FIGURE 5.10
(a) Structure of the (0001) planes of wurtzite. Light circles represent the B′ planes and the gray circles represent the A planes. Heavy lines show the offset hexagonal rings in the two levels. Bonds between the light and gray atoms have been omitted for clarity. (b) Stacking sequence of a wurtzite structure showing the AA′BB′ stacking. Light circles represent Zn and gray circles S. Heavy lines represent bonds (some omitted for clarity).

5.4.4 Wurtzite (B4)

A less common ZnS ore, known as wurtzite, crystallizes in the hexagonal form (space group P6$_3$mc). Here the stacking sequence of the hexagonal planes is A(Zn) A′(S) B(Zn) B′(S), etc., as shown in Figure 5.10. The AA′ and BB′ planes are both open hexagonal structures in which the atoms in the BB′ planes lie on every other tetrahedral site of the AA′ planes. The Zn atoms in the A plane lie directly over the S atoms in the A′-plane as do the Zn and S atoms in the BB′ planes. In this manner each atom is tetrahedrally coordinated with atoms of the opposite species. Note that there are no bonds between the atoms making up the hexagonal rings in the various planes.

Crystals with the wurtzite structure include AlN, BeO, CdSe, GaN, InN, MgTe, SiC, ZnO, and ZnS. As one might imagine, there is little energy difference between the wurtzite and sphalerite structure, so many of the III–V and II–VI compounds can have either structure.

5.4.5 Graphite (W9)

Graphite is formed from sp^2 bonded carbon creating a planar network of benzene-like hexagons (space group P6$_3$mmc). The remaining electron becomes dislocated and can move throughout the basal plane making the material a good electric and thermal conductor in the direction of the plane. The layers are staggered with the trijunction of one layer centered in the middle of the hexagon in the adjacent layers to stack in the ABABAB manner as seen in Figure 5.11. The layers are bonded only by van der Waals forces, which accounts for the soft, slippery feel of graphite making it useful as a dry lubricant. (Moly-disulfide [MoS$_2$] has a similar layered structure with van der Walls bonding between the layers and is also used as a dry lubricant.) Although weak in shear between the layers, the strength of graphite within the basal plane is very great because of the strong covalent

A

B

A

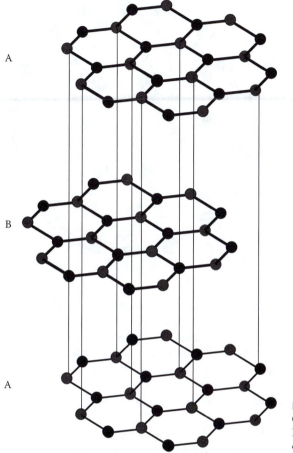

FIGURE 5.11
Graphite showing the ABAB layered structure.
Each lattice point in the A planes lies above the
center of a hexagon in the B plane and vice versa.

sp^2 bonds. These planes can be rolled up like rolls of chicken wire to form graphite fibers that are used to strengthen various composites.

It is now possible to make sheets of highly ordered pyrolytic graphite that have a thermal conductivity of seven times that of aluminum. This material is finding its way into thermal management systems ranging from heat sinks in laptop computers, to industrial heat exchangers, to avionic components for high performance aircraft.

5.4.6 Fullerenes, Fullerites, and Fullerides

In 1985, Richard Smalley discovered that carbon could exist in the form of C_{60}, a large molecule in the form of the geodesic structure invented by Buckminster Fuller. The 60 carbon atoms are connected by sp^2 bonded trijunctions as shown in Figure 5.12. These molecules are appropriately referred to as bucky balls and this class of carbon structures is now known as "fullerenes." The original stimulus for this discovery was an attempt to explain the observed absorption spectra in interplanetary dust, thought by the astrophysicists to be caused by clusters of carbon and chains of polycyclic aromatic hydrocarbons. However, there seems to be no correlation of the C_{60} absorption with the astrophysical observations. Nonetheless, this discovery prompted an intense interest in understanding how C_{60} and other carbon molecules form and in determining their properties.

After it became possible to produce macroquantities of C_{60}, it was discovered that these molecules assembled into fcc close-packed structures of solid C_{60} (with a small percentage

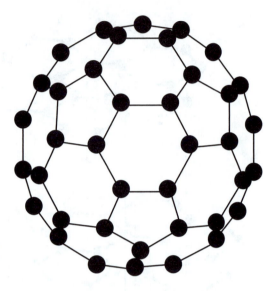

FIGURE 5.12
A bucky ball is basically a soccer ball with C atoms on
each of its 60 vertices forming a series of pentagons and
hexagons.

of C_{70}) known as fullerites. These fullerites are harder and have a higher bulk modulus and sound velocity than diamond.

It is possible to add interstitial atoms to fcc C_{60} to form A_3C_{60} compounds known as fullerides. Since there are 12 A-atoms in the unit cell, all tetrahedral and all octahedral sites will be occupied. If A is potassium, the materials will be superconducting with a critical temperature of 18 K. Going up the group 1 monovalent metals increases the critical temperature (T_c) until A_3 is Rb_2Cs, which has a T_c of 28 K. Interestingly, Rb_3C_{60} is not superconducting.

Because the C_n molecule is like a hollow shell with a diameter ranging from 0.4 to 1 nm as n goes from 60 to 240, other atoms or molecules can be placed inside of this protective cage to form a new class of materials called endohedral fullerenes. Dipolar salt molecules such as LiF, LiCl, and NaCl encapsulated in C_{60} exhibit paramagnetic ordering. Molecular hydrogen can be encapsulated in C_{60}. Metal atoms from group II and III encaged by C_{82} form endohedral mettalofullerenes. Electron transfer takes place between the metal atom and the C cage to form a super atom such as $Y^{3+}@C_{82}{}^{3-}$ (the @ indicates the Y is inside the C_{82} molecule).

5.4.7 Carbon Nanotubes

The discovery of carbon nanotubes (CNTs) is generally attributed to Sumio Iijima who in 1991 discovered tubular forms of rolled-up graphene sheets (sheets of hexagonal sp^2-bonded C atoms) with bucky ball-like end caps. Iijima's paper brought definitive news of such structures to the western world where it set off a flurry of research activity.

CNTs can now be produced as single wall nanotubes (SWNTs), double wall nanotubes (DWNTs), or multiwall nanotubes (MWNTs). Individual SWNTs are the strongest materials known with yield strengths some 50 times that of the strongest steels, elastic moduli approaching a TeraPascal, and specific strengths (strength/mass) several hundred times that of any other material. All of these properties are attributed to the incredible strength of the sp^2 bond. Even more remarkable in view of these mechanical properties is the flexibility of CNTs. They can be bent at 90°, formed into a torus, and even knotted.

SWNTs have very unusual properties that depend on their chirality (the sense and degree of twist in their structure). The chirality is measured by the chiral vector C_h defined

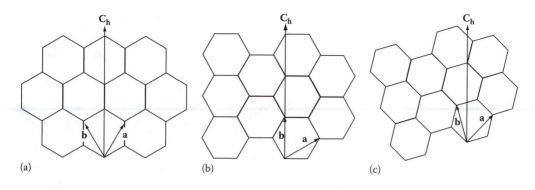

FIGURE 5.13

(a) $m = n$, armchair structure. (b) $m = 0$, $n = 1$, zigzag structure. (c) $m = 1$, $n = 2$, chiral structure.

as $C_h = am + bn$, where **a** and **b** are unit vectors defined as shown in Figure 5.14 and m and n are integers. The chiral vector is taken to be parallel to the circumference of the nanotube. If $m = n$, the "armchair" structure in Figure 5.13a results. If m or $n = 0$, the zigzag structure in Figure 5.13b results. Structures with other values of m or n as seen in Figure 5.13c are referred to as "chiral."

The electric and thermal conductivity of metallic SWNTs can be very high; electron trajectories are almost ballistic-like with little scattering from the lattice atoms. Current densities can be a thousand times greater than in metals. The electrical properties of SWNTs are determined by the quantum confinement of the electrons around the circumference of the SWNT and therefore are dependent on the diameter and chirality of the structure. The SWNTs are metallic or conductive if $(2m + n)/3 =$ an integer. Otherwise the SWNT is a semiconductor whose bandgap depends on the diameter as well as the m and n values. Inserting various endohedral metallofullerenes to form "pea-pod" structures can alter bandgaps and other electronic properties thus allowing SWNTs to function as diodes and transistors. Clearly SWNTs will play important roles in the future of nanoelectronics.

The development of ways to make SWNTs and control their structure as well as ways to manipulate them and incorporate them into devices is at the forefront of research in nanotechnology. Recent developments in CVD growth of nanotubes have resulted in the ability to grow bundles of CNTs that can be harvested and spun into fibers to make a super thread. The individual CNTs in a super thread are bonded by van der Waals forces, but it is possible to alter their mechanical and electrical properties by heat treating and irradiation. The sides of CNTs have been functionalized to bind with epoxies and polymers to form composites and their tips have been functionalized to serve as chemical sensors, atomic force microscope (AFM) tips, ion and electron emitters, and for other novel applications.

5.4.8 Other Hexagonal Ring Structures

α-Boron nitride (B_k) also forms a hexagonal ring structure similar to graphite except the rings have alternating B and N atoms as shown in Figure 5.14 (space group $P6_3/mmc$). The stacking is also ABAB so that each B atom has an N atom directly above and below it and vice versa giving each atom a coordination number of 5. The layers are bonded by a mixture of ionic and covalent bonds instead of van der Walls bonds, making α-BN very hard and inert.

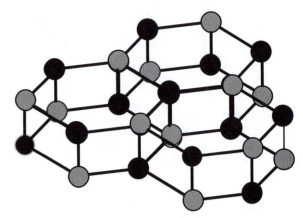

FIGURE 5.14
B_k ring structure of α-boron nitride. Unlike graphite, the layers sit on top of each other are bonded together by a combination of ionic and covalent bonds to make a very strong and inert structure.

Like graphite, MoS_2 (C7), or molydisulfide as it is commonly known, is an excellent dry lubricant that gets its slippery feel by inert planes sliding over one another, bonded only by van der Waals forces. However, in this case the inert planes are sandwiches of hexagonal ABA planes as shown in Figure 5.15. The A plane consists of S-atoms that sit on every other vertex of the hexagonal layer of Mo that forms the B plane with space group $P6_3/mmc$. Since all bonds are satisfied, the ABA planes are inert leaving only van der Waals forces between the planes.

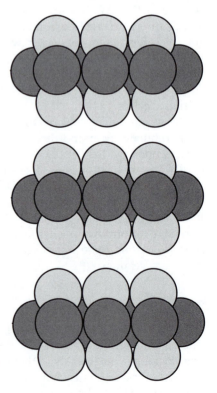

FIGURE 5.15
Layered C7 structure of Mo_2S. The layers consist of S atoms (light spheres) on either side of the Mo atoms (grey spheres) in an hcp arrangement. Since all bonds are satisfied, the layers slip over each other, attracted only by van der Walls forces.

5.4.9 Quartz

The silicates, which are thought to have a mixture of ionic and covalent bonds, can form a vast variety of structures. The basic building block is the orthosilicate ion, SiO_4^{-4}, formed by a Si atom in the center of a tetrahedron of four O atoms as seen in Figure 5.16. The tetrahedra can join with other tetraherda by sharing the bridging O atoms at the corners of the tetrahedra to form silica or SiO_2. Because there is little barrier to rotation, these bonds allow a random network structure (fused quartz), which is the basis for the silicate glasses, as well as ordered crystalline structures to be formed.

Crystalline quartz is polymorphic, meaning it can have more than one crystal structure. In fact nine different structures of quartz are classified in mineralogy tables, several of which are metastable forms, meaning they are only stable at high temperatures and pressures. The stable form under ambient conditions is α-quartz or low quartz, which is stable up to 573°C. It has a trigonal structure with space group $P3_221$. Above 573°C, β-quartz, or high quartz (C8), is stable under normal pressure. It is hexagonal with space group $P6_222$. Above 867°C, hexagonal β-tridymite (C10) forms with space group $P6_3/mmc$. Finally, β-cristobalite (C9) forms above 1470°C before melting at 1713°C. The Si atoms in β-cristobalite form an $Fd\bar{3}m$ diamond structure of Si atoms with the O atoms inserted between them. The structure of these different polymorphs gets progressively more open as the temperatures of their stable phase increases. At high pressures, a monoclinic form called coesite can exist and at even higher pressures, tetragonal stishovite ($P4/mmm$) forms. These latter two metastable forms are found in impact craters.

If crystalline quartz is compressed, say by forcing the top ion toward the bottom three ions, the Si^{4+} ion will be displaced from the center of the charge distribution resulting in a displacement current. Thus quartz is a piezoelectric material. Conversely, an applied electric field will cause the structure to distort. More details about this highly useful property will be discussed in Chapter 23.

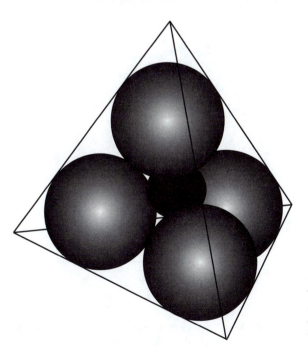

FIGURE 5.16
The SiO_4^{-4} tetrahedron, the basic building block of the silicates. The small Si^{4+} ion in the center can barely keep the large O^{2-} anions from touching one another.

5.4.10 Other Silicate Structures

The basic SiO_4^{4-} orthosilicate ion can combine with two Mg^{2+} ions to form Mg_2SiO_4 (fosterite) or with two Fe^{2+} to form Fe_2SiO_4 (Figure 5.17a). These two compounds form solid solutions known as otrthosilicates or olivines. The $Si_2O_7^{6-}$ pyrosilicates (Figure 5.17b) can combine with other metal ions to form pyrosilicates such as akermanite ($Ca_2MgSi_2O_7$). A wide variety of minerals are formed from the chain and ring structures (Figure 5.17c and d). Enstatite ($MgSiO_3$) has a chain structure. Wollastanite ($CaSiO_3$) is formed from SiO_3^9 rings and beryl ($Be_3AlSi_6O_{18}$) is formed from a $Si_6O_{18}^{12-}$ rings.

The SiO_4^{4-} units can also combine to form sheet structures as shown in Figure 5.18. These sheets form the substrate for various clays and layered minerals such as muscovite (mica). An $Al_2(OH)_4^{2+}$ layer can form ionic/covalent bonds with the SiO_4^{4-} layer to form a tightly bonded neutral sheet. Stacks of these sheets make up the clay kaolinite ($Al_2Si_2O_5(OH)_4$). Since these sheets are neutral and all primary bonds are saturated, only van der Waals forces are available to hold these sheets together, which accounts for the fact that such clays are easily deformable. The montmorillonite clay sandwiches two of these silicate sheets between an $Al_2(OH)_4^{2+}$ layer to form $Al_2(Si_2O_3)_2(OH)_2$. Other minerals based on this sheet structure include $Mg_2(Si_2O_3)_2(OH)_2$ (talc) and $KAl_3Si_3O_{10}(OH)_2$ (muscovite or mica).

5.4.11 Zeolites

Another important class of the aluminosilicate family is the zeolites. Zeolites have porous open structures that can accommodate various alkali metal ions that can easily be exchanged, making them useful as molecular sieves and as powerful catalysts.

There are 48 natural occurring zeolites and many more have been synthesized for specific purposes. They are widely used in gas and water purification and as sorption pumps in vacuum systems. They are used as desiccants because of their ability to adsorb

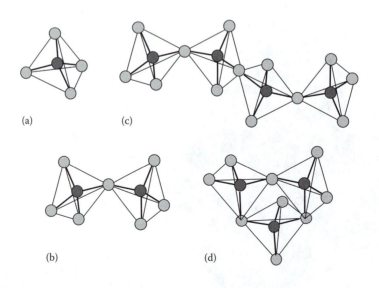

FIGURE 5.17
Structures formed by combining the basic SiO_4^{-4} orthosilicate unit (a). Combining two of the units forms the $Si_2O_7^{6-}$ backbone for the pyrosilicates (b). Longer chains (c) and rings (d) are the backbone for a large number of minerals.

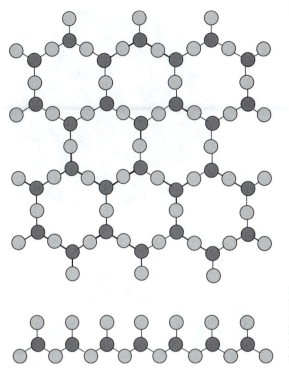

FIGURE 5.18
Bottom view (upper) and edge view (lower) of a $(Si_2O_5)^{2-}$ sheet structure. Hydroxides of Al, K, and Mg can bond with the O ions on the top of this layer to form electrically neutral sheets that make up layered structures such as clays, talc, and mica.

large amounts of water. Added to soils, they can store water, making plants more drought resistant. As solid-state acids, they are used in the petrochemical industries for cracking. The nuclear industry uses them for trapping and immobilizing radioactive waste.

5.4.12 Electron-Deficient Solids

Elements on the right-hand side of the periodic do not have electrons to give up to form a metallic bond, so they must share what they have to form covalent bonds. As a result, they are known as electron-deficient solids. Boron forms icosahedrons that are interconnected to form the rhombohedral or trigonal structures seen in Figure 5.19. As, Sb, and Bi form sheets of puckered hexagonal rings. The puckering results from π-bonds that form nearly 90° angles causing every other atom to be above or below the median plane. Like graphite, the planes are bonded by van der Waals forces. The unit cells for these structures are rhombohedral or trigonal. The group VI elements, S, Se, and Te, form π-bonded puckered rings of eight atoms shown in Figure 5.20. Their structures are complex with a number of polymorphs possible as these clusters of rings can condense to form a variety of structures.

5.5 Structure of Glass

Technically, a glass is a material that can be cooled from the melt without crystallizing. Since it does not crystallize, there is no distinct melting point. Instead, the viscosity

FIGURE 5.19
Icosahedral cluster of boron atoms (T-50). These clusters occupy five symmetry sites in the primitive cell of the $P4_2/nnm$ space group.

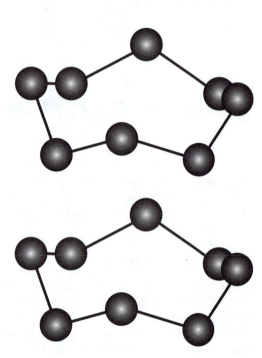

FIGURE 5.20
The 8-atom molecular rings of sulfur (A-16). The space group is Fddd.

increases as the temperature is lowered until the glass transition temperature is reached. Above the glass transition temperature, the material is in the state of an undercooled melt. Below the glass transition temperature, the molecules are locked into place just as a crystallized solid and the viscosity of the glass is virtually infinite. However, since the molecules in a glass form a random network instead of an ordered solid, the volume of a glass is slightly larger than its crystalline counterpart.

Since the glass consists of a random network of molecules, it lacks the long-range order of a crystalline solid, hence a glass does not show sharp x-ray diffraction peaks. Instead low, broad peaks in a continuum can be seen that reflect only the short range order of the molecule. Since glass is essentially transparent to x-rays, glass tubes are used to contain powder x-ray samples. We are accustomed to thinking of glass as being transparent to visible light because most glasses are formed form ceramics whose bandgap is higher than the top of the visible spectrum. Being transparent to visible light is by no means a property of a glass; metallic glasses are as reflective as crystalline metals.

In 1932, Zachariasen proposed a continuous random network model that predicted that oxide glasses of the form A_nO_m could exist with configurational energies that were only slightly higher than the crystalline state if the following conditions were met:

1. O atom is not bonded to more than two A-atoms.
2. Number of O atoms surrounding an A-atom must be no more than three in a triangle or four in a tetrahedron.
3. O polyhedra share corners with one another but not edges or faces.
4. At least three corners of each O polyhedra must be shared with the other O polyhedra to assure a 3D network.

Examples of these rules are illustrated in Figure 5.21 (see Chapter 14 for more details on glass formation).

5.6 Structure of Polymers

5.6.1 Polymerization Processes

Most polymers consist of linear covalently bonded chains of repeating mer units. The chains can self-assemble by the process of addition or chain polymerization. An initiator, usually some form of a free radical (a reactive compound containing an unpaired electron designated by R•), opens a double bond and attaches itself to a monomer molecule. The resulting open bond transfers the unpaired electron to the monomer and allows a second monomer to attach as shown in Figure 5.21. The process continues until the chain is terminated by another initiator molecule, by joining end to end with another growing chain, or until little monomer is left. Under carefully controlled conditions, ionic polymers can retain their charged chain ends until all the monomer has reacted. Such polymers are called living polymers. Polymerization can be continued by adding new monomer, which may be different from the original. This process is used to create block copolymers. A block copolymer simply means a chain of one type of repeating mer units interspersed with chains of a second type of repeating mer units.

Instead of the addition polymerization described above, some polymers are formed by a chemical reaction between two monomers that produces the repeating units together with a byproduct. This process is called condensation or step reaction polymerization. An example would be the reaction between dimethyl terephthalate and ethylene glycol to produce polyethylene terephthalate (PET) and methyl alcohol as a byproduct. The nylons and polycarbonates are linear chain polymers that also polymerize by the condensation process.

5.6.2 Linear Polymers

Although polymer chains are represented by a straight line of repeating single bonded units, remember that the sp^3 bonds are 109° apart, so it is more realistic to think of the backbone of the chain in terms of a zigzag structure as depicted in Figure 5.23. Also recall that there is no barrier to rotation about the σ-bonds between the C atoms in the chain so each mer unit can rotate anywhere about a 109° cone about the bond with the previous

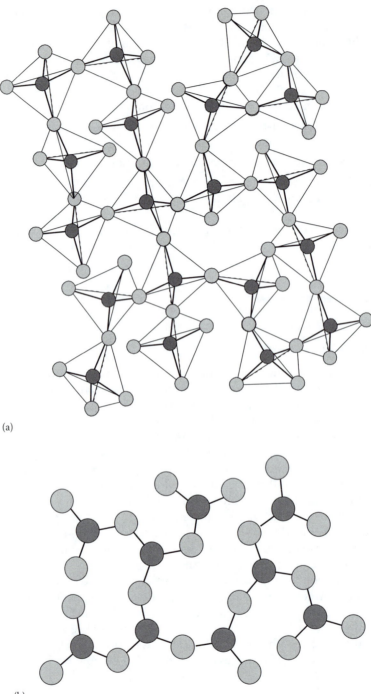

(a)

(b)

FIGURE 5.21
Two different ways to form a random net structure according to Zachariasen's rules. (a) represents SiO_2 in which the shared bridging O atoms sit at the vertices of the tretrahedra and (b) represents B_2O_3 in which the repeating units are triangles.

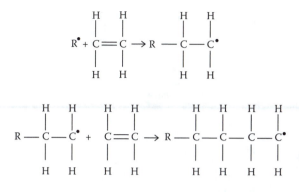

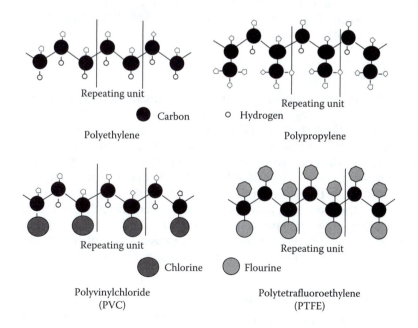

FIGURE 5.22
Illustration of addition or chain polymerization. A free radical $R^\bullet$ opens a double bond of an ethylene molecule, transferring the unpaired electron to its opposite end where it can open the double bond of another polyethylene molecule and so on.

atom. Thus the chains actually form random coils and their average length is $N^{1/2}$ times the length of the repeating unit, where N is the number of repeating units.

The chains are held together by secondary bonds and can slip by each other in response to stress, which accounts for the plastic response of polymers. This response can be greatly varied depending on the structure of the polymer chains. Increasing the molecular weight of the polymer (by adding more mer units) increases the stiffness and yield strength of the polymer and raises its glass transition temperature (the temperature at which chain motion is no longer possible and the material exhibits brittle rather than plastic behavior). Under favorable circumstances, the chains can fold back and forth on themselves to form a crystalline structure. Crystallinity combines strength and stiffness with ductility. A variety of different properties can be obtained by replacing a hydrogen in polyethylene with a large anion such as Cl^- to form polyvinyl chloride, or with blocky side groups such as CH_3^- to form polypropylene, or with an aromatic ring to form polystyrene. These structures are illustrated in Figure 5.23.

FIGURE 5.23
Examples of linear chain polymers shown as a zigzag chain to emphasize that the bond angles on the carbon atoms in the chain are 109°.

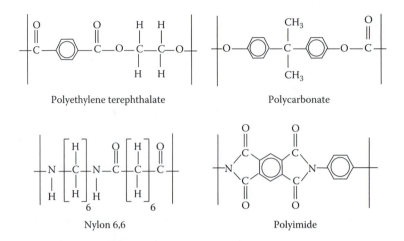

Polyethylene terephthalate   Polycarbonate

Nylon 6,6   Polyimide

FIGURE 5.24
Repeating units of selected polymers.

A variety of different repeating units may be used instead of ethylene, which gives the polymer scientist much freedom and versatility in the development of new materials. A few examples are shown in Figure 5.24.

5.6.3 Branched Polymers

Another chain may attach itself to one of the sites on the primary chain, which is normally occupied by an H atom or a side group, to form a branched polymer. Adding side chains keeps the primary chains from getting close together, which lowers the density and prevents crystallization. This results in a lower glass transition temperature and a softer plastic. Low-density polyethylene (LDPE) is an example.

5.6.4 Net Polymers

Cross-linking the primary chains by covalent bonds greatly strengthens the material but limits the ductility. Noncross-linked chain polymers will generally soften, melt, and resolidify reversibly. The polymers in this class are referred to as thermoplastics. Another class of polymers have their primary structures covalently bonded or heavily cross-linked. In this case it is necessary to break the covalent bond in order to melt the material and generally the materials will decompose before this happens. These polymers are called thermosetting plastics or thermosets. Bakelite (phenol formaldehyde), shown in Figure 5.25, is an example of a thermosetting polymer. As may be seen from its structure, it is trifunctional, meaning that it has three points where repeating units may be attached allowing it to form a 3D network like the B_2O_3 glass network discussed previously. Other phenolics, as well as epoxies, urethanes, and cross-linked linear polymers are thermosets.

5.6.5 Ziegler–Natta Catalysts and Stereoisomerism

In the 1950s Karl Ziegler developed a catalyst for making linear or high-density polyethylene (HDPE). Shortly thereafter, Giulio Natta discovered that Ziegler-type catalysts could polymerize propylene with all of the methyl groups on the same side of the chain. This configuration is called isotactic. Natta was also able to find a catalyst that would cause

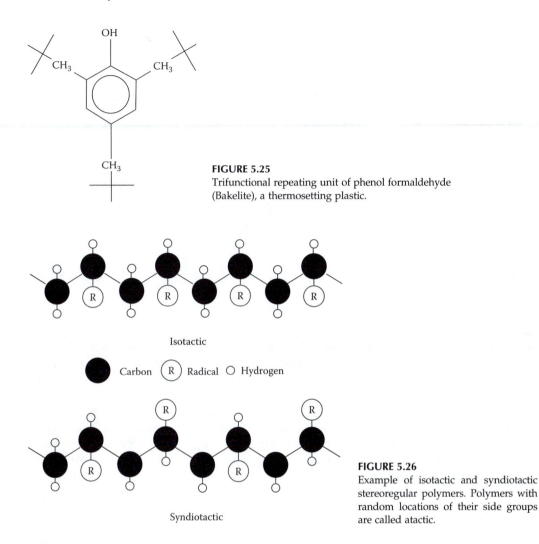

FIGURE 5.25
Trifunctional repeating unit of phenol formaldehyde (Bakelite), a thermosetting plastic.

Isotactic

● Carbon ⓡ Radical ○ Hydrogen

Syndiotactic

FIGURE 5.26
Example of isotactic and syndiotactic stereoregular polymers. Polymers with random locations of their side groups are called atactic.

the side groups to alternate from one side to the other in regular fashion. This configuration is call syndiotactic. If the distribution of the side groups is random, the configuration is atactic. The isotatic and syndiotactic configurations are shown in Figure 5.26. On a one-dimensional (1D) diagram such as Figure 5.24, it would appear that one could change the position of a side group by a simple bond rotation. However, as can be seen in Figure 5.26, such a rotation is not possible without breaking bonds.

Isotactic and syndiotactic polymers are referred to as stereoregular. Being able to control the positions of the side groups so that they repeat in a regular fashion is important to make high-strength materials because the uniform structure leads to close packing of the polymer chains and a high degree of crystallinity. The catalyst systems employed to make stereoregular polymers are now referred to as Ziegler–Natta catalysts.

5.6.6 Geometrical Isomerism

When a repeating unit contains a double bond, the position of the side group relative to the carbon atoms forming the double bond has a great influence on the properties of the polymer. Consider the isoprene polymers shown in Figure 5.27. The *cis* and *trans* refer

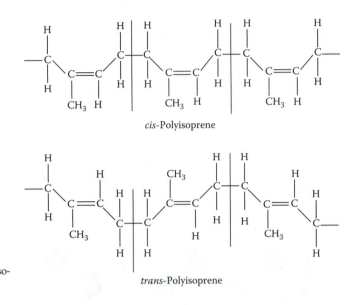

cis-Polyisoprene

trans-Polyisoprene

FIGURE 5.27
Comparison of the *cis*- and *trans*-iso-
mers of the isoprene repeating unit.

to which side of the double bond the side groups are located on. The *cis*-configuration places both side groups on the same side while the *trans*-configuration places them on opposite sides. Since atoms forming a double bond cannot be rotated, the effect is to place the CH_3 groups on the same side in *cis*-polyisoprene and on alternating sides in *trans*-polyisoprene.

Both *cis*- and *trans*-polyisoprene are natural products that come from different trees. The *cis*-polyisoprene is natural rubber while *trans*-polyisoprene is something called gutta percha, a tough leathery substance that was originally used for golf balls and for underwater insulation. It has now been largely replaced by synthetics although it is still used for packing teeth after a root canal.

5.6.7 Vulcanization

Natural rubber was too soft to be of much use. It was called rubber by Joseph Priestly, who found it useful for rubbing out pencil marks. The vulcanization process, developed in 1839 by Charles Goodyear, added sulfur to open the double bonds in the isoprene molecule and form covalent sulfur bonds that cross-linked the chains together as illustrated in Figure 5.28. The tougher product immediately found use in bicycle tires and later in automobile tires. The shortage of natural rubber during World War II prompted the development of synthetic rubbers with properties far superior to the best natural rubbers.

5.7 Summary

Because there is no directionality in the metallic bond and because the charge on the ion cores is screened by the intervening sea of electrons, metals tend to form close-packed structures. The two close-packed structures that can fill all space are the fcc and the hcp. The two structures differ in the packing sequence of hexagonal layers, hcp packs as ABAB,

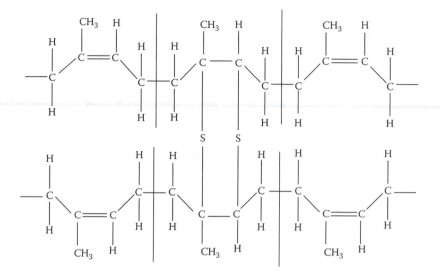

FIGURE 5.28

Vulcanization of *cis*-polyisoprene. The addition of sulfur opens the carbon double bond and forms a covalent sulfur bridge between two adjacent chains which greatly strengthens the final product.

whereas fcc packs as ABCABC. The coordination number (number of nearest neighbors) for both systems is 12 and they both have the same atomic packing factor, 0.74. The fcc has a higher symmetry and more high atomic fraction planes which can act as slip planes, making fcc metals more ductile (less brittle) than hcp metals.

Other metals form bcc structures with eight nearest neighbors and six second nearest neighbors, which are closer than the second nearest neighbors in fcc or hcp. These 14 close neighbors apparently produce nearly the same bond energy as the two close-packed systems because nearly as many metals form this structure as those forming either the fcc or hcp structures. All the monovalent alkali metals form fcc structures and the nonmagnetic transition metals tend to follow a pattern pairwise that goes hcp → bcc → hcp → fcc. Calculation of the free energies of these metallic systems requires solving the Schrödinger equation for a many electron system. Recent advances in this field have been facilitated by the development of the density functional theory (DFT) by Walter Kohn (Nobel Prize, 1998).

Ionically bonded systems must obey Pauling's rules for ionic crystal formation which, among other things, requires each ion to be surrounded by a polyhedron of counterions and also that the cation must be large enough to at least touch each of the counterions. These requirements prevent ionically bonded systems from forming close-packed structures and the ratio of the cation to anion size will play a role in determining the coordination number. In addition, the unit cell containing the anions and cations must be electrically neutral and must be able to fill all space by translations along its lattice vectors.

Covalently bonded systems, because of the directionality of the covalent bond, tend to form open systems with low coordination numbers. Like the metals, they can form cubic and hexagonal systems by altering their stacking sequence.

The various covalent and ionic structures are summarized in Table 5.2.

TABLE 5.2

Summary of Common Ionic and Covalent Structures

Structure	Type	Lattice	Unit Cell Description	Examples
Diamond	A	fcc $Fd\bar{3}m$	A-atoms on fcc sites and on every other tetrahedral site	C, Si, Ge, and α-Sn
Sphalerite (zinc blende)	AB	fcc $F\bar{4}3m$	A-atoms on fcc sites and B atoms on every other tetrahedral site	ZnS, SiC, and most III–V
Wurtzite	A(Zn) B(s)	hex $P6_3mc$	Zn atoms on hexagonal sites to form A layer. S atoms over Zn atoms to form A′ layer. Zn atoms on every other tetrahedral site to form B layer. S atoms over Zn atoms to form B′ layer resulting in alternating stacks of AA′BB′	ZnS, some III–V and II–VI
α-BN	AB	hex $P6_3mc$	Planar hexagonal rings alternating A and B covalently bonded together	α-BN
Cesium chloride	AX	sc $Pm\bar{3}m$	A-atoms on corners of cube and B atom at center of cube	Cs–Cl, AgZn CsBr, and BeCu
Rock salt (NaCl)	AX	fcc $Fm\bar{3}m$	A-atoms on fcc sites and B atoms on every octahedral site	NaCl, MgO, and FeO
Fluorite	AX_2	fcc $Fm\bar{3}m$	A-atoms on fcc sites and B atoms on every tetrahedral site	CaF_2, UO_2, and ThO_2
Perovskite	ABX_3	sc $Pm\bar{3}m$	A in the center of the cube, B on the corners, and C on the faces	$BaTiO_3$, $SrZrO_3$, $SrSnO_3$,
Spinel	AB_2X_4	fcc $Fd\bar{3}m$	X on fcc sites, A on a tetrahedral site and B on two adjacent octahedral sites	Al_2O_3 MgO Al_2O_3 FeO
Inverse spinel	AB_2X_4	fcc $Fd\bar{3}m$	X on fcc sites, one B on a tetrahedral site. Other B and A on octahedral sites	Fe_2O_3 FeO

The stable form of carbon is graphite in which the sp^2 bonds form hexagonal sheets bonded together by weak van der Waals forces. However, carbon atoms can also form a new class of materials called fullerenes which consists of molecules in the form C_n ($n = 60$–140) as well as cylindrical structures called CNTs. The C_{60} molecules, known as bucky balls, can assemble in an fcc close-packed array to form fullerites, a solid form of carbon that is harder than diamond and has many other remarkable properties, especially when metal atoms are intercalated into the structure to form fullerides. The CNTs also have remarkable mechanical and electronic properties ranging from super-strong composites to nanoelectronics.

Most polymers consist of covalently bonded long chain molecules that make secondary bonds with each other. The physical properties are dependent on the length of these chains and on the type and arrangement of the side groups that are placed on the chains. Linear chains with regularly oriented side groups promote crystallinity and such polymers are stronger, more dense, and have higher glass transition temperatures than polymers with randomly placed side groups or side branches.

Most chain polymers will soften and become more plastic when heated above their glass transition temperature and their chain molecules are able to move past one another. Such polymers are called thermoplastics. Chain polymers can be hardened by cross-linking while other polymer systems form covalently bonded networks. Such systems do not soften and primary bonds must be broken in order to melt. Such systems are called thermosetting plastics or thermosets.

Bibliography

Papers Dealing with Methods for Computing Lattice Energies

Hohenberg, P. and Kohn, W., *Inhomogeneous Electron Gas, Phys. Rev.* B864, 1964, 136.

Koch, W. and Holthausen, M.C., *A Chemist's Guide to Density Functional Theory*, 2nd edn., Wiley-VCH, Weinheim, 2002.

Kohn, W. and Sham, L.J., *Phys. Rev.* A1133, 1965, 140.

Kohn, W., *Electronic Structure of Matter—Wave Functions and Density Functionals*, Nobel Lecture, Jan. 28, 1999.

Other References Involving Structures

Barsoum, M.W., The MAX phases: Unique new carbide and nitride materials, *American Scientist*, 89/4, July–Aug. 2001, 334.

Barsoum, M.W., *Fundamentals of Ceramics*, Institute of Physic Publishing, Bristol and Philadelphia, 2003.

Billmeyer, F.W. Jr., *Textbook of Polymer Science*, 3rd edn., John Wiley & Sons, New York, 1984.

Kingery, W.D., Bowen, H.K., and Uhlmann, D.R., *Introduction to Ceramics*, 2nd edn., John Wiley & Sons, 1976.

Villars, P., Factors Governing Crystal Structures, in *Intermetallic Compounds, Principles and Practice*, ed. J.H. Westbrook and R.L. Fleischer, John Wiley & Sons, UK, 1994, p. 227.

Crystal structures may be viewed from various angles on the website provided by the Center for Computational Materials Science of the United States Naval Research Laboratory, Crystal Lattice Structures, see http://cst-www.nrl.navy.mil/lattice/.

For More Information on Intermetallic Compounds

Westbrook, J.H. and Fleisher, R.L., Eds., *Intermetallic Compounds, Volume 1, Crystal Structures*, 1st ed., John Wiley & Sons, U.K. 2000.

Books and Papers on Fullerenes

Dresselhaus, M.S., Dresselhaus, G., and Eklund, P.C., *Science of Fullerenes and Carbon Nanotubes: Their Properties and Applications*, Academic Press, London, 1966.

Dresselhaus, M.S., Dresselhaus, G., and Avouris, P., Eds., *Nanotubes: Synthesis, Structure, Properties and Applications* (Foreword by R.E. Smalley), Springer, Berlin, 2001.

Harris, P.J.F., *Carbon Nanotubes and Related Structures*, Cambridge University Press, Cambridge, 2002.

Iijima, S., Helical microtubules of graphitic carbon, *Nature*, 354, 56, 1991.

Kadish, K.M. and Ruoff, R.S., Eds., *Fullerenes: Chemistry, Physics, and Technology*, Wiley-Interscience, Pennington, NJ, 2000.

O'Connell, M.J., Ed., *Carbon Nanotubes: Properties and Applications*, CRC Press, Boca Raton, FL, 2006.

Saito, R., *Physical Properties of Carbon Nanotubes*, World Scientific Publishing, Singapore, 1998.

Stephens, P.W., Ed., *Physics and Chemistry of Fullerenes*, World Scientific Publishing, Singapore, 1994.

Problems

1. Show that the ideal c/a ratio for an hcp structure is 1.633.

2. Verify the cation/anion radius for the various coordination numbers shown in Table 5.1.

3. Write the basis for the following structures: (a) hcp metal, (b) sphalerite, and (c) wurtzite.

4. Pd can absorb 900 times its volume of H under standard conditions. Write a formula for this compound.

6

Reciprocal Lattice and X-Ray Diffraction

Most readers are probably familiar with the Bragg formula for x-ray diffraction (XRD), $2d \sin \theta = n\lambda$, and the selection rules that tell us that the sum of the Miller indices for body-centered cubic (bcc) crystals must be even and the indices for face-centered cubic (fcc) crystals must be all even or all odd in order to produce a diffraction peak. The intent of this chapter is to give a more general derivation of the Laue conditions leading to Bragg reflections of electrons and phonons that take place in three-dimensional (3D) crystals. These reflections from lattice planes within the crystal are fundamental to the understanding of the thermal and electronic properties of materials. In the process, the reciprocal lattice will be introduced which will be widely used in subsequent chapters to describe how materials behave thermally, electronically, magnetically, and photonically.

6.1 Reciprocal Lattice

Much of solid-state physics is carried out in reciprocal space, sometimes called Fourier space. The reasons for this will become more apparent as we progress. One of the more obvious reasons is that it is easier to represent the momentum of photons and phonons as well as particles such as electrons and neutrons and thus their interactions in reciprocal space. Recall that the momentum of either a massless particle such as a photon or a particle with mass can be written as

$$\mathbf{p} = \frac{h}{\lambda} = \frac{2\pi\hbar}{\lambda} = \hbar \mathbf{k}, \tag{6.1}$$

where $\mathbf{k} = 2\pi/\lambda$ is sometimes called the propagation vector. (For mass particles, λ is the deBroglie wavelength.)

6.1.1 Fourier Expansion of the Electron Density

Since the lattice has a periodicity with a repeat distance $\mathbf{a}$ in one dimension, we would expect the electron density surrounding the atoms in any unit cell to have the same periodicity. Thus we should be able to represent the electron density of the solid by a Fourier series of the form

$$n(x) = \sum_{h} \widetilde{A}_h e^{2\pi i h x / |\mathbf{a}|}. \tag{6.2}$$

The tilde on Fourier coefficients is to remind us that they are complex.

The electron density should be invariant under any translation operation $\mathbf{T}$ such that $n(x + \mathbf{T}) = n(x)$, where $\mathbf{T} = m\mathbf{a}$. In other words, the electron density in one cell should look the same as in any other cell. We can see that this is the case from

$$n(x + \mathbf{T}) = \sum_h \tilde{A}_h e^{2\pi i h x / |\mathbf{a}|} e^{2\pi i h m} = \sum_h \tilde{A}_h e^{2\pi i h x / |\mathbf{a}|} = n(x), \tag{6.3}$$

since $\exp(2\pi i h m) = 1$ and h and m are integers.

6.1.2 Reciprocal Lattice Vector

Now let us generalize this to three dimensions. We must expand $n(\mathbf{r})$ in Fourier series.

$$n(\mathbf{r}) = \sum_G \tilde{A}_G e^{i(\mathbf{G} \cdot \mathbf{r})} \tag{6.4}$$

such that $n(\mathbf{r} + \mathbf{T}) = n(\mathbf{r})$. To accomplish this, we must find a vector $\mathbf{G}$ that expresses the periodicity of the lattice in reciprocal space since $\mathbf{G}$ must have the dimension of $1/r$. It may be seen that in one dimension, $\mathbf{G} = 2\pi h / \mathbf{a}$, where h is an integer and $r = a$.

In a 3D space, $\mathbf{T} = m\mathbf{a} + n\mathbf{b} + p\mathbf{c}$, where $\mathbf{a}$, $\mathbf{b}$, $\mathbf{c}$ are the lattice vectors and m, n, and p are integers. To enforce $n(\mathbf{r} + \mathbf{T}) = n(\mathbf{r})$, Equation 6.3 must be written as

$$n(\mathbf{r} + \mathbf{T}) = \sum_G \tilde{A}_G e^{i(\mathbf{G} \cdot \mathbf{r})} e^{i(\mathbf{G} \cdot \mathbf{T})} = n(\mathbf{r}). \tag{6.5}$$

This requires $\exp(i\mathbf{G} \cdot \mathbf{T}) = 1$, which means that $\mathbf{G} \cdot \mathbf{T} = 2\pi \times$ integer.

To facilitate the construction of the vector $\mathbf{G}$, we define $\mathbf{G}$ as a translation vector in reciprocal space:

$$\mathbf{G}_{hk\ell} = h\mathbf{A} + k\mathbf{B} + \ell\mathbf{C}, \tag{6.6}$$

where
  $\mathbf{A}$, $\mathbf{B}$, and $\mathbf{C}$ are unit cell vectors in reciprocal space
  h, k, and ℓ are integers.

(You may note we use the same notation for these integers as we did for the Miller indices. This is no accident—you will find later that they are the Miller indices.)

Using the definition of $\mathbf{G}$ as the translation vector in reciprocal space,

$$\mathbf{G} \cdot \mathbf{T} = (h\mathbf{A} + k\mathbf{B} + \ell\mathbf{C}) \cdot (m\mathbf{a} + n\mathbf{b} + p\mathbf{c}) = 2\pi \times \text{integer}.$$

To meet this condition, we must find $\mathbf{A}$, $\mathbf{B}$, $\mathbf{C}$ such that

$$\begin{matrix} \mathbf{a} \cdot \mathbf{A} = 2\pi & \mathbf{a} \cdot \mathbf{B} = 0 & \mathbf{a} \cdot \mathbf{C} = 0 \\ \mathbf{b} \cdot \mathbf{A} = 0 & \mathbf{b} \cdot \mathbf{B} = 2\pi & \mathbf{b} \cdot \mathbf{C} = 0 \\ \mathbf{c} \cdot \mathbf{A} = 0 & \mathbf{c} \cdot \mathbf{B} = 0 & \mathbf{c} \cdot \mathbf{C} = 2\pi \end{matrix}.$$

Therefore, we must construct $\mathbf{A}$ so that it is perpendicular to both $\mathbf{b}$ and $\mathbf{c}$, $\mathbf{B}$ perpendicular to $\mathbf{c}$ and $\mathbf{a}$, and $\mathbf{C}$ perpendicular to $\mathbf{a}$ and $\mathbf{b}$. We do this by taking vector cross products,

$$\mathbf{A} = \text{const } \mathbf{b} \times \mathbf{c}, \quad \mathbf{B} = \text{const } \mathbf{c} \times \mathbf{a}, \quad \mathbf{C} = \text{const } \mathbf{a} \times \mathbf{b}. \tag{6.7}$$

To evaluate the const, we dot **a** with the first of the above equations and set it equal to 2π.

$$\mathbf{a} \cdot \mathbf{A} = \text{const } \mathbf{a} \cdot \mathbf{b} \times \mathbf{c} = 2\pi, \tag{6.8}$$

from which const $= 2\pi / \mathbf{a} \cdot \mathbf{b} \times \mathbf{c}$. Therefore, the inverse lattice vectors are given by

$$\mathbf{A} = 2\pi \frac{\mathbf{b} \times \mathbf{c}}{\mathbf{a} \cdot \mathbf{b} \times \mathbf{c}}, \quad \mathbf{B} = 2\pi \frac{\mathbf{c} \times \mathbf{a}}{\mathbf{a} \cdot \mathbf{b} \times \mathbf{c}}, \quad \mathbf{C} = 2\pi \frac{\mathbf{a} \times \mathbf{b}}{\mathbf{a} \cdot \mathbf{b} \times \mathbf{c}}, \tag{6.9}$$

and the reciprocal lattice translation vector $\mathbf{G}_{hk\ell} = h\mathbf{A} + k\mathbf{B} + \ell\mathbf{C}$ plays the same role in reciprocal space as the translation vector **T** in direct space and the lattice points in reciprocal space represent planes in direct space.

Thus we have constructed $\mathbf{G}_{hk\ell}$ so that it is perpendicular to the plane with Miller indices h, k, ℓ and it can be shown (proof left to the student) that the distance from the origin to lattice plane $(hk\ell)$ is given by

$$d_{hk\ell} = \frac{2\pi}{|\mathbf{G}_{hk\ell}|}. \tag{6.10}$$

6.1.3 Lattice Types in Reciprocal Space

6.1.3.1 Simple Cubic Direct Lattice

For a simple cubic lattice, $\mathbf{a} = a\widehat{x}$, $\mathbf{b} = a\widehat{y}$, $\mathbf{c} = a\widehat{z}$ and

$$\mathbf{A} = \frac{2\pi(\mathbf{b} \times \mathbf{c})}{\mathbf{a} \cdot \mathbf{b} \times \mathbf{c}} = \frac{2\pi}{a}\widehat{x}, \quad \mathbf{B} = \frac{2\pi}{a}\widehat{y}, \quad \text{and} \quad \mathbf{C} = \frac{2\pi}{a}\widehat{z}. \tag{6.11}$$

The reciprocal lattice vectors form a simple cubic lattice in reciprocal space with a lattice constant given by $2\pi/a$.

6.1.3.2 Body-Centered Cubic Direct Lattice

Next consider a bcc lattice. A set of primitive bcc lattice vectors is

$$\mathbf{a}' = \frac{a}{2}\begin{pmatrix} 1 \\ 1 \\ -1 \end{pmatrix}, \quad \mathbf{b}' = \frac{a}{2}\begin{pmatrix} -1 \\ 1 \\ 1 \end{pmatrix}, \quad \mathbf{c}' = \frac{a}{2}\begin{pmatrix} 1 \\ -1 \\ 1 \end{pmatrix}. \tag{6.12}$$

We use the primitive vectors for the bcc lattice to distinguish its special symmetry from the nonprimitive bcc lattice and use the prime symbol to distinguish these primitive lattice vectors from the lattice vectors **a**, **b**, and **c**. The choice of the primitive vectors is arbitrary. Any set will produce similar results. Using the set of relations in Equation 6.9 to transform these vectors to reciprocal space,

$$\mathbf{A}' = \frac{2\pi}{a}\begin{pmatrix} 1 \\ 1 \\ 0 \end{pmatrix}, \quad \mathbf{B}' = \frac{2\pi}{a}\begin{pmatrix} 0 \\ 1 \\ 1 \end{pmatrix}, \quad \mathbf{C}' = \frac{2\pi}{a}\begin{pmatrix} 1 \\ 0 \\ 1 \end{pmatrix}. \tag{6.13}$$

These are the Cartesian coordinates of a set of primitive translation vectors for an fcc lattice.

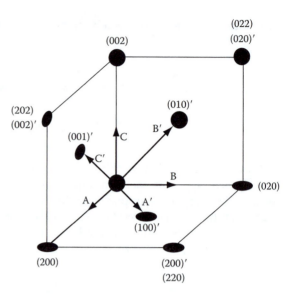

FIGURE 6.1
Primitive and nonprimitive vectors in an fcc reciprocal lattice in which the primitive vectors and their translation sites are denoted by prime symbols. Translations along these primitive vectors produce a nonprimitive fcc reciprocal lattice with a basis of (000), (110), (101), (011). Notice that primitive translations (100)' are equivalent to (110), (200)' to (220), etc.

Thus the primitive translation vector in reciprocal space, $G' = h'\mathbf{A}' + k'\mathbf{B}' + \ell'\mathbf{C}'$ for a bcc direct lattice (fcc reciprocal lattice) is

$$\mathbf{G}'_{h'k'\ell'} = \frac{2\pi}{a} \begin{pmatrix} h' + \ell' \\ h' + k' \\ k' + \ell' \end{pmatrix}, \tag{6.14}$$

where h', k', ℓ' specify the translations along the primitive vectors. These translations along the primitive vectors will map all of the lattice points of the fcc lattice, but the coordinates of the lattice points will be different from the nonprimitive coordinates as may be seen in Figure 6.1.

The nonprimitive reciprocal lattice vectors $\mathbf{A}$, $\mathbf{B}$, and $\mathbf{C}$ are given by Equation 6.9 using the nonprimitive lattice translation vectors $\mathbf{a}$, $\mathbf{b}$, $\mathbf{c}$. These nonprimitive reciprocal lattice vectors are along the x, y, z axes and form a cubic unit cell with length $2\pi/a$ and a basis of (000), (110), (101), (011) as may be seen in Figure 6.1. Note that the primitive reciprocal lattice vector $|\mathbf{A}'| = \sqrt{2}|\mathbf{A}|$.

6.1.3.3 Face-Centered Cubic Direct Lattice

Starting with a set of primitive vectors of an fcc direct lattice and using Equation 6.12 to transform them into a reciprocal lattice, we obtain primitive vectors for a bcc reciprocal lattice given by

$$\mathbf{G}'_{h'k'\ell'} = \frac{2\pi}{a} \begin{pmatrix} h' - k' + \ell' \\ h' + k' - \ell' \\ -h' + k' + \ell' \end{pmatrix}. \tag{6.15}$$

Similarly to the previous case, $h'k'\ell'$ translations of these primitive reciprocal lattice vectors will map out a bcc nonprimitive reciprocal lattice with length $2\pi/a$ and a basis of (000), (111).

6.2 Diffraction Conditions

6.2.1 Laue Conditions

Let a wave with propagation vector **k** be incident on a crystal as shown in Figure 6.2. Some of the wave is scattered in the direction **k'** by the electron density surrounding the atom at position **r**. Let **R** be the vector from the source to a reference point on the lattice and **P** the vector from the reference point to the observation point P. The phase of the wave when it reaches the atom at **r** is given by $\exp[i\mathbf{k}\cdot(\mathbf{R}+\mathbf{r})]$ and, when it arrives at the observation point P, the phase is $\exp[i\mathbf{k}\cdot(\mathbf{R}+\mathbf{r})+i\mathbf{k'}\cdot(\mathbf{P}-\mathbf{r})]$. The phase of the wave that is scattered from the reference atom when it arrives at the observation point P is given by $\exp[i\mathbf{k}\cdot\mathbf{R}+i\mathbf{k'}\cdot\mathbf{P}]$. Here it is assumed that the $|\mathbf{P}|\gg|\mathbf{r}|$ so that **k'** is parallel to both $\mathbf{P}-\mathbf{r}$ and **P**. Therefore, the phase shift between the incident and scattered wave is $\exp[-i(\Delta\mathbf{k}\cdot\mathbf{r})]$, where $\Delta\mathbf{k}=\mathbf{k'}-\mathbf{k}$.

The amplitude of the scattered wave is proportional to the electron density $n(\mathbf{r})$ times the phase shift integrated over the crystal, or

$$\mathbf{E}(\mathbf{k'}) \sim \int_{crystal} dV n(\mathbf{r})\exp[-i(\Delta\mathbf{k}\cdot\mathbf{r})]. \tag{6.16}$$

Putting the Fourier expansion for $n(\mathbf{r})$ from Equation 6.4 into Equation 6.16,

$$\mathbf{E}(\mathbf{k'}) \sim \int_{crystal} dV \sum_{hk\ell} \tilde{A}_{\mathbf{G}_{hk\ell}}\exp[i(\mathbf{G}_{hk\ell}-\Delta\mathbf{k})\cdot\mathbf{r}]. \tag{6.17}$$

The wave scattered in direction **k'** can only have significant amplitude when the waves scattered from the same point in each cell arrive in phase, which requires

$$\mathbf{G}_{hk\ell} = \Delta\mathbf{k}. \tag{6.18}$$

This is the Laue condition for Bragg reflection. The Laue diffraction condition stated in the Equation 6.18 is a necessary, but not sufficient condition, for a diffraction peak. The intensity of the peak, or whether the peak may be cancelled out, depends on the contents of the unit cell as discussed later.

The Laue condition can also be written as

$$\mathbf{k} + \mathbf{G}_{hk\ell} = \mathbf{k'}. \tag{6.19}$$

Squaring both sides and making use of the fact that the amplitude of the scattered wave is unchanged for the incident wave (elastic scattering), we can write

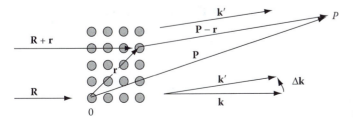

FIGURE 6.2
Schematic of x-rays being diffracted.

$$2\mathbf{k} \cdot \mathbf{G} + \mathbf{G}^2 = 0. \tag{6.20}$$

If we dot the Laue condition $\mathbf{G}_{hk\ell} = \Delta\mathbf{k}$ with the lattice vectors $\mathbf{a}$, $\mathbf{b}$, and $\mathbf{c}$, we obtain the Laue equations:

$$\mathbf{a} \cdot \Delta\mathbf{k} = 2\pi h, \quad \mathbf{b} \cdot \Delta\mathbf{k} = 2\pi k \quad \mathbf{c} \cdot \Delta\mathbf{k} = 2\pi\ell. \tag{6.21}$$

Each equation describes a cone of possible $\Delta\mathbf{k}$s about that particular lattice vector that could satisfy the Laue condition. All three equations must be satisfied simultaneously in order to meet the Laue condition. This requires that each cone must have a common line of intersection with the other cones. Note that if $\mathbf{G}$ satisfies the Laue equations (Equation 6.21), $-\mathbf{G}$ will also $(-h, -k, -\ell$ denote the same plane as $h, k, \ell)$. Therefore, the Laue condition Equation 6.20 can also be written as $2\mathbf{k} \cdot \mathbf{G} = \pm\mathbf{G}^2$.

6.2.2 Ewald Construction

Ewald devised a construction in reciprocal space that provides a good way to visualize how the Laue condition may be met. An Ewald sphere with radius $\mathbf{k} = 2\pi/\lambda$ is drawn around the crystal. A vector $\mathbf{k}$ is drawn from the crystal in the direction of the incoming x-ray beam to a reciprocal lattice point on the Ewald sphere which is taken as the origin of the reciprocal lattice (000) as shown in Figure 6.3. If another reciprocal lattice point falls on the Ewald sphere, as is the case for the (100) point shown in Figure 6.3 (remember, reciprocal lattice points denote planes in the direct lattice), the Laue condition $\mathbf{G} = \mathbf{k}' - \mathbf{k}$ is met by a vector $\mathbf{k}'$ drawn from the crystal to that point. The diffracted beam $\mathbf{k}'$ makes angle 2ϕ with the incident beam and ϕ with the $\mathbf{G}_{100}$ vector which is normal to the (100) plane. Recall from Equation 6.10 that $|\mathbf{G}_{100}|$ equals $2\pi/d_{100}$ where d_{100} is the spacing between the (100) planes in the direct lattice. From Figure 6.4, it may be seen that $|\mathbf{G}_{100}| = 2|\mathbf{k}|\sin\theta$ and since $|\mathbf{k}| = 2\pi/\lambda$,

$$|\mathbf{G}_{100}| = 2\left(\frac{2\pi}{\lambda}\right)\sin\theta = \frac{2\pi}{d_{100}} \tag{6.22}$$

from which the more familiar form of Bragg's law, $2d_{100}\sin\theta = \lambda$ is obtained. Figure 6.4 illustrates this relationship in the direct lattice.

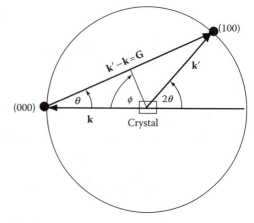

FIGURE 6.3
Ewald construction for an x-ray with wave vector $\mathbf{k}$ incident at angle ϕ relative to the (100) plane. The reflection spot will be in the direction of $\mathbf{k}'$. A simple geometrical argument shows this construction satisfies Equation 6.20.

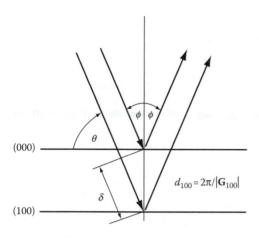

FIGURE 6.4
More familiar illustration of Bragg reflection. If the path difference of the two rays $2\delta = 2d_{100} \sin\theta = \lambda$, constructive interference results.

What if the reciprocal lattice point corresponding to the (200) plane was sitting on the Ewald sphere? The reciprocal lattice vector would now be $|G_{200}| = 2\pi/d_{200}$. Bragg's law would be written as $2d_{200} \sin\theta = \lambda$. The d_{200} is the spacing between the (200) planes, which falls halfway between the (100) planes. But what if there were no atoms on the (200) plane to diffract the x-rays? Reflections from the (200) plane can be thought of as second-order reflections from the (100) planes in which the path difference between the two rays, $2\delta = 2d_{100} \sin\theta = 2\lambda$ or $2d_{100} \sin\theta = n\lambda$, where $n = 2$.

A useful feature of the Ewald construction is the ability to see which planes will be able to meet the Laue condition. Consider the reciprocal lattice superimposed on the Ewald sphere shown in Figure 6.5a. In this case, the **k** is 2.4 times larger than the reciprocal lattice vector, which means the direct lattice spacing a is 2.4 times larger than the wavelength λ. None of the reciprocal lattice points lie on the Ewald sphere (at least in the

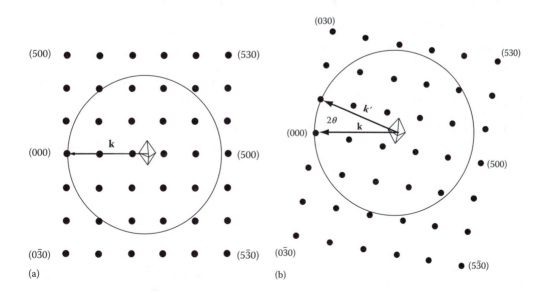

FIGURE 6.5
(a) Ewald sphere lattice projected on a simple cubic reciprocal lattice. The **k** is 2.4 times larger than the reciprocal lattice vector. None of the lattice points satisfy the Laue condition. (b) The crystal is rotated by 13° to satisfy the Laue condition for the (010) reflection. The angle 2θ between **k** and **k'** is 26°

two-dimensional [2D] projection) when the radiation is normal to the (100) face of the crystal. However rotating the crystal also rotates the reciprocal lattice about the (000) origin, and all of the points for which $|G_{hk\ell}| < 2|k|$ will eventually intersect the Ewald sphere.

If polychromatic radiation is being used, as in the Laue method, the Ewald circle represents the short wavelength limit (highest energy photon). Now all planes whose representative points lie inside the circle represent possible reflections and the symmetry of the reflections is the symmetry of the reciprocal lattice relative to the incident radiation. Thus the Laue method is useful for determining the orientation of the axes in a single crystal.

6.2.3 Brillouin Zones

Any wave-like particle traveling through a crystal, e.g., x-ray photon, electron, neutron, or phonon, can be reflected from lattice planes as they move through the solid just like x-rays; therefore, the Bragg law applies equally to the electrons, neutrons, and phonons in the crystal as it does to x-ray scattering. Recall that the Laue condition for scattering can be expressed as $G = \Delta k = k' - k$. For elastic scattering with k in the opposite direction of G, the scattered wave vector k' must be equal and opposite k. Thus the Laue condition is met if $k = G/2$. It is easy to show that this condition is also met everywhere along a perpendicular bisector of any reciprocal lattice vector G (proof left to the student).

DEFINITION

The smallest polyhedron centered at the origin, (000) in k-space, that is enclosed by the perpendicular bisectors of G to each nearest reciprocal lattice point is called the first Brillouin zone. *Similarly, the smallest polyhedron enclosed by the perpendicular bisectors of the next set of reciprocal lattice vectors is the* second Brillouin zone, *etc.*

This definition can be illustrated by constructing the first two Brillouin zones for a 2D oblique lattice as shown in Figure 6.6.

A similar construction called a Wigner–Seitz cell can be formed in direct space by taking perpendicular bisectors between each lattice point. The polyhedra formed will fill the 3D space and contain the volume associated with that particular lattice point. From this analogy, the Brillouin zones can be considered Wigner–Seitz cells in reciprocal space. The 3D Brillouin zones for the bcc, fcc, and hexagonal close-packed (hcp) reciprocal lattices are shown in Figure 6.7.

The Γ denotes the origin and the other letters indicate directions of high symmetry. We encounter these designations later in mapping electron energy bands (although the choice of letters may vary).

FIGURE 6.6
Construction of the first two Brillouin zones in a 2D oblique reciprocal lattice by taking perpendicular bisectors of the reciprocal lattice vectors G drawn from the origin to each lattice point. Since the Laue condition is met everywhere along a perpendicular bisector, the Brillouin zone surfaces can be thought of as mirrors that reflect radiation traversing the crystal for which $2(k \cdot G) + G^2 = 0$.

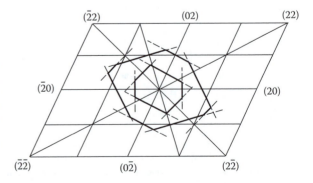

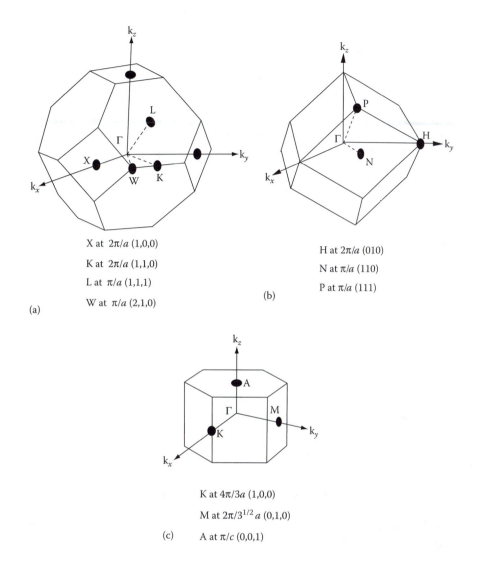

X at $2\pi/a$ (1,0,0)

K at $2\pi/a$ (1,1,0)

L at π/a (1,1,1)

W at π/a (2,1,0)

(a)

H at $2\pi/a$ (010)

N at π/a (110)

P at π/a (111)

(b)

K at $4\pi/3a$ (1,0,0)

M at $2\pi/3^{1/2}a$ (0,1,0)

A at π/c (0,0,1)

(c)

FIGURE 6.7

Brillouin zones for fcc direct lattice (bcc reciprocal) (a), bcc direct lattice (fcc reciprocal) (b), and hexagonal direct lattice (hexagonal reciprocal) (c). The letter Γ designates the origin and the other letters designate directions of high symmetry. This notation will be seen later in the band structure of various materials.

6.3 Diffraction Intensity

A primitive structure that contains only one atom per unit cell will produce a diffraction peak for every plane that satisfies the Laue condition. However, the presence or absence as well as the intensity of the reflections from a structure containing multiple atoms per unit cell will be determined by the contents of the unit cell.

Let $\boldsymbol{\rho}_j$ be a vector from the origin of the m, n, p unit cell to the nucleus of the jth atom in the unit cell and let $\boldsymbol{\rho}$ be a vector from the nucleus to a point in its electron cloud. Let $\psi(\boldsymbol{\rho})$ be the radial electron density function of the atom. The vector $\mathbf{r}$ extends from the crystal origin to $\boldsymbol{\rho}$ so that $\mathbf{r} = \mathbf{T}_{mnp} + \boldsymbol{\rho}_j + \boldsymbol{\rho}$ as shown in Figure 6.8.

FIGURE 6.8

Geometry for analyzing scattering of x-rays from electron density located at $\boldsymbol{\rho}$ relative to the jth atom in the unit cell translated by $\mathbf{T}_{mnp}$ from the origin.

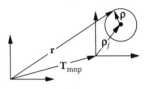

The electron density in the vicinity if the jth atom is given by $n_j(\boldsymbol{\rho}) = \int dV \psi_j(\boldsymbol{\rho})\psi_j^*(\boldsymbol{\rho})$. The electron concentration $n(\mathbf{r})$ in the crystal may be written as

$$n(\mathbf{r}) = \sum_{mnp}^{MNP} \sum_{j=1}^{s} \int dV \psi_j(\boldsymbol{\rho})\psi_j^*(\boldsymbol{\rho}). \tag{6.23}$$

The scattering amplitude is proportional to the electron concentration times the appropriate phase factor:

$$\int dV n(\mathbf{r}) \exp^{-i\mathbf{r}\cdot\Delta k} = \sum_{mnp}^{MNP} \sum_{j=1}^{s} \int dV \psi_j(\boldsymbol{\rho})\psi_j^*(\boldsymbol{\rho}) \exp^{-i\mathbf{r}\cdot\Delta k}. \tag{6.24}$$

6.3.1 Atomic Form Factor

The contribution from the jth atom to a diffraction spot may be obtained by setting $\Delta\mathbf{k} = \mathbf{G}$ and $\mathbf{r} = \mathbf{T}_{mnp} + \boldsymbol{\rho}_j + \boldsymbol{\rho}$ in Equation 6.24 and writing

$$\int dV \psi(\boldsymbol{\rho})\psi^*(\boldsymbol{\rho}) e^{-i\boldsymbol{\rho}\cdot\mathbf{G}} e^{-i(\boldsymbol{\rho}_j + \mathbf{T}_{mnp})\cdot\mathbf{G}} = f_j e^{-i(\boldsymbol{\rho}_j + \mathbf{T}_{mnp})\cdot\mathbf{G}}, \tag{6.25}$$

where we define $f_j = \int dV \psi(\boldsymbol{\rho})\psi^*(\boldsymbol{\rho}) e^{-\boldsymbol{\rho}\cdot\mathbf{G}}$ as the atomic form factor for atom j.

We can make a simple estimate of the atomic form factor by assuming a spherical distribution of electrons with radius R_j surrounding the jth atom and considering the phase shift from the scattering across the region.

$$f_j = 2\pi n_j \int_0^{R_j} \rho^2 \, d\rho \int_0^{\pi} e^{-i\rho G \cos\alpha} \sin\alpha \, d\alpha = 4\pi n_j \int_0^{R_j} \rho^2 \, d\rho \, \frac{\sin(\rho G)}{\rho G}, \tag{6.26}$$

where n_j is the electron density in the cloud surrounding the jth atom. We assume G satisfies the Bragg equation (Equation 6.22) for a beam with wavelength λ being scattered at θ and write $G\rho$ as

$$G\rho = \frac{4\pi\rho}{\lambda} \sin(\theta) = x \sin(\theta) \tag{6.27}$$

and write Equation 6.26 as

$$f_j = \frac{4\pi R_j^3 n_j}{3} \int_0^{x_{oj}} \frac{3x^2}{(x_{0j})^3} \frac{\sin[x\sin(\theta)]}{x\sin(\theta)} \, dx = \frac{4\pi R_j^3 n_j}{3} I(\theta, x_{0j}). \tag{6.28}$$

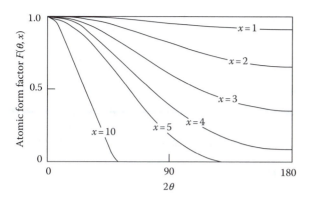

FIGURE 6.9
Angular part of the atomic form factor for various values of $x = 4\pi$ particle size/wavelength. X-rays with wavelengths much shorter than the particle size will be scattered in the forward direction. As the wavelength increases, the scattering becomes more omnidirectional.

where $x_{0j} = 4\pi R_j/\lambda$. Thus we see that the atomic form factor is just the total charge on the jth atom times an angular scattering factor that depends on the ratio of atomic radius to wavelength. Plots of this scattering function for different values of the radius to wavelength ratio are shown in Figure 6.9. If the wavelength of the radiation is short compared to the size of the atoms, the scattering will be more intense in the forward direction. As the wavelength becomes comparable to the size of the atoms, the scattering becomes more omnidirectional and backscattering can be observed.

Actual values for the atomic form factors for various atoms as a function of scattering angle can be found in the *International Tables for X-Ray Crystallography* or from the online NIST database http://physics.nist.gov/PhysRefData/FFast/Text/cover.html.

6.3.2 Structure Factor

We define the structure factor as

$$S_G = \sum_{j=1}^{s} f_j e^{-i\boldsymbol{\rho}_j \cdot \mathbf{G}}. \tag{6.29}$$

If the Laue condition is met, $\mathbf{G} = \Delta\mathbf{k}$ in Equation 6.24 and the scattering amplitude will be proportional to the sum of the contributions from each of the j atoms in all of the M^3 unit cells, or

$$\mathbf{E} \sim \sum_{mnp}^{MNP} \sum_{j=1}^{s} f_j e^{-i(\boldsymbol{\rho}_j + \mathbf{T}_{mnp}) \cdot \mathbf{G}} = S_G \sum_{mnp}^{MNP} e^{-i\mathbf{T}_{mnp} \cdot \mathbf{G}} = S_G MNP = S_G N_{\text{unit cells}}. \tag{6.30}$$

Thus the scattered E-wave is directly proportional to the structure factor S_G, which contains details of the unit cell, and to the sum over all of the lattice phase factors, which is just the number of unit cells. The scattered intensity I_{sc} will be then be proportional to

$$I_{sc} \sim |\mathbf{E}|^2 = EE^* = \frac{E_0^2 N_{\text{unit cells}}^2 S_G S_G^*}{R^2}, \tag{6.31}$$

where R is the distance from the crystal to the observation point. Since the atomic form factors in the structure factors are proportional to the Z of their atoms, the intensity of the diffraction peak will be proportional to Z^2. This fact makes if more difficult to use XRD techniques with very low Z material.

Expanding $\boldsymbol{\rho}_j \cdot \mathbf{G} = (x_j\mathbf{a} + y_j\mathbf{b} + z_j\mathbf{c}) \cdot (h\mathbf{A} + k\mathbf{B} + \ell\mathbf{C}) = 2\pi(x_jh + y_jk + z_j\ell)$. The structure factor becomes

$$S_G = \sum_{j=1}^{s} f_j e^{-2\pi i(x_jh + y_jk + z_j\ell)}, \tag{6.32}$$

where $x_j + y_j + z_j$ are the positions of the jth atom in the unit cell. For a simple cubic structure with a single atom at 0, 0, 0; $S_{hk\ell} = f_1$ for all h, k, ℓ and all reflections permitted by the Laue criterion will occur.

For a CsCl structure, which can be represented by a simple cubic lattice with a basis of (0,0,0) and (1/2,1/2,1/2), the structure factor is

$$S_{hk\ell} = f_1 + f_2 e^{-\pi i(h+k+\ell)} = f_1 \pm f_2 \quad (- \text{ if } h + k + \ell = \text{odd}, + \text{ if even}) \tag{6.33}$$

and all reflections will occur unless $f_1 = f_2$.

However, for a bcc structure, which can be represented by a simple cubic unit cell with atoms at (0,0,0) and (1/2,1/2,1/2), both atoms are the same and the structure factor reduces to

$$S_{hk\ell} = f_1 \left[1 + e^{-\pi i(h+k+\ell)} \right]. \tag{6.34}$$

In this case, $S_{hk\ell} = 0$ if $(h + k + \ell)$ is odd and $S_G = 2f_1$ if $(h + k + \ell)$ is even. Hence reflections from the (100), (111), (210), (300), etc., planes will be absent.

The basis for an fcc structure is (0,0,0), (1/2,1/2,0), (1/2,0,1/2) and (0,1/2,1/2). Now the structure factor becomes

$$S_{hk\ell} = f_1 \left[1 + e^{-\pi i(k+\ell)} + e^{-\pi i(h+\ell)} + e^{-\pi i(h+k)} \right]. \tag{6.35}$$

It can be seen that now $S_{hk\ell} = 4f_1$ if h, k, ℓ are either all odd or all even and $S_{hk\ell} = 0$ if they are mixed. Thus the (100), (101), (210), (211), etc., lines will be absent.

The disappearance of the (100) line in the case of both the bcc and fcc structure can be easily understood by the fact that in both cases, there are atoms on the (200) planes halfway between the (100) planes. When the path difference between (100) planes is λ, meeting the Laue condition for (100) reflections, the path difference between the (200) planes is $\lambda/2$ and destructive interference results. Thus the reflections from the (200) planes cancel the reflections from the (100) planes. When the path difference between (200) planes is λ, meeting the Laue condition for (200) reflections, the reflections from the (100) planes (which technically are also (200) planes) reinforce the reflections from the (200) planes and the intensities are enhanced.

Note that we could have represented the bcc and fcc structures by their primitive lattices and since these primitive unit cells contain single atoms, there is no interference and all $h'k'\ell'$ reflections would occur. Now if we take the primitive $h'k'\ell'$ indices for the fcc reciprocal lattice shown in Figure 6.1 and map these primitive indices into the nonprimitive fcc reciprocal lattice as shown in Figure 6.10a, we generate points at (200), (110), (220), etc. in the first layer ($\ell = 0$) and (101), (211), (301), etc. in the second layer ($\ell = 1$) layer. These points correspond to the allowed reflections of a bcc direct lattice in accordance with the structure factor calculations (Equation 6.34), which requires $hk\ell$ to be net even. Similarly, one can see in Figure 6.10b that the primitive vectors in the bcc reciprocal lattice will map out (200), (220), (222), (400), etc. in the even-numbered ℓ planes and (111), (311), (331), etc.

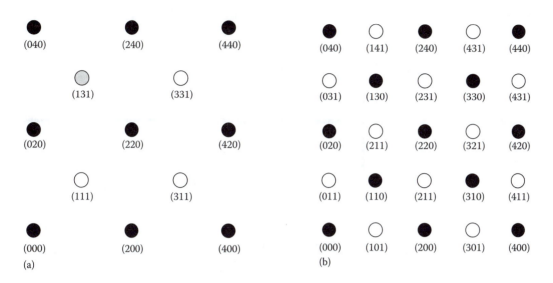

FIGURE 6.10
(a) First two layers of the reciprocal lattice points for an fcc reciprocal lattice generated by translating the primitive reciprocal lattice vectors shown in Figure 6.1. These are the Laue indices of the allowed reflections for a bcc direct lattice. Note that the net sum is even. (b) First two layers of reciprocal lattice points of a bcc reciprocal lattice generated by translating the primitive bcc reciprocal lattice vectors. These are the Laue indices of the allowed reflections for an fcc direct lattice. Note that the indices are all even or all odd.

in the odd-numbered ℓ planes, which are the allowed reflections in the fcc direct lattice in accordance with the structure factor calculations (Equation 6.34), which requires that $hk\ell$ be all even or all odd.

6.4   Methods and Uses of X-Ray Diffraction

6.4.1   Debye–Scherrer or Powder Method

One of the most fundamental tools for studying materials is X-ray diffraction (XRD) using the Debye–Scherrer or powder method. The sample is prepared as a finely ground powder and placed in a glass ampoule. The sample is illuminated with monochromatic (single wavelength) radiation. Those tiny crystallites that happen to be oriented such that they meet the Laue conditions will diffract at the angle that satisfies the Laue condition. Thus the diffracted radiation comes off in a series of cones whose axes are concentric with the incident radiation. The results are recorded on film or by an electronic counter as a function of angle relative to the incident beam. By indexing the peaks to determine their $hk\ell$ sequence, the presence or absence of certain peaks gives some indication of the structure, i.e., fcc, bcc, simple cubic, or other. Knowing the wavelength, the spacing between the $hk\ell$ planes may be found and from these spacing, the lattice parameters may be determined. Since the lattice parameters are unique to the various elements, this measurement is extremely useful for materials identification. Modern x-ray diffractometers are equipped with a computer library with the fingerprints of all of the common elements and can automatically do a material analysis by matching these known diffraction patterns with the observed spectrum. A typical powder diffraction pattern is shown in Figure 6.11.

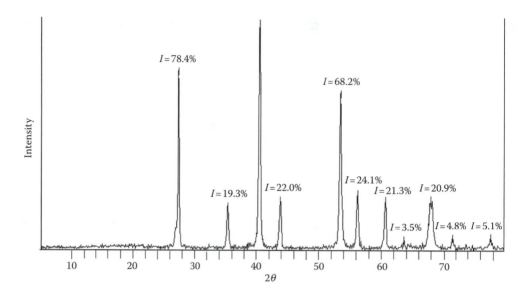

FIGURE 6.11

XRD pattern for MgF taken with CuK-α radiation. (Photograph courtesy of M. Banish, University of Alabama, Huntsville.)

When equipped with a hot stage, the powder method is also useful for observing solid-state phase changes or peritectic reactions in which a solid reacts with a liquid to form a new solid.

6.4.2 Laue Method

Another type of XRD analysis uses the Laue technique. In this application, a single crystal is illuminated with a polychromatic beam. The observed pattern, usually taken in two dimensions on a film or an imaging plate, reflects the symmetry of the crystal. This method is primarily used for determining the orientation of the crystal. It can also be used to judge the quality of the crystal. If the crystal is strained or has a high dislocation density, the spots appear smeared instead of sharp. Figure 6.12 is a Laue image of a GaAs crystal taken along its [111] axis. The spots with sixfold symmetry are the reflections from 6 of the 24 {642} planes that are included in the Ewald sphere. The spots with threefold symmetry are reflections from 3 of the 12 {422} planes in the Ewald sphere.

6.4.3 Rocking Curves

A variation of the Laue method technique uses monochromatic radiation to image a particular reflection spot on the detector and then the crystal is rocked back and forth over a small angle. The intensity versus the rocking angle is called a rocking curve and is used to quantify the quality of the crystal. If the crystal is strain-free, the angle of the reflected beam from each plane of atoms will be the same and the rocking curve will be very narrow. Internal strains from dislocations will cause the Bragg condition in various parts of the crystal to be met at slightly different angles, causing a spread in the rocking curve. Thus the width of the rocking curve (full width at half maximum) is a measure of the perfection of the crystal. Figure 6.13a shows a simulated rocking curve of a good single crystal.

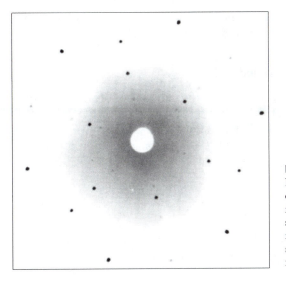

FIGURE 6.12
Back reflection Laue photograph along the [111] axis of GaAs. The spots showing sixfold symmetry are from 6 of the 24 {642} planes that lie in the Ewald sphere. The spots showing threefold symmetry are from 4 of the 12 {422} planes that lie in the Ewald sphere. (Picture courtesy of D. Gillies, private communication.)

Strain in the crystal would broaden the width of the rocking curve and twist, tilt, or small angle grain boundaries would show up as multiple peaks as illustrated in Figure 6.13b.

6.4.4 Rotating Crystal Method

Referring back to the Ewald construction, a diffraction pattern obtained by illuminating a single crystal with a polychromatic radiation (Laue method) contains the spots that lie between two concentric spheres: the inner sphere corresponds to the smallest k-value (longest wavelength) and the outer sphere corresponds to the largest k-value (shortest wavelength) of the polychromatic radiation used. Illuminating a single crystal with monochromatic radiation will only produce spots from reciprocal lattice points that happen to lie on the Ewald sphere corresponding to the wavelength being used. For the determination of the content of the unit cell, many spots must be precisely measured as discussed in Section 6.4.5. One method for accomplishing this is to rotate the crystal. Such an operation

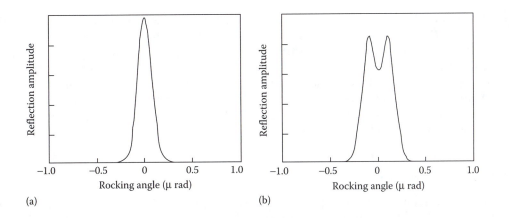

FIGURE 6.13
(a) Simulated rocking curves of an ideal crystal and (b) crystal with a small angle grain boundary.

will cause every spot within $2k$ of the origin of the Ewald diagram to cross the Ewald sphere, thus producing rows of spots that can be indexed and measured.

6.4.5 Obtaining Structure from X-Ray Diffraction Data

As stated previously, the Laue conditions determine the circumstances under which reflections might occur, but the intensity of the reflections is determined by the content of the unit cell through the structure factor. Recall that in Equation 6.24 when we integrated the electron density function over the volume to get the E-vector amplitude of the spot pattern, we obtained

$$\mathbf{E_G} \sim S_G N_{\text{unit cells}} = \int dV n(\mathbf{r}) e^{-i\mathbf{r} \cdot \mathbf{G}} \tag{6.36}$$

when the Laue condition $\mathbf{G} = \Delta\mathbf{k}$ is satisfied. Also, recall that the complex Fourier transform that transforms a time varying signal $f(t)$ to the frequency domain is given by

$$F(\omega) = \frac{1}{\sqrt{2\pi}} \int\limits_{-\infty}^{\infty} dt f(t) e^{-i\omega t}. \tag{6.37}$$

It is easy to see by analogy that the integral over the electron density function times the phase shift is just a complex 3D Fourier transform that transforms a periodic function in direct lattice space to reciprocal lattice space. Therefore, one may think of the diffraction pattern as a Fourier transform of the direct lattice. In principal, one should then be able to measure the intensities and locations of the reflections in the diffraction pattern, take the inverse Fourier transform, and recover the electron density function. The inverse transform is obtained by

$$n(x,y,z) = \frac{1}{V} \sum_h \sum_k \sum_l S_{G(h,k,\ell)} e^{2\pi i (hx + ky + \ell z)}. \tag{6.38}$$

6.4.5.1 Phase Problem

However, there is a major difficulty in carrying out this inversion process. The electric field vector $\mathbf{E}(h,k,\ell)$ of the scattered x-rays is directly proportional to $S_{G(h,k,\ell)}$, but at the present there is no method for directly measuring this E-field. For x-rays, we are limited to the intensity I as the observable, which is given by $I \sim |\mathbf{E}|^2$. The $\mathbf{E}$ is a complex function involving both amplitude and phase, but the phase information is lost unless one uses a coherent source to record the phase information as is done in holography. Since we do not yet have x-ray lasers suitable for this purpose, this is known as the phase problem in crystallography (see McPherson, 1999).

6.4.5.2 Direct Method

There are several methods for approaching the phase problem. The direct method or brute force approach simply treats the coordinates of the n atoms in the unit cell as $3n$ unknowns and solves the overdetermined set of nonlinear equations relating the observed intensities $I(h,k,\ell)$,

$$I(h,k,\ell) \sim S_G S_G^* = \left(\sum_j f_j e^{-2\pi i(x_j h + y_j k + z_j \ell)} \right) \left(\sum_j f_j e^{2\pi i(x_j h + y_j k + z_j \ell)} \right), \qquad (6.39)$$

assuming the number of measured $I(h,k,\ell) \gg 3n$. (This is why a large number of spots must be measured.) There are algorithms for solving a set of nonlinear overdetermined equations, but they are time-consuming and the above approach is feasible only for small molecule crystals where $n < 100$ atoms per unit cell.

6.4.5.3 *Isomorphous Substitution*

The other approach involves the use of isomorphous substitutions. Since the scattered intensity $\sim Z^2$, the substitution of a heavy metal for one of the atoms in the molecule will give a distinctive difference in intensity of the various spots on the recorded diffraction pattern. These spots can serve as points of zero phase and by comparing diffraction patterns with and without the heavy metal ions, it is possible to unravel the phase problem and determine the structure of a complex macromolecule such as a protein.

The pioneering work of Perutz and Kendrew (1962 Nobel Laureates in Chemistry), who used XRD to obtain the first 3D structures of simple proteins such as hemoglobin and myoglobin, has revolutionized biology. Biologists for the first time are now able to understand the actions of enzymes and other proteins as well as DNA, viruses, and cellular components at the molecular level. The protein or biomacromolecule must first be crystallized (not a trivial task) and then XRD data sets are taken using a monochromatic x-ray beam, often from a synchrotron source. A single crystal in a monochromatic beam may not fulfill the Laue conditions, so the crystal is rotated around different axes to bring the various reflection planes into the proper angles to meet the Laue conditions. These spots are indexed and the intensities measured and the inverse Fourier transform is computed as described previously to obtain the 3D electron density map. Generally the amino acid sequence of the protein is known from other techniques, so the electron density map allows the structural biologist to construct a 3D structure of these building blocks that best fits the observed electron density.

A limitation of XRD in obtaining biological structures is that the positions of the hydrogen atoms cannot be determined except by inference because they do not scatter the x-rays due to their low Z. Neutron diffraction can solve this problem since neutrons are strongly scattered by small atoms. The preceding mathematical analysis applies to neutrons as well as to x-rays. However, neutron diffraction analysis generally requires much larger crystals to obtain useful results, which has been a limiting factor in its use.

6.5 Summary

Much of the wave nature of solids is carried out in reciprocal space because the momentum vector of photons, phonons, and particles such as electrons and neutrons can be expressed directly in terms of their vector $\mathbf{k}$ as $|k| = 2\pi/\lambda$. The reciprocal lattice vectors $\mathbf{A}$, $\mathbf{B}$, and $\mathbf{C}$ are constructed such that they are perpendicular to the direct lattice vectors $\mathbf{a}$, $\mathbf{b}$, and $\mathbf{c}$ and their lengths are proportional to the inverse length of the direct lattice vectors. The lattice points in reciprocal space correspond to planes in direct space and a reciprocal translational vector is defined as $\mathbf{G}_{hk\ell} = h\mathbf{A} + k\mathbf{B} + \ell\mathbf{C}$, where $hk\ell$ are the Miller indices of these planes. It is shown that $\mathbf{G}_{hk\ell}$ is perpendicular to the $(hk\ell)$ plane and the distance of this plane from the origin is given by $d_{hk\ell} = 2\pi/|\mathbf{G}_{hk\ell}|$.

TABLE 6.1

Applications of XRD

Method	Radiation	Sample	Information Obtained
Debye–Scherrer	Monochromatic	Fine powder	Lattice parameters Material identification Solid-state phase change
Laue	Polychromatic	Single crystal	Crystal orientation Estimate of perfection
Rocking curve	Monochromatic	Single crystal	Quantify perfection
Rotating crystal	Monochromatic	Single crystal	Structure determination

The Laue condition for diffraction (including x-ray, electron, and neutron diffraction) can simply be stated as $\mathbf{G}_{hk\ell} = \Delta\mathbf{k}$, where $\Delta\mathbf{k}$ is the vector difference between the incident and scattered wave vector. For elastic scattering, the Laue condition can also be written as $2\mathbf{k} \cdot \mathbf{G} = \pm\mathbf{G}^2$ which is shown to be equivalent to the more familiar Bragg equation, $2d_{hk\ell} \sin\theta = \lambda$.

The Laue condition is satisfied everywhere on a perpendicular bisector to $\mathbf{G}_{hk\ell}$. Thus planes that are perpendicular bisectors of the various $\mathbf{G}_{hk\ell}$ vectors act as diffraction mirrors for waves or particles that meet the Laue condition. The Brillouin zones, polyhedra bounded by these perpendicular bisecting planes, will play an important role in the electron theory of metals as will be shown in later chapters.

Satisfying the Laue (or Bragg) condition is a necessary but not sufficient condition for a diffraction spot. The presence or absence of the $hk\ell$ diffraction peak, as well as its intensity, is determined by the structure factor $S_{hk\ell} = \sum_{j=1}^{s} f_j e^{-2\pi i(x_j h + y_j k + z_j \ell)}$ where x_j, y_j, z_j are the coordinates of the jth atom in the unit cell, f_j is the atomic form factor for the jth atom and is proportional to the number of electrons surrounding the atom, and the sum is taken over all of the atoms in the unit cell. Therefore, the reflected intensity is proportional to the square of the number of unit cells in the beam and to the square of the atomic numbers of the atoms in the cells.

For primitive systems with only one atom per unit cell, all peaks are visible. But for systems with multiple atoms per unit cell, many of the peaks cancel out. For a bcc structure, the sum of $hk\ell$ must be even to produce a diffraction peak. For an fcc structure, $hk\ell$ must either be all odd or all even to produce a diffraction peak.

There are a number of different ways in which XRD is used to study materials. A summary of these methods is given in Table 6.1.

Bibliography

Ashcroft, N.W. and Mermin, N.D., *Solid State Physics*, Brooks Cole, Philadelphia, 1976.

Christman, J.R., *Fundamentals of Solid State Physics*, John Wiley and Sons, Inc., New York, 1988.

Ibach, H. and Lüth, H., *Solid State Physics*, 3rd edn., Springer-Verlag, New York, 1990.

Kittel, C., *Introduction to Solid State Physics*, 7th edn., John Wiley and Sons, Inc., N.Y., 1966.

McPherson, A., Crystallization of Biomolecular Molecules, Cold Spring Harbor Laboratory Press, MA, 1999.

Problems

1. Show that the h, k, ℓ components of the **G** vector are the Miller indices.
2. Show that the spacing between lattice planes is given by $d_{hk\ell} = 2\pi/|\mathbf{G}_{hk\ell}|$ for any lattice.
3. Show that the Bragg condition is met everywhere on a perpendicular bisector of any reciprocal lattice vector **G**.
4. Show that the Bragg condition is met by any reciprocal lattice point that falls on the Ewald sphere.
5. Diamond structure is fcc with every other tetrahedral interstitial site occupied. Write the structure factor. Write the Laue indices for the first five diffraction peaks.
6. GaAs has the zinc blende structure, which is the diamond structure with one type of atom on the fcc sites and the other on the interstitial sites. Write the structure factor and the Laue indices for the first five diffraction peaks.
7. Find the reciprocal lattice vectors for a hexagonal lattice and verify the size of the hexagonal Brillouin zone shown in Figure 6.7c.
8. Basis for the hcp system is $(0,0,0)$ and $(a/2, a/2\sqrt{3}, c/2)$. Find the structure factor and identify a criterion for determining whether or not a peak would be visible.
9. Assume the nearest neighbor distances for the fcc and hcp form of a given material are the same. Find the distinguishing features of their XRD patterns that would identify which form you have.

7

Theory of Elasticity

The study of elastic behavior is fundamental to the understanding of the mechanical properties of materials. The elastic moduli are a property of the strength of the chemical bond of the material and are not affected by the various strengthening mechanisms employed to increase the yield or tensile strength of the material. Elastic deformations involve the stretching but not the breaking of bonds, so that the solid returns to its original configuration when stresses are relaxed.

7.1 Elastic Coefficients

We start by enumerating the various forces acting on the faces of an elastically deformable cube in Figure 7.1.

In Figure 7.1, the capital letter indicates the direction of the stress (force/area) and the subscript the face the force acts on. The forces perpendicular to the three faces, X_x, Y_y, Z_z, are the tensile stresses, denoted as T_x, T_y, T_z. In equilibrium, there can be no net torques; therefore, the stresses acting along the three faces must balance out according to $(Y_x \, \Delta z \, \Delta y)$ $\Delta x - (X_y \, \Delta x \, \Delta z)\Delta y = 0$, or $Y_x = X_y$, etc. We define the shear stress normal to the z-axis, $S_z = Y_x = X_y$. Similarly, $Z_x = X_z = S_y$ and $Z_y = Y_z = S_x$.

7.1.1 Stress Tensor

The stresses may be represented as a symmetric tensor of rank 2,

$$\mathbf{S}_{ij} = \begin{pmatrix} X_x & X_y & X_z \\ Y_x & Y_y & Y_z \\ Z_x & Z_y & Z_z \end{pmatrix} = \begin{pmatrix} T_x & S_z & S_y \\ S_z & T_y & S_x \\ S_y & S_x & T_z \end{pmatrix}. \tag{7.1}$$

7.1.2 Strain Tensor

The above solid is anchored at the origin so that it cannot translate. The deformation produced by the stress tensor can be represented by ξ, η, ζ, which are the dimensional changes along each of the respective coordinate axes. The tensile strains are the change in length per unit length or $e_{xx} = \partial \xi / \partial x$, $e_{yy} = \partial \eta / \partial y$, $e_{zz} = \partial \zeta / \partial z$.

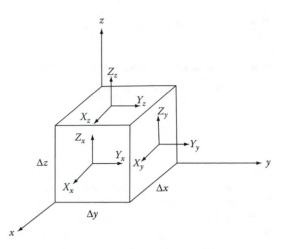

FIGURE 7.1
Stresses acting on the faces on a stationary elastic cube anchored at the origin.

The shear strains are derived from Figure 7.2.

The shear strains are defined as $e_{xy} = \partial\eta/\partial x + \partial\xi/\partial y = e_{yx}$, $e_{xz} = \partial\zeta/\partial x + \partial\xi/\partial z = e_{zx}$, and $e_{yz} = (\partial\eta/\partial y) + (\partial\zeta/\partial y) = e_{zy}$. Note that $(\partial\eta/\partial x) = \theta_y$ and $(\partial\xi/\partial y) = \theta_x$; therefore, $e_{xy} = \theta_x + \theta_y = \theta_{xy}$, etc.

Thus we can also write the strain as a symmetric tensor of rank 2,

$$\mathbf{e}_{k\ell} = \begin{pmatrix} e_{xx} & e_{xy} & e_{xz} \\ e_{xy} & e_{yy} & e_{yz} \\ e_{xz} & e_{yz} & e_{zz} \end{pmatrix}. \tag{7.2}$$

7.1.3　Elastic Coefficient Tensor

The stress and strain can be related by a coefficient tensor of rank 4, i.e.,

$$\mathbf{S}_{ij} = \mathbf{C}_{ijk\ell}\mathbf{e}_{k\ell}, \tag{7.3}$$

where each subscript runs from 1 to 3, giving the coefficient matrix 81 components. However, since $\mathbf{S}_{ij}$ and $\mathbf{e}_{k\ell}$ each have only six unique components, it is more convenient to represent them as vectors with six components and write

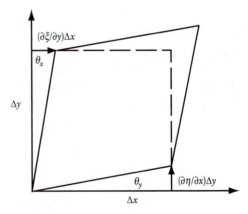

FIGURE 7.2
Deformation of elastic cube under shear stress S_z.

$$
\begin{pmatrix} T_x \\ T_y \\ T_z \\ S_x \\ S_y \\ S_z \end{pmatrix} = \begin{pmatrix} C_{11} & \cdot & \cdot & \cdot & \cdot & C_{16} \\ \cdot & \cdot & & & & \\ \cdot & \cdot & & & & \\ \cdot & \cdot & & & & \\ \cdot & \cdot & & & & \\ C_{61} & \cdot & \cdot & \cdot & \cdot & C_{66} \end{pmatrix} \begin{pmatrix} e_{xx} \\ e_{yy} \\ e_{zz} \\ e_{yz} \\ e_{zx} \\ e_{xy} \end{pmatrix}, \tag{7.4}
$$

where the C_{ij} are the modulus of elasticity or stiffness coefficients. It is easy to show that the C_{ij} elements are symmetric by the following argument. Consider an increment of work δU where

$$
\delta U = T_x \delta e_{xx} + T_y \delta e_{yy} + T_z \delta e_{zz} + S_x \delta e_{yz} + S_y \delta e_{zx} + S_z \delta e_{xy}.
$$

Since $\partial U/\partial e_{xx} = T_x$ and $\partial U/\partial e_{yy} = T_y$, $\partial T_x/\partial e_{yy} = \partial^2 U/\partial e_{xx}\partial e_{yy} = \partial T_y/\partial e_{xx}$. Since $T_x = C_{11}e_{xx} + C_{12}e_{yy} + \cdots$ and $T_y = C_{21}e_{xx} + C_{22}e_{yy} + \cdots$, $(\partial T_x/\partial e_{yy}) = C_{12}$ and $(\partial T_y/\partial e_{xx}) = C_{21}$; hence, $C_{12} = C_{21}$. Similarly it can be shown that $C_{ij} = C_{ji}$ for all i and j. Since the coefficient matrix is symmetric, the number of unique components is reduced to 21.

The above results are general and apply to any coordinate system—not necessarily an orthogonal system. However, if we have cubic symmetry, $C_{11} = C_{22} = C_{33}$, $C_{44} = C_{55} = C_{66}$, $C_{12} = C_{13} = C_{23} = C_{21} = C_{31} = C_{32}$. Since there are mirror planes perpendicular to each of the axes, shears in opposite directions must have opposite signs. But we have already shown that all of the matrix elements are symmetric; hence, the off-diagonal terms must vanish. Now the coefficient matrix can be written as

$$
\begin{pmatrix} T_x \\ T_y \\ T_z \\ S_x \\ S_y \\ S_z \end{pmatrix} = \begin{pmatrix} C_{11} & C_{12} & C_{12} & 0 & 0 & 0 \\ C_{12} & C_{11} & C_{12} & 0 & 0 & 0 \\ C_{12} & C_{12} & C_{11} & 0 & 0 & 0 \\ 0 & 0 & 0 & C_{44} & 0 & 0 \\ 0 & 0 & 0 & 0 & C_{44} & 0 \\ 0 & 0 & 0 & 0 & 0 & C_{44} \end{pmatrix} \begin{pmatrix} e_{xx} \\ e_{yy} \\ e_{zz} \\ e_{yz} \\ e_{zx} \\ e_{xy} \end{pmatrix} \tag{7.5}
$$

and only three elastic moduli or stiffness coefficients are needed to characterize the response of the material.

7.2 Properties of Crystals with Cubic Symmetry

Simplifying the stiffness coefficient matrix by assuming a material with cubic symmetry allows the various bulk elastic properties to be expressed in terms of the three elastic coefficients and various interrelations among them may be derived as illustrated in the following examples.

7.2.1 Shear Modulus

The shear modulus is defined as the ratio of shear stress to shear strain. For cubic symmetry, the off-diagonal coefficients vanish, $G = C_{44}$.

7.2.2 Bulk Modulus

The compressibility K is defined as negative change in volume per unit volume in response to a given hydrostatic pressure, $K = -(1/V)(\Delta V/\Delta p)$. The bulk modulus B is the resistance to compression, or $B = 1/K = -V(\Delta p/\Delta V)$. Under hydrostatic pressure, $-\Delta p = \Delta T_x = \Delta T_y = \Delta T_z$. Since $\Delta T_x = C_{11}\Delta e_{xx} + C_{12}(\Delta e_{yy} + \Delta e_{zz})$, $\Delta T_y = C_{11}\Delta e_{yy} + C_{12}(\Delta e_{xx} + \Delta e_{zz})$, and $\Delta T_z = C_{11}\Delta e_{zz} + C_{12}(\Delta e_{xx} + \Delta e_{yy})$, we see that $e_{xx} = e_{yy} = e_{zz}$. The $\Delta V/V = e_{xx} + e_{yy} + e_{zz} = 3e_{xx}$. Therefore, we can write

$$B = \frac{(C_{11} + 2C_{12})e_{xx}}{3e_{xx}} = \frac{1}{3}(C_{11} + 2C_{12}). \tag{7.6}$$

7.2.3 Young's Modulus

The Young's modulus is defined as the ratio of axial stress to strain or $E = T_x/e_{xx}$. From the stiffness matrix, we obtain the following relations:

$$T_x = C_{11}e_{xx} + C_{12}(e_{yy} + e_{zz})$$
$$0 = C_{11}e_{yy} + C_{12}(e_{xx} + e_{zz})$$
$$0 = C_{11}e_{zz} + C_{12}(e_{xx} + e_{yy}).$$

Using the last two equations to eliminate e_{yy} and e_{zz} in the first equation, we can write

$$E = C_{11} - \frac{2C_{12}^2}{C_{11} + C_{12}}. \tag{7.7}$$

7.2.4 Poisson's Ratio

The Poisson's ratio, ν, is defined as $-e_{yy}/e_{xx} = -e_{zz}/e_{xx}$. Using the last two equations as before to express e_{xx} in terms of e_{yy}, we can write

$$\nu = \frac{C_{12}}{C_{11} + C_{12}}. \tag{7.8}$$

7.3 Measurement of Elastic Coefficients

For single crystals, the elastic coefficients may be determined by acoustic velocity measurements. For example, it may be shown that the velocity of a longitudinal wave propagating in the [100] direction is given by

$$v_L[100] = \sqrt{C_{11}/\rho} \tag{7.9}$$

and the velocity of a transverse wave in the same direction is given by

$$v_T[100] = \sqrt{C_{44}/\rho}. \tag{7.10}$$

The velocity of a longitudinal wave moving in the [110] direction can be shown to be given by

$$v_L[110] = \sqrt{(C_{11} + C_{12} + 2C_{44})/2\rho}. \tag{7.11}$$

If the material is isotropic, the velocity in the [100] direction will be the same as in the [110] direction; hence

$$C_{11} = C_{12} + 2C_{44}. \tag{7.12}$$

This is not true for single crystals because they are generally anisotropic, i.e., the velocity of propagation is not the same in the [100] as in the [110] directions. However, polycrystalline materials and glasses are isotropic because of the random nature (or absence) of their grain structure. Therefore, the above relationship holds for bulk polycrystalline solids. This allows a number of relationships between B, E, G, and v to be derived. Using the relation between the Poisson ratio and C_{11} and C_{12} to eliminate C_{12}, the bulk modulus may be related to Young's modulus by

$$E = 3B(1 - 2v). \tag{7.13}$$

Using Equation 7.12 to relate C_{44} to C_{11} and C_{12} for isotropic materials, we can write

$$E = 2G(1 + v) \tag{7.14}$$

and

$$G = \frac{3B}{2}\left(\frac{1 - 2v}{1 + v}\right). \tag{7.15}$$

7.4 Bond Energy—Elastic Coefficients Relationships

A number of simple empirical potential functions were presented in Chapter 3. We will now attempt to use some of these functions in order to derive relationships between the various elastic properties of materials.

7.4.1 Bulk Modulus

We will concentrate primarily on the bulk modulus because, as the resistance to being compressed, it is the most fundamental of the elastic relationships. Also, by symmetry, it lends itself from being described by a simple central force potential. The resistance to being compressed or crushed arises from the Pauli principle that forbids any two electrons from being in the same state. As a material is being compressed, strong repulsion arises to resist the overlap of similar wavefunctions. At high pressures, when the valence electrons have been essentially collapsed into the inner core, the bulk modulus is no longer material dependent. But until this point is reached, the bulk modulus tells us a great deal about the mechanical properties of material.

The bulk modulus is defined as

$$B = -V\left(\frac{\partial^2 U(r)}{\partial V^2}\right)_{r_0} = -V\left(\frac{\partial^2 U(r)}{\partial r^2}\right)_{r_0}\left(\frac{\partial r}{\partial V}\right)^2. \tag{7.16}$$

Let $V = cr^3$, $dV = 3cr^2\,dr$. The bulk modulus can be written as

$$B = cr_0^3 U_0''\left(\frac{1}{3cr_0^2}\right)^2 = \frac{U_0''}{9cr_0}; \quad U_0'' = -\left(\frac{\partial^2 U(r)}{\partial r^2}\right)_{r_0} \tag{7.17}$$

and r_0 is the nearest neighbor distance.

7.4.2 Elastic Coefficients

Remember that the $U(r)$ is a central force potential and must be modified in order to obtain a unidirectional force. A unit cell with a NaCl structure contains 54 nearest neighbor bonds that would have to be stretched to expand this cell in three dimensions. However, to expand it in one dimension only requires stretching 18 bonds. Therefore, since only 1/3 the number of bonds need to be stretched in a linear expansion, the stress can be written as

$$\sigma(x) = \frac{F(x)}{A} = -\frac{\partial U(x)}{A\partial x} = -\frac{\partial U(r)}{3A\partial r}. \tag{7.18}$$

The elastic coefficient C_{11} is defined as the ratio of one-dimensional (1-D) stress to strain at r_0 and can be found from

$$C_{11} = \left(\frac{\partial \sigma(x)}{\partial(x/x_0)}\right)_{x_0} = -\frac{x_0}{A}\frac{\partial}{\partial x}\left(\frac{\partial U(x)}{\partial x}\right)_{x_0} = -\frac{x_0}{3A}\frac{\partial}{\partial r}\left(\frac{\partial U(r)}{3\partial r}\right)_{r_0} = \frac{U_0''}{9br_0} = B\frac{c}{b}, \tag{7.19}$$

where

$x_0 = r_0$
$A = br_0^2$

From Equation 7.19 we see that $C_{11} = cB/b$ and from Equation 7.6

$$C_{12} = \frac{1}{2}(3B - C_{11}) = \frac{C_{11}}{2}\left(3\frac{b}{c} - 1\right) \tag{7.20}$$

and from Equation 7.8

$$\nu = \frac{C_{12}}{C_{11} + C_{12}} = \frac{3b - c}{3b + c}. \tag{7.21}$$

For isotropic media, the shear modulus C_{44} may be estimated from

$$C_{44} = \frac{C_{11} - C_{12}}{2} = \frac{3C_{11}}{4}\left(1 - \frac{b}{c}\right). \tag{7.22}$$

7.4.3 Thermal Expansion

Most solids expand when heated. Their length increases according to

$$L(T) = L_0(1 + \alpha T), \tag{7.23}$$

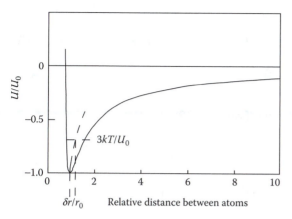

FIGURE 7.3
Equilibrium position of an atom in the bond-energy diagram at $3kT/U_0$ (dashed line). The asymmetry of the bond energy causes this equilibrium position to shift to larger values as the temperature increases, which is responsible for thermal expansion $\delta r/r_0$. Note that this shift is highly exaggerated in this drawing. Recall that the melting point occurs at only ~4% of the bond energy for most materials.

where α is the coefficient of linear thermal expansion (CTE) which is O (10^{-6} K^{-1}). A similar expression can be written for volume expansion and it is easy to show that since α is small, the coefficient of volumetric expansion is 3α. Thermal expansion occurs because of the asymmetry in the bond-energy function. This asymmetry produces anharmonicity in the oscillation of the atoms in the lattice; i.e., instead of oscillating with equal amplitude on both sides of their equilibrium position, they can swing farther to the soft side of the bond-energy function than toward the steep side. As a result, the average positions of the atoms are displaced from the equilibrium position at the 0 K location on the bottom of the bond-energy curve as seen in Figure 7.3. The average displacement of the lattice atoms can be found from

$$\langle \delta r \rangle = \frac{\int\limits_0^\infty \delta r e^{-\beta U(\delta r)}\, d\delta r}{\int\limits_0^\infty e^{-\beta U(\delta r)}\, d\delta r}, \tag{7.24}$$

where $U(\delta r)$ is the bond energy as a function of the displacement δr from the equilibrium position and $\beta = 1/kT$.

The bond-energy curve can be approximated by $U(\delta r) = C\delta r^2 - D\delta r^3 + \cdots$ in which C represents the harmonic part of the potential and D is the anharmonic part. It is assumed that $\beta D x^3 \ll 1$ so that $\exp(-\beta D x^3)$ can be expanded as $1 + \beta D x^3 + \cdots$. Keeping only terms through second order, the integrals can then be written as

$$\langle \delta r \rangle = \frac{\int\limits_{-\infty}^\infty e^{-\beta C\delta r^2}\left(\delta r + \beta D\delta r^4 + \cdots\right) d\delta r}{\int\limits_{-\infty}^\infty e^{-\beta C\delta r^2}\left(1 + \beta D\delta r^3 + \cdots\right) d\delta r} = \frac{3DkT}{4C^2}. \tag{7.25}$$

The coefficient of thermal expansion can be expressed as

$$\alpha = \frac{dr}{dT} = \frac{3RD}{4r_0 C^2}, \tag{7.26}$$

where R is the gas constant (assuming U is expressed in J/mol). If we expand our $U(\delta r)$ in a Taylor's series, we can identify $C = U_0''/2$ and $D = -U_0'''/6$. Thus the coefficient of thermal expansion becomes

$$\alpha = -\frac{RU_0'''}{2r_0 U_0''^2}. \tag{7.27}$$

7.4.4 Mie Potential

We shall now use the Mie potential to calculate the derivatives of $U(r)$. From Equation 3.36

$$U(r) = \frac{U(r_0)}{n-m}\left[n\left(\frac{r_0}{r}\right)^m - m\left(\frac{r_0}{r}\right)^n\right] = -\frac{U_0}{n-m}\left[n\left(\frac{r_0}{r}\right)^m - m\left(\frac{r_0}{r}\right)^n\right], \tag{7.28}$$

where U_0 is a positive quantity. Differentiating Equation 7.28

$$U'(r) = \frac{U_0mn}{(n-m)r_0}\left[\left(\frac{r_0}{r}\right)^m - \left(\frac{r_0}{r}\right)^n\right], \tag{7.29}$$

$$U''(r) = \frac{U_0mn}{(n-m)r_0^2}\left[-(m-1)\left(\frac{r_0}{r}\right)^m + (n-1)\left(\frac{r_0}{r}\right)^n\right], \tag{7.30}$$

from which

$$U_0'' = \frac{U_0mn}{r_0^2} \quad \text{and} \quad U_0''' = -\frac{U_0mn}{r_0^3}(n+m+3). \tag{7.31}$$

7.4.5 Ionically Bonded Systems

We used the measured lattice energy in Chapter 3 to obtain the values of n for various ionic compounds with the NaCl structure. Setting $m=1$ in the Mie potential Equation 7.28,

$$U(r) = -\frac{U_0}{n-1}\left[n\left(\frac{r_0}{r}\right) - \left(\frac{r_0}{r}\right)^n\right]. \tag{7.32}$$

For the NaCl structure, $c=8$, and $b=4$. From Equations 7.17 and 7.31 we have

$$B = -\frac{U(r_0)''}{9cr_0} = \frac{nU_0}{9cr_0^3} = \frac{n(U_0/V)}{9}. \tag{7.33}$$

From Equation 7.19

$$C_{11} = B\frac{c}{b} = 2B, \tag{7.34}$$

from Equation 7.20

$$C_{12} = \frac{C_{11}}{2}\left(3\frac{b}{c} - 1\right) = \frac{C_{11}}{4}, \tag{7.35}$$

from Equation 7.21

$$\nu = \frac{3b-c}{3b+c} = \frac{1}{5}, \tag{7.36}$$

and from Equation 7.22

$$C_{44} = \frac{3}{4}\left(\frac{c-b}{c}\right)C_{11} = \frac{3C_{11}}{8}. \tag{7.37}$$

TABLE 7.1

Elastic Properties of Covalent Compounds

	U_0/V (GJ/m³)	n	B Calculated Values (GPa)	B Measured Values (GPa)	C_{11} Calculated Values (GPa)	C_{11} Measured Values (GPa)
LiF	102.9	6.23	71.23	67.1	134.2	124.6
LiCl	40.66	8.3	37.50	29.8	59.6	49.4
NaCl	28.34	8.98	28.28	24	48	48.7
KF	34.45	8.21	31.43	30.2	60.4	65.6
KCl	18.39	9.46	19.33	17.4	34.8	40.5
KBr	15.33	9.92	16.90	14.8	29.6	34.6
KI	11.74	10.74	14.01	11.7	23.4	33.8

Source: Data taken from Kittel, C., *Introduction to Solid State Physics*, John Wiley & Sons, New York, 1966.

Gilman points out that the shear modulus cannot be estimated from a simple nearest neighbor central force potential. He represented the cross section of a rod using nine atoms connected by springs of one stiffness for nearest neighbors and springs of a different stiffness for second nearest neighbor. His conclusion was that the shear modulus for ionic compounds could be approximated as $C_{44} = G = 3/5B$. Using Equations 7.34 and 7.37, we estimated $G = 3/4B$ for ionic compounds. However, we did not just use a nearest neighbor central force potential.

We now check our predicted elastic constants against their measured values in Table 7.1.

There is reasonably fair agreement between the predicted and measured values of B and C_{11}. We will now examine the other predictions in Table 7.2.

The values for B calculated from Equation 7.6 may be slightly different from the measured values quoted in Table 7.1, but were used to be consistent with other values that could be computed from the elastic constants in this table.

We see from Table 7.2 that the predicted values for $C_{12}/C_{11} = 0.25$, $B/C_{11} = 0.50$, and $v = 0.20$ are somewhat lower than the mean observed values, but are well within the standard deviations of the observed values.

TABLE 7.2

Predicted Elastic Properties of Covalent Compounds

	C_{11} Measured Values (GPa)	C_{12} Measured Values (GPa)	B Equation 7.6 (GPa)	C_{12}/C_{11}	B/C_{11}	v Equation 7.8
LiF	113	47.7	69.5	0.42	0.61	0.3
NaF	96.8	24.2	48.4	0.25	0.5	0.2
KF	65.6	14.6	31.6	0.22	0.48	0.18
LiCl	49.4	22.8	31.7	0.46	0.64	0.32
NaCl	48.7	12.4	24.5	0.25	0.5	0.2
KCl	40.5	6.2	17.6	0.15	0.44	0.13
RbCl	36.3	6.2	16.2	0.17	0.45	0.15
AgCl	60	36	44	0.6	0.73	0.38
KBr	34.6	5.8	15.4	0.17	0.45	0.14
NaBr	33	13	19.7	0.39	0.6	0.28
AgBr	56	33	40.7	0.59	0.73	0.37
KI	27	4.3	11.9	0.16	0.44	0.14
MgO	286	87	153.3	0.3	0.54	0.23
				0.32 ± 0.15	0.55 ± 0.10	0.23 ± 0.09

Note: Measured values for C_{12} and C_{11} were taken from a variety of texts including Ashcroft and Mermin (1976), Gersten and Smith (2001), and Kittel (1966; 2nd and 7th eds.).

TABLE 7.3

Estimated Thermal Expansion fo Covalent Compounds

Material	U (kJ mol^{-1})	N	α_{Cal} (K^{-1})	α_{Meas} (K^{-1})
LiF	1012.6	6.23	6.74×10^{-6}	9.20×10^{-6}
LiCl	831.4	8.30	6.83×10^{-6}	12.20×10^{-6}
NaCl	765.6	8.98	7.83×10^{-6}	11.00×10^{-6}

Source: Measured values are taken from Barsoum, M.W., *Fundamentals of Ceramics*, Institute of Physic Publishing, Bristol and Philadelphia, (2003).

Recall in Figure 3.3 we saw that the specific energy (U/V) exhibited approximately a fourth power relationship with the nearest neighbor distance. Since the bulk modulus and the elastic coefficients are directly related to (U/V), one would expect a similar fourth power dependence of these quantities on the nearest neighbor distance.

We now investigate the ability to estimate thermal expansion from Equations 7.23 and 7.28.

$$\alpha = -\frac{RU_0'''}{2U_0''^2} = \frac{R(m+n+3)}{2mnU_0} = \frac{R(n+4)}{2nU_0}. \tag{7.38}$$

The results are shown in Table 7.3 for a few selected compounds. The agreement is only fair.

7.4.6 Simple Metals

We now turn our attention to the s and s–p bonded metals, or the elements in groups IA–IVA. We include the IVA elements even though they are not metals because their elastic properties behave similar to the simple metals because of their s and p bonds. Lacking a rational model for their attractive potential, such as the Coulomb potential model that can be used for the ionically bonded salts, the best that can be done is to try to fit one of the empirical potential functions such as the Morse potential, the Buckingham potential, or the Mie potential that were discussed in Chapter 3 to their measured properties. However, since it is necessary to use most of the observed properties to calculate the parameters needed to fit the potential, there is little left to predict. Also, it should be mentioned that the relation between the 1-D and 3-D central force potential, based on the number of bonds between the ions in the NaCl structure, would not be the same for fcc, hcp, or bcc structures. Hence the relations develop for the NaCl system, Equations 7.34 through 7.37 are not valid for other systems.

Rather than pursue the previous course, we will look for unifying relations between the structure of these metals and their elastic properties. Because the bulk modulus is a measure of the resistance of the valence electrons as they are forced together in response to the Pauli principle, one might expect that the bulk modulus should be related to the valence electron density. Indeed this is the case as may be seen in Figure 7.4. The bulk modulus for the simple metals and group IVA elements scales as the 1.22 power of the electron density. Similar power law expressions may be obtained for cohesive energy and elastic coefficients of the individual groups of the simple metals, but the curves for these individual groups do not coalesce into a single curve as in the case of the bulk modulus.

7.4.7 Transition Metals

Because of the saturation of the covalent bonds formed by the d electrons, the elastic properties of the transition metals are not directly related to their valence electron density

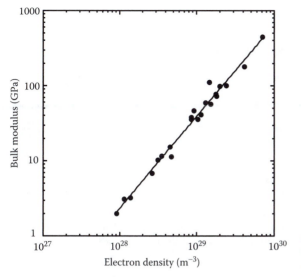

FIGURE 7.4
Bulk modulus versus electron density for simple
metals. The slope of the line is 1.22.

as was the case for the simple metals. Instead, as seen in Figure 7.5, they more closely
follow the same behavior as their specific energies shown in Figure 3.15.

The elastic properties of selected transition metals are tabulated in Table 7.4.

Note that the ratio of B/U_0 tends to increase as one moves down and to the right of the
periodic table.

The M-shell transition metals have magnetic interactions that weaken their bonds
because some of the d-electrons are locked in spin states, which prevents them from
forming covalent bonds, (discussed in Chapter 3). The bulk moduli of the N- and O-shell
elements show a very strong (inverse eighth power dependence) on their atomic radii
(Figure 7.6). Recall from Figure 3.16 that there is a very small difference between the atomic
radii of the N- and O-shell transition metals, which would account for the strong dependence on their radii.

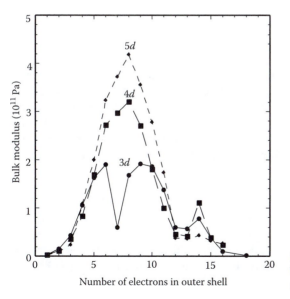

FIGURE 7.5
bulk modulus of the 3d, 4d, and 5d transition
metals.

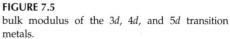

TABLE 7.4

Elastic Properties of Transitions Metals

Element	Atomic Number	Group	U_0 (GJ/m^3)	B (Gpa)	C_{11} (Gpa)	B/C_{11}	B/U_0
V	23	IVB	61.42	161	229	1.41	2.64
Ta	73	VB	72.02	200	267	1.33	2.78
Ni	28	VIIB	65.01	186	261	1.41	2.86
Fe	26	VIIB	58.28	168	234	1.39	2.89
Cu	29	VIIB	47.24	137	168	1.22	2.90
W	74	VIB	89.82	323	501	1.54	3.60
Mo	42	VIB	70.14	272	412	1.51	3.88
Ag	47	IB	24.43	100	131	1.30	4.12
Pd	46	VIIB	43.0	180	234	1.30	4.21
Pt	78	VIIB	61.93	278	366	1.32	4.49
Au	79	IB	36.01	173	186	1.07	4.81

Source: Data from Kittel, C., *Introduction to Solid State Physics*, 7th ed., John Wiley & Sons, New York, 1966; Ashcroft, N. and Mermin, N.D., *Solid State Physics*, Brooks Cole, Belmont, MA, 1976.

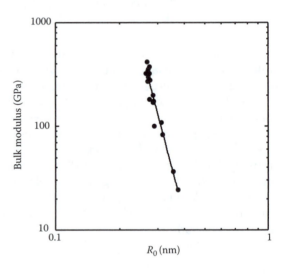

FIGURE 7.6
Bulk moduli of the N- and O-shell transition metals as a function of their atomic radii. Note the approximate eighth power dependence on their radii.

7.5 Theoretical Strength

It is useful to try to estimate the theoretical strength of materials from fundamental principles in order to understand what physical properties contribute to the strength, to find how far away we are from the maximum possible strength, and to get some insight into how to improve the performance of materials. There are several models for the theoretical strength that depend on the assumed failure mechanism of the material. We shall first consider the cohesive model which assumes failure by simply separating the bonds that hold the atoms together.

7.5.1 Cohesive Models

One way of estimating the theoretical strength of a material would be to compute the maximum force/area required to separate the atoms. We assume the potential as a function of a linear displacement x can be represented by the Morse potential. The Morse potential and its derivatives can be written as

$$U(x) = U_0 \left[e^{-2(x-x_0)/\xi} - 2e^{-(x-x_0)/\xi} \right], \quad U'(x) = -\frac{2U_0}{\xi} \left[e^{-2(x-x_0)/\xi} - e^{-(x-x_0)/\xi} \right],$$

$$U''(x) = \frac{2U_0}{\xi^2} \left[2e^{-2(x-x_0)/\xi} - e^{-(x-x_0)/\xi} \right], \quad U(x_0)'' = \frac{2U_0}{\xi^2}. \tag{7.39}$$

The second derivative has a root at $x_{max} = x_0 + \xi \ln 2$, so the maximum stress will be given by

$$\sigma_{max} = \frac{U'(x_{max})}{bx_0^2} = \frac{U_0}{2bx_0^3(\xi/x_0)} = \frac{cU_0/V}{2b(\xi/x_0)}. \tag{7.40}$$

If there are no defects in the crystal, the elastic bonds theoretically will stretch until x reaches x_{max}. Beyond this point the system energy is lowered by increasing the separation. In other words, if the stress is released at $x < x_{max}$, the system will recover elastically, but if strained beyond x_{max} the bonds will be broken and the system will separate into two parts.

Since from Equation 7.19

$$C_{11} = \frac{U(x_0)''}{9bx_0} = \frac{2U_0}{9bx_0\xi^2} = \frac{2c(U_0/V)}{9b(\xi/x_0)^2}, \tag{7.41}$$

the maximum stress can be obtained in terms of C_{11} by substituting Equation 7.41 into Equation 7.40 to give

$$\sigma_{max} = \frac{9C_{11}(\xi/r_0)}{4}. \tag{7.42}$$

The ξ/x_0 must be found from Equation 7.41 using the measured values of C_{11} and U_0/V. Putting this into Equation 7.42 yields

$$\sigma_{max} = \sqrt{\frac{9cC_{11}U_0/V}{8b}}. \tag{7.43}$$

For Al, $U_0/V = 32$ GJ/m^3, $C_{11} = 114.3$ MPa, and c/b is $2/\sqrt{2}$ for fcc. Using these values, the maximum cohesive strength for Al is predicted to be 76.3 GPa. The tensile strength of pure annealed 1100-0 Al is only 90 MPa, almost three orders of magnitude lower than the theoretical cohesive strength.

Often the Young's modulus is used instead for C_{11} and the theoretical strength for brittle fracture is generally estimated as $\sim E/10$ (see Section 9.5). For Al, the Young's modulus is 69 GPa, which gives an estimated maximum brittle strength of 6.9 GPa. The tensile strength of 7075 T-6 Al alloy is 572 MPa which is more than an order of magnitude lower than the theoretical maximum.

7.5.2 Shear Models

We will learn in the subsequent chapters that ductile metals yield by shearing, so now we will attempt to estimate their theoretical yield strength, or what will be called their critical resolved shear stress (crss). Shearing requires planes of atoms to slide over each other, rather like shearing a deck of cards as illustrated in Figure 7.7a. For cubic crystals, Equation 7.5 gives $S_z = C_{44}e_{xy} = G\theta_{xy} = Gx/d$ where G is the shear coefficient and $x \ll d$.

Assume a cubic crystal that has N planes with thickness δ and nearest neighbor distance R_0. In order to shear the crystal in the absence of dislocations, the shear force must move N planes of atoms over a distance R_0 as illustrated in Figure 7.7b. The maximum shear force

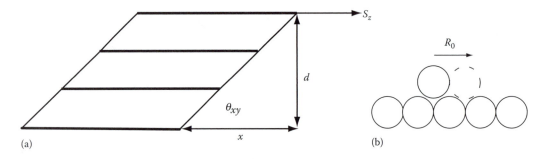

FIGURE 7.7
(a) Geometry of a crystal of thickness d sheared by distance x. (b) Illustration of the barrier in moving planes of crystal over one another.

will occur at $x = R_0/2$ as the planes of atoms are at the top of each other. Therefore, the theoretical maximum critically resolved shear force is given by

$$\sigma_{crss} = NG\left(\frac{R_0}{2d}\right) = \frac{G}{2}\left(\frac{R_0}{\delta}\right). \tag{7.44}$$

For fcc, $\delta = \sqrt{2}R_0$ and for bcc, $\delta = 2/\sqrt{3}R_0$. As a general rule the theoretical shear strength is estimated to be $G/4$.

The shear modulus for Al is ~40 GPa which translates into an estimated theoretical shear strength of 10 GPa. The actual yield strength of 7075 T-6 Al is ~500 MPa, about 1/20 of its estimated theoretical value. For pure Al (1100-0 annealed) the yield strength is only 34 MPa.

It should be understood that the elastic properties we have been discussing are a function only of the bonds and are not affected by any of the strengthening mechanisms (except for alloying) that will be discussed in the later chapters. The elastic properties of alloys will generally follow the rule of mixtures (see Section 10.3.1).

7.6   Summary

The stress and strain tensors can be written as six-component vector (three tensile and three shear) that are related by an elastic coefficient matrix, which can be reduced from 36 to 21 components by symmetry arguments. For systems with cubic symmetry, the elastic coefficient matrix can be further reduced to only three elastic constants, C_{11}, C_{12}, and C_{44}. These constants can be determined by measuring the velocity of sound in the medium in different directions in a single crystal. Once these elastic constants are known, other properties such as the bulk modulus B, the Young's modulus E, Poisson's ratio ν, and the shear modulus G may be derived, i.e.,

$$B = \frac{1}{3}(C_{11} + 2C_{12}), \quad E = C_{11} - \frac{2C_{12}^2}{C_{11} + C_{12}}, \quad \nu = \frac{C_{12}}{C_{11} + C_{12}}, \quad \text{and} \quad G = C_{44}$$

along with a number of interrelationships between these properties.

Using the simple bond-energy model and potential functions introduced in Chapter 3, we attempt to estimate the bulk modulus, the thermal expansion coefficient, and the linear elastic coefficient with varying degrees of success.

Using similar relations, we then estimated the theoretical or maximum tensile and yield strength of a material from its bond strength and elastic coefficients. For Al, the estimated theoretical yield and tensile strengths proved to be more than three orders of magnitude greater that the strengths of pure Al in the annealed state. Strengthening mechanisms (to be discussed in subsequent chapters) increase the actual strengths by about an order of magnitude, but they are still an order of magnitude short of the theoretical values for brittle fracture. So we have much room for improvement in developing stronger materials.

Bibliography

Ashcroft, N. and Mermin, N.D., *Solid State Physics*, Brooks Cole, Belmont, MA, 1976.

Barsoum, M.W., *Fundamentals of Ceramics*, Institute of Physic Publishing, Bristol and Philadelphia, 2003.

Gersten, J.I. and Smith, F.W., *The Physics and Chemistry of Materials*, John Wiley & Sons, New York, 2001.

Gilman, J.J., *The Electronic Basis of the Strength of Materials*, Cambridge University Press, UK, 2003.

Hosford, W.F., *Physical Metallurgy*, Taylor & Francis, CRC Press, Boca Raton, FL, 2005.

Kingery, W.D., Bowen, H.K., and Uhlmann, D.R., *Introduction to Ceramics*, 2nd ed. John Wiley & Sons, New York, 1976.

Kittel, C., *Introduction to Solid State Physics*, 7th ed., John Wiley & Sons, New York, 1966.

McMillan, N.H., Review: The theoretical strength of solids, *J. Mat. Sci.* 7, 1972, 239–254.

Russell, A.M. and Lee, K.L., *Structure–Property Relations in Nonferrous Metals*, John Wiley & Sons, New York, 1966.

Theoretical Strength of Materials, Prepared by the Materials Advisory Board, Publication MAB-221-M, National Academy of Sciences–National Research Council, Aug. 1966 (available online).

Problems

1. Show that for an isotropic solid the Young's modulus E is related to the shear modulus G and the Poisson ratio ν by $E = 2G(1 + \nu)$. Also show that $E = 9\,BG/(G + 3B)$.

2. If the density of a material remained constant while it was stretched elastically, what would the Poison ratio have to be? What would the bulk modulus have to be? Is it realistic to assume that the density remains constant while a material is deformed elastically?

3. What would be the theoretical maximum elongation of Al? What would be the stored energy?

4. The Materials Advisory Board (Aug. 1966) introduced a simple bond-energy function, $U(\rho) = -U_0(1 + \rho/D)\exp(-\rho/D)$ where ρ is the displacement from equilibrium and D is an adjustable parameter. Use this function to compute the theoretical maximum cohesive strength at rupture, and the strain at rupture. How does this result compare with the predictions using the Morse potential.

8

Defects in Crystals

As we shall see, nature does not allow a crystal to be perfect, but even if it did, the performance of a perfect crystal would not necessarily be improved. In fact the defect structure very much determines the mechanical, electrical, thermal, optical, and magnetic properties of a material and our ability to understand the role of defects and to be able to control their formation is key to the development of useful materials.

8.1 What Are Defects?

Generally speaking, defects are disruptions in the long-range order of the crystal that was discussed in Chapter 4. Such disruptions can range from an atom out of its place to gross defects such as voids or inclusions in the crystal. The mechanical properties of a material are largely influenced by its defect structure and such defects are often engineered into the material to improve its properties. On the other hand, certain types of defects degrade the electronic and optical performance of materials used for these purposes and are to be avoided. Therefore, it is important to develop an understanding of how various types of defects arise in crystal and how to control them.

The defects we will be concerned with can be classified into four categories: point defects, line defects, surface defects, and volume defects. The formation of these defects and their relation to the mechanical properties of the material are treated in the following sections.

8.2 Point Defects

As the name implies, point defects involve atoms that are missing, out of place, impurities that were purposely added or those that crept in.

8.2.1 Vacancy Defects

No crystal is perfect. No matter which solidification process is used or how careful one is in controlling the process, every crystal will have, at the very least, point defects known as vacancies.

Vacancies arise spontaneously to minimize the free energy at the local temperature. Since it requires energy to take an atom out of a lattice site to form a vacancy, one might ask, why should vacancies exist at equilibrium? The answer lies in the fact that at equilibrium, the Gibbs energy (or free energy) G is minimized—not the internal energy U. The Gibbs energy is $G = U + pV - TS$. (Note: For solids at atmospheric pressure, the pV contribution is negligible and the Helmholtz energy $F = U - TS$ is often used in place of G.)

Minimizing the free energy requires $dF = dU - TdS = 0$. If there are N_v vacancies, the change in internal energy is just $N_v Q$, where Q is the activation energy per atom to form a vacancy. Now we have to find a way to determine dS.

Despite what you may have learned about entropy in your undergraduate thermo-dynamics, there is a simple definition of entropy that you should always remember:

$$S = k \ln W, \tag{8.1}$$

where
  k is Boltzmann's constant
  W is the number of accessible states the system can be in for a given temperature, pressure, and volume.

Clearly, if all of the sites were occupied with atoms, there would be only one accessible state. However, if there were N_v vacancies and if the energy U did not depend upon which sites were vacant and which were occupied, there would be many more accessible states. The S would increase until the $T \, dS$ term is balanced by the increase in U.

Now let us quantify the number of accessible states. If we have N sites and N_v vacancies, just as there are 52! ways of arranging a deck of cards, there are $N!$ ways of arranging N sites with N_v vacancies. However, since atoms are indistinguishable, rearranging the $N - N_v$ atoms among the sites they occupy does not constitute different states. Thus we must divide the $N!$ ways of arranging the N sites by $(N - N_v)!$ to eliminate the redundancies from the fact that atoms are indistinguishable. Similarly, since interchanging one vacant site for another does not constitute a different state, we must also divide $N!$ by $N_v!$ to remove this redundancy. Thus the number of accessible states is given by

$$W = \frac{N!}{(N - N_v)! N_v!}. \tag{8.2}$$

As a sanity check, one can see that if all the sites were occupied, $W = 1$ (remember that $0! = 1$). If there were only 1 vacancy, there would be N places that vacancy could occur. If there are 2 vacancies, $W = N(N - 1)/2$ etc.

However, we must take the natural log of W. Since for macroscale systems both the number of sites and the number of vacancies are very large, we can employ Stirling's approximation $\ln x! = x \ln x - x$, which holds for large x. Thus

$$\ln W = N \ln N - (N - N_v) \ln(N - N_v) - N_v \ln N_v. \tag{8.3}$$

To minimize G with respect to N_v, we must take $d(\ln W)/dN_v$.

$$\frac{d \ln W}{dN_v} = \ln(N - N_v) - \ln N_v = \ln\left(\frac{N - N_v}{N_v}\right). \tag{8.4}$$

Since $dU/dN_v = Q$, the free energy is minimized with respect to N_v when

$$Q = kT \ln\left(\frac{N - N_v}{N_v}\right) \tag{8.5}$$

or

$$\frac{N_v}{N - N_v} = \frac{N_v}{N_{atoms}} = e^{-Q/kT}. \tag{8.6}$$

This is just Boltzmann's law, which says that the ratio of atoms (in this case, vacancies) occupying a state with energy Q to those in the ground state at temperature T is given by the Boltzmann factor, $\exp(-Q/kT)$. This type of behavior is also said to follow the Arrhenius law. If one plots the $\log(N/N_v)$ vs. $1/T$, a straight line should result and the activation energy can be determined from its slope. Conversely, any property whose log plots a straight line against $1/T$ is said to have Arrhenius-like behavior, meaning that there is an activation energy involved. Diffusion and viscosity coefficients are two other materials properties that exhibit Arrhenius-like behavior.

Atoms or ions leaving a lattice site to form a vacancy can either jump into an interstitial site or move to the surface of the solid. Other atoms or ions may move into the site they left behind; thus vacancies may meander or diffuse through the solid, a process known as vacancy diffusion. In metals, single-atom vacancies are possible, but in ionically bonded compounds, vacancies must occur in pairs in order to maintain charge balance.

8.2.2 Frenkel and Schottky Defects

In an ionic solid, a cation may move into an interstitial site leaving a cation vacancy. This is known as a Frenkel defect and the cation interstitial-cation vacancy is known as a Frenkel pair (Figure 8.1). An anion could also move into an interstitial to form an anion interstitial-anion vacancy, but since anions are generally larger than cations, the former is more probable.

It is also possible for a cation and an anion to move to the surface, forming an electrically neutral cation–anion vacancy. This type of vacancy is known as a Schottky defect and the pair of vacancies is called a Schottky vacancy pair.

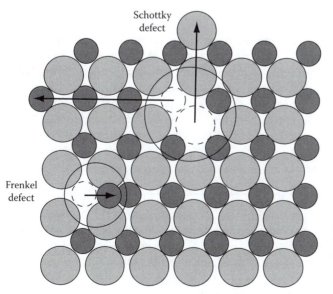

FIGURE 8.1
Illustration of Frenkel and Schottky defects in an ionic crystal. Larger light gray circles represent anions and smaller darker gray circles represent cations.

8.2.3 Impurity Defects

Another type of point defect is the incorporation of an impurity atom. Impurity atoms may replace host atoms in the regular crystal structure, in which case they are called substitutional defects, or they may occupy an interstitial site as interstitial impurities. Impurities are often purposely introduced in a lattice to strengthen it (solid solution hardening) or to otherwise alter its properties, e.g., doping a semiconductor to tailor the number and sign of charge carriers. However, as seen later, it is virtually impossible to completely eliminate unwanted impurity atoms.

8.2.4 Nonstoichiometric Compounds

If the ratio of cations to anions in a solid reflects the chemical formula, the system is said to be stoichiometric. For example, if the ratio of Na^+ to Cl^- in NaCl is 1:1, the system is stoichiometric. Nonstoichiometry arises when a cation or anion with a different valency is substituted into an ionic crystal. Anion or cation vacancies must then result in order to maintain charge neutrality. For example, if a Ca^{2+} were substituted for Na^+ in NaCl, a cation vacancy would be needed to accommodate the extra charge. If O^{2-} was substituted for Cl^-, an anion vacancy would have to occur. Iron can exist with a valence of +2 or +3. If a trivalent Fe^{3+} was substituted for Fe^{2+} in FeO, one Fe^{2+} vacancy would have to be formed for each pair of Fe^{3+} incorporated. Remember, Mother Nature is pretty serious about charge neutrality.

   Some ceramic compounds have a wide phase field of stability and can be quite nonstoichiometric. Spinel ($MgAl_2O_4$), for example, is shoichiometric at 50 At% MgO and 50 At% Al_2O_3, but is stable over quite a range of compositions. Here again a Mg^{2+} vacancy must occur for each additional pair of Al^{3+} above the stoichiometric value. On the other hand, one O^{2-} vacancy must occur for each pair of Mg^{2+} substituted for Al^{3+} above the stoichiometric number.

8.3 Line or One-Dimensional Defects

8.3.1 Slip in Metallic Crystals

Metallic crystals seldom deform plastically under pure tensile stress. They are much more likely to yield under shear stress. When a shear stress is applied to a metallic crystal, the crystal is found to deform more readily along certain directions than others. The planes along which shear deformation is most easily accomplished are called slip planes with the highest planar fraction or where the atoms are closest together. Also directions within these planes in which the atoms are closest together (highest linear fraction) are called slip directions. A specific slip direction together with its slip plane constitutes a slip system. (Think of the atoms as ball bearings. It is much easier to move an object over a surface when the ball bearings are close together.)

8.3.2 Resolved Shear Stress

When a tensile stress is applied to a crystal, it may be resolved into two perpendicular shear components, one making angle ϕ with the normal to the slip plane and the other making angle λ with the slip direction. The resolved shear stress is then

$$\tau_{rss} = \tau_{tensile} \cos\phi \cos\lambda. \tag{8.7}$$

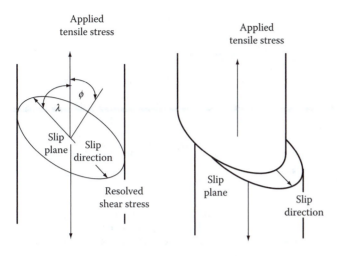

FIGURE 8.2
When the resolved shear stress from the applied tensile stress exceeds the crss, the material yields by shearing along a slip direction in a slip plane.

When the resolved shear stress exceeds the critical resolved shear stress τ_{crss} (a material property), slip will occur as shown in Figure 8.2. The yield stress can then be written as

$$\tau_{yield} = \frac{\tau_{crss}}{\cos \phi \cos \lambda}, \tag{8.8}$$

which is known as Schmid's law. The $\cos \phi \cos \lambda$ is called the Schmid factor. The maximum Schmidt factor is 0.5 and occurs when $\phi = \lambda = 45°$. The slip system having the largest Schmid factor will be the first to yield. It should be noted that if the Schmidt factor is zero, the yield strength is infinite. In a polycrystalline system, there will always be grains with a nonzero Schmidt factor; they will slip, and the materials will yield through a combination of this slip combined with grain boundary slip. However, it is possible to orient a single crystal so that there is no resolved stress in the direction of the applied tensile stress. This is the reason for making single crystal turbine blades for high performance jet engines.

8.3.3 Edge Dislocations

In Chapter 7, the theoretical shear strength was estimated to be $\sim G/4$, where G is the shear modulus. $G = 118$ GPa for Fe and $G = 125$ GPa for Ni. Therefore, the theoretical yield strength (the crss where material begins to deform plastically) is ~ 30 GPa. The actual yield for pure Fe and Ni is 130 and 138 MPa, respectively, approximately 240 times smaller than their theoretical limits. Similarly for Al, $G = 28$ GPa and the yield strength is 35 MPa, 200 times smaller than the theoretical estimate. Alloying and various hardening methods to be discussed are able to increase the yield strength to as high as 700 MPa for some stainless steels and Ni-based superalloys have yield strengths as high as 1600 MPa. 7075-T6 Al has a yield strength of 500 MPa, but these values are still only 1%–5% of the theoretical limit. (Since the elastic properties are determined by the chemical bond in the host material, alloying and hardening do not change the elastic properties significantly.) Why do these metals not reach their theoretical performance?

In 1934 Sir Geoffrey Taylor, together with Michael Polanyi and Egon Orowan, suggested a model that explained why metals failed to achieve their theoretical strength. They postulated that if a half plane of atoms were to be inserted into the lattice, the crystal could be deformed under shear by breaking bonds only along a line of atoms instead of throughout a plane of atoms. This would be equivalent to moving a carpet by creating a

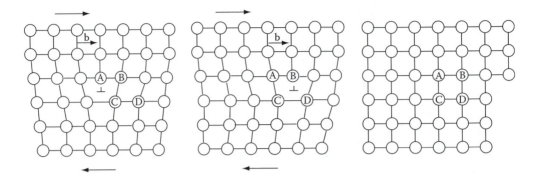

FIGURE 8.3
Mechanism for dislocation movement. Under a shear stress, bonds between B and C are broken and reform between C and A. This moves the dislocation line to between C and D. The process continues with bonds reforming between D and B until the dislocation moves through the edge of the solid causing a displacement of one lattice vector.

kink at one end and moving the kink across the carpet rather than trying to move the entire carpet at once.

The line along the core where the extra half plane of atoms terminates is called a dislocation. If this line emerges perpendicularly to the face of the crystal, it is called an edge dislocation. The symbol ⊥ denotes an edge dislocation with the vertical part in the direction of the extra half plane. The mechanism by which crystals might deform by the motion of edge dislocations is shown in Figure 8.3.

8.3.4 Burgers Vector

A dislocation is characterized by its Burgers vector **b,** which is determined by the step needed to complete the Burgers loop about the dislocation. Starting from an arbitrary point, move from lattice point to lattice point around a square loop containing the dislocation, making the same number of steps in each direction. The additional step required to complete the loop is the Burgers vector as may be seen in Figure 8.3. Define a unit vector $\hat{t}$ tangent to the dislocation line; the Burgers vector will be perpendicular to $\hat{t}$ for an edge dislocation.

There is lattice strain in the region of a dislocation. The region with the extra half plane is under compression while the region with the missing half plane is under tension. These strain fields can produce mutual attraction or repulsion on other dislocations depending on the signs of their Burgers vectors. If they have the same sign, meaning the strain fields are the same on either side of the slip plane, they will repel each other. Opposite signs will attract and the two opposite dislocations can annihilate each other.

8.3.5 Screw Dislocations

Another type of dislocation is the screw dislocation. It is best illustrated by cutting the crystal halfway through and shearing the cut halves by one lattice spacing before rejoining the two pieces as shown in Figure 8.4. Using the Burgers loop as before, it is seen that the Burgers vector is parallel to the tangent vector for a screw dislocation.

8.3.6 Mixed Dislocations

A mixed dislocation can contain a component of an edge as well as a screw dislocation and may curve through the crystal, entering one face as an edge dislocation and emerging

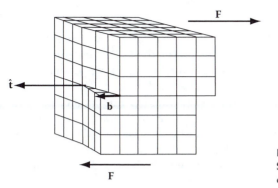

FIGURE 8.4
Schematic showing a screw dislocation on the face of a crystal.

through a perpendicular face as a screw dislocation as illustrated in Figure 8.5. In such a process, the Burgers vector is conserved in direction and magnitude. The screw component is given by $(\mathbf{b} \cdot \hat{\mathbf{t}})\hat{\mathbf{t}}$ and the edge component is given by $\mathbf{b} - (\mathbf{b} \cdot \hat{\mathbf{t}})\hat{\mathbf{t}}$, so the vector sum is always $\mathbf{b}$.

The direction taken around the dislocation loop to define the Burgers vector is arbitrary, but must be consistent with the requirement that $\mathbf{b}$ is invariant. For example, a clockwise Burgers loop was taken to define the Burgers vector in Figure 8.3. If this dislocation had emerged on the opposite side of the crystal, the Burgers loop would have to be taken in the anticlockwise direction to maintain the direction of $\mathbf{b}$. It follows that the actual sign of the vector $\mathbf{b}$ is arbitrary as long as it is assigned consistently.

Dislocations cannot begin or end inside a crystal. They can begin and end on free surfaces, on grain boundaries, or on other dislocations. They can also close on themselves to form dislocation loops within the crystal. Dislocations can move along any plane containing both $\hat{\mathbf{t}}$ and $\mathbf{b}$ in the direction of the resolved shear force. Their mobility is highest along slip planes and in slip directions where the atoms are closest together. After a dislocation has passed through the crystal as illustrated in Figure 8.4, that portion of the crystal will have been displaced or sheared by the magnitude of the Burger's vector by only breaking bonds along one line of atoms instead of the entire plane of atoms. The implications of this process in the strength of materials are dealt with in greater detail in later chapters.

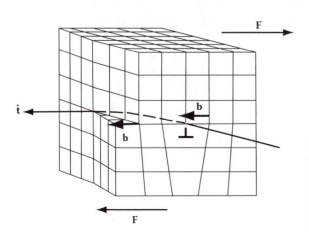

FIGURE 8.5
An example of a mixed dislocation. The dislocation line enters the crystal as an edge dislocation on one face, curves through the crystal, and emerges as a screw dislocation. The magnitude and direction of the Burgers vector is conserved throughout.

8.3.7 Observing Dislocations

Dislocations may be observed in metals using transmission electron microscopes, but the sample must be carefully prepared and can only be ~5 μm thick. Because the region around where a dislocation emerges from a surface is strained, it is more vulnerable to chemical attack and the dislocation can be revealed by chemical etching. The dislocation density is defined as the number of dislocation lines intersecting per unit area of material and can be estimated by counting the number of etch pits in a known area. The etch pit density (epd) can range from 10^{10} to 10^{12} cm^{-2} in heavily strained samples down to a few 100 cm^{-2} in carefully grown crystal as shown in Figure 8.6. It is now possible to grow dislocation-free silicon using the floating zone process in which there is no wall contact with the growing crystal.

8.3.8 Creating Dislocations

What causes dislocations? There are a number of mechanisms that can produce dislocations, most of which have to do with internal stresses. Such stresses can be generated from differential expansion as a material is cooled from the melt or by cold-working the material. Frank and Reed proposed a mechanism whereby a dislocation can be pinned at two points, either by impurity atoms or other dislocations. As stress is applied, the dislocation line becomes increasingly bowed, until at last it folds back on itself and pinches off to form a dislocation loop, which continues to expand under the applied stress. The remainder of the dislocation loop after it pinches off now repeats the process as seen in Figure 8.7. This mechanism is called a Frank–Reed source.

Large numbers of vacancies are present when a material first solidifies from the melt. If the material is cooled faster, then the number of vacancies cannot reach their equilibrium value; they become quenched in and can then migrate and coalesce into dislocation loops as seen in Figure 8.8.

8.3.9 Screw Locations and Crystal Growth

If a screw dislocation is present on the growth interface of a crystal growing from the vapor, the kink provides an ideal location for the new atoms or molecules to attach to the

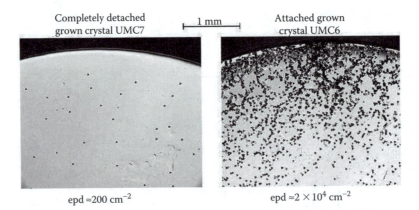

Completely detached grown crystal UMC7   ⊢ 1 mm ⊣   Attached grown crystal UMC6

epd ≈200 cm^{-2} epd ≈2 × 10^4 cm^{-2}

FIGURE 8.6
Etch pits in Ge single crystals grown with and without wall contact. Notice the 100-fold decrease in dislocation density that results from eliminating wall contact during the growth. (From Schweizer, M., et al. *J. Crystal Growth*, 235, 161, 2002. With permission.)

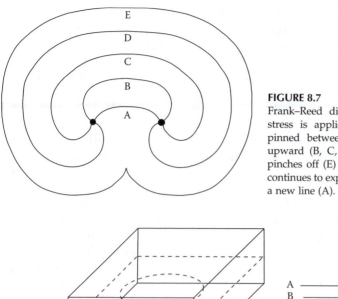

FIGURE 8.7
Frank–Reed dislocation generating mechanism. As stress is applied, a dislocation line (A), becomes pinned between two points, moves progressively upward (B, C, D) until it folds back on itself and pinches off (E) to form a dislocation loop. This loop continues to expand while the process starts over with a new line (A).

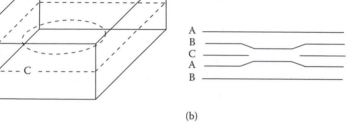

(a)   (b)

FIGURE 8.8
Coalescence of vacancies into dislocation loop. Vacancies coalesce in the plane C to form the empty region in the diagram to the right (b). This produces the half plane vacancy in the region bounded by the dislocation loop shown on the left (a).

growing crystal since they can satisfy more bonds. As they attach, the spiral keeps advancing, always providing additional low energy attachment sites for new atoms or molecules to be incorporated. Superstrength single crystal whiskers may be grown from the vapor in this manner. Since the screw axis runs along the length of the crystal, the Schmid factor is zero along this axis and such whiskers can achieve near theoretical strength.

8.4 Two-Dimensional or Planar Defects

8.4.1 Interfacial Energy

Any surface of a crystal can be considered as a planar defect. Atoms sitting on the surface of a crystal have a different environment from those deep inside. For one thing, their bonds are only partly satisfied and they are not as tightly bonded as the atoms on the inside. These unsatisfied bonds create a surface excess energy meaning that the atoms on the surface have a higher (less negative) energy than those buried inside the crystal. This situation is analogous to atoms or molecules in a drop of liquid. Liquids, in order to minimize their surface energy, minimize their surface. It requires a force to extend the surface and this force per unit length is called surface tension, which is expressed in N/m. We could just as well speak of the work required to extend the surface over some area as interfacial energy and express it as J/m^2. In other words, this is the energy it costs to make new surface, which is the same as the surface excess energy. Surface tension and interfacial energy are numerically as well as dimensionally the same. Interfacial energies for metals are on the order of 1–10 J/m^2.

Some crystals can be cleaved along planes that are parallel to faces of their growth habit. These are generally planes of minimum interfacial energy since the growing crystal likes to surround itself with surfaces that cost the least amount of energy in order to lower the total free energy. Another way of saying this is that the growth atoms tend to attach to higher energy surfaces where the force of attraction is the greatest. It is often said that you never see the fastest growing surface of a crystal because these surfaces grow themselves out, leaving the slower growing surfaces exposed, which are generally the planes with the highest atomic density.

The basal planes of mica or annealed pyrolytic graphite can easily be separated with a razor blade. Ordinary table salt has a cubic habit and can be cleaved easily along the {100} faces by tapping a suitably aligned razor blade. However, attempts to cleave along a {110} will cause the crystal to shatter. The {100} faces have the highest area density which means the ions are closer together, hence their bonds are stronger and the excess surface energy will be minimized. The {110} unit cell faces have the same number of atoms as the {100} faces but have approximately 40% more area, so they have a much higher surface energy (see Problem 3.5). The {111} faces also have a high area density, but they are not electrically neutral. All the ions on these faces have the same charge; therefore, the face would have an extremely high surface excess energy.

8.4.2 Estimating Surface Energy for Ionic Solids

We can compute the interfacial energy for ionically bonded compounds by computing the lattice sum for a plane of atoms (see the Appendix of Chapter 3.) For the (100) face of the rock salt structure, we find the Madelung sum A_{plane} to be 1.615542. The Madelung constant for the lattice is 1.748. The Madelung constant for an ion pair on the surface will be given by $A_{interface} = A_{plane} + A_{below}$. The sum of the three contributions $A_{above} + A_{plane} + A_{below}$ must equal the total lattice sum A. If the plane is taken somewhere near the middle of the solid as in Figure 8.9 so that the contributions from the atoms near the top or bottom are negligible, we can set $A_{above} = A_{below}$.

Therefore,

$$A_{plane} = A - A_{above} - A_{below} = A - 2A_{below} \tag{8.9}$$

$$A_{interface} = A_{plane} + A_{below} = A_{plane} + \frac{A - A_{plane}}{2} = \frac{A + A_{plane}}{2}. \tag{8.10}$$

Since A_{plane} is smaller than A, ion pairs at the interface will have a higher energy (not as deep in the energy well) as those in the interior. This excess energy is the interfacial energy. Using Equation 3.7, the surface excess energy is given by

FIGURE 8.9
Schematic for estimating the surface energy of an ionic solid.

$$\gamma = -(A_{\text{interface}} - A)\frac{N_s e^2}{4\pi\varepsilon_0 r_0}\left(\frac{n-1}{n}\right) = \frac{(A - A_{\text{plane}})}{2}\frac{e^2}{8\pi\varepsilon_0 r_0^3}\left(\frac{n-1}{n}\right), \qquad (8.11)$$

where the surface density $N_s = 2$ ion pairs per $4r_0^2$. The nearest neighbor distance, for NaCl, $r_0 = 0.282$ nm and n was found to be 8.98 (from Table 7.1). Using these values, the interfacial energy of the [100] face is found to be 0.306 J/m^2. This is close to the observed value of 0.32 cited in Barsoum (2003, p. 103).

8.4.3 Grain Boundaries

In polycrystalline materials, grain boundaries interrupt the periodicity and the orientation of the crystalline structure and are also considered planar or surface defects. They have surface excess energies on the order of 1–3 J/m^2 but are not completely free surfaces since some of the bonds are satisfied. As mentioned previously, they are also more subject to chemical attack because of the unsatisfied bonds and, like dislocations, they can be made visible for analysis by cutting, polishing, and chemical etching. The reflectivity of metals is a function of the crystal orientation that provides the contrast needed to distinguish the different grains when viewed under a metallurgical microscope (Figure 8.10).

8.4.4 Tilt and Twist Boundaries

A low angle tilt boundary is a special type of grain boundary in which a portion of the grain is tilted as a result of a periodic array of edge dislocations as shown in Figure 8.11. Similarly, a low angle twist boundary can be formed by an ordered array of screw dislocations.

8.4.5 Stacking Faults

Stacking faults are another form of surface defects. They occur from mistakes during crystal growth, often from growing too fast where the ad atoms have not had a chance to reach equilibrium at the surface. They can also be caused by stresses or by vacancy

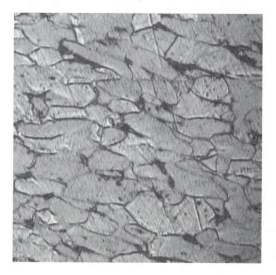

FIGURE 8.10
Grain boundaries of a polished and etched steel nail at ×60. Note the elongated grains that result from the drawing process. (Photograph courtesy of Dr. Michael Banish, University of Alabama in Huntsville. Personal communication.)

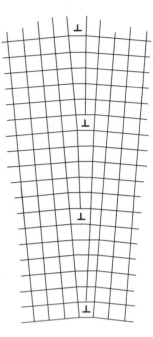

FIGURE 8.11
Tilt boundaries. Formation of a low-angle tilt boundary
through a series of dislocations.

coalescence. A stacking fault from a missing plane, such as ABCBCABC, is referred to as an intrinsic stacking fault; whereas adding an extra plane such as ABABCABAB or ABCBABC is called an extrinsic stacking fault. Since the coordination numbers are the same in either of these planes, the stacking fault energy can be very small, ranging for 0.01–0.3 J/m^2.

8.4.6 Twinning

A twin is formed when a crystal grows in a mirror image of itself about a plane which is not a normal mirror plane. In the simple cubic and body-centered cubic systems, the {111} planes are possible twin planes (they are not mirror planes). Similarly, in the face-centered cubic system, ABCABCBACBA, the middle C (also a {111} plane) is a twin plane. In the hexagonal close-packed system, ABABCBABA, C would be a twin plane. Twining is caused by stress during cooling and complex vacancy and dislocation mechanisms. The energy involved in forming twins is about the same as in producing stacking faults. In space crystal growth experiments in which crystals were grown in the absence of hydro-static pressure and wall contact, both the dislocation density and number of twins were dramatically reduced.

8.5 Volume or Three-Dimensional Defects

Volume defects consist of inclusions or precipitates of a second phase material or voids. Voids can be formed by vacancy clusters or from the nucleation of bubbles from dissolved gases or from components with high vapor pressures. Such defects can range in size from microscopic to gross. Bear in mind that not all such defects are unwanted. Many are purposely introduced into the final solid to tailor certain electrical, optical, and magnetic properties, or to serve as strengthening mechanisms. These topics are discussed in later chapters.

8.6 Diffusion

Atoms can move through solids by diffusion just as they are able to do in liquids and gases, except the process is much slower. Mechanisms by which atoms are able to move through solids include vacancy exchange diffusion, in which atoms fill vacancies creating new vacancies; interstitial diffusion, in which atoms can move through the interstitial sites; grain boundary diffusion, in which atoms can move along grain boundaries; and pipe diffusion, in which atoms move along dislocations.

Diffusion can be considered to be a random process in which an impurity atom is just as likely to jump one way as the other. However, think of two herds of sheep, a small herd and a larger herd separated by a fence. The sheep start jumping over the fence. The sheep in either herd are just as likely to jump the fence, but more sheep from the larger herd will actually jump the fence, simply because there are more of them. So the diffusion of sheep will be driven by the concentration gradient of the herd. Eventually, there will be the same number of sheep on each side of the fence and they will be jumping back and forth at the same rate. We would say an equilibrium has been reached.

So it is in a solid. If there are more atoms of one variety in one part of the solid, they will tend to spread out until a uniform concentration is reached throughout the solid by the mechanism of solid-state diffusion; however, this process can be extremely slow, especially at ambient temperatures.

8.6.1 Diffusion Coefficient

The rate of diffusion is governed by the diffusion coefficient in conjunction with Fick's laws of diffusion. Because diffusion is thermally activated, the diffusion coefficients follow the Arrhenius behavior and it is a common practice to express their thermal dependence in terms of a pre-exponential coefficient D_0 and an activation energy Q,

$$D(T) = D_0\, e^{-Q/RT}. \tag{8.12}$$

There are no general rules governing diffusion coefficients; therefore, they must be determined experimentally. Each of the diffusion processes has its own activation energy, which depends not only on the process involved but also on the host and the diffusing material. At higher temperatures, vacancy diffusion is generally prevalent since it has the highest activation energy. Since the activation energy for vacancy diffusion requires the breaking of bonds, one would expect metals with higher melting points to have higher activation energies, thus lower diffusion coefficients. Grain boundary and pipe diffusion have lower activation energies and become more important at lower temperatures. For very small molecules, interstitial diffusion will generally dominate. Naturally one would expect a small atom such as carbon to diffuse through metals faster than other atoms with sizes comparable to their host. Gases diffuse through metals much faster than atoms of other metals, which is why vacuum chambers are constructed from thick walls of stainless steel. He diffuses faster than H_2 because of its smaller molecular size. Values of the D_0 and the activation energy for different diffusion couples are found in various handbooks (see bibliography).

8.6.2 Kirkendall Effect

In the process of vacancy diffusion at a junction between two metals, one component may diffuse faster than the other. This imbalance of flow can cause the interface to move and create vacancies and even voids in the region being depleted by the faster diffusing component. This effect has been implicated in the failure of solder joints and Al–Au interconnects as they age, which is sometimes referred to as the purple plague. However, researchers at the Max Planck Institute of Microstructure Physics have exploited the Kirkendall effect to make hollow spinel ($ZnAl_3O_4$) nanotubes by coating ZnO nanowires with Al_2O_3 and letting the Zn diffuse into the Al_2O_3 coating (*Nature Materials*, Aug. 2006).

Because the diffusion rates of two materials may be different, it is necessary to define an effective diffusivity $\widetilde{D}_{BA} = xD_{AB} + (1-x)D_{BA}$, where x is the mole fraction of component B in A, D_{AB} is the diffusion coefficient of A into B, and D_{BA} is the diffusion of B into A. This distinction is only important for nondilute alloy systems. For dilute systems, x is small and $\widetilde{D}_{BA} \rightarrow D_{BA}$.

8.6.3 Fick's Laws

The rate of diffusion is governed by Fick's laws. Fick's first law simply states that the diffusive flux (number of atoms of component B crossing unit area per unit time) is proportional to the concentration gradient C_B of B atoms in the host of A atoms; or

$$J_B = -D_{BA}\left(\frac{\partial C_B}{\partial x} + \frac{\partial C_B}{\partial y} + \frac{\partial C_B}{\partial z}\right) = -\widetilde{D}_{BA}\nabla C_B. \tag{8.13}$$

Fick's first law is useful for solving simple problems in which the concentration is known and the instantaneous flux is required, but most problems in diffusion require solving the time-dependent partial differential equation known as Fick's second law (as modified by Darken), i.e.,

$$\frac{\partial C_B}{\partial t} = \frac{\partial}{\partial x}\left(\widetilde{D}_{BA}\frac{\partial C_B}{\partial x}\right) + \frac{\partial}{\partial y}\left(\widetilde{D}_{BA}\frac{\partial C_B}{\partial y}\right) + \frac{\partial}{\partial z}\left(\widetilde{D}_{BA}\frac{\partial C_B}{\partial z}\right) = \nabla\left(\widetilde{D}_{BA}\nabla C_B\right). \tag{8.14}$$

The $\widetilde{D}_{BA}$ is included inside the second derivative to take into account a possible spatial variation in the diffusion coefficient. If there is no spatial variation, Fick's second law simplifies to

$$\frac{\partial C_B}{\partial t} = \widetilde{D}_{BA}\left(\frac{\partial^2 C_B}{\partial x^2} + \frac{\partial^2 C_B}{\partial y^2} + \frac{\partial^2 C_B}{\partial z^2}\right) = \widetilde{D}_{BA}\nabla^2 C_B. \tag{8.15}$$

Note the similarity between this equation and Fourier's equations for heat flow:

$$\frac{\partial T}{\partial t} = \kappa\left(\frac{\partial^2 T}{\partial x^2} + \frac{\partial^2 T}{\partial y^2} + \frac{\partial^2 T}{\partial z^2}\right) = \kappa\nabla^2 T.$$

Heat flow is also by thermal diffusion in solids and κ in the above equation is the thermal diffusivity. Both coefficients have unit of m^2/s.

8.6.4 Example Problem

The simplest solution to a time-dependent diffusion problem is the case of one-dimensional (1-D) diffusion into a semi-infinite solid. The plane surface of the solid has a constant concentration of solute and we wish to know the concentration profile of the solute at various times. The boundary conditions may be written as

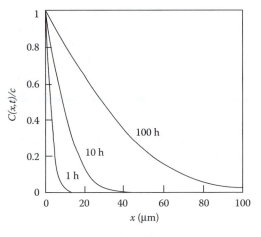

FIGURE 8.12
Diffusion profile at different times for a diffusion coefficient $D = 2.5 \times 10^{-15}$ m^2/s.

$$C = C_0 \quad \text{for } 0 \leq x \leq \infty \text{ and } t = 0$$
$$C = C_s \quad \text{for } x = 0 \text{ and } t \geq 0.$$

The solution is

$$C(x,t) - C_0 = (C_s - C_0)\, \text{erfc}\left(\frac{x}{2\sqrt{Dt}}\right), \tag{8.16}$$

where erfc(z) is the complimentary error function equal to $1 - \text{erf}(z)$. The error function is defined as

$$\text{erf}(z) = \frac{2}{\sqrt{\pi}} \int_0^z e^{-y^2}\, dy \tag{8.17}$$

and is tabulated in most mathematical tables. What is important to remember is that erf$(0) = 0$, erf$(0.5) = 0.5202$, and that erf$(\infty) = 1$. Several diffusion profiles are shown in Figure 8.12 for a diffusion coefficient of 2.5×10^{-15} m^2/s.

A great deal of effort on the part of nineteenth century (and some twentieth century) mathematicians was devoted to solving the above diffusive heat flow equation for every conceivable boundary condition for which a closed form solution was possible. The solutions are compiled in *Conduction of Heat on Solids* (Carslaw and Jaeger 1959). A similar compilation of solutions to the diffusion equation was compiled by Crank. If you find yourself confronted with finding a solution to either of these equations, before you resort to a numerical solution, I would suggest trying to find an analytical solution in one of these two references. Even if the problem for which an analytical solution is known is not exactly the same as the problem you want to solve, you can often find a solved problem in these references that is close enough to serve as a sanity check on your numerical solution. In fact, it is a good idea to check your numerical code by solving one of these solved problems and comparing the solutions. This procedure has been known to turn up unknown flaws in stock codes that were thought to be completely reliable.

8.6.5 Useful Approximation

Note in Equation 8.16 that the solute concentration $(C(x,t) - C_0)$ has increased to approximately $0.5(C_s - C_0)$ when or $x/\sqrt{Dt} = 1$ or $x = \sqrt{Dt}$. This provides a handy means for estimating the diffusion depth. If, for example, you wanted to diffuse As into Si in order to make an *n*-type layer *d* μm thick by heating the Si in the presence of As vapor with concentration C_s. You know the diffusion coefficient of As in Si at the temperature you are operating is 2.5×10^{-15} m^2/s and you want to know how long it would take for the diffusion front ($C \sim 0.5C_s$) to advance 10 μm into the Si. Using the approximation, $t^2 \approx x/D = 3 \times 10^{-6}/2.5 \times 10^{-15} = 1.2 \times 10^{-9}$ s^2 or 9.6 h. This estimate agrees quite well with the actual solution shown in Figure 8.12.

This same approximation is useful for estimating the temperature at depth *d* after time *t* in a material whose surface temperature is T_s. Substituting the thermal diffusivity κ for the chemical diffusion coefficient *D*, the temperature *T* at depth *d* will be $\sim 0.5(T_s - T_0)$ when $t = d^2/\kappa$.

8.6.6 Expansions for the Error Function

Often in diffusion problems, it is convenient to be able to approximate the error function for very small or very large arguments. For such purposes, the following relations are useful:

$$\text{erf}(x) = \frac{2}{\sqrt{\pi}} \left[x - \frac{x^3}{3} + \cdots \right], \quad x \ll 1 \tag{8.18}$$

and

$$\text{erf}(x) = 1 - \frac{e^{-x^2}}{x\sqrt{\pi}} \left[1 - \frac{1}{2x^2} + \cdots \right], \quad x \gg 1. \tag{8.19}$$

8.7 Summary

Defects play a major role in the performance of materials, some wanted and some unwanted. Therefore, it is necessary to understand what they do, how they form, and how to control them. Defect are categorized by their dimensionality: point defects (zero dimensional), line defects (1-D), planar defects (2-D), and volume defects (three dimensional [3-D]).

Point defects include vacancies, impurity or substitutional, and interstitial. Vacancy defects are the result of a thermodynamic equilibrium and are unavoidable. Substitutional and interstitial defects may contain unwanted impurity atoms, or may be engineered to improve the material's properties. If impurity atoms are added that have different valences from the host atoms, the material is nonstoichiometric and vacancies must form to maintain charge neutrality.

Dislocations are line defects that form when a half plane of atoms is inserted (or removed) causing a mismatch in the number of atoms above and below a certain plane. Such defects can be edge, screw, or mixed, depending on how the material is distorted. They are characterized by the Burgers vector, which remains invariant along the dislocation line. The Burgers vector is defined as the missing step in a square loop drawn around the dislocation line. Dislocations allow the material to slip along slip planes and in slip directions that have high atomic density by only breaking bonds along a line rather than along an entire surface and are thus responsible for materials being far weaker than their theoretical strength. Strengthening mechanisms generally focus on preventing the motion of dislocations.

Any surface, whether it be the face of a crystal or a grain boundary in a polycrystalline solid, is considered a defect because the atoms on the surface have unsatisfied bonds, which give them a higher energy that those atoms deep inside the crystal. This interfacial energy plays a major role in the surface properties of solids. Stacking faults and twin boundaries are other forms of planar defects.

Volume defects in the form of a second phase that is purposely added or precipitated from a supersaturated solid solution are often used to improve the mechanical properties. Voids due to the clustering of vacancies or from trapped gases are examples of unwanted 3-D defects.

Diffusion is a process by which different components can move through their host material. Diffusion mechanisms include vacancy exchange diffusion, diffusion along grain boundaries, through interstitials, and along dislocation lines. Diffusion in solids is a very slow process, especially at low temperatures. Being a thermally activated process, the diffusion coefficients follow an Arrhenius law and are generally expressed in the form $D(T) = D_0 e^{-Q/RT}$, where the D_0 and the activation energy Q are determined experimentally. The dimensions of the diffusion coefficient are expressed in m^2/s. The diffusion flux is given by Fick's first law that says the rate of atoms crossing unit area is the product of the diffusion coefficient and the concentration gradient of the diffusing component. Time-dependent diffusion is governed by Fick's second law that is a second-order partial differential equation, similar to Fourier's heat flow equation, which usually requires a numerical solution. However, it is often possible to obtain the information needed from the diffusion length, $d = \sqrt{Dt}$ where d is the distance the diffusion front has moved in time t.

Bibliography

Ashcroft, N.W. and Mermin, N.D., *Solid State Physics*, Brooks Cole, Philadelphia, 1976.

Barsoum, M.W., *Fundamentals of Ceramics*, Institute of Physics Publishing, Bristol/Philadelphia, PA, 2003.

Callister, W.D., *Materials Science and Engineering: An Introduction*, 7th edn., John Wiley & Sons, New York, 2007.

Carslaw, H.S. and Jaeger, J.C., *Conduction of Heat on Solids*, 2nd edn., Oxford University Press, Oxford, 1959.

Crank, J., *The Mathematics of Diffusion*, 2nd edn., Oxford University Press, Oxford, Reprinted in 1976.

Glicksman, M.E., *Diffusion in Solids: Field Theory, Solid-State Principles, and Applications*, John Wiley & Sons, New York, 2000.

Hosford, W.F., *Physical Metallurgy*, Taylor & Francis, CRC Press, Boca Raton, FL, 2005.

Kingery, W.D., Bowen, H.K., and Uhlmann, D.R., *Introduction to Ceramics*, 2nd edn., John Wiley & Sons, 1976.

Kittel, C., *Introduction to Solid State Physics*, 7th edn., John Wiley & Sons, New York, 1966.

Schaffer, J.P., et al., *The Science and Design of Engineering Materials*, Richard D. Irwin, 1995.

Schweizer, M., et al., Defect density characterization of detached-grown germanium crystals, *J. Crystal Growth*, 235, 161–166, 2002.

Problems

1. Activation energy for vacancies in Cu is 1.28 eV. Find the fractional number of vacancies at the melting temperature. Estimate the contribution of vacancies to the thermal expansion at the melting temperature.

2. In close-packed metals, we generally can ignore Frenkel defects. Given this assumption, devise a method for determining the concentration of vacancies as a function of tem-

perature. Is it necessary to be able to ignore Frenkel defects for your scheme to work? Why?

3. Activation energy for diffusion of C into Ni is 146 kJ/mol. By what factor does the steady-state flux of C into Ni increase if the temperature increases from 25°C to 50°C?

4. I want to diffusion-bond a Si chip to a Au-coated substrate and I want the bond to penetrate to roughly 0.1 μm. Au has a diffusion coefficient in Si of 7×10^{-7} m²/s at 1100°C and 10^{-9} m²/s at 800°C. Create a time–temperature plot for this process.

5. Diffusion coefficient of As in Si is 10^{-15} m²/s at 1400°C and is 10^{-18} m²/s at 1000°C. I have a wafer of intrinsic Si that I need to make *n*-type with a carrier concentration of 10^{18} electrons/m³ at a depth of 1 μm (which means a concentration of 10^{18} As at 1 μm). I propose to achieve this by exposing the wafer to As vapor. Devise a practical method for achieving this doping level in terms of the temperature and pressure of the As gas and the exposure time required. (Assume the As can be considered as an ideal gas.)

6. Find the interfacial energy of the (110) face of NaCl. (see Problem 3.5).

9

Mechanical Properties of Materials

In previous chapters, we saw how materials are structured, chemically bonded, and how their elastic properties are related to their structure and their bonds. Now we must deal with how materials actually perform and how we can improve their performance.

9.1 Stress–Strain Relationships

9.1.1 Tensile Test

Perhaps the most basic measurement to characterize a material's mechanical properties is the tensile test. A sample of the material to be tested is prepared according to American Society for Testing and Materials (ASTM) standards as illustrated in Figure 9.1. One end of the sample is clamped and fixed in position while the other end is attached to a crosshead that is moved at constant speed by a mechanical or hydraulic mechanism. A load cell in the pulling device records the force being exerted as the sample deforms while the change in the gage length of the sample is recorded by an extensometer placed on the sample. The measured force is converted to engineering stress (force/area) in the sample based on its original diameter and the displacement is converted to engineering strain ($\Delta L/L$) based on the sample's original gage length. The reader is referred to ASTM Standards E8, E9, and E143 for more details.

A typical engineering stress–strain diagram for a polycrystalline metal is shown in Figure 9.2. Initially the strain is proportional to the applied stress. The Young's modulus is the ratio of stress per strain and is represented by the slope of the line; the stiffer the material, the steeper the slope. Since strain is defined as $\Delta L/L$, it is dimensionless and the Young's modulus must have the dimensions of stress or force per area. In the International system, the Young's modulus is expressed in megapascals (MPa) or gigapascals (GPa) (1 pascal $= 1$ N/m^2). Unfortunately the English system is still widely used in this country and Young's modulus as well as other strength properties are expressed in pounds per square inch (1 MPa $= 145$ lb/in.^{2} or psi).

Deformations along the linear portion of the stress–strain curve are elastic and involve stretching of the bonds. If the strain is less than the elastic limit, e_{elastic} (usually expressed in terms of percent elongation), the unloading curve will be along the loading curve. In other words, the specimen will return to its original configuration when stress is relieved. The stress at the elastic limit is called the yield strength. Generally materials are chosen for specific applications to operate at some fraction of their yield stress to provide a margin of safety. The area under the curve bounded by the elastic limit represents the maximum

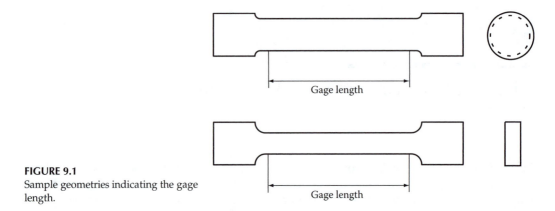

FIGURE 9.1
Sample geometries indicating the gage length.

amount of elastic energy the material can store and is termed the resiliency. The resiliency can be expressed as

$$\frac{\sigma_{\text{yield}} \, e_{\text{elastic}}}{2} = \frac{Ee_{\text{elastic}}^2}{2} = \frac{\sigma_{\text{yield}}^2}{2E},$$ (9.1)

where E is the Young's modulus. Materials with high resiliencies are used for springs and other applications where elastic energy storage is required.

The departure from the linear relationship between stress and strain indicates an irreversible change in the structure. Bonds are now being broken as individual grains undergo slip by motion of dislocations as their resolved stress exceeds their critically resolved shear stress. Individual grains begin to move past each other by grain boundary slip. Once a material is stressed beyond its yield point, it will suffer a permanent deformation when stress is relieved, as is indicated by the dashed unloading line.

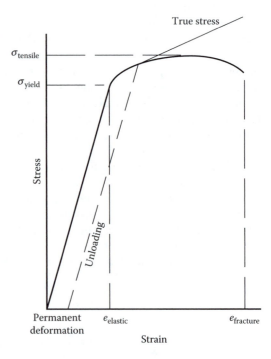

FIGURE 9.2
Typical stress–strain curve for a ductile material.

9.1.2 Tensile Strength

The peak of the stress–strain curve is the tensile strength or fracture strength of the material. The downturn in the curve around this point is a result of the specimen necking down in the middle as it yields. Since the load cell is recording only force and the stress is inferred assuming a constant area of the specimen, the actual stress being applied is actually increasing, first because of the reduced cross-sectional area as the sample is being stretched according to the Poisson's ratio before the neck begins to form, and next as the cross section decreases during the necking process. As the material begins to neck, it is actually getting stronger because of work hardening as indicated by the dashed line labeled true stress in Figure 9.2.

Brittle materials fail at or near the yield point without forming a neck. A crack opens and propagates rapidly across the specimen, leaving a smooth surface. A highly ductile viscoelastic material will continue to neck down to a point and simply pinch off when it fractures. Ductile metals will form a neck but voids will begin to open in the interior of the neck. The voids will coalesce and eventually form an elliptical crack which then propagates across the area of the neck. Finally the material near the surface of the neck fails by shearing at a 45° angle relative to the applied stress, forming a cone and cup fracture. The reduction in area (ratio of area at the point of fracture to the original area) is used to compute the true stress at the point of fracture.

The area under the curve bounded by the entire stress–strain curve to the fracture elongation is a measure of toughness of the material. The ductility of a material is defined as the percent elongation at the fracture point. Brittle materials fracture at or near their yield strength. Ductile materials bend before they break and are generally tougher. A really tough material has a large Young's modulus, high yield strength, and an extended elongation before fracture.

The yield strength, ductility, and tensile strength can be altered by various strengthening mechanisms, but, as was shown in Chapter 7, the Young's modulus is a function only of the bond energy and is not affected by work hardening, dispersion hardening, or grain refinement. The Young's modulus of an alloy will, of course, be dependent on the bond energies of the alloy's components. The dramatic effect of the various hardening mechanisms to be discussed in the following sections can be seen in Table 9.1 (see Chapter 14 for descriptions of the alloys listed in Table 9.1).

The modest change in the Young's modulus between the pure material and the alloys is quite evident. In the case of the steels, the lowest carbon steel (since iron must contain a minimal amount of carbon to be considered a steel) is used as a basis. The enormous

TABLE 9.1

Strength and Ductility of Selected Alloys

Alloy	E (GPa)	σ_{yield} (MPa)	$\sigma_{tensile}$ (MPa)	e_{yield} (%)	$e_{fracture}$ (%)
1100-0 Al	69	34	90	0.04	40
7075-T6 Al	71	504	572	0.7	11
Ti (pure annealed)	103	170	240	0.16	30
Ti6Al4V	114	760	900	0.65	14
Ti6Al4V (age hardened)	114	1103	1172	0.9	10
1020 low carbon steel	207	295	395	0.14	36.5
17-7PH SS	204	1310	1450	0.64	3.5
440A SS (tempered)	200	1650	1790	0.8	5
VIT-001	93	1863	1863	2.0	2.0

Source: From Callister, W.D. *Materials Science and Engineering: An Introduction*, 7th edn., John Wiley & Sons, New York, 2007.

increase in both yield strength and tensile strength by the various hardening techniques is quite apparent, but note that this increase comes at the expense of ductility (the strain at fracture). The elastic limit is obviously increased along with the yield strength. The VIT-001 is a bulk metallic glass and will be discussed in Section 9.3.6.

9.1.3 True Stress and True Strain

Corrections to the engineering stress and strain (what is seen on the stress–strain diagram) can be made to obtain true stress and strain. True stress is simply $\sigma_t = F/A$, where A is the cross-sectional area at the neck of the sample.

Since the testing machine is measuring the displacement ΔL as the material is being stretched, the true stain is defined as

$$e_t = \int_{L_0}^{L_f} \frac{dL}{L} = \ln\left(\frac{L_f}{L_0}\right) = \ln\left(1 + \frac{\Delta L}{L_0}\right) = \ln(1 + e) \approx e. \tag{9.2}$$

After the material begins to yield, the true stress can be empirically related to true strain by

$$\sigma_t = K e_t^n = K[\ln(1 + e)]^n, \tag{9.3}$$

where
  K is a constant
  n is the strain-hardening exponent ($n < 1$)

However, for most applications, it is a common practice to simply use the engineering stress in evaluating materials.

9.1.4 Bend Test

For materials that are not suited for standard tensile testing, an alternative method for assessing the elastic modulus as well as yield and fracture stress is the three-point bend test. The sample in the form of a small bar d cm wide and b cm thick is supported on one side at two points spaced L cm apart while a force is applied at the midpoint on the opposite side. The elastic modulus as well as the yield and fracture stress can be calculated from the force vs. bending data along with the geometry of the system. The flexure strength is given by

$$\sigma_f = \frac{3F_f}{2bd^2}. \tag{9.4}$$

The procedure is outlined in ASTM Standard C1161.

Bend tests are useful for very hard ceramics that cannot be machined to the shapes required for use in a standard tensile testing machine. It has also been applied to measure the elastic modulus and strength of material that is too soft and fragile to be clamped in the testing machine. The elastic modulus and strength of very delicate protein crystals, only a millimeter or so in length, have been measured from which their bond energy was determined.

9.1.5 Hardness Testing

Hardness is measured by applying a given force to an indenter, usually in the form of a small sphere or diamond point, and measuring the depth or diameter of the resulting dent.

Arbitrary hardness scales have been established that allow some correlation between hardness and yield strength. There are a variety of different methods including Brinell, Rockwell, Vickers, and Knoop, each with its own hardness scale. The reader is referred to ASTM Standards E10, E18, E92, and E384 for details. The primary value of hardness testing is that it provides a nondestructive method for assessing the strength of a material and is useful for certifying that the actual materials used for specific components in a system meet the required material specifications.

9.2 Relationship between Lattice Type and Ductility

9.2.1 Slip Systems in Metals

The ductility of a metal depends on the number and type of slip systems. The slip systems for the various lattice types are shown in Table 9.2. Note that in the {111} family there are four different planes (111), ($\bar{1}$11), (1$\bar{1}$1), and (11$\bar{1}$). (Other permutations of the indices are planes parallel to these four distinct planes.) The slip directions for the (111) must lie in the plane, hence their dot product with the vector normal to the plane must be zero. Thus the three slip directions in the (111) plane will be [$\bar{1}$10], [$\bar{1}$01], and [0$\bar{1}$1]. (Remember this only works only for cubic systems.) Again, other permutations of these indices are directions parallel to these three directions.

Even though the bcc system has many more slip systems than the fcc, fcc metals tend to be more ductile than bcc metals because their slip planes have a larger area density fraction. These could be considered as primary slip systems. Recall from Chapter 4 that the area density fraction of the fcc (111) plane was 0.907 while the area density fraction of the bcc (110) plane was only 0.833. The higher index bcc slip systems have even smaller area densities, as may be seen in Figure 9.3 and can be considered as secondary slip systems. The lower area density slip systems in the bcc structure do not operate as effectively at low temperatures, which can cause a ductile-to-brittle transition as the temperature is lowered.

The basal plane in the hcp system is the only high-density plane in that system. The other planes are usually not active at room temperature and are not even mentioned in some texts. Dislocations that become blocked from moving in one plane can move in an intersecting plane, a process known as cross slip. But since the primary hcp slip planes are parallel to each other, cross slip cannot occur in hcp structures. Consequently, hcp metals tend to be more brittle than fcc or bcc metals.

9.2.2 Slip Systems in Ionically Bonded Ceramics

The slip systems in ionically bonded ceramic systems are generally much less effective than those in metallic systems. To begin with, it is much more costly to create a dislocation in an

TABLE 9.2

Slip Systems

Crystal System	Slip Planes	Slip Directions	Number of Slip Systems
fcc ($Fm\bar{3}m$)	{111}	$\langle\bar{1}10\rangle$	$4 \times 3 = 12$
bcc ($Im\bar{3}m$)	{110}	$\langle\bar{1}11\rangle$	$6 \times 2 = 12$
	{211}	$\langle\bar{1}11\rangle$	$12 \times 1 = 12$
	{321}	$\langle\bar{1}11\rangle$	$24 \times 1 = 24$
hcp ($P6_3/mmc$)	{0001}	$\langle11\bar{2}0\rangle$	$1 \times 3 = 3$
	{10$\bar{1}$0}	$\langle11\bar{2}0\rangle$	$3 \times 1 = 3$
	{10$\bar{1}$1}	$\langle11\bar{2}0\rangle$	$6 \times 1 = 6$

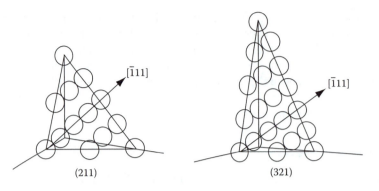

FIGURE 9.3
Higher index bcc slip systems.

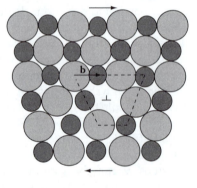

FIGURE 9.4
Edge dislocation in the (100) face of a rock salt structure. The Burger's circuit is shown by dashed line and the Burger's vector **b** has a magnitude of $(r_1 + r_2)$ in the [010] direction, which makes (001) the slip plane. Dislocations cannot propagate to the right because they are blocked by the strong coulomb repulsive force from the large anions that must move past each other.

ionically bonded system because a dislocation cannot form with the insertion of a single extra half plane of atoms. This would result in like atoms touching like atoms. Dislocations can form in ionic crystals by inserting two complimentary half planes of atoms as shown in Figure 9.4. Of course this will be more costly in energy than forming dislocations in metals because of the increased lattice strain. In addition, it is more difficult for dislocations to move in an ionic system. Consider the (100) plane of the rock salt structure as shown in Figure 9.4. The large anion is blocked from moving to the left and filling the vacancy left by the dislocation as it comes into contact with another anion by the Coulomb repulsion. In the [110] plane, it is possible for an ion to move without direct contact with other ions of the same sign, but this is not a high-density plane. The extra energy required to form dislocations in ionic systems, together with the inhibited motion of dislocations due to ionic repulsion, accounts for the hardness and brittleness of ionic compounds.

9.3 Strengthening Mechanisms

The strength of a material is related to its defect structure as well as to its bond energy. As stated previously, dislocations can begin to move through the grains when the resolved shear stress reaches or exceeds the critical resolved shear stress. The primary strengthening mechanism then involves inhibiting the motion of dislocations. This can be accomplished in a variety of ways as will be shown.

9.3.1 Work Hardening

It is difficult for a dislocation to move past another dislocation, so one strengthening mechanism involves simply creating more dislocations. As the dislocations pile up on one another, it becomes very difficult for additional dislocations to move through the material and metal becomes much harder. One way to create more dislocations is to work harden the material by flexing, hammering, or cold rolling it.

A dramatic example of work hardening can be demonstrated by taking a piece of soft copper wire, such as a short length of #8 grounding wire. Bend it sharply in half and try to straighten it. Continued flexing will cause the wire to become harder and brittle until it eventually breaks from fatigue.

9.3.2 Grain Refining

Since dislocations cannot move through grain boundaries, another strengthening mechanism is the creation of finer grains, a process known as grain refining. The yield strength of a material is empirically related to the average grain size d by

$$\sigma_{\text{yield}} = \sigma_0 + k/\sqrt{d}, \tag{9.5}$$

which is known as the Hall–Petch relationship. The k is a constant for a given material. Methods for controlling the grain structure will be discussed in Chapter 14. Work hardening and grain refining strengthening mechanisms are effective in elemental systems as well as in alloys.

9.3.3 Solid Solution Hardening

Alloying the metals with other components offers other strengthening mechanisms. Adding a second metal as substitutional atoms in the lattice is known as solid solution hardening. The size difference between the solvent (host) atoms and the solute (impurity) atoms strains the lattice and makes it difficult for dislocations to move. Adding smaller atoms that can go into the interstices produces a similar hardening in the lattice. This is the role that carbon plays in the strengthening of steel.

9.3.4 Precipitation Hardening

Second phase particles can be made to precipitate from a supersaturated solute. By carefully controlling the time and temperature, these precipitates can ripen by solid-state diffusion until they reach the optimum size in which their lattice nearly matches the host lattice, but their mismatch puts enough strain in the lattice to block the motion of dislocations. This process is called precipitation hardening or age hardening because the size of the precipitate is controlled by the time the alloy is held at a temperature the precipitate can grow by solid-state diffusion. The methods for forming such precipitates will be discussed in Chapter 14.

9.3.5 Dispersion Hardening

Adding inert second phase particles results in dispersion hardening. Submicron oxide or carbide particles act as pinning sites that block the motion of dislocations. Intermetallic phases can be made to form, which not only block the motion of dislocations, but also help to stabilize grain boundaries. The hardening effect of a dispersed phase is not as effective as a coherent precipitated phase, but dispersion hardening can be used in systems that do not lend themselves to precipitation hardening. It also has the advantage of remaining effective throughout long exposures to elevated temperatures without the overaging problem encountered in precipitation-hardened systems.

9.3.6 Amorphous Structure

Slip systems can be eliminated by taking away the long-range order of the crystalline lattice. Glasses are inherently stronger than metals for this reason, but since there is no mechanism for them to deform plastically, they are quite brittle. Recent developments have made it possible to form bulk metallic glasses with an amorphous microstructure that have exceptional strength and resiliency (see Chapter 14).

9.4 Creep

Creep is a thermally activated phenomenon in which a material slowly becomes elongated in response to applied stress. The onset of creep usually begins at around 30% of the absolute melting temperature and can become significant above 50% of the absolute melting point. Naturally, this phenomenon becomes of great concern in the design of rotating machinery such as the turbine blades in a gas turbine or in a jet engine that must operate at high temperatures.

There are three stages of creep. In the initial or primary stage, creep is fairly rapid but diminishes with time as the material strain hardens. During the secondary stage, creep continues at a constant rate as the rate of strain hardening is balanced by the rate of thermal softening or recovery (to be discussed in Chapter 11). In the final stage, the creep velocity accelerates rapidly as gross defects begin to form just before rupture.

There are several mechanisms that give rise to creep. At high temperatures there are more vacancies as well as enhanced diffusion which allows the vacancies to migrate. In the Nabarro-Herring creep mechanism, atoms migrate in the direction of applied stress while vacancies migrate to the lateral surfaces, causing a net elongation in the stress direction. Coble creep is similar except the vacancies migrate along grain boundaries. Another creep mechanism is dislocation climb. Atoms in the extra half plane of a dislocation migrate into vacancies, which moves the dislocation line in the direction of the applied stress (Figure 9.5). Other creep mechanisms include grain boundary sliding in response to the resolved shear stress.

Ceramics are much more resistant to creep than are metals because of the difficulties in forming vacancies and dislocations as well as to the lack of mobility of vacancies and

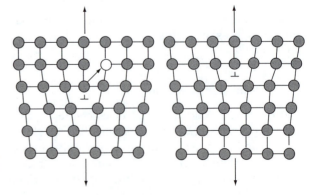

FIGURE 9.5
Mechanism of dislocation climb. Atoms from dislocation plane migrate to vacancies making the specimen longer and thinner.

dislocations due to the charges on the atoms. However, creep can become significant even in ceramics at very high temperatures through the diffusional mechanisms as well as through grain boundary slippage.

Some ceramics may have a glass phase connecting the individual crystallites. Above the glass transition temperature, glasses deform by viscous flow that increases exponentially with temperature. Therefore, these ceramics as well as metallic glasses and polymers can creep by viscous flow.

Also, it should be noted that creep is not just a high temperature phenomena found in high performance gas turbine engines, but can occur at ambient temperatures in polymers and solders.

9.5 Fracture Mechanics

In the nineteenth century, designers of bridges, ships, locomotives, and other machinery did attempt to calculate the distribution of loads on their structures in a rather general way, but paid little attention to the detailed geometry of the structure, particularly the effects of windows and doors or holes on the stress patterns. They compensated by applying large factors of safety or more precisely, factors of ignorance, in their designs. Despite factors of safety typically as large as 5 or 6 (and even up to 18 in locomotives), bridges fell, ships broke apart, machinery failed, and wheels and axels broke on locomotives with alarming regularity, often with catastrophic results and loss of life. Even though the fundamental theory of fracture mechanics was developed by A.A. Griffith in 1920, his work was largely ignored until the third Comet passenger jet mysteriously disintegrated in midair in 1954.

9.5.1 Stress Concentration

When a specimen is stressed, one can think of continuous lines of force running through the material; the stronger the force, the more the lines. If the specimen has constant cross section, the stress, which can be represented as the number of line of force per area, will be constant throughout the specimen. If the specimen narrows, the lines of force become more concentrated and the stress increases in that region. If there is a hole in the specimen, the lines will tend to flow around the hole and become more concentrated in its vicinity. These lines become even more concentrated in the vicinity of a sharp discontinuity such as a square hole or a crack. (This is why portholes and hatches in the bulkheads of ships are round or oval. The windows in the Comet aircraft were square.)

In 1898, G. Kirsch showed that stress concentration in the vicinity of a round hole increased by a factor of 3. G.V. Kolosoff solved the problem of stress concentration in the vicinity of an elliptical hole in 1910, but his paper was published in Russian and was virtually unknown in the West. In the West, the seminal paper on the subject was published by C.E. Inglis in 1913 in which he showed that the stress concentration in the vicinity of the tip of an elliptic hole was given by

$$K = 1 + 2\sqrt{\frac{L}{r}}, \qquad (9.6)$$

where
  2*L* is the length of the hole
  r is the tip radius.

One can appreciate the effectiveness of the stress concentration in a crack whose tip radius is on the order of an atomic dimension. No doubt the reader is familiar with how much easier it is to tear open a plastic packet if there is a small indentation or crack to start the tear. It is also a common practice to drill a small hole at the end of a crack in a metal plate to "blunt the crack" by increasing the tip radius to relieve the stress concentration.

One can make a simple model for fracture by estimating the stress at the tip of the crack by multiplying Equation 9.6 by the applied stress and equating this to the theoretical tensile strength, estimated to be $E/10$,

$$\frac{E}{10} = \sigma_{\text{applied}}\left(1 + 2\sqrt{\frac{L}{r}}\right) \approx 2\sigma_{\text{applied}}\sqrt{\frac{L}{r}} \tag{9.7}$$

or

$$\sigma_{\text{fail}} \approx \frac{E}{20}\sqrt{\frac{r}{L}}. \tag{9.8}$$

However, the Inglis stress concentration by itself may not be sufficient to cause a crack to self-propagate even though the applied stress is greater than σ_{fail} in Equation 9.8 because energy is required to do the work of fracture, i.e., the energy to break bonds and create new surfaces. Since the applied stress is doing no work on the system, this energy must come from somewhere else if the crack is to continue to propagate.

9.5.2 Griffith's Theory of Brittle Fracture

Griffith proposed that the theoretical tensile strength of a brittle material should be obtained by equating the work required to separate two atomic planes to the surface energy required to make two new surfaces. We have shown previously that the energy per volume of a stressed brittle solid is $\sigma^2/2E$. If the specimen has cross-sectional area A and must be deformed by distance x in order to separate a plane on atoms, the work required is $\sim\sigma^2 Ax/2E$. The energy required to make two new surfaces is $2A\gamma$, where γ is the interfacial energy. Equating the two yields

$$\sigma_{\text{theoretical}} = 2\sqrt{\frac{\gamma E}{x}}. \tag{9.9}$$

Correcting for the decrease in the force between the planes of atoms as they are separated and integrating over x yields

$$\sigma_{\text{theoretical}} = \sqrt{\frac{\gamma E}{x}}. \tag{9.10}$$

Griffith wanted to test his theory against the actual strength of materials. However, the γ was not known and he did not have a way to measure it directly for metals. So Griffith turned to glasses. The measurement of the surface tension of a liquid is straightforward. Since a glass is essentially a frozen-in liquid, Griffith reasoned that he could extrapolate the measured surface tension of molten glass to the room temperature value, which he found to be 0.56 N/m. A simple tensile test determined the Young's modulus of his glass samples to be 62 GPa. Taking the distance of separation to be 1 nm, the theoretical tensile strength was estimated to be 13 GPa. (Note that this is the same order of magnitude as estimated in Chapter 7 as the stress required to break bonds by pulling planes of atoms apart.)

Beginning with 1 mm diameter glass rods, Griffith found they broke at 0.17 GPa, almost two orders of magnitude below the theoretical estimate. However, when he tested smaller diameter rods, he found them to be progressively stronger as the rods decreased in diameter. A rod of 0.0025 mm, the smallest he could make and measure accurately, exhibited a tensile strength of 3.4 GPa and extrapolation suggested that the theoretical strength would be attained at vanishingly small diameters.

Why should strength increase with decreasing diameters? Griffith postulated that the fracture of the brittle rods was initiated by microscopic cracks in their surface. The strain energy per unit volume, we recall, is $\sigma^2/2E$. The strain energy that is released in the vicinity of a crack with length L in a material with thickness d is approximately $\pi d L^2 \sigma^2/2E$. The energy required to make new surfaces is $2Ld\gamma$. The energy of the system under stress σ is

$$U = V\frac{\sigma^2}{2E} - \frac{\pi d L^2 \sigma^2}{2E} + 2Ld\gamma, \tag{9.11}$$

where V is the volume. Initially the energy increases with increasing L as energy is required to make new crack surfaces. But eventually the energy released in the vicinity of the crack more than compensates for that required to create the new surface. The maximum energy is found by differentiating with respect to L and is

$$L_C = \frac{2\gamma E}{\pi\sigma^2}. \tag{9.12}$$

Cracks shorter than the critical L_C cost more surface energy than they release in order to propagate and are therefore stable. Those longer than L_C release more strain energy than it costs to make their surfaces and will spontaneously propagate. The smaller the diameter of the glass rods used by Griffith, the smaller the crack they could contain and therefore they could withstand higher tensile stresses before whatever cracks they may have had would become critical.

You may have noticed that glass cutters scribe the surface of a glass plate they wish to cut with a diamond point and then apply a thin coating of kerosene or light oil that wets the surface. The scratch serves as the stress concentrator and crack initiator and the wetting agent reduces the interfacial energy making it easier for the crack to propagate through the glass.

9.5.3 Orowan–Griffith Theory

Griffith's work applied only to materials that fail by brittle fracture because he did not consider any plastic deformation that would occur before the fracture of ductile materials. Orowan extended the work of Griffith to ductile materials by introducing an effective surface energy which contained both the energy required to make new surface and the plastic deformation during the fracture of ductile material. This new material property became known as the work of fracture G or toughness. The modified form of Griffith's Equation 9.12 becomes

$$L_C = \frac{GE}{\pi\sigma^2}. \tag{9.13}$$

For brittle materials, the work of fracture G reduces to 2γ.

9.5.4 Fracture Toughness

During the late 1950s, George Irwin improved on the Griffith–Orowon model by introducing a stress intensity parameter, K, for predicting failure in structures regardless of their size and shape or crack geometry. The definition of this parameter varies with the crack geometry and the geometry of the stress, but for a horizontal crack length, L, in a thin sheet similar to the previous geometry described by Equation 9.13, $K_1 = \sigma\sqrt{\pi L}$. The K_1-values for other geometries are available in engineering books. For each material, a material toughness parameter K_{1C} is determined such that fractures will occur when K_1 equals or exceeds K_{1C}. Rewriting Equation 9.13, failure occurs when

$$\sigma\sqrt{\pi L_C} = \sqrt{GE}, \tag{9.14}$$

from which we identify the fracture toughness as $K_{1C} = \sqrt{GE}$. Notice that fracture toughness has the rather peculiar units of Pa $m^{1/2}$.

There are a number of ways to measure fracture toughness such as loading specimens with carefully prepared flaws. Impact testing is a widely used method for comparative testing. A notched sample is prepared according to ASTM standards. In the Charpy impact test, the sample is held at the ends while a heavy pendulum with a knife-edge is allowed to impact the sample on the side opposite the notch. The work of fracture is obtained by taking the difference in height of the pendulum before and after the impact. Figure 9.6 shows the dramatic change in work of fracture in different steels as a function of temperature obtained from impact testing described herein. Such tests clearly demonstrate the ductile–brittle transition as the available slip systems in the bcc system become inactive as the temperature is lowered.

The most important thing to remember about Griffith cracks is that the critical length is absolute; that is, the critical crack length does not scale with the size of the structure. This fact can have dramatic consequences in the scale-up of designs. It is not enough to simply design a structure to accept stresses equal to the tensile strength divided by some safety factor; one must also consider the critical crack length which is the same for the main wing spar of a 747 as well as for a small component. Rolls Royce learned this lesson the hard way when they developed the RB211 jet engine with carbon fan blades. The work of fracture of these large blades was not sufficient to withstand the impact of small birds or other debris and the engine had to be redesigned with metal blades. This not only resulted in loss of

FIGURE 9.6
Impact testing data for pearlitic steels as a function of temperature for various wt% carbon content. High carbon steels are stronger but become brittle at ambient temperatures, whereas low carbon steels are ductile until −50°C, when they suddenly become brittle. (From Reinbolt, J.A. and Harris, W.J., *Trans. ASM*, 43, 1951. Reprinted with permission of ASM International. All rights reserved.)

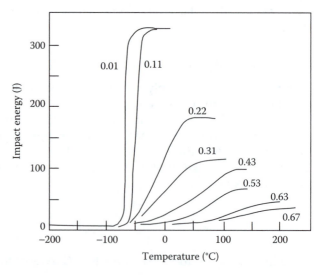

performance of the engine, but caused also Rolls Royce to declare bankruptcy in 1971 because they could not meet their delivery schedule. This delay also seriously affected Lockheed who was waiting for these Rolls Royce engines for their new L1011 passenger jet. The delay in getting the L1011 into service costs Lockheed considerable market share and was one of the factors that caused them to abandon their commercial passenger business. (The above discussion on Griffith cracks was taken from Gordon (1988).)

The best example of a self-propagating crack is a balloon when punctured by a pin. Instead of allowing the air to be slowly released through the pinhole, the high stress in the skin results in a very small critical crack length. Similarly, designers of pressurized containers, including the monocoque design of modern passenger jets that utilizes the stressed skin for rigidity, need to be more concerned with critical crack length than just yield strength. If one wants to design such a vessel that will leak before bursting, the stresses in the skin should be limited to a critical stress intensity factor in which the critical crack length is equals to or greater than the skin of the vessel.

9.5.5 Ductile-to-Brittle Transformation

It was mentioned in Chapter 4 that the lower atomic density slip systems in bcc materials become inactive at low temperatures and a ductile-to-brittle transition occurs. From Figure 9.6 it is seen that for some high-carbon steels, this transition occurs well above the freezing point of water. A number of the early Liberty ships that operated in the North Atlantic during World War II were apparently made from the wrong kind of steel and broke apart without warning.

9.5.6 Toughening Methods

Glass fibers are easily abraded and are often clad with a thin layer of a polymer or other protective surface to prevent crack formation in fiberglass or other applications that use glass rods or fibers for strengthen components.

The safety of plate glass is often increased by tempering, a process by which the outer surfaces are cooled by blasts of air while the inner portion is still soft. As the inner regions cool and contract, the outer surfaces, which have already cooled and contracted somewhat, are put into compression. This compression tends to seal any small surface crack to keep it from propagating.

Ceramics may be toughened in a similar manner by a technique known as transformation toughening. Zirconia (ZrO_2) is stable in the monoclinic phase under ambient conditions. It can be stabilized in a more dense metastable cubic or tetragonal form by the addition of calcia. Particles of partially stabilized tetragonal zirconia are added to the host matrix, usually Al_2O_3. An advancing crack will cause the tetragonal zirconia to transform to the monoclinic form and the ensuing volume increase seals the crack.

Transition induced plasticity (TRIP) steels are some of the newest and most exciting developments in steel making. TRIP steels retain some of the fcc austenite in addition to the bcc ferrites and banites normally found in steels. This retained austenitic phase makes the steel more ductile at a given strength than other steels and the transformation of the retained austenite to martensite upon plastic deformation gives it a very high work hardening rate (see Chapter 14 for more details).

Adding planes of weakness perpendicular to the direction of the propagation of the crack can act as crack stoppers. Wood has such planes in their grain structure, which tend to resist radial crack propagation

Clearly if the size of the structure is less than the critical crack length, cracks cannot propagate. Therefore, smaller structures have a major advantage because they can design for a smaller critical crack length, which allows the maximum applied stress σ in Equation 9.14 to be larger. Subdividing structures can significantly increase their resistance to tensile failure. A stranded cable is an excellent example. Smaller individual strands can sustain higher stress and remain stable than a single thicker strand because of their smaller critical crack length. In addition, the stranded cable has the advantage of redundancy: several smaller strands can break while the rest carry the load if there is no mechanism for transferring the release energy from one strand to another. This same principle carries over to structures with plates that are welded together vs. plates that riveted or bolted together. The tendency for modern aircraft to use a monocoque design in which the stresses are carried predominantly by the skin improves their strength-to-weight ratio, but makes them more susceptible to catastrophic failure from a crack exceeding its critical length. Presumably this is what happened to the Comet passenger jets that disintegrated in midair.

9.5.7 Fatigue

Fatigue is a primary cause of structure failure. Perhaps the most spectacular failure was the sudden loss of the skin over the forward passenger cabin of a Boeing 737–200 used on Aloha Airlines Flight 243 in 1988. The only casualty was a flight attendant who was blown overboard. Miraculously, the pilot was able to land the plane safely with the passengers strapped in their seats. The investigation found that metal fatigue, exacerbated by stress corrosion, was the reason for the failure. Similar incidents have occurred on other aircraft, trains, and structures.

Fatigue can be thought of as a gradual wearing out of a material due to cumulative internal structural damage brought about by repeated stresses and strains such that the material eventually fails at a stress that is well below the normal tensile strength of the material. Repeated bending of a tab on a drink can or paper clip will cause a fracture at a small fraction of the material's tensile strength. Why? The first bending generates dislocations that intersect one another and work hardens the material. (Recall the copper wire bending experiment.) Repeated bending creates more dislocations that continue to pile up making the material more brittle because the dislocations can no longer move. Eventually, the lattice becomes so disordered that cracks begin to form and propagate, resulting in brittle fracture.

Metals are tested for fatigue by repeated loading or flexing at a given stress level until they fail. The test is then repeated at different stress levels until enough points are generated to construct what is termed an S–N (stress vs. number of cycles to failure) curve illustrated schematically in Figure 9.7. Since at low stress, as many as 10^7 cycles may be required for a single datum point, it can be appreciated that fatigue tests can be time-consuming and

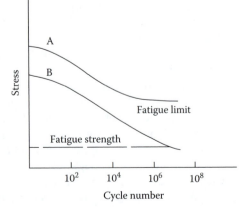

FIGURE 9.7
Typical S–N fatigue curve. Material A exhibits a fatigue limit and can withstand an unlimited number of cycles if the stress is held below this level. Material B does not have a fatigue limit. Its fatigue strength is defined as the average stress that results in failure after 10^7 cycles.

expensive. To make things worse, there will be significant statistical scatter in the measured fatigue life, the average number of cycles before failure at a given stress. This means that enough tests must be run to determine the average number of cycles to failure with some level of confidence. Stresses that cause some plastic deformation cause failure at much fewer cycles $O(10^4–10^5)$ and are termed low cycle fatigue. If the deformations are purely elastic, the fatigue life is greatly extended in what is termed high cycle fatigue.

Notice that curve A approaches a lower stress limit called the fatigue limit or endurance limit below which fatigue failure does not occur, where in curve B the fatigue stress continues to decrease with increasing cycles. In this case the fatigue strength is defined as that strength for which failure occurs at 10^7 cycles. Steels tend to have a fatigue limit, which favors their use for applications such as train wheels, axels, crankshafts, connecting rods, and other components of reciprocating machines. Aluminum, copper, and other fcc metals do not exhibit an endurance limit.

Surface treatments to prevent surface cracks forming in steels include case hardening, which consists of carburizing or nitriding in which carbon or nitrogen is diffused into the interstities of the bcc lattice to put the surface into compression. Another surface treatment used on connecting rods and other components of high performance engines is shot peening. Small diameter shot (0.1–1 mm diameter) is impinged at high velocity on the component. The small dents from the impacts work harden the surface and seal incipient surface cracks. Both of these techniques raise the S–N curve as well as the fatigue limit.

Ceramics are not generally subject to fatigue since they do not deform plastically because dislocation motion is inhibited.

9.6 Mechanical Properties of Polymers

The mechanical behavior of polymers is quite different from metals and ceramics and depends greatly on their structure and operating temperature. Below their glass transition temperature, T_g (the temperature at which their covalent bonded chains can no longer move relative to one another), they are quite brittle and exhibit glass-like behavior. Above their T_g, they behave plastically. In this sense, they are similar to metals that exhibit ductile-to-brittle transitions, but for entirely different reasons.

9.6.1 Semicrystalline Polymers

Above their T_g, semicrystalline polymers at low strains have a metal-like behavior as may be seen in Figure 9.8. The linear portion of their stress–strain diagram results from

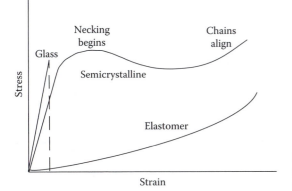

FIGURE 9.8
Schematic stress vs. strain behavior of different classes of polymers.

stretching or bending of covalent bonds in their crystalline chains and reversible displacements of the molecules in their amorphous regions. At the onset of plastic deformation, the covalent chains in the amorphous regions start to move and align themselves along the direction of the applied stress and the specimen begins to form a neck. Next the crystallized regions begin to align themselves in the direction of the stress and separate into segments of crystal linked by the covalent strands. At this point the material actually becomes stronger as all of the stress is being borne by aligned covalent chains. This can be demonstrated by taking the plastic (polyethylene) band from a six pack of drinks and stretching it. Observe how the material forms a neck of aligned fibers as it stretches and how much stronger the aligned material in the neck is compared to the original material.

9.6.2 Elastomers

The class of polymers known as elastomers do not crystallize. Instead their randomly coiled covalent chains are loosely cross-linked. Initially they are easily deformed, but as their chains uncoil and become aligned, their stiffness increases giving their stress–strain behavior a sort of "lazy-J" appearance as shown in the sketch in Figure 9.8.

Stretching an elastomer increases the order in the system as the strands become aligned. When the stress is relieved, the second law of thermodynamics requires the entropy to increase, causing the covalently bonded chains to form random coils again and, guided by the loose cross-links, the elastomer returns to its original shape.

9.6.3 Viscoelastic Behavior

A crystalline material has a distinct melting point where there is a distinct entropy change in going from a disordered melt to a more ordered solid. Since an amorphous material remains disordered as it is cooled, it has no distinct melting point. Instead, its viscosity continues to increase until molecular motion ceases and it becomes a glass. Likewise, an amorphous polymer goes from a viscous liquid to a viscoelastic solid to an elastic brittle glass as it is cooled from its melt.

An elastic solid responds to stress like a spring. It responds instantly by stretching in proportion to the applied stress, and recovers completely when the stress is removed. A viscous liquid, on the other hand, responds like a dashpot or shock absorber. It deforms with a velocity that is proportional to the stress and does not recover when the force is removed. A viscoelastic material combines the two behaviors and can be modeled as a spring and dashpot in series or parallel as shown in Figure 9.9. Newtonian fluids have pure

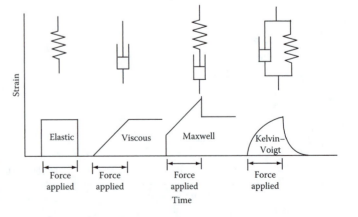

FIGURE 9.9
Time response of different rheological systems to applied forces. The Maxwell model gives steady creep with some post stress recovery, representative of a polymer with no cross-linking. The Kelvin–Voigt model gives a retarded viscoelastic behavior expected from a cross-linked polymer.

viscous behavior. Many polymer melts are non-Newtonian and have to be modeled by various combinations of springs and dashpots.

In the Maxwell model of the viscoelastic system, initially the entire force is transmitted through the dashpot causing the spring to stretch in proportion to the force. As the dashpot begins to move, the force is divided between that required to move the fluid and to stretch the spring allowing the spring to begin to relax. The partial relaxing of the spring accounts for the bending of the response curve and limited recovery when the force is removed. In the Kelvin–Voigt model, the spring response is damped by shock absorber as in an automobile suspension, which results in a delayed elastic response. Adding additional elements gives more variables to fit the behavior of any type of viscoelastic material.

9.6.4 Viscoelastic Relaxation Modulus

The time dependency of the stress–strain of viscoelastic materials presents a difficulty in testing this class of materials. In order to develop a meaningful test, a time-dependent relaxation modulus $E_R(t)$ introduced is defined as

$$E_R(t) = \frac{\sigma(t)}{\varepsilon_0}.$$
(9.15)

A specimen is rapidly strained by an amount ε_0 and then the stress $\sigma(t)$ required to hold that strain is measured as a function of time. For an elastic material, $E_R(t)$ is the Young's modulus and is constant in time. For a purely viscous material, $E_R(t)$ quickly goes to zero. To obtain the temperature behavior of a viscoelastic material, $E_R(t)$ is evaluated at some particular time, usually 10s, during the relaxation process. The values of $E_R(10)$ are then plotted against temperature as shown in Figure 9.10.

All thermoplastic polymers start out with a glassy behavior below the glass transition temperature T_g. As T is increased above T_g, they all become more ductile or leathery and later become softer or more rubber-like. A semicrystalline isotactic polymer will retain its strength longer than an amorphous polymer because its crystalline structure permits more elastic deformation while the others deform viscoelactically. Eventually, semicrystalline isotactic polymer as well as the amorphous polymer will turn into a viscous liquid as they eventually melt. A cross-linked polymer softens somewhat at the glass temperature but does not melt because of the strong covalent cross-link bonds. It will decompose before these bonds are broken.

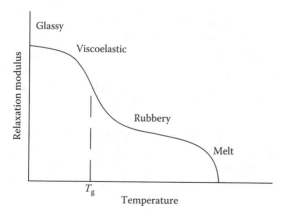

FIGURE 9.10
Schematic of the relaxation modulus plotted against temperature for a viscoelastic polymer.

9.7 Summary

At low stresses, metals tend to deform elastically. The strain is proportional to the stress (Hooke's law), no bonds are broken, and there is complete recovery when the stress is relieved. At a certain stress level (the yield strength), ductile metals will began to deform plastically, bonds are being broken, and a permanent deformation will result when the stress is relieved. As more stress is applied, ductile metals will begin to form a neck and eventually rupture. The maximum recorded stress is the tensile strength of the material. However, the recorded stress at rupture will usually be less than the maximum recorded stress because of the area reduction as the neck forms. The strain at rupture is a measure of the ductility, usually expressed in percent elongation. These engineering stress and strain measurements can be converted into true stress and strain by taking into account the area reduction when the neck is formed. The total area under the stress–strain curve is a measure of the toughness of the material. The area under the stress–strain diagram at the elastic limit is the elastic work stored in the material and is called the resiliency and is given by $\sigma_{yield}\, e_{elastic}/2 = E e_{elastic}^2/2 = \sigma_{yield}^2/2E$, where E is the Young's modulus.

Brittle metals and ceramics undergo little or no plastic deformation before rupturing. Rupture occurs at stress well below the theoretical strength because of surface cracks that open and eventually exceed their critical length and propagate through the material leaving a planar surface. Very ductile materials tend to neck down and simply pinch off at stresses well below their theoretical strength. Less ductile metals tend to form voids due to vacancy coalescence during the necking process. These voids form an elliptical crack that propagates across the specimen, leaving a characteristic cup and cone fracture.

Metals deform by dislocations moving along slip planes, planes with high atom densities (atoms/area), in directions that have high linear atom densities. The ductility of a metal depends on the number of active slip systems. Face-centered lattices have the most number of active systems and are generally the most ductile. Body-centered systems actually have more slip systems than fcc systems, but the slip planes have lower atom densities, hence are not as active. Some of these slip systems become inactive at low temperatures causing a ductile-to-brittle transition to occur. Hexagonal close-packed systems have only one active slip system. Since their slip planes are parallel, cross-slip cannot occur and hcp systems are generally more brittle than other metals.

Ceramics tend to be very brittle because (1) it is more difficult to form dislocations because a dislocation requires two half planes of ions to be inserted; and (2) it is difficult for the dislocations to move because of the Coulomb repulsion as large anions must slip past each other.

Since plastic deformation involves dislocation motion, most strengthening methods seek to inhibit the motion of dislocations. There are essentially five ways of accomplishing this: work hardening, solid solution hardening, grain refining, dispersion hardening, and precipitation hardening. Glasses can be strengthened by tempering, which is accomplished by chilling the exterior to put it into compression and sealing off any insipient surface cracks. Metals can also be hardened by diffusing small molecules such as C or N into their interstities or by shot peening to seal any surface cracks. Steels and ceramics may be transition toughened by incorporation of a metastable phase that will undergo a transition under stress to seal cracks.

Creep is a thermally activated process under which materials will slowly stretch under stress when the temperature exceeds 50% of the melting temperature. Creep mechanisms include dislocation climb, vacancy migration, and grain boundary slip. Creep is especially important in gas turbine engines and a class of Ni-based superalloys has been developed to minimize its effects.

Modern fracture mechanics was pioneered by A.A. Griffith who showed that brittle materials could fail catastrophically by cracks that become self-propagating even at stresses much lower than their tensile strength. Griffith's theory was expanded to include ductile materials and has led to the definition of a material property called fracture toughness that allows the prediction of critical crack lengths.

The mechanical properties of polymers are quite different from those of metals and ceramics and depend on the structure of the polymer and temperature. Semicrystalline polymers initially deform elastically much like metals by the stretching of their covalent bonds. Plastic deformation occurs when the covalent chains start to move and align themselves with the applied stress. This alignment of the strong covalent chains causes an increase in strength before rupture.

Elastomers have tightly coiled chains that can easily be deformed by uncoiling the chains. Driven by entropy and guided by a light cross-linkage, the chains recoil to their original state when stress is relieved.

Amorphous polymers deform viscously above their glass transition temperature. Polymers generally exhibit non-Newtonian behavior (viscosity is not constant) and their time-dependent behavior can be modeled mechanically by a combination of springs and dashpots.

Bibliography

Barsoum, M.W., *Fundamentals of Ceramics*, Institute of Physic Publishing, Bristol/Philadelphia, PA, 2003.

Callister, W.D., *Materials Science and Engineering: An Introduction*, 7th edn., John Wiley & Sons, New York, 2007.

Gordon, J.E., *The Science of Structures and Materials*, Scientific American Library, New York, 1988.

Hosford, W.F., *Physical Metallurgy*, Taylor & Francis, CRC Press, Boca Raton, FL, 2005.

Kingery, W.D., Bowen, H.K., and Uhlmann, D.R., *Introduction to Ceramics*, 2nd edn., John Wiley & Sons, New York, 1976.

Lawn, B., *Fracture of Brittle Solids*, 2nd edn., Cambridge University Press, UK, 1993.

Reinbolt, J.A. and Harris, W.J., *Trans. ASM*, 1951.

Russell, A.M. and Lee, K.L., *Structure–Property Relations in Nonferrous Metals*, John Wiley & Sons, New York, 1966.

Problems

1. Simply plotting the observed stress against the observed strain in the elastic region does not give the true value for the Young's modulus because it does not take into account the narrowing of the sample due to the Poisson's ratio. Derive the true Young's modulus taking into account true stress and true strain. How much error is incurred in using the engineering value?

2. Explain how you could derive the modulus of elasticity from a bend test using the relation given by Equation 9.4.

3. Find the area density fraction for the {211} and {321} families of planes for the bcc system.

4. Find the area density fractions for the $(10\bar{1}0)$ and the $(10\bar{1}1)$ planes in the hcp system. How do these compare the (211) and (321) planes in the bcc system?

5. Comment on Griffith's choice of 1 nm for the length at fracture. Compare his estimate of the theoretical strength of his glass rods against the discussion on theoretical strength in Chapter 7.

6. You have an automobile with 16 in. Al alloy wheels. Should you start worrying about fatigue failure of these wheels at 100,000 mi.? Estimate the safe lifetime of these wheels.

7. You are asked to design a pressure sphere of 0.5 m in. diameter that will hold 100 atmospheres with minimum weight. A safety factor of 2 is required below ultimate strength. What material would you chose? How thick would the walls be? What would the weight be?

8. The requirements in Problem 5 have changed. The customer wants the pressure sphere to leak before it ruptures. How would this change your design?

9. Silly putty bounces when thrown against the ground but slumps when left sitting. How would you model this response?

10

Composites

The ability to join dissimilar materials has greatly expanded the use of composite materials in applications where great strength, light weight, and dimensional stability are required. This chapter will review the present state of the development and the use of composites with some insight into how they are able to increase the performance of materials.

10.1   History of Composites

Often it is possible to combine dissimilar materials to obtain the desirable properties of both while minimizing the undesirable properties of each. An example would be the combination of a ceramic, which is hard and strong but brittle with a metal that is weaker but ductile to form a cermet (a metal/ceramic composite) which has both strength and ductility. Nature has developed a number of composite systems such as wood that consists of lamia of cellulose, which is soft and flexible, with lignin, which is hard and strong. As a result, trees can bend in wind without breaking. Similarly bone consists of a combination of hard apatite and flexible collagen to obtain the necessary strength and fracture toughness. Eutectics form alternating lamina or rods of two phases to form *in situ* composites. Pearlitic steels derive their strength by combining the hardness of the cementite phase with the ductility of the ferrite phase. Precipitation hardened alloys are another example of *in situ* composites.

However, we will confine our discussion to composites that do not naturally occur, but result from the purposeful joining together of two or more dissimilar materials to take advantage of desirable properties of each (or to lessen the undesirable properties of either).

Man's use of composites dates back to the ancient Egypt where straw was added to mud bricks (a form of ceramic/polymer composite) to strengthen them. The builders of Rome used concrete and mortar (a ceramic/ceramic composite) to build the ancient city. Cloisonné jewelry (a metal/ceramic composite) has been around for centuries. A number of composites have found their way into the market place through the early to mid-twentieth century. Plywood and corrugated paper boxes (polymer/polymer composites) have been around for many years. Because of a shortage of metal during World War II, aircraft were fabricated by laying thin strips of wood over a concrete form and gluing them together. The famous deHaviland Mosquito was built entirely of plywood as was the Howard Hughes "Spruce Goose" seaplane.

Steel-belted tires (a polymer/metal composite) and later aramid-reinforced tires (a polymer/polymer composite) greatly improved the safety of tires against blowouts. Carbon in the form of lampblack is added to vulcanized rubber (a polymer/ceramic assuming carbon is considered a ceramic) as an inexpensive filler that strengthens and toughens it as well as making it less susceptible to abrasion and tearing. Fiberglass (a polymer/ceramic composite) has found its way into cars, boats, and baths.

With the advent of the space age in the 1960s, there arose a need for new materials that were lighter and stronger. Considering that it costs several thousand dollars a pound to put something in orbit, any small weight saving would pay for a lot of development and production cost of a lighter material that would do the job. As electronics got more sophisticated, the need for materials for better thermal management and for heat rejection arose. Also as a satellite goes in and out of the Earth's shadow 18 times a day, its skin temperature and the temperature of its components such as dish antennas and masts can fluctuate by more than 100°C, so control of thermal expansion becomes an issue. Ways to maintain dimensional stability of the trusses supporting the large optical systems of the great observatories had to be found. Finding materials for reentry vehicles that can stand the hypersonic loads and searing temperatures is a major challenge.

Meanwhile, the Air Force was beginning to explore the use of composite materials in their aircraft. The advent of airborne radar required new nonmetallic materials to house it in the noses of supersonic aircraft. The Minuteman II third stage motor casing was filament-wound using a new S-glass developed by Owen Corning Fiberglass to improve its performance. Carbon and boron fibers became available and composites began to find their way into the F-15 and F-16 fighters. Composites became indispensable in the development of low-observable aircraft such as the B-2 bomber and F-22 Raptor. The Strategic Defense Initiative in the mid-1980s provided an additional need for lighter and better materials to support their space-based missile defense system.

Composites were introduced into nonmilitary aircraft by pioneers such as Bert Rutan, whose Voyager airplane that flew nonstop around the world without refueling and his SpaceShipOne that won the X-prize built almost entirely from laminated composites. Rutan's company, Scaled Composites is known for its use of nonmetal, fiber-composite materials in the development of concept aircraft, and prototype fabrication processes.

The builders of commercial aircraft were a little slower to start using composites in the structures of their passenger aircraft. First the need for weight saving, though important, was not as compelling as it was for space-borne systems. There was also the issue of inspection and service life expectancy. Both Boeing and Airbus are now making extensive use of composites in their new 787 and A380 aircraft. Composites account for about 9% of the structural weight of Boeing's 777, compared to ~3% of their previous commercial aircraft and their new 787 reportedly will use as much as 50% composite materials. The Airbus 340's structural weight consisted of 15% composites and the 380 will use ~25% composite materials.

One of the composites used by Airbus is a GLAss-REinforced (GLARE) fiber metal laminate that was first tested on the A310. It consists of alternating layers of thin Al sheets interspersed with glass fibers in an epoxy matrix. The glass fibers may be oriented along the directions of maximum stress, which allows a thinner, lighter material to replace a thicker sheet of Al that would have the same strength in that direction. In addition to saving weight, the use of the composite inhibits crack propagation (remember Griffith's theory).

Some of the composites have also been in the consumer market for some time. In addition to plywood, homebuilders and furniture makers are making extensive use of particle board (woodchips in a polymer matrix) because of its superior dimensional stability. Often it is laminated to a wood veneer for use as furniture. Decorative vinyl-clad steel or aluminum panels are used in office buildings. Generally, the more advanced composites such fiber-reinforced metals and ceramics have been too expensive for consumer use although we are now seeing aluminum metal matrix composites (ALMMCs) appearing in bicycle frames and in the drive shafts of drag racers and Corvettes. Carbon fiber-reinforced silicon carbide (SiC) is now used to eliminate brake fade in the brake rotors used by Porsche as well as in Formula 1 race cars and in some aircraft. Other manufacturers are considering the use of these high-performance ceramics for brake rotors and clutch plates in less exotic vehicles.

10.2 Types of Composites

There are three basic types of engineered composites: (1) laminates, (2) particle-reinforced composites, and (3) fiber-reinforced composites. In particle-reinforced composites, one can make the distinction between small (submicron) particle composites, where the particles are incorporated in the microstructure, vs. large particle composites, where the particles themselves actually do the work or carry the load. The reinforcing fibers can be discontinuous or continuous. The fibers in discontinuous fiber-reinforced composites can be randomly oriented to provide isotropic properties or aligned to enhance a specific property in a specific direction. Continuous fiber composites are generally designed for their unidirectional properties but can be crisscrossed to obtain multi-directional property enhancement such as in a filament-wound pressure container. All possible permutations of metal, ceramic, and polymer are found in the laminated as well as in the reinforced composites.

10.2.1 Laminated Composites

10.2.1.1 Sandwich Panels

One of the motivations for making laminated composites is to make very light panels that are extremely stiff. Recall that the stiffness modulus in the bend test is proportional to the square of the thickness of the specimen. The goal here is to increase the thickness of a panel without increasing its weight. One way to make a very rigid lightweight panel is to bond two thin sheets to a thick, very low density core. To bend the panel, the outer sheet is put in tension while the inner sheet is in compression and failure occurs only when the compressive strength of the core is no longer able to keep the inner sheet from buckling. Poster board (or foam board) and corrugated board are examples of materials that use this principle to maintain their stiffness. The performance of such panels can be improved immensely by replacing the core with a honeycomb made from thin Al or aramid fiber. The bulk density of such honeycomb cores is typically 0.05 g/cm^3. The elevators and fins of high-performance aircraft are often fabricated in this manner.

Another reason for laminating panels is to improve dimensional stability. Wood has a propensity to warp and crack or split along the grain. However, by gluing two or more pieces of wood together with their grain rotated at 90° prevents splitting along the grain direction and makes the panel dimensionally stable against warping.

Decorative panels are made by laminating vinyl or wood veneer to structural panels which can be steel, aluminum, plywood, or particle board. Foam cores between the structural panels offer stiffness, sound deadening, and thermal insulation. Metal–polymer laminates are used as decorative panels in buildings. A viscoelastic core can be added between the metal sheets for sound damping.

10.2.1.2 Metal–Metal Laminates

Bonding two thin metal strips together that have different coefficients of thermal expansion (CTE) produces a laminated composite that responds to a change in temperature. These bimetallic strips are widely used in thermostats and other temperature sensing devices.

Many coins are now a laminated composite of different metals. US coins minted after 1964 have a Cu interior and a Ni–alloy cladding. This preserves the appearance and noncorrosive properties of Ni at much less expense than making the coin from solid Ni or Ag.

Explosive bonding, developed at the Colorado School of Mines, is a method for joining dissimilar metals by using the force of a controlled explosion to weld two or more sheets of metals together. A sheet of plastic explosive is placed above the stack of metals to be bonded. When the explosive is detonated, the metals are forced together to form a strong metallic bond. The process is considered a form of cold welding that preserves the native properties of the metals without forming intermetallic phases that could cause brittleness. This process was used to make clad coins, from 1965 to 1967.

10.2.1.3 Metal–Graphite Laminates

As mentioned in Chapter 3, graphite has a remarkably high thermal conductivity ($\sim$1500 W/m-deg.) in its (100) plane. As the thermal conductivity of graphite is several times higher than the thermal conductivity of aluminum, Ktechnology Corporation developed a method for manufacturing sheets of annealed pyrolytic graphite (APG) in which the basal planes are nearly aligned. The sheets of APG are somewhat frangible so Ktechnology developed a method for encapsulating them with thin sheets of Al or other metals. These Al/Gr/Al sandwiches have five times the thermal conductive of pure Al. Later Ktechnology was able to replace the Al encapsulant with aligned graphite fibers to improve the unidirectional thermal conductivity. The space and defense industries are the primary users of this technology, now sold under the trade name K-Core, to extract heat from batteries and avionics as well as other thermal management systems.

10.2.1.4 Ceramic/Metal Laminates

Chobham armor, developed by the British at the Chobham Commons proving grounds, is an example of a ceramic/metal laminated composite. The ceramic panel (BC, TiB_2, SiC, AlN, etc.) is designed to break up the projectile and adsorb most of the energy through the breaking of bonds and making new surfaces. The metal backing is designed to catch the ceramic particles that spall off the back of the ceramic from the reflected shockwave. Ultra-high strength polymers such as Kevlar, Spectra, or M-5 fibers are often used to supplement or in place of the metal spall shield in personnel armor. Elastomers are often included in the design to isolate the ceramic plates to prevent the shock waves from the panel that was hit from penetrating into the other panels.

10.2.2 Particle-Reinforced Composites

Particle-reinforced composites can be classified as either small particle composites, in which the particles sometimes represent only a small volume fraction of the composite and interact on an atomic scale, or as large particle composite where the particles represent a significant volume fraction and operate on a macroscopic scale.

10.2.2.1 Small Particle Composites

Generally, the role of the particles in small particle composites is to dispersion strengthen the host material. A relatively small volume fraction of very small (submicron sized) insoluble particles such as various carbides or oxides in a metal hinders the motion of dislocations and thereby hardens and strengthens the alloy as discussed in Chapter 9. Dispersion hardening is an effective method for preventing creep in alloys and superalloys used in high-temperature rotating machinery such as gas turbines, jet engines, or other applications where maintaining dimensional stability at high temperatures is important. There are many examples of dispersion-strengthened materials. A few examples are listed below.

10.2.2.1.1 Dispersion-Hardened Alloys

Fine flakes of Al immediately form an oxide layer (Al_2O_3). When this powder is used to make sintered alloys, the oxide is retained as a dispersoid. This oxide prevents grain growth as well as dislocation motion, thus producing a high strength, high creep resistant sintered aluminum powder (SAP) alloy.

Copper electrodes are dispersion hardened with fine particles of Mo, B, or boron carbide (BC) to improve their strength at high temperatures.

Similarly, tungsten electrodes used in tungsten inert gas welding and filaments used in light bulbs and vacuum tubes are strengthened by dispersions of thoria (ThO_2) particles (thoriated tungsten). The presence of thoria in W also lowers work function and is used to improve the efficiency of electron emitters in x-ray tubes and other electron beam devices.

10.2.2.1.2 Other Small Particle Composites

As discussed in Chapter 9, small particles of partially stabilized tetragonal zirconia are added to a ceramic matrix to toughen it. If a crack starts to develop, these particles swell and seal the crack.

The addition of carbon black (nanometer-sized carbon particles or soot) to vulcanized rubber to strengthen and toughen it was also mentioned earlier. Not only is carbon black an inexpensive filler, it also makes the tires less susceptible to abrasion and tearing. As a result, automobile tires and other rubber products may consist of up to 30% carbon black as the reader may have discovered from the black on his/her hands after changing rubber washers and other plumbing fittings.

10.2.2.2 Large Particle Composites

10.2.2.2.1 Concrete and Mortar

The most widely used particle-reinforced composite is concrete. Unlike the previous systems, concrete is a large particle composite consisting of sand, gravel, and cement. The cement is a calcia–silica powder formed by heating clay and lime to 1400°C and grinding the resulting material into a fine powder. When water is added, a hydration reaction occurs resulting in a hydrated calcium silicate which is a very hard ceramic material that binds the sand and gravel aggregate together. It should be remembered that concrete does not dry; the water is taken up in the hydration reaction in the curing process. For this reason concrete can be cured under water.

Concrete has great compression strength because of the hardness of the gravel in the aggregate. The sand particles fill in the spaces between the larger gravel pieces, transferring the load between them without allowing any cracks to propagate. However, concrete is very brittle and its tensile and shear strength is many times less than its compression strength. For this reason steel reinforcing bars or wire nets are used to help hold concrete slabs from cracking. The high pH of the concrete prevents the steel from rusting. To provide the necessary strength for concrete beams used for highway bridges and over-passes, the beams are cast with a number of small cylindrical holes running the length of the beam. After the beam is cured, steel cables are threaded through these holes, tensioned hydraulically, and swaged at both ends. This prestressing keeps the beam in compression thus taking advantage of the high compressive strength of the concrete.

Brick mason's mortar is basically concrete without the gravel. The sand acts as a filler and the cement binds the sand together with the bricks by penetrating into small pores in the brick. The strength of the mortar depends greatly on how well the cement is able to penetrate into the brick, which is determined by the porosity of the brick and the moisture content at the brick interface.

10.2.2.2.2 Particle Board

Particle board is a composite of wood chips and sawdust and a thermosetting resin that is used as a less expensive alternative to plywood. Although not as attractive as plywood, it is remarkably strong and dimensionally stable. For this reason, it is often used in furniture that is to be covered with a veneer.

10.2.2.2.3 Cermets

Cermets are a form of metal matrix composites (MMCs) usually composed of a large volume percent of ceramic in a metal host phase. Cutting tools consisting of very hard ceramics such as WC, TiC, or even industrial diamond, embedded in a metal matrix combine the cutting ability of the ceramic with the fracture toughness as well as the heat dissipation provided by the metal. For example, carbide drills and saw blades are often made from a tough cobalt matrix with tungsten carbide particles inside. The metal isolates the hard ceramic particles from each other and prevents cracks from propagating from one particle to another.

10.2.2.2.4 Graphite Particle-Reinforced Aluminum

Metal Matrix Cast Composites, Inc. (MMCC) produces graphite particle-reinforced alumi-num composites with a range of thermal conductivities and CTEs that can be optimized for specific applications.

10.2.2.2.5 Sawdust-Reinforced Ice

Though of little or no practical value, as an example of how effective large particle composites can be in toughening materials, the British were seriously considering using sawdust dispersed in ice to build aircraft carriers to protect their convoys in the North Atlantic from German U-boats. In a demonstration of the strength of this composite, Lord Montbatten was reportedly asked to fire his pistol into a large block of ice, which of course shattered. When he fired into the sawdust-reinforced ice, the bullet simply bounced off.

10.2.3 Fiber-Reinforced Composites

As the technology for making very high strength fibers from polymers (Kevlar), glasses (E-glass), metals (whiskers and continuous boron fibers), and more recently, carbon (graphite and carbon nanotubes [CNTs]) has matured, much use is being made of using such fibers to reinforce metal, ceramic, and polymer composites.

The use of fiber reinforcing can take many forms: continuous vs. discontinuous, and aligned vs. randomly oriented. The general idea is to take advantage of the high strength of the fiber by transferring the load from the matrix to the fiber. In order to do this effectively, a critical length of the fiber is required which depends on the strength of the fiber as well as the bond between the fiber and the matrix. This relationship as well as other mechanical properties will be discussed in Section 10.3. Here we will briefly survey some of the applications of fiber-reinforced composites.

10.2.3.1 Fiber-Reinforced Metal Matrix Composites

An MMC consists of a metal or alloy host phase reinforced with metal or ceramic particles, whiskers, or fibers as the dispersed phase. Graphite is of particular interest as a reinforcing fiber because it has a negative CTE. Thus it becomes possible to engineer the CTE of the composite by selecting the amount and orientation of graphite fiber to be incorporated into metal matrix. Since graphite tends to form unstable carbides in the presence of metals at high temperatures, one of the major technical hurdles that had to be overcome was the development of a coating for the graphite fiber that is both protective from and is wet by the matrix metal.

Discontinuous fiber MMCs are isotropic and are superior in terms of specific strength, elastic modulus, toughness, and thermal conductivity compared to conventional alloys. They are used for brackets and fittings that are subject to multiaxial loads. The ability to engineer the CTEs of discontinuously reinforced aluminum (DRA) composites makes it possible to save substantial weight by replacing Invar or Kovar in critical design situations where a low CTE material is needed.

Continuous fiber MMCs have very high strength and other mechanical and thermal properties in the longitudinal direction but are relatively weak in the transverse direction as might be expected. 6061 Al reinforced with 40 vol % P100 continuous graphite fiber has a Young's modulus of ~350 GPa and an ultimate strength of ~900 MPa in the longitudinal direction compared to 69 GPa and 350 MPa for 6061T-6 Al. Of course, the transverse values drop to ~35 GPa and 25 MPa, respectively, as would be predicted by the law of mixtures (see Section 10.4). The thermal conductivity is increased from 180 to 320 W/m-deg. in the longitudinal direction, but drops to 72 W/m-deg. in the transverse direction. The CTE in the longitudinal direction is actually $-0.49 \times 10^{-6}/°C$ because of the negative CTE of the graphite fibers. The same Al reinforced with B fibers is somewhat stronger, but also slightly more dense and does not have as high thermal conductivity or negative CTE. With these highly anisotropic properties, the use of continuous fiber-reinforced composites is restricted to applications where the directions of the expected loads (or thermal paths) are uniaxial and predictable, such as in trusses or other structural components.

SCS6-Ti (Ti reinforced with coated SiC fibers) composites are of great interest for potential use in many aerospace applications where lightweight materials are required that keep their strength at high temperature. These MMCs are also being investigated for possible use as structural components in hypersonic aircraft and missiles. However, there is some reluctance on the part of designers to incorporate these composites into their designs because it is felt that they need to be better understood in order to be put into widespread usage. Of particular concern is their response to fatigue and a significant research effort is underway to understand why some of the TiMMCs do not meet their theoretical fatigue limit.

Because the high cost of these MMCs had made them unattractive to the world of commercial products, the Aluminum MMC Consortium was created in 1997 whose mission was "to facilitate expanded application of reinforced aluminum and other metal matrix composite products by developing manufacturing process technology, design confidence, and increased awareness." As a result, much of the technology developed by the military and space industries that had been too costly for use commercially is beginning to see the use in specialty consumer products. ALMMCs with discontinuous fibers (DRAs) are finding many applications that include drive shafts, pistons, brake rotors, bicycle frames, and sporting goods. Al MMC drive shafts are advertised to be 30% stronger than 6061-T6 shafts.

10.2.3.2 Fiber-Reinforced Ceramic Matrix Composites

Because of their hardness and their ability to withstand high-temperature oxidizing environments, ceramics would be the ideal material of choice for many applications, but until recently, their usefulness had been limited because of their lack of toughness. For example, gas turbine engines could operate several hundred degrees higher at much greater efficiency if their superalloy blades were replaced with ceramic blades.

A collaborative effort involving private companies, universities, and national laboratories was begun by the Department of Energy in 1992 to develop continuous fiber ceramic composites (CFCCs) with the goal of making ceramics tougher and more resistant to thermal shock. Nonoxide ceramics such as BC, aluminum nitride (AlN), silicon nitride (Si_3N_4), and SiC are suitable for the matrix phase of CFCCs. AlN as the highest thermal conductivity, Si_3N_4 has the highest strength, but SiC is the most widely used because of the combination of low density, good strength, and thermal conductivity and is the best understood. Oxide ceramics such as alumina and mullite can be used as the matrix when a high tolerance to corrosive or oxygen environments are required, but are generally inferior in terms of strength to the nonoxide ceramics. Among continuous fibers of glass, mullite, alumina, carbon, and SiC, again SiC fibers seem to be the material of choice because of their high strength, stiffness, and thermal stability. Unfortunately, the cost of some of these continuous fibers can be as high as \$5000/lb, which limits the scope of industrial applications for the product. This cost is projected to come down to ~\$100/lb as demand increases and fiber production goes from the pilot to large-scale production. Also special techniques are required to infuse the matrix material through the fibers. Presently, the cost of CFCC products is greater than \$1000/lb compared to ~\$50/lb for discontinuous fiber-reinforced ceramic composites.

Typical properties for a SiC/SiC CFCC are Young's modulus ~100 GPa, tensile strength ~260 MPa, and a strain at failure of 0.4%, making it semiductile. Applications include burners, heat exchangers, hot gas fans and filters, and gas turbine combustion liners and shrouds. They have demonstrated 1,000 h without failure as turbine blades in gas turbine engines and 10,000 thermal cycles as radiant burner screens without failure.

Discontinuous fiber ceramic composites have a large cost advantage over CFCCs in that they can be produced by the same methods used to process ordinary ceramics and do not require the expensive continuous fibers. However, they remain quite brittle and have not achieved the ductility of CFCCS. Discontinuous fibers can be whiskers, nanotubes, or chopped fibers of glass, mullite, alumina, carbon, or SiC. The main applications of discontinuous fiber ceramic composites have been special tools for cutting very hard materials, in abrasives, and liners for chutes in the mining industry. As mentioned in Section 10.1, fiber reinforced ceramic composites are now available to high-end consumers in the form of high-performance clutch plates and brake systems for high-performance cars. In addition to being smaller and lighter than the conventional systems, a ceramic clutch can transmit substantially more torque. Ceramic brakes are not only fade resistant, but can reduce the unsprung weight of wheels.

10.2.3.3 Carbon–Carbon Composites

Carbon–carbon (C–C) composites are light, very strong, and can withstand temperatures in excess of 2000°C. For these reasons they are used in rocket nozzles, high performance aircraft, high-temperature molds, and heat shields. The nose and wing leading edges on the space shuttles are C–C composites. These composites owe their high strength to the sp^2 bond in graphite. The carbon filaments are made by oxidizing and pyrolyzing polyacrylonitrile (PAN). By heating to ∼3000°, the fiber turns to almost pure graphite with a tensile strength of ∼5 GPa and an elastic modulus of ∼500 GPa. Layers of these continuous carbon fibers are impregnated with a resin, usually a phenolic, and allowed to cure. The material is then pyrolyzed or heated to the destruction of the binder, leaving only a carbon matrix reinforced by carbon fibers.

One difficulty in using C–C composites in space is that they can be degraded by exposure to atomic oxygen, which is present in near-earth orbit. For this reason a protective layer of SiC must be added to the portions of the C–C materials that are exposed to the space environment.

10.2.3.4 Polymer Matrix Composites

Polymer matrix composites (PMC) consist of a fiber-reinforced polymer matrix that is usually a polyester, polyimide, polyvinyl, or epoxy. The fibers are generally glass, carbon, graphite, boron, or a high-strength polymer such as an aramid-type fiber (Kevlar, Spectra, or M-5). PMCs are both strong and light, but are generally restricted to operating temperatures below 200°C–400°C where the matrix begins to soften.

Fiberglass is by far the most widely used PMC, having been around since the 1950s. The glass fiber is usually E-glass, a type of glass that is strong, easily drawn into fibers, fairly ductile, chemically inert and, best of all, it is relatively inexpensive. Typically the fibers are a few microns in diameter (remember, smaller is stronger). Great care must be taken not to nick, scratch, or abrade the surface during the processing for fear of weakening the fiber and often the fibers are given a protective coating. The glass fibers may be continuous and aligned, or chopped and randomly oriented, depending of the design requirements. The Corvette has had a fiberglass body since its inception in 1953. The hulls of most small boats are now constructed from fiberglass, which avoids the problems of leaks, dry rot, and the periodic need for repainting that plagues wooden hulls. Fiberglass is now widely used in the manufacture of bathtubs, shower stalls, and other household items.

Aligned continuous fiber PMC tubes combine strength and rigidity to make very light structural members and trusses. Porsche and other automobile manufacturers are experimenting with carbon fiber PMC tubular frames to lighten their cars. The Porsche Carrera GT is built with a carbon fiber panel monocoque body on a carbon PMC subframe with practically no metal parts in the structure.

A distinction is made between carbon and graphite fibers. Both are made by oxidizing and pyrolyzing PAN. Carbon fibers form initially with some ordered graphite, but heating to ∼3000°C is required to fully graphitize the fiber. Graphite fibers may be thought of as graphite sheets rolled up like chicken wire. The high strength and thermal conductivity approaches that of graphite sheets along the basal plane. The negative CTE of the graphite fiber can be balanced against the positive CTE of the matrix to provide a near zero CTE. The metering truss that supports the optics on the Hubble Space Telescope had extremely stringent requirements for rigidity and dimensional stability over a wide temperature range. This challenging design problem was solved by fabricating the truss from a graphite/epoxy composite. To compensate for the large negative CTE of the graphite fibers, it was necessary to add powdered Al to the matrix to achieve the near-zero CTE required by the design.

It is possible to fabricate graphite fibers with thermal conductivities as high as 1200 W/m-deg. These fibers are available in a polymer prepreg that can be cured by baking at 125°C. The high thermal conductivity, low density, and high stiffness make this composite ideal for fabricating highly efficient radiators for deep space missions that use solar/electric of nuclear/electric power. Rejecting the large amount of waste heat from such power systems is a major design challenge calling for highly efficient radiators. One such prototype design was fabricated by ATK Space Systems using multiple layers of a prepreg containing K13D2U in EX1551 cyanate ester to produce a layered tapered fin. Wrapping the prepreg around a Ti/H_2O heat pipe effectively conducts the heat from the heat pipe to the fin where it can radiate to space. The assembly was laid up in a graphite mold with a grove to accept the Ti tube. Crossing the fibers at different angles in the laminated fin provided stiffness and controls the CTE. The finished assembly is then cured under pressure in an autoclave. Since the radiator must operate over a range from $-100°$ to 550 K, it is important to match the CTE of the graphite fiber-reinforced polymer composite with the Ti heat pipe so it does not delaminate during thermal cycling.

With the discovery of CNTs and their remarkable properties came great interest in trying to incorporate them into the matrix material as well as use them in the fiber to improve the strength of fiber-reinforced composites. Originally the nanotubes were prohibitively expensive to experiment with on the scale needed for such experimentation and their lengths were shorter than the critical length needed to achieve good coupling to the matrix. Also, ways need to be found to functionalize the nanotubes in order for them to bond with the matrix material.

Recent developments in synthesis and assembling CNTs into continuous fibers have opened the doors to more widespread research in the effort to develop the ultimate fiber (see Section 5.4.7). When subjected to high pressure, some of the sp^2 bonds can convert to sp^3 bonds, which may provide a means for cross-linking the CNT fibers into a super-strong fabric—carbon threads joined by diamonds.

10.3 Modeling the Performance of Composites

10.3.1 Rule of Mixtures

It is difficult to model the mechanical properties of particle-reinforced composites because of the complexity of the structure. However it is possible to place limits on some of the properties of a composite using a system of springs as an analog. If two springs are placed in parallel, the applied force F_A must be shared by the two springs and the displacement Δx will be the same in both springs. The force on spring 1 is $F_1 = k_1 \Delta x$ and the force on the second spring is $F_1 = k_1 \Delta x$. The net spring constant

$$k = F_A/\Delta x = (F_1 + F_2)/\Delta x = k_1 + k_2. \tag{10.1}$$

Substituting the net Young's modulus times the area of the composite for $k\Delta x$ and the Young's modulus times the areas of each component for $k_1\Delta x$ and $k_2\Delta x$, the net modulus becomes

$$E = \frac{E_1 A_1}{A_C} + \frac{E_2 A_2}{A_C} = E_1 V_1 + E_2 V_2 = E_{\text{Matrix}}(1 - V_{\text{Particle}}) + E_{\text{Particle}} V_{\text{Particle}}, \tag{10.2}$$

where it is assumed that the volume fractions V_1 and V_2 of the components are given by their cross-section area divided by the total cross-section area.

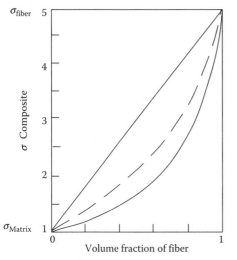

FIGURE 10.1
Upper and lower bounds of the performance of a particle-reinforced composite set by the rule of mixtures. In this illustration the strength of the fiber is five times the matrix. The dashed line in the middle is the estimated performance of a randomly oriented discontinuous fibers (1/3 aligned with the applied stress, 2/3 normal to the applied stress).

The other extreme is represented by placing the springs in series. Now the F_A is the same on each spring, but the displacements will be given by $\Delta x_1 = F_A/k_1$ and $\Delta x_2 = F_A/k_2$. Now the net displacement is given by $\Delta x = \Delta x_1 + \Delta x_2$ and the net spring constant becomes

$$k = \frac{F_A}{\Delta x_1 + \Delta x_2} = \frac{1}{1/k_1 + 1/k_2} = \frac{k_1 k_2}{k_1 + k_2}. \tag{10.3}$$

If we substitute the Young's modulus times the area of the composite for $k\Delta x$ and identify the strains $\varepsilon_1 = F_A A_1/E_1$ and $\varepsilon_2 = F_A A_2/E_2$

$$E_A A_C = k\Delta x = \frac{F_A \Delta x}{\Delta x_1 + \Delta x_2} = \frac{F_A}{\varepsilon_1 + \varepsilon_2} = \frac{F_A}{F_A A_1/E_1 + F_A A_2/E_2} = \frac{E_1 E_2}{E_1 A_2 + E_2 A_1}. \tag{10.4}$$

Now by assuming that $A_1/A_C = V_1$ and $A_2/A_C = V_2$, we obtain

$$E = \frac{E_1 E_2}{E_1 V_2 + E_2 V_1} = \frac{E_{Matrix} E_{Particle}}{E_{Matrix}(V_{Particle}) + E_{Particle}(1 - V_{Particle})}. \tag{10.5}$$

One can see in Figure 10.1 that the second model will give the lowest effective modulus since the weakest spring in a series will stretch the most.

These derivations, while far from rigorous, do set upper and lower limits on the combined effect of the two components. The net thermal conductivity and CTE can also be estimated in this manner.

10.3.2 Critical Fiber Length

As mentioned in 10.2.3, super-strong polymeric, ceramic, metallic, and carbon fibers are now used to reinforce various types of composites. However strong the fiber may be, the composite is no better than the bond between the fiber and its matrix. Consider a fiber of diameter D and length L, half of which is imbedded in a matrix along the z-axis. If a longitudinal force F_z is applied that tries to pull the fiber out of the matrix, the matrix will exert a restraining force given by $dF_z = \pi D\,\sigma_B dz$, where σ_B is the shear strength of the bond between the fiber and the matrix. The stress in the fiber is

$$\sigma(z) = \frac{4}{\pi D^2}\int_0^z dF_z = \frac{4}{D}\int_0^{L/2}\sigma_B dz; \quad \sigma < \sigma_F, \tag{10.6}$$

where σ_F is the tensile strength of the fiber. When the stress in the fiber reaches σ_F, the maximum load has been transferred to the fiber. Thus there is a critical fiber aspect ratio given by

$$\left(\frac{L}{D}\right)_C = \frac{\sigma_F}{2\sigma_B}. \tag{10.7}$$

If the aspect ratio of the fibers is less than the critical value, the bond between the fibers and the matrix will fail before the fibers break and full advantage of the fiber's strength will not be realized.

10.3.3 Continuous Aligned Fiber-Reinforced Composites

In continuous aligned fiber-reinforced composites, it is assumed that the fibers are much longer than the critical length so that the maximum load can be transferred to the fiber. The elastic modulus for a longitudinal force is given by the parallel spring model developed in the previous section, and the elastic modulus of the composite is given by Equation 10.2. The yield and fracture strengths depend on the plastic behavior of the matrix as may be seen in Figure 10.2 where it is assumed that the fibers fail by brittle fracture and that the matrix undergoes ductile fracture. The yield strength of the composite is determined by the strain at the elastic limit of the matrix e_{YM} and, from Equation 10.2,

$$\sigma_{YC} = \sigma_{YM}(1 - V_M) + E_F(V_F). \tag{10.8}$$

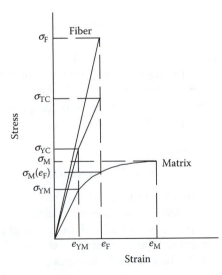

FIGURE 10.2
Schematic stress–strain behavior of aligned composites. It is assumed that the matrix is ductile and the fibers fail by brittle fracture. The yield strength of the composite is σ_{YC} and its tensile strength is σ_{TC}.

As additional stress is applied, the matrix will continue to transfer load to the fibers as it yields plastically until the strain at fracture e_F of the fiber is reached. The tensile strength of the composite is given by

$$\sigma_{TC} = \sigma_M(e_F)(1 - V_M) + \sigma_{TF}(V_F), \tag{10.9}$$

where $\sigma_M(e_F)$ is the stress in the matrix at e_F.

If the force is applied in the transverse direction relative to the aligned fibers, the series spring model applies given by Equation 10.5, which corresponds to the lower curve in Figure 10.1.

10.3.4 Discontinuous Fibers

If the fibers are aligned but shorter than the critical length L_C (the length corresponding to the critical aspect ratio given by Equation 10.7), the modulus is given by

$$E = E_F V_F(L/2L_C) + E_M(1 - V_F), \quad L < L_C. \tag{10.10}$$

The factor of 2 in the denominator comes about since the average stress in a fiber $\geq$ the critical length is half the maximum stress. For aligned fibers $\geq L_C$, the modulus is given by

$$E = E_F V_F(1 - L_C/2L) + E_M(1 - V_F), \quad L > L_C. \tag{10.11}$$

When the fiber length $L \sim 15\ L_C$, the fibers are considered to be fully effective as reinforcements to the matrix.

Aligned discontinuous fibers offer a negligible advantage in the transverse direction. To obtain isotropic performance, the fibers may be randomly distributed. The modulus may be written as

$$E = K E_F(L/2L_C)V_F + E_F(1 - V_F), \tag{10.12}$$

where K is a fiber efficiency factor that varies from 1 for aligned fibers to $\sim$0.2 from randomly distributed fibers.

10.4 Summary

Both nature and man have made extensive use of composite materials in which two or more different materials are joined in such a manner that they maintain their identity but work together to add their strengths and decrease their weaknesses. Composites can be classified into three categories: (1) Laminates, in which sheets of different materials are laminated together; (2) particle-reinforced composites, in which particles of one material are imbedded in a matrix of a second material; and (3) fiber-reinforced composites, in which fibers of one material are encapsulated in a matrix of a second material. Particle-reinforced composites can be subdivided into small particle composites, where the particles are incorporated into the microstructure, such as dispersion-hardened alloys, and large particle composites, where the matrix simply supports the particles. Fiber-reinforced composites may have continuous versus discontinuous fibers and aligned versus randomly oriented fibers, which can provide anisotropic versus isotropic properties. Composites combine all combinations of metals, ceramics, and polymers into MMCs, where a metal is the matrix material, ceramic matrix composites (CMCs), and PMCs. Sometimes all three materials are combined to form hybrid composites.

There is a critical aspect ratio of the reinforcing fibers, which goes as the ratio of the fiber strength to the strength of the bond between the fiber and the matrix. If the fibers do not exceed this critical aspect ratio, the matrix will separate from the fiber before the fibers break and the full strengthening effect of the fibers will not be attained.

Upper and lower bounds on the properties of a particle- or fiber-reinforced composite can be estimated using the law of mixtures. For aligned composites the upper bound corresponds to the longitudinal properties and the lower bound to the transverse properties.

The development of super-strong wires and continuous fibers of boron, graphite, and polymers such as Kevlar, Spectra, and M-5 have led to families of advanced composites. Of these, perhaps graphite is the most interesting because of its high thermal conductivity and negative CTE as well as its strength. Graphite fibers in MMCs, CMCs, and PMCs are being used as highly efficient thermal conductors and radiators. Their negative CTE allows the tailoring of composites to match the CTE of other materials as well as the construction of light, strong, dimensionally stable structures.

Bibliography

Agarwal, B.D. and Broutman, L.J., *Analysis and Performance of Fiber Composites*, 2nd edn., John Wiley & Sons, New York, 1990.

Callister, W.D., *Materials Science and Engineering: An Introduction*, 7th edn., John Wiley and Sons, New York, 2007.

Clyne, T.W. and Withers, P.J., *An Introduction to Metal Matrix Composites*, Cambridge University Press, Cambridge, 1993.

Petersen, R.C., Discontinuous fiber-reinforced composites above critical length, *J. Dent. Res.*, 84(4), 365–370, 2005.

Rawal, S., Metal matrix composites for space applications, *JOM* 53/4, 2001, 14–17.

Stark, N.M. and Rowlands, R.E., Effects of wood fiber characteristics on mechanical properties of wood/polypropylene composites, *Wood Fiber Sci.*, 35(2), 167–174, 2003.

Suresh, S., Mortensen, A., and Needleman, A., *Fundamentals of Metal Matrix Composites*, Butterworth-Heinemann, Boston, MA, 1993.

Problems

1. Sketch a stress–strain diagram for a composite with a smaller volume fraction of the fibers to that shown in Figure 10.2 in which the yield strength of the composite is lower than the fracture strength of the matrix.

2. Write expressions for the yield strength and the tensile strength of the composite in Problem 1.

11

Phase Equilibria in Single Component Systems

Phase transformations such as evaporation, sublimation, melting, and solidification play an essential role in the processing of materials. We will begin the study of how and why such transformations occur by first considering the melting and solidification of systems with only a single component such as a pure metal.

11.1 Definition of a Phase

Strictly speaking, a phase is defined as a region in a material in which the structure and composition are homogeneous. This assumes that the system is in chemical equilibrium, a situation which would eventually occur because of diffusion. However, for reasons that will become clear shortly, such a situation rarely occurs in a multicomponent solid, even if the system is isomorphous (all components completely soluble) because of the slow inter-diffusion of components within a solid. Therefore, practically speaking, we generally define a phase as a region in which the crystal structure is the same and the components are the same but allow for the fact that chemical equilibrium may not yet have been reached.

For example, if we dope a semiconductor with an impurity by diffusion, there will be a higher concentration of dopant atoms near the surface, but we do not consider this a separate phase unless we exceed the solid solubility with the concentration of dopant atoms causing them to form a precipitate. The precipitate would be considered a second phase. Similarly, if we solidify a solid solution (isomorphous) system, the resulting poly-crystalline alloy is considered a single phase because all of the grains have the same crystal structure, even though the first-to-freeze grains may have a different composition than the last-to-freeze. On the other hand, a liquid or gas may have the same composition as a solid, but clearly have different structures and therefore must be considered as separate phases.

11.2 Solidification of Pure Systems

11.2.1 Gibbs Phase Rule

The Gibbs phase rule can be stated as the number of phases present + degrees of free-dom = number of components + number of variables. A single component system such as a gas has three thermodynamic coordinates, generally p, V, and T. The equation of state relating these three variables reduces the number of independent variables to two. If a single phase is present, the Gibbs phase rule tells us there are two degrees of freedom, i.e., we must specify two variables to define the state of the system. If two phases are present, we have

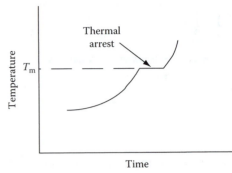

FIGURE 11.1
Typical melting curve showing the thermal arrest.

only one degree of freedom. Specifying any one variable automatically fixes the other two along the solid–vapor line, the liquid–solid line, or the liquid–vapor line on the phase diagram. If three phases are present, there are no degrees of freedom and three phases can exist in equilibrium only at a single fixed value of p, V, and T, namely the triple point.

In dealing with materials, we generally operate at constant pressure so there is only one independent variable. A single component system would then have only one degree of freedom with a single phase present, and no degrees of freedom with two phases present. Therefore, a melt can only be in equilibrium with its solid at a fixed temperature, i.e., the equilibrium melting temperature T_M. (Note the emphasis on equilibrium.) Solids can coexist with their melts indefinitely at the equilibrium melting temperature T_M.

11.2.2 Enthalpy of Fusion

Crystalline solids begin to melt as their temperature is raised to their T_M as shown in Figure 11.1. The flat region is called the thermal arrest during which the temperature remains the same as the applied heat goes into the enthalpy of fusion (latent heat of melting). This latent heat or enthalpy of melting represents the difference in binding energy between the molten phase and the solid phase.

It is interesting to compare the energy difference between the melt and the solid with the cohesive energy of the lattice. One can see from Figure 11.2 that, with the exception of the covalently bonded group IV elements, the enthalpy of fusion ΔH is only ~4% of the cohesive energy. This suggests that there is little energy difference between the solid and the melt, or in other words, there is still a very large bonding energy involved in the melt. This can be understood from the fact that the change in coordination number as a metal melts is generally small, going from 12 to ~11 in close-packed metals.

11.2.3 Entropy of Fusion

It is also instructive to look at the entropy change associated with melting as the ordered solid goes to a disordered liquid. Again, as may be seen in Figure 11.3, in most metals there is only ~2.3 cal/mol-deg. entropy difference between the melt and the solid.

11.2.4 Gibbs Free Energy

What determines this equilibrium melting temperature? To understand the underlying mechanisms governing phase change, we introduce a thermodynamic coordinate, the Gibbs energy (sometimes called Gibbs free energy or simply free energy), defined as

$$G \equiv U + pV - TS = H - TS. \tag{11.1}$$

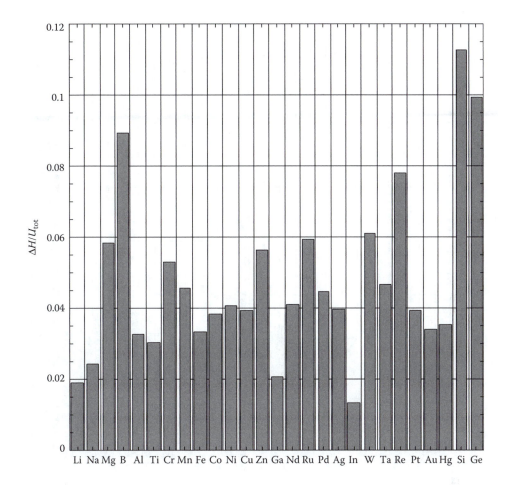

FIGURE 11.2
Ratio of the enthalpy of fusion ΔH to the binding energy of various crystalline solids. This ratio is only ~4% for most metals.

The Gibbs free energy is constructed such that it will be a minimum when the system is at equilibrium, or we might say that nature tends to want to minimize the free energy of a system. Whichever phase has the lowest free energy will be the stable phase.

Let $\Delta G = G_L - G_S = \Delta H - T\Delta S$ be the difference in Gibbs energy between the liquid phase and the solid phase. The change in enthalpy $\Delta H = H_L - H_S > 0$ since the solid is more tightly bound than the liquid. At low temperatures where $T\Delta S < \Delta H$, $G_S < G_L$ and the solid is the stable phase. The slope of the free energy of the liquid, $\partial G_L/\partial T = \partial H_L/\partial T - S_L - T\partial S_L/\partial T \sim -S_L$ since the first and last terms tend to cancel each other. Similarly, the slope of the solid free energy, $\partial G_S/\partial T = \partial H_S/\partial T - S_S - T\partial S_S/\partial T \sim -S_S$. The free energies of both the solid and the melt decrease with increasing temperatures because of the $-TS$ term, but the free energy of the melt decreases more rapidly because the entropy of the melt is greater than the solid ($S_L > S_S$) since there are more accessible states in the melt (the molecules are free to move about). Eventually the melt free energy curve crosses and falls below the free energy curve of the solid at the equilibrium melting temperature T_M, as seen in Figure 11.4, and the liquid becomes the more stable phase at temperatures $>T_M$. A similar argument applies to the transition between liquid and vapor phase at even higher temperatures.

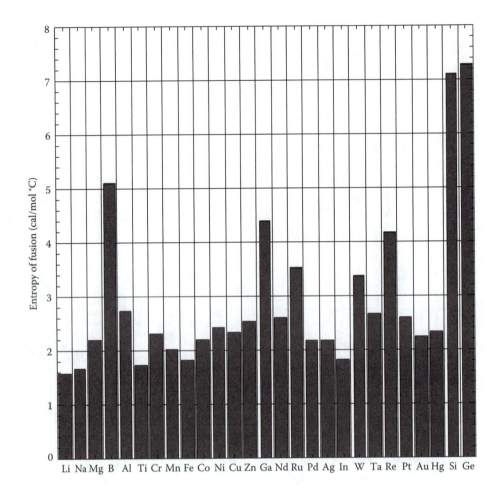

FIGURE 11.3
Entropy of fusion for various crystalline solids. For most metals, ΔS is only a few calories per mole.

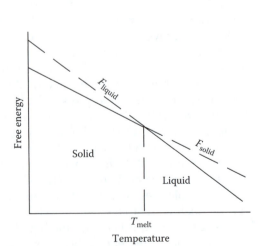

FIGURE 11.4
Schematic of the free energy whose slope becomes discontinuous during a first-order phase transition.

11.2.5 First- and Second-Order Phase Transitions

Notice the slope of the free energy of the system in Figure 11.4 is discontinuous at the T_M, where $\Delta G = 0$ and the shallower slope of the solid goes over to the steeper slope of the melt. A discontinuity in the first derivative of the free energy signals a first-order phase transition, which is marked by a distinct change in entropy (and molar volume). Since $\Delta G = \Delta H - T_M \Delta S = 0$, $\Delta H = T_M \Delta S = Q$, where Q is the heat of fusion that must absorbed to break the bonds of the more tightly bound solid in order for it to melt. In later chapters, we will encounter phase changes involving magnetic ordering, superconducting transitions, and other phenomena in which the first derivative of the free energy will remain continuous but the second derivative is discontinuous. Such transitions are second-order phase changes. There is no entropy change in second-order transitions, hence no heat of transition, but there will be a discontinuity in the slope of the entropy, hence a discontinuity in heat capacity since $C = T(\partial S / \partial T) = -T(\partial^2 G / \partial T^2)$.

11.3 Solidification Process

11.3.1 Undercooling

As stated previously, a melt can remain in equilibrium with its parent solid indefinitely, but melting and solidification are not equilibrium processes. To cause the sample to melt further, additional heat must be provided to supply the enthalpy needed to break the solid bonds in order to form the melt. To drive additional solidification, it is necessary to increase the difference between the free energy of the liquid and the solid, which is usually accomplished by lowering the temperature of the melt below the equilibrium melting temperature, a process known as undercooling. Define $\Delta T = T_M - T$. Since $\Delta S \approx \Delta H / T_M$ in the vicinity of T_M,

$$\Delta G(\Delta T) = G_L - G_S = \Delta H - T\Delta S = \Delta H - (T_M - \Delta T)\Delta S = \Delta T \Delta S = \frac{\Delta H \Delta T}{T_M} \tag{11.2}$$

so increasing the undercooling ΔT increases the thermodynamic driving force to advance the solidification front. For a pure melt solidifying on to its parent solid, very little undercooling is required (typically only a few millikelvin) and we say the process is governed by fast surface kinetics, meaning that the atoms can quickly be incorporated into the growing crystalline solid.

11.3.2 Recalescence

Incipient melting begins when a solid is raised to its equilibrium melting temperature T_M. The reverse is not always true when a melt is lowered to the T_M. If there is no crystalline material in contact with the melt, a nucleation event is required to initiate solidification and the temperature can fall well below T_M before nucleation occurs. (Fahrenheit first noticed this when he was trying to establish the zero point on his temperature scale.) When nucleation does occur, the heat of fusion (ΔH) is released causing the melt to return to T_M before cooling further. This sudden increase in temperature, illustrated in Figure 11.5, is called recalescence. Recalescence can be observed as a sudden brightening of the melt. If the ΔT is so large that the ΔH is not sufficient to bring the recalescence temperature back up to T_M, the melt is said to be hypercooled. This large undercooling can occur because of a significant ΔG barrier to initiating the solidification from a pure melt as shown

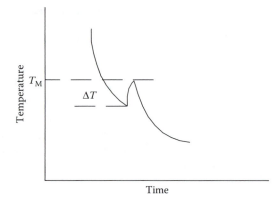

FIGURE 11.5
Typical cooling curve during solidification of a metal showing the recalescence as the release of latent heat causes a brief increase in temperature.

in Figure 10.7. It should be noted that solidification of an undercooled melt after recalescence can be very rapid since it has already given up its latent heat.

11.3.3 Effect of Pressure on Melting Point

To understand how such a nucleation barrier comes about, it is necessary to understand how curvature of a particle affects its melting temperature, which in turn requires a review of the effect of pressure on the melting point. We start by writing the differential of G as

$$dG_L = V_L dp - S_L dT \tag{11.3}$$

$$dG_S = V_S dp - S_S dT \tag{11.4}$$

where V_L and V_S are the molar volumes of the liquid and solid.

At equilibrium, $dG_L = dG_S$. Subtracting the above equations yields

$$(V_L - V_S)dp - (S_L - S_S)dT = 0, \tag{11.5}$$

which can also be written as

$$\frac{dT_M}{dp} = \frac{(V_L - V_S)}{(S_L - S_S)} = \frac{T_M^0}{\Delta H}(V_L - V_S). \tag{11.6}$$

You should recognize this as the Clapeyron equation, which gives the slope of the melting curve at some reference T_M^0, which in this case is taken as the equilibrium melting point of a planar surface. Since $\Delta H > 0$, materials that contract upon freezing will have $dT_M/dp > 0$. Materials with open structures such as group IV materials with diamond structure (e.g., Ge, Si), group III–V and II–VI compounds with zinc blende or wurtzite structures (e.g., GaAs, ZnS, etc.), and, of course, H_2O, expand when they freeze; hence $dT_M/dp < 0$.

11.3.4 Effect of Curvature on Melting Point

The pressure difference in a small droplet of liquid in equilibrium with it own vapor is given by Laplace's equation, $\Delta p = 2\gamma\kappa$, where γ is the surface tension and κ is the curvature defined as $\kappa = 1/2(1/r_1 + 1/r_2)$, r_1 and r_2 being the principle radii of curvature, respectively. (Note: For a sphere, $r_1 = r_2$; for a cylinder $r_2 = \infty$.) This Δp may be understood from Figure 11.6. If the pressure in the sphere is Δp, the force tending to separate the two hemispheres $F = \pi r^2 \Delta p$. The force must be balanced by the stress in the walls of the shell $F = 2\pi\sigma r d$. Equating the forces, $\Delta p = 2\sigma d/r$. The quantity σd is equivalent to the surface tension γ, so the pressure inside the drop is $2\gamma/r$ greater than the ambient pressure.

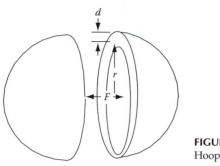

FIGURE 11.6
Hoop stress in a sphere with a pressure difference.

A small solid sphere in its melt also feels a similar pressure difference except γ now refers to the interfacial energy between the solid and melt. Just as in the case of surface tension, this interfacial energy comes about because of unsatisfied or partially satisfied bonds at the interface which give rise to a net radial inward force on the surface atoms.

From Equations 11.3 and 11.4, the differential Gibbs energies can be written

$$\delta G_S = -S_S \delta T + V_S \delta p = -S_S \delta T + \frac{2\gamma V_S}{r} \tag{11.7}$$

and

$$\delta G_L = -S_L \delta T. \tag{11.8}$$

Note that the pressure term is absent in the liquid because the pressure is only felt in the solid. Again equating the ΔG terms to obtain phase equilibrium, we find

$$\delta T_M = \frac{V_S \delta p}{(S_S - S_L)} = \frac{2\gamma V_S}{r(S_S - S_L)} = -\frac{2\gamma T_M^0 V_S}{r\Delta H}. \tag{11.9}$$

Since $\Delta H = H_L - H_S > 0$, we find that the local melting temperature for the solid sphere is lowered by an amount that is inversely proportional to its radius. This lowering of the local melting point by interfacial curvature is sometimes known as the Gibbs–Thompson effect. Notice that it does not depend on whether the material expands or contracts when freezing. In either case, the effect of positive curvature is to lower the melting point. Furthermore, we see that the equilibrium melting temperature only applies to a planar interface. Small particles will therefore melt before larger ones and larger particles will grow at the expense of the smaller particles in a constitutionally undercooled melt (a melt whose local freezing temperature is lower than the local temperature because of compositional differences). This phenomenon, known as Ostwald ripening, is important in grain growth and precipitation hardening and is discussed in greater detail in Chapter 13. The lowering of the melting temperature with curvature plays an important role in sintering by lowering the melting point at the sharp corners of the grains which allows fusion of the material to take place below the normal melting temperature.

It is also important to understand the difference between the lowering of the melting temperature predicted by Equation 11.9 and that predicted by the Clapeyron equation (Equation 11.6). The Clapeyron equation applies when both the solid and the liquid phase are under the same pressure. It predicts a raising of the freezing temperature with pressure of a substance such as a metal that contracts when it freezes. Equation 11.9 applies when only the solid is subjected to pressure such as occurs at the point of contact between powdered metal grains as they are being compressed by a process known as hot isostatic pressing (HIP). This local pressure, applied only to the solid particles, depresses the local freezing point below the ambient temperature and causes localized melting. Similarly, the lowering of the local melting point may play a role in friction or stir welding in which a rotating tool is pressed with great force against the surfaces to be joined.

11.4 Classical Homogeneous Nucleation Theory

11.4.1 Nucleation Barrier

From Equation 11.9, for an embryonic nucleus of radius r to coexist with its melt, the melt must be undercooled by an amount equal to ΔT_M. Or the critical size for a viable nucleus for a given undercooling is given by

$$r_{critical} = \frac{2\gamma T_M^0 V_S}{\Delta H \Delta T}.$$ (11.10)

(Note that $\delta T = -\Delta T$.)

The total Gibbs energy difference between the solid with radius r and its melt is expressed by

$$\Delta G^*(r) = \frac{4}{3}\pi r^3 \frac{-\Delta G}{V_S} + 4\pi r^2 \gamma,$$ (11.11)

where ΔG refers to the difference in Gibbs energy per unit volume between the melt and solid for a plane interface and the second term is the added surface energy. From Equation 11.2, this can be written as

$$\Delta G^*(r) = -\frac{4}{3}\pi r^3 \frac{\Delta H \Delta T}{T_M^0 V_S} + 4\pi r^2 \gamma.$$ (11.12)

Figure 11.7 is a typical plot of $\Delta G^*(r)$ as a function of r. The function initially rises because of the positive r^2 term, but the negative r^3 term eventually dominates and reduces $\Delta G^*(r)$.

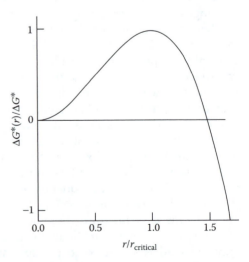

FIGURE 11.7
Barrier to nucleation. Embryonic nuclei with $r < r_{critical}$ decrease their free energy by melting; hence are not viable. Nuclei with $r > r_{critical}$ decrease their free energy by growing; hence become stable.

The peak ΔG^* at $r_{critical}$ is the barrier to nucleation. An embryonic nucleus with $r < r_{critical}$ can reduce its free energy by dissolving, hence it is not viable. On the other hand, if $r > r_{critical}$, the embryonic nucleus can reduce its free energy by growing and becomes a stable nucleus.

Differentiating Equation 11.11 with respect to r to find $r_{critical}$ yields

$$\frac{d}{dr}\Delta G^*(r) = -4\pi r^2 \frac{\Delta H \Delta T}{T_M^0 V_S} + 8\pi r \gamma = 0, \tag{11.13}$$

from which

$$r_{critical} = \frac{2\gamma T_M^0 V_S}{\Delta H \Delta T}, \tag{11.14}$$

which is the same result as found previously in Equation 11.10 using a different but equivalent argument. Putting this value for $r_{critical}$ back into Equation 11.6 to find the barrier to nucleation,

$$\Delta G^* = \frac{16}{3}\frac{\pi \gamma^3 \left(T_M^0\right)^2 V_S^2}{\Delta H^2 \Delta T^2}. \tag{11.15}$$

11.4.2 Nucleation Rate

The nucleation rate is given by

$$I_V = I_0 \exp\left(-\frac{\Delta G^*}{K_B T}\right), \tag{11.16}$$

where I_0 contains the number of atoms per unit volume, an activation energy for an atom to move from one site to another, and some frequency term. There have been several methods proposed in an attempt to derive I_0.

David Turnbull, a pioneer in the study of nucleation in metals, estimated $I_0 = N(kT/h)\exp(-\Delta F/kT)$, where N is the number of atoms/cm^3, the kT/h is an estimate of the fluctuation frequency based on the uncertainty principle, and ΔF is the transport activation energy. According to Turnbull, $\exp(-\Delta F/T)$ is the on the order of 10^{-2} at typical solidification temperatures. Altogether Turnbull estimates $I_0 = 10^{33\pm1}$ cm^{-3} s^{-1}.

Turnbull conducted extensive undercooling experiments on very small ($\sim$10 μm in diameter) metallic droplets and observed that the maximum undercooling was $\sim$20% of the T_m for a large number of metals. He further observed a correlation between the interfacial energy and the heat of fusion if the interfacial energy is expressed as a molar quantity, suggesting that the energy of an atom at the solid–liquid interface has a certain fraction λ of the difference in enthalpy between the two states or $\gamma_m = \lambda \Delta H$.

The molar interfacial energy γ_m can be related to γ by setting $\gamma_m/A = \gamma$, where A is the area of a mole of atoms with a thickness of one atomic layer. Let d be the diameter of the atom in question. The molar volume is $V_s = N_A d^3$ and $A = N_A d^2$ from which $A = N_A^{1/3} V^{2/3}$ where N_A is Avogadro's number. Turnbull found $\lambda = 0.45$ for most metals, and $= 0.32$ for semimetals and water, generally systems with a more open structure.

Substituting this relationship for γ into Equation 11.15,

$$\Delta G^* = \frac{16}{3}\frac{\pi \left(\lambda \Delta H N_A^{-1/3} V_s^{-2/3}\right)^3 \left(T_M^0\right)^2 V_S^2}{\Delta H^2 \Delta T^2} = \frac{16}{3}\frac{\pi \lambda^3 \Delta H T_M^{02}}{N_A \Delta T^2}. \tag{11.17}$$

We can set $\Delta H = \Delta S T_M^0$ and write

$$\frac{-\Delta G^*}{kT} = -\frac{16}{3} \frac{\pi \lambda^3 \Delta S (T_M^0)^3}{RT \Delta T^2} = -\frac{16}{3} \frac{\pi \lambda^3 \Delta S}{R \Theta (1 - \Theta)^2}, \tag{11.18}$$

where Θ is T/T_M^0. As discussed previously, the entropy of fusion for most metals is approximately 2.3 cal/K mol. Thus we can write a general Boltzmann factor for the nucleation of metals as

$$\frac{-\Delta G^*}{kT} = -\frac{16}{3} \frac{\pi (0.45^3) 2.3}{R \Theta (1 - \Theta)^2} = -\frac{1.767}{\Theta (1 - \Theta)^2} \tag{11.19}$$

and the critical radius can be written as

$$r_{critical} = \frac{2 \gamma T_M^0 V_S}{\Delta H \Delta T} = \frac{2 \lambda V_S^{1/3}}{N_A^{1/3}} \frac{T_M^0}{\Delta T}. \tag{11.20}$$

Using Turnbull's estimate for the pre-exponential term to be $10^{33 \pm 1}$ in $cm^{-3} \, s^{-1}$, the nucleation rate is given by

$$I_V = 10^{33 \pm 1} \exp\left(-\frac{1.767}{\Theta (1 - \Theta)^2}\right) cm^{-3} \, s^{-1}. \tag{11.21}$$

Flemings estimates I_0 by starting with the number of critical nuclei n^* in equilibrium with N atoms $n^* = N \exp(-\Delta G^*/K_B T)$. He then takes the number of atoms adjacent to a critical nucleus, $4\pi (r_{critical})^2/a_0^2$ and estimates the nucleation rate as the rate at which these atoms join the critical nucleus to push it over the energy barrier. The rate at which the adjacent atoms join a critical nucleus is given by D_L/a_0^2, where D_L is the liquid diffusion coefficient and a_0 is an atomic radius. If one takes $N \sim 10^{22}$ atoms cm^{-3}, $r_{critical}/a_0 \sim 4$, and $D_L \sim 10^{-4} \, cm^2 \, s^{-1}$; $4\pi (r_{critical})^2/a_0^2 \sim 10^2$, $D_L/a_0^2 \sim 10^{11} \, s^{-1}$ and $I_0 \sim 10^{35} \, cm^{-3} \, s^{-1}$.

Kingery et al. suggest a model similar to Flemings' except they estimate the frequency that atoms cross from the melt to the embryonic solid as $kT/3\pi a_0^3 \eta$, where η is the absolute viscosity. For typical metallic melts, $\eta \sim 0.01 \, g \, cm^{-1} \, s^{-1}$. For $T \sim 1000 \, K$, the jump frequency $D_L/a_0^2 \sim 10^{11} \, s^{-1}$ which is the same order of magnitude as estimated by Flemings.

As it turns out, the pre-exponential term is relatively unimportant because the nucleation rate is dominated by the exponential term in Equation 11.21. As Θ increases past 0.8, the absolute value of the argument increases rapidly, presenting a very large barrier to nucleation. The exponential of a very large negative number becomes so small that the nucleation rate is nil regardless of the pre-exponential factor.

Perhaps the best way to represent the nucleation rate is through the use of Poisson statistics. For a Poisson distribution, the probability $P(n)$ of observing n events, when the expected number of events is n_{exp}, is given by

$$P(n) = \frac{n_{exp}^n \exp(-n_{exp})}{n!}. \tag{11.22}$$

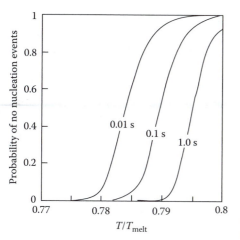

FIGURE 11.8
Probability of no nucleation events in a 5 μm droplet as a function of T/T_{melt} in 0.01, 0.1, and 1.0 s using Turnbull's model. There is virtually no chance that a particle can avoid nucleation when undercooled below 0.80 of its melting temperature.

(Poisson statistics apply when n_{exp}, is small. As n_{exp} becomes large, the Poisson distribution becomes a Gaussian distribution.) We can now use Equation 11.22 to compute the probability of observing no nucleation event in a volume V (cm^3) after t s.

$$P(0) = \exp[I_V V t] = \exp\left[-10^{33\pm1} V t \exp\left(-\frac{1.767}{\Theta(1-\Theta)^2}\right)\right]. \qquad (11.23)$$

Turnbull used very small drops to maximize the chance of eliminating impurities that could serve as nucleation sites. Assume a 5 μm drop. Taking the pre-exponential term as 10^{33}, the probability of observing no nucleation events in 0.1 s as a function of the dimensionless temperature is shown in Figure 11.8. From this plot, it can be seen that half of the drops would nucleate at $\Theta = 0.7895$ or undercool by 21.05%. Practically all of the drops would undercool by at least 20% and virtually none of the drops would undercool by more than 21.5%. To illustrate the insensitivity of the degree of undercooling to the pre-exponential factor, curves were also plotted for $t = 0.01$ and 1.0 s. Varying the pre-exponential factor by two orders of magnitude only shifts the undercooling for $P(0)$ from 0.784 to 0.795 or ±0.7%. In other words, a two-order magnitude variation in the pre-exponential factor only shifts the expected amount of undercooling from 20.5% to 21.6% of the melting temperature.

Equation 11.23 can also be used to find the average time a given droplet size can be held at a given Θ before it will be nucleate homogeneously.

$$t_{ave} = \frac{-\ln(0.5)}{10^{33\pm1} V} = \frac{0.693 \times 10^{-33\pm1}}{V} \exp\left(\frac{1.767}{\Theta(1-\Theta)^2}\right). \qquad (11.24)$$

Figure 11.9 illustrates the time before nucleation as a function of undercooling for different size molten droplets using Turnbull's theory. Droplets of virtually any size can be held more or less indefinitely at $\Delta T/T = 0.9\%$ or 10% undercooling. However, the time before nucleation decreases rapidly as the undercooling approaches 20%, especially for larger droplets. It became generally accepted that the maximum undercooling that could be achieved in a melt was 20% of the melting temperature and this became known as the Turnbull limit for undercooling.

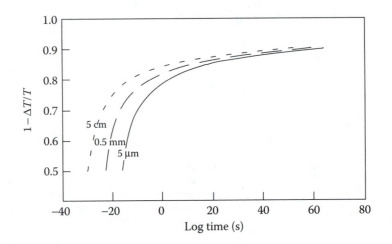

FIGURE 11.9
Time before probability of nucleation reaches 0.5 for 5 μm droplets (solid), 0.5 mm (dashed), and 5 cm (dotted) computed from Turnbull's model. Any size droplet can be undercooled by 10% more or less indefinitely. Only very small droplets can be undercooled by ~20%.

11.5 Heterogeneous Nucleation

The above theory of homogeneous nucleation applies only to melts in which there are no heterogeneous nucleation sites. In practice, it is very difficult to avoid heterogeneous nucleation sites that decrease the ΔG^* barrier height and limit the amount of undercooling that can be achieved. In most cases container walls act as nucleation sites, especially if they have a crystalline structure similar to the materials being undercooled. Nucleation can be delayed if the walls are chosen to be nonwetted by the melt and/or by making them amorphous. Still minute cracks or crevices will promote nucleation. Other nucleation sites may be in the form of microscopic second phase impurities with higher melting points.

A modification to the nucleation barrier to account for heterogeneous nucleation developed by Volmer simply multiplies the energy barrier ΔG^* in Equation 11.15 by $f(\theta)$, the ratio of the volume of a spherical cap sitting a surface with contact angle θ as shown in Figure 11.10, to the volume of a sphere with the same radius.

$$f(\theta) = \frac{(4\pi r^3/3)(1/4)(2 + \cos\theta)(1 - \cos\theta)^2}{4\pi r^3/3} = \frac{1}{4}(2 + \cos\theta)(1 - \cos\theta)^2. \qquad (11.25)$$

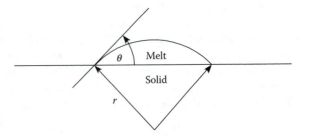

FIGURE 11.10
Molten sessile drop making contact angle θ with a solid surface.

Since a melt completely wets its own solid, the contact angle goes to zero and the barrier to nucleation disappears once the solid has formed. Nucleation can be inhibited by surrounding the melt with nonwetting surfaces. As the contact angle approaches 180°, the $f(\theta)$ approaches unity and the nucleation barrier approaches the ΔG^* for homogeneous nucleation.

11.6 Recent Developments in Undercooling Experiments

Turnbull was very limited in the techniques available to him during the 1950s when he did his pioneering work on the undercooling of melts at the General Electric Research Laboratory. He worked with an array of small droplets, typically 10–100 μm in diameter, arrayed on a flake of freshly blown quartz placed on top of a Pt strip heater and used a microscope to observe the solidification, either by the change of appearance of the droplet at low temperatures, or by the recalescence blick at high temperatures. He worked with small particles because there was less likely to be a foreign particle that could catalyze nucleation in an individual particle so the chance that it would undercool would be greater. He had no means of suspending the particles, so he had to rely on the large contact angle between metallic melts and quartz to limit heterogeneous nucleation. Since this angle is not 180°, it is doubtful that any of his particles actually nucleated homogeneously. In later experiments, he encapsulated Ge in molten B_2O_3, a glassy material that acts as a flux to remove surface contaminants and prevent oxidation, and was able to undercool Ge by 0.34 T_M^0. Subsequently, molecular dynamics simulations suggested that the maximum undercooling possible would be 0.33 T_M^0. It should be understood that if nucleation can be delayed until the viscosity of the melt becomes high enough to prevent crystal growth, a glass will result.

High-temperature refractory metals are highly reactive and require containerless techniques if they are to be undercooled. A 100 m drop tube was put in operation at the Marshall Space Flight Center (MSFC) in Huntsville, Alabama, followed by a 47.1 m drop tube in the Centre d'Etudes Nucléaires de Grenoble, France. Pendant metal droplets were melted from a wire using an electron beam and allowed to fall down the evacuated tube. Infrared detectors observed the recalescence flash and the amount of undercooling was determined by the heat lost to radiation at the time the flash was observed. Among other things, these facilities were used to study the microstructure and metastable phases in rapidly solidified metals.

An electromagnetic levitation facility developed through a joint effort by the German Aerospace Center DLR and ESA was flown on the second International Microgravity Lab (IML-2) and again on the Material Science Lab (MSL-1) Spacelab flights. Samples as large as 7–8 mm in diameter could be repeatedly melted and undercooled in a controlled environment without physical contact. Novel noncontact methods for measuring temperature and other thermophysical properties such as surface tension, viscosity, conductivity, thermal expansion, etc. were developed. An international team used this facility to study nucleation in a repeatable manner and to measure thermophysical properties in an undercooled melt that had not been possible before. Knowing such properties is very important in the design of bulk metallic glasses.

Meanwhile, a group at the Jet Propulsion Lab developed an electrostatic levitator that could levitate drops as large as 2 mm in diameter in an ultrahigh vacuum or a controlled atmosphere using an active feedback system to position the sample. A similar instrument was acquired by the MSFC and was equipped with lasers to melt the nonmetal as well as metal samples and with diagnostics for measuring properties of molten materials in their

undercooled state. A similar facility was later constructed to operate in the beam line of the Advanced Photon Source at Argonne National Laboratory where it is being used to study the structure of materials during the nucleation process.

11.7 Summary

In dealing with a single component system, the Gibbs phase rule tells us the melting point is a function of pressure only. Since most of the processes we will be dealing with take place at ambient pressure we will concern ourselves with T_M^0, the temperature at which a plane front solid can remain indefinitely with its melt.

The Gibbs free energy is defined as $G = H - TS$. Nature always strives to minimize the free energy of a system. The liquid phase has a higher (less negative) enthalpy than a solid because bonds have been broken. It also has higher entropy than a solid (more disorder, more available states). At low temperatures, the lower enthalpy of the solid outweighs the larger $-TS$ term of the liquid and the solid is the stable phase. As the temperature is increased, the $-TS$ of the melt eventually dominates and liquid becomes the stable phase. The T_M^0 occurs when the Gibbs energy of the solid equals that of the melt.

There is a discontinuity in the slope of the free energy when plotted against temperature at the melting point as the solid free energy curve goes over to the steeper liquid free energy curve. A phase transition in which there is a discontinuity in the first derivative of the free energy is called a first-order phase transition and is accompanied by a release (or absorption) of the enthalpy of fusion. A second-order phase transformation has a discontinuity in the second derivative of the free energy and there is a discontinuity in the heat capacity rather than in the enthalpy.

If both the solid and melt are under the same pressure, the dependence of the melting point with pressure is determined by the Clapeyron equation and depends on whether the solid expands or contracts on freezing. However, if only the solid is under pressure, the melting point is always lowered by increasing the pressure on the solid. A solid with a convex solid–liquid interface is under pressure due to its interfacial energy; therefore, the local melting point is lowered (Gibbs–Thompson effect). Very small (nanometer-size) particles can have melting points well below the T_M^0 of a plane face solid. These effects become important in understanding grain growth and other ripening phenomena as well as in nucleation theory.

When a heated crystal reaches its melting temperature, its temperature remains constant until the enthalpy of fusion is absorbed and the crystal is completely melted before the temperature will continue to rise, causing a thermal arrest. When a melt is cooled to its melting temperature, it does not immediately solidify. In order to drive solidification, the free energy of the solid must be lowered, which is generally done by lowering the temperature below the melting point, which is called undercooling. Very little undercooling is required for a melt in the presence of its solid, but in the absence of a crystalline particle or surface to act as a nucleation site, melts can be undercooled by as much as ~33% of their absolute melting temperature before nucleation occurs. When nucleation does occur, the release of the latent heat (enthalpy of fusion) brings the melt back to the melting temperature before additional cooling can occur. This sudden increase in temperature is called recalescence.

Even though the release of latent heat raises the temperature of an undercooled melt back to the melting point, this heat has already been removed so that solidification is very rapid. Therefore, the ability to undercool a melt is very useful in developing rapid solidification processing for producing metastable states or other novel microstructures.

In the absence of any nucleation site, homogeneous nucleation will eventually occur as the atoms cluster together to form embryonic nuclei. There is a critical radius for these

nuclei to become viable and grow into a solid, which is obtained from the Gibbs–Thompson equation. The free energy required to create a viable nucleus forms a nucleation barrier. Thus homogeneous nucleation is a stochastic process in which we can only estimate the probability of a thermal fluctuation large enough to surmount the nucleation barrier for a cluster of atoms to form a viable nucleus.

Several different models have been proposed for estimating the nucleation rate and Poisson statistics can be used to estimate the average amount of undercooling one would expect for a given system, or one can estimate the length of time a given system can be held before nucleation occurs. These models contain simplifying assumptions and have large uncertainties, but still are useful in developing processes for glass formation where it is necessary to undercool bulk melts to the point where the viscosity of the melt becomes high enough to prevent crystal formation.

Nucleation events are found in many aspects of materials processing besides the solidification of a melt. A nucleation event is required for precipitates to form in precipitation-hardened alloys, for the decomposition of immiscible systems, for new grains to form in a casting, for graphite flakes to form in cast iron, phase selection in alloy solidification and in crystal growth, etc. Understanding of the nucleation process is key to being able to control it in order to achieve the desired result in a process.

Bibliography

Bayuzick, R.J. et al., Review on long drop towers and long drop tubes, Collins, E.W and Koch, C.C., Eds., *Undercooled Alloy Phases*, ASM International, 1986, 207.

Chalmers, B., *Principles of Solidification*, Robert E. Krieger, Huntington, New York (Reprinted by arrangement from 1964 version published by John Wiley & Sons).

Flemings, M.C., *Solidification Processing*, McGraw Hill, New York, 1974 (McGraw Hill Series *in Materials Science and Engineering*).

Gocken, N.A. and Reddy, R.G., *Thermodynamics*, 2nd edn., Springer, Berlin, 1996.

Ragone, D.V., *Thermodynamics of Materials*, Vol. II, John Wiley & Sons, New York, 1995.

Rathz, T.J., et al., The Marshall space flight center drop tube facility, *Rev. Sci. Instrum.*, 61/12, 1990, 3846.

Rhim, W.K., An electrostatic levitator for high-temperature containerless materials processing in 1-G, *Rev. Sci. Instrum.*, 64/10, 1993, 2961–2970.

Schmelzer, J.W.P., Ed., *Nucleation Theory and Applications*, Wiley-VCH, Weinheim, 2005.

Turnbull, D., Formation of crystal nuclei in liquid metals, *J. Appl. Phys.*, 21, 1950, 1022–1027.

Problems

1. We generally ignore the atmospheric pressure when dealing with materials. (a) Find the difference between the melting point of Fe at 1 atm (atmospheric pressure $\sim$1 bar $= 10^5$ Pa) and in a vacuum. (b) What would be the melting point of Fe deep inside the Earth where the pressure is 5 Mbar. Assume the solid is immersed in its melt. (c) How would we know if the Fe was molten or not at that pressure?

2. Powdered metals are sintered together using a hot isostatic press (HIP) typically operating at 2 kbars. If 1 μm Fe particles are pressed at 2 kbars, how much would their melting temperature be lowered by the applied pressure? (Assume the particles have only solid–solid contact.) How much would the local melting point be lowered by their tip radius?

3. Use the Poisson probability distribution to find the number of times you would expect heads (or tails) to occur in 100 coin tosses.

4. How many times can you expect to roll a pair of dice without a 7 showing up?

5. Calculate the critical radius of an Fe nuclei in a melt that is undercooled by 0.33 of its normal melting temperature. How many atoms would be contained in such a cluster?

6. Show that the volume of a spherical cap sitting on a surface with contact angle θ is given by $(4\pi r^3/3)(1/4)[(2 + \cos\theta)(1 - \cos\theta)]$.

12

Phase Equilibria in Multicomponent Systems

Phase diagrams are the basic maps required for determining the melting points of alloys or compounds and for designing processes to obtain a desired composition of a system to be solidified from the melt. In this chapter we will develop not only a working knowledge of how to use phase diagrams but also an understanding of how the various phase diagrams evolve from free energy considerations.

12.1 Gibbs Phase Rule

The Gibbs phase rule tells us that two phases can coexist in a two-component system at constant pressure with one degree of freedom, i.e., the phase line will be a function of both temperature and composition. There are no degrees of freedom when three phases coexist; therefore, this can occur only at a specific temperature and composition called an invariant point.

12.2 Entropy of Mixing

For solidification processes carried out at ambient pressure (~ 1 bar), the Helmholtz free energy, $F = U + TS$, is sometimes used interchangeably with G. The term "free energy" is often used without specifying Gibbs or Helmholtz energy which, for all practical purposes, are synonymous for solidification processes at atmospheric pressure because the pdV contribution is negligible at this pressure. Two additional terms will have to be included in the Gibbs free energy function: heat of mixing and entropy of mixing.

Let us first consider the entropy of mixing. A mixture has more accessible states than a pure substance, hence a higher entropy. This can be quantified in the following manner.

Let N_A be the number of A atoms and N_B be the number of B atoms. The total number of atoms $N = N_A + N_B$. The N atoms can be arranged in $N!$ different ways, but since atoms are indistinguishable, N_A and N_B of these are redundant. Thus the number of unique ways of arranging N atoms is

$$W = \frac{N!}{N_A! N_B!} = \frac{N!}{(N - N_B)! N_B!}. \tag{12.1}$$

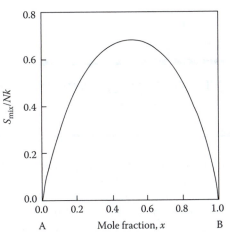

FIGURE 12.1
Dimensionless entropy of mixing vs. mole fraction.

Since the Ns are very large, the factorials can be represented by Stirling's approximation, $\ln x! \approx x \ln x - x$. The entropy of mixing can then be expressed as

$$S_{mix} = k \ln\left(\frac{N!}{(N - N_B)! N_B!}\right) = k\left[N_B \ln\left(\frac{N - N_B}{N_B}\right) - N \ln\left(\frac{N - N_B}{N}\right)\right]. \qquad (12.2)$$

It is more convenient to work with the mole fraction $x \equiv N_B/N$. Introducing this in the above expression,

$$S_{mix} = -Nk\left[\ln(1 - x) + x \ln\left(\frac{1 - x}{x}\right)\right] = -Nk[(1 - x)\ln(1 - x) + x \ln(x)]. \qquad (12.3)$$

This function is symmetrical about $x = 0.5$ as may be seen in Figure 12.1. Note also that the slope of S approaches infinity at $x = 0$ and minus infinity at $x = 1$ (which can easily be verified by differentiating Equation 12.3). This behavior has significant implications in the contamination of melts by their containers (see Figure 12.18).

12.3 Heat of Mixing

Next we consider the heat of mixing using a simple nearest neighbor model. Let u_{AA} be the nearest neighbor binding energy per atom for pure A, u_{BB} for pure B, and u_{AB} for an A atom next to a B atom. For mole fraction x of B atoms, the probability of atom B next to atom A is x and the total energy for the A atoms is

$$U_A = N_A \frac{p}{2}[(1 - x)u_{AA} + xu_{AB}], \qquad (12.4)$$

where p is the number of nearest neighbors. The factor $1/2$ is included since each bond is shared between two atoms. Similarly, the total energy for the B atoms is

$$U_B = N_B \frac{p}{2}[(1 - x)u_{AB} + xu_{BB}]. \qquad (12.5)$$

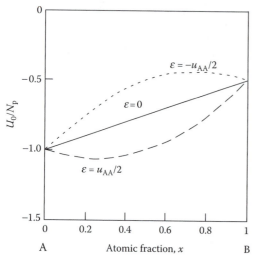

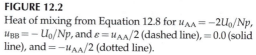

FIGURE 12.2

Heat of mixing from Equation 12.8 for $u_{AA} = -2U_0/Np$, $u_{BB} = -U_0/Np$, and $\varepsilon = u_{AA}/2$ (dashed line), $= 0.0$ (solid line), and $= -u_{AA}/2$ (dotted line).

Since the number of A atoms is $N(1-x)$ and the number of B atoms is Nx, the total energy is

$$U_{tot} = N\frac{p}{2}\left[(1-x)^2 u_{AA} + x^2 u_{BB} + 2x(1-x)u_{AB}\right].$$ (12.6)

Note that if u_{AB} is the average of u_{AA} and u_{BB}, Equation 12.6 reduces to

$$U_{tot} = N\frac{p}{2}\left[(1-x)u_{AA} + xu_{BB}\right].$$ (12.7)

If u_{AB} is written as $\langle u_{AA} + u_{BB}\rangle/2 + \varepsilon$, where ε is the excess heat of mixing per atom, Equation 12.6 can be written as

$$U_{tot} = N\frac{p}{2}\left[(1-x)u_{AA} + xu_{BB} + 2x(1-x)\varepsilon\right].$$ (12.8)

Note that Equation 12.8 plots as a straight line in Figure 12.2 for $\varepsilon = 0$ (no excess heat of mixing), curves upward for $\varepsilon > 0$ (positive excess heat of mixing), and curves downward for $\varepsilon < 0$ (negative excess heat of mixing).

12.4 Free Energy

The total free energy can be written as $F(T,x) = U_{tot} - TS_{mix}$. From Equations 12.3 and 12.8, we obtain

$$F(x,T) = \frac{Np}{2}\left[(1-x)\mu_{AA} + x\mu_{BB} + 2x(1-x)\varepsilon\right] + NkT[(1-x)\ln(1-x) + x\ln(x)].$$ (12.9)

The free energy per atom as a function of temperature is shown in Figure 12.3 for different excess energies. Taking the second derivatives of Equation 12.9

$$\frac{\partial^2 F(T,x)}{\partial x^2} = -2Np\varepsilon + \frac{NkT}{x(1-x)}.$$ (12.10)

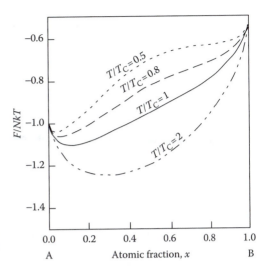

FIGURE 12.3
Free energy plot for various values of T/T_C, where the consolute temperature T_C is defined by Equation 12.12. The mixed phase is stable for all x if $T = T_C$ (solid line) but the system will become segregated if $T < T_C$ as will be explained in Section 12.4.1.

The curvature of the free energy will continue to be positive (curve upwards) so long as

$$kT - 2p\varepsilon x(1 - x) \geq 0 \qquad (12.11)$$

and the mixed phase will be stable over the entire range of x. Clearly this will always be the case if the excess energy $\varepsilon < 0$ (the atoms prefer their counterparts to their own kind). However, if $\varepsilon \geq 0$, the free energy curve may exhibit an inflection point if Equation 12.11 becomes 0. The lowest temperature where this can happen is called the consolute temperature T_C, where for this system

$$kT_C = p\varepsilon/2. \qquad (12.12)$$

Also note that no matter how large the heat of mixing may be, the initial free energy curves always have an initial drop at $x = 0$ and 1. This is because the entropy of mixing has an infinite slope at $x = 0$ and 1 as was shown previously. This implies that no matter how dissimilar the two components may be, there will always be a limited region of solid phase solubility, even though it may be vanishingly small. This will have significant consequence in the purification of materials, especially when we later discover that certain trace impurities at concentrations far less than our ability to detect chemically (less than parts per billion) can have a dramatic influence of the electrical properties of semiconductors.

12.4.1 Miscibility

Systems in which the bonding between atoms A and B are different, e.g., a polar substance such as water and a nonpolar substance such as oil, will have $\varepsilon > 0$. Solids in which the A atoms have substantially different sizes than B atoms will not form bonds between each other that are as strong as bonds between themselves; hence $\varepsilon > 0$. However in these cases, a single phase will form at $T > T_C$ because of the lowering of the free energy due to the $-TS$ term, but for $T < T_C$ there will be a range of compositions over which the system will separate into two phases. This region is termed as a miscibility gap and can be present in either the solid or the liquid phase, or both.

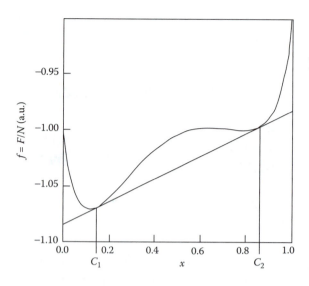

FIGURE 12.4
Free energy curve at some arbitrary temperature for a binary system with an immiscibility
gap. The free energy of the system is minimized
when the system separates into two phases, one
with $x = C_1$ (α phase) and the other with $x = C_2$
(β phase). The compositions are determined by
constructing a line that is mutually tangent to
the lowest regions of the free energy curve. The
compositions of the two phases correspond to
the x-values of the two tangent points.

12.4.2 Method of Tangents

To illustrate the method for mapping out multiple phase systems, an arbitrary free energy
curve for $T < T_C$ is shown in Figure 12.4. In order to find the compositions of the two
segregated phases, a line is drawn that is mutually tangent to the two lowest regions of the
free energy curve. The composition of each of the two phases is given by x-values of the
two tangent points. The phase with the smallest x-value is usually called the α phase and,
for the case illustrated by Figure 12.4, it would have a composition A-13.2 At% B at the
temperature corresponding to this particular free energy curve. The second phase is called
the β phase, and in this particular case, the composition is A-85.9 At% B (or B-14.1 A). At%
refers to atomic percent or mole percent. Compositions are also frequently specified in
terms of weight percent (wt%). The method for conversion is given in elementary texts
(see Problem 12.3).

Notice that the free energy curve in Figure 12.4 is in the units of $f = F/N$ which is the
average free energy per particle. To find the total free energy of the segregated mixture, we
can write

$$F = N_\alpha f(x_\alpha) + N_\beta f(x_\beta), \tag{12.13}$$

where
  N_α and N_β are the number of atoms in the α and β phases, respectively
  $f(x_\alpha)$ and $f(x_\beta)$ are the respective values of the free energy curve at x_α and x_β

We shall proceed to show that the free energy of the segregated system is lower than the
free energy of the homogeneous system in the miscibility gap region. We shall also
demonstrate that the segregated system with the composition found by the method of
tangents is in equilibrium. But first we have to find the amount of the two phases present,
namely N_α and N_β.

12.4.3 Lever Rule

The phase diagram tells you directly the compositions of the two phases that are in
equilibrium, but to get the relative amounts of the material in each of the two phases,

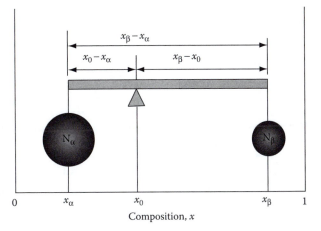

FIGURE 12.5
Illustration of the lever law. The fulcrum represents the starting composition x_0N = number of B atoms in the system. N_α is the number of atoms in the α phase and $x_\alpha N_\alpha$ is the number of B atoms in the α phase. Likewise, N_β is the number of atoms in the β phase and $x_\beta N_\beta$ is the number of B atoms in the β phase.

we first specify the average composition x_0 of the system. Since $N_\alpha + N_\beta = N$, and $x_\alpha N_\alpha + x_\beta N_\beta = N_B$ (total number of B atoms) $= x_0N$, we can solve these two equations simultaneously to obtain

$$\frac{N_\alpha}{N} = \frac{x_\beta - x_0}{x_\beta - x_\alpha}, \quad \frac{N_\beta}{N} = \frac{x_0 - x_\alpha}{x_\beta - x_\alpha}. \tag{12.14}$$

This simple but important result is known as the lever rule because it has the same form as the equation for balancing a lever arm as shown in Figure 12.5.

Assume the left end of the lever arm is located at x_α, the right end is at x_β, and x_0 is the fulcrum. If x_0 is closer to x_α than to x_β, N_α must be greater than N_β to balance the system.

The lever rule also applies to phase diagrams in weight fraction or wt% if the xs in Equation 12.14 are expressed in terms of weight fraction or wt% and mole fractions N_α/N and N_β/N are replaced by weight fractions W_α/W and W_β/W, respectively.

Putting these results into Equation 12.13, the free energy of the segregated system becomes

$$\hat{f}(x) = \frac{F}{N} = \frac{1}{x_\beta - x_\alpha} \left[(x_\beta - x)f(x_\alpha) + (x - x_\alpha)f(x_\beta) \right], \quad x_\alpha \le x \le x_\beta. \tag{12.15}$$

This function is linear in x and intersects $f(x_\alpha)$ at x_α and intersects $f(x_\beta)$ at x_β, which is just the mutual tangent curve we constructed previously in Figure 12.5. Since the free energy of the segregated system $\bar{f}(x)$ is lower than the homogeneous system $f(x)$ between x_α and x_β, the stable configuration of the system will be two distinct phases with compositions x_α and x_β.

12.4.4 Chemical Potential

In order to demonstrate that the compositions x_α and x_β are in equilibrium, we need to introduce the chemical potential, μ, of a particular species defined as the change in free energy resulting from the addition of one atom of that species, all other species held constant. Symbolically, this is written as

$$\mu_i \equiv \left(\frac{\partial G}{\partial N_i} \right)_{T,p,N_1,N_2,\dots,N_{i,i\ne j}} = \left(\frac{\partial F}{\partial N_i} \right)_{T,V,N_1,N_2,\dots,N_{i,i\ne j}} \tag{12.16}$$

even though, strictly speaking, the functions are not continuous, hence are not differentiable. For the system to be in equilibrium, each component must have the same chemical potential in the coexisting phases; otherwise, the free energy could be reduced by a net migration of components between phases. To show that the above system is indeed in equilibrium with compositions x_α and x_β, we must show that $\mu_A(x_\alpha) = \mu_A(x_\beta)$ and $\mu_B(x_\alpha) = \mu_B(x_\beta)$.

To proceed, we introduce $f = F/N$, where f is the average free energy per particle as used in Figure 12.5. For a two-component system

$$\mu_A = \left(\frac{\partial Nf}{\partial N_A}\right)_{T,p,N_B} = f + N\frac{\partial f}{\partial N_A} = f + N\frac{\partial f}{\partial x}\frac{\partial x}{\partial N_A} = f + N\frac{\partial f}{\partial x}\left(-\frac{x}{N}\right),$$

which reduces to

$$\mu_A = f - x\frac{\partial f}{\partial x}. \tag{12.17}$$

Similarly, it can be shown that

$$\mu_B = f + (1 - x)\frac{\partial f}{\partial x}. \tag{12.18}$$

Eliminating f in Equations 12.17 and 12.18, we can write

$$f' = \mu_B - \mu_A. \tag{12.19}$$

Since $f'(x_\alpha) = f'(x_\beta)$ because of the mutual tangent line, $\mu_B(x_\alpha) - \mu_A(x_\alpha) = \mu_B(x_\beta) - \mu_A(x_\beta)$, from which $\mu_B(x_\alpha) = \mu_B(x_\beta)$ and $\mu_A(x_\alpha) = \mu_A(x_\beta)$, QED.

12.4.5 Constructing a Phase Diagram from Free Energy Curves

The free energy curves at the top of Figure 12.6 were computed for an arbitrary system in which $u_{AA} = -1$, $u_{BB} = -0.9$, and $\varepsilon = 1$. Therefore, the $kT_C = 1$. Given this family of free energy curves for different temperatures, the phase diagram may be constructed by drawing mutual tangents to the lower extremities of each of the free energy curves. The points of intersection of each free energy curve with its mutual tangent line are then projected onto the corresponding isotherm (family of horizontal lines on the phase diagram). The locus of these point maps out the coexistence region for the two phases on a temperature vs. composition plot which constitutes the phase diagram as shown in Figure 12.5. For $T > T_C$, a single phase exists for all compositions. For $T < T_C$, the system segregates into two immiscible phases as indicated by L_1 and L_2. Although the system is labeled as two immiscible liquid phases, the same argument applies to immiscible solid phases.

12.4.6 Spinodal Decomposition

We just found the extent of the immiscible region using the method of tangents to map out the points where these tangents intersected the lowest portions of the free energy curve. However, the portions of the free energy curve between the end points and the inflections have positive curvature, which means they would be stable if taken individually. As a result, the region between the intersection of the mutual tangent and the inflection point is metastable, meaning that so long as the second derivative of the free energy remains

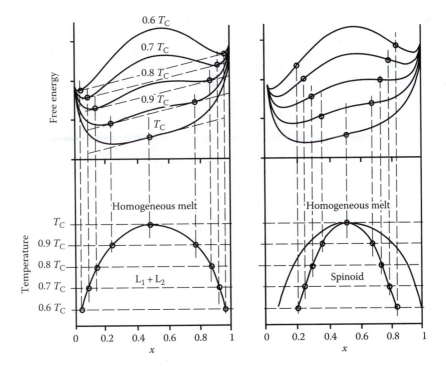

FIGURE 12.6
Phase diagram (left hand side) and spinodal (right hand side) for a completely immiscible system derived from free energy considerations. Mixtures cooled into the spinodal separate spontaneously whereas mixtures between the spinodal and the region of immiscibility are metastable and require a nucleation event to separate.

positive, a nucleation barrier exists that must be overcome in order to form the segregated phase. However, in the region between the two inflection points where the second derivative of the free energy is negative, there is no barrier to nucleation and decomposition occurs spontaneously. This spontaneous decomposition is known as spinodal decomposition and can occur in the solid as well as in the liquid phase.

The spinodal is located within the miscibility gap and can be mapped by projecting the points of inflection of the free energy curves (where $\partial^2 F/\partial x^2 = 0$) on to their corresponding isotherms as shown in Figure 12.6.

12.5 Phase Diagram for Ideal (Isomorphic) Systems

12.5.1 Solid Solutions and the Hume–Rotherty Criteria

Complete solid solubility for a binary system means that atoms of one constituent can be substituted for the other over the entire range of compositions in both the solid and liquid state without introducing a different phase. Such alloys are said to be isomorphous. Hume–Rothery developed a set of criteria that must be met in order for a system to exhibit complete solid solubility:

- The atoms must differ in size by no more than ~15%.
- The two components must crystallize with the same lattice type.

- The two components should not differ in electronegativity by $>\sim 0.4$ (otherwise intermediate phases are likely to form).
- The two components must have the same valence. (The reason for this rule involves Fermi surfaces and Brillouin zones and will be discussed in Chapter 19.)

Given the similarity of atomic radii between the 4d and 5d transition metals, one would expect those pairs in the same column to be isomorphous. This is the case for Hf–Zr, Mo–W, Ru–Os, and Au–Ag. Si–Ge is also isomorphous as are several compound systems such as HgTe–CdTe, Al_2O_3–Cr_2O_3 that exhibit pseudobinary isomorphous phase diagrams.

Some mixed valency systems such as Cu–Ni and Ag–Pd are isomorphic. Other systems meet all of the Hume–Rothery criteria but are immiscible in the liquid state (e.g., Nb–Ta) or are completely miscible in the melt but decompose spinodally in the solid state (e.g., Pd–Pt). One must conclude that there is nothing hard and fast about the Hume–Rothery rules and they should be used only as a guide to estimating which systems may be isomorphous.

We will now set out to compute the free energies for the solid and liquid phases of an ideal isomorphous binary system and the resulting phase diagram using a very simple model. If we add the entropy of mixing (Equation 12.3) to the chemical potentials of pure A and pure B solid μ_A^0 and μ_B^0, we can write the free energy of the solid as

$$F_S(x \cdot T) = U_S(x) - TS_S = (1 - x)\mu_A^0 + x\mu_B^0 + RT(x\ln(x) + (1 - x)\ln(1 - x)), \quad (12.20)$$

where x is the mole fraction of component A. The free energies of the pure liquids are $\mu_A^0 + \Delta S(T_A - T)$ and $\mu_B^0 + \Delta S(T_B - T)$, where T_A and T_B are the melting temperatures of A and B, respectively. Taking the pure solid A and B and the standard states, we may set their chemical potentials to 0. Assuming no excess heat of mixing, the free energies of the solid and liquid then become

$$F_S(x \cdot T) = RT(x\ln(x) + (1 - x)\ln(1 - x)) \quad (12.21)$$

and

$$F_L(x \cdot T) = RT(x\ln(x) + (1 - x)\ln(1 - x)) + \Delta S[x(T_B - T) + (1 - x)(T_A - T)]. \quad (12.22)$$

Taking $\Delta S = 2.3$ cal mol^{-1} = 9.6 J mol^{-1}, $T_A = 1683$ K, and $T_B = 1210$ K (melting points for Si–Ge), the free energy curves were calculated for different temperatures in Figure 12.8. At the top are the free energies at 1210 K. Note that the F_S and F_L meet at $x = 1$ (pure Ge) and that the F_S curve is always lower than the F_L curve, indicating that the solid phase is stable below 1210 K for all x. As the temperature is increased above the melting point of Ge, notice that the F_L curves drop because of the $-T\Delta S$ term and move through the F_S curve until at $T = 1683$ K, the F_L curve is always lower than the F_S curve (except where they meet at 1683 K), indicating that the liquid phase is present at temperatures above 1683 K for all x. At each temperature between the two melting points, mutual tangents can be drawn between the two free energy curves near the point of their intersection. The x-values of these points of intersection can be projected onto a T versus x plot, which maps out the phase diagram for the Si–Ge binary alloy seen in Figure 12.7. This may be compared to the actual Ge–Si phase diagram in Figure 12.8.

12.5.2 Segregation or Partition Coefficient

Note that even ideal alloys melt and freeze incongruently. Unless there is a negative heat of mixing, the lower melting point component B will be rejected as the first-to-freeze A-rich crystal grows. As a result, the composition of the melt will be different from the solid at a

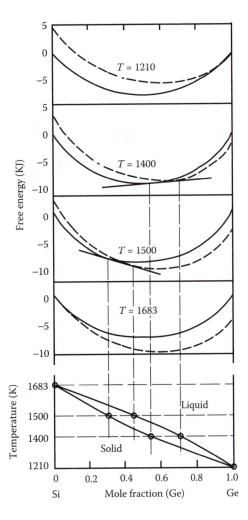

FIGURE 12.7
Construction of a phase diagram for the Si–Ge system, which is assumed to be isomorphic with no excess energy of mixing in either phase. The solid lines represent the free energy of the solid phase and the dashed line represent the free energies of the liquid phase. In the phase diagram, the line above which everything is liquid is the liquidus line. The line below which everything is solid is the solidus line. Liquid and solid coexist in the region between these two lines.

given temperature. In the liquid–solid coexistence region, an isotherm in the form of a horizontal tie-line is drawn through the two-phase region as shown in Figure 12.9. The intersections of this tie-line with the boundaries of the two-phase region dictate the composition of the melt and solid that are in equilibrium with each other. For example, if a melt with a starting composition C_0 is cooled to temperature T_1 where it first contacts the liquidus, grains of solid with composition C_1 begin to form (assuming there is no nucleation barrier) which are in equilibrium with melt composition C_0. The ratio of the composition of the solid that is in equilibrium with the melt is called the segregation or the partition coefficient, $k = C_S/C_L$.

12.5.3 Equilibrium Solidification

As the temperature is lowered to T_2, the A atoms that have gone into the solid have been removed from the melt, shifting its composition from C_0 to C_3. Meanwhile some of the B atoms have been incorporated into the growing solid giving it composition C_2 so it is in equilibrium with the liquid at C_3. The relative amounts of the solid and melt at this temperature can be found from the lever law, $N_S(C_0 - C_2) = N_L(C_3 - C_0)$. Since $N_S + N_L = N$

$$N_S = N\frac{C_3 - C_0}{C_3 - C_2}; \quad N_L = N\frac{C_0 - C_2}{C_3 - C_2}. \tag{12.23}$$

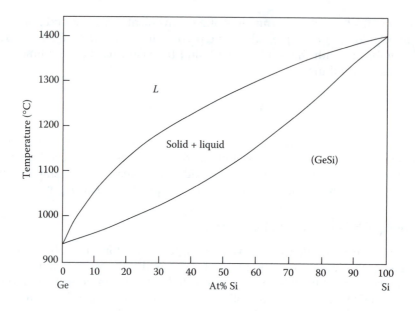

FIGURE 12.8

Phase diagram for the Ge–Si system. The Ge is on the left in this diagram and the liquid–solid two-phase region is somewhat fatter than in the calculated case, possibly due to small departures from ideality of the system. (Reprinted from Massalski, T.B., *Handbook of Binary Alloy Phase Diagrams*, Vols 1, 2, and 3, ASM, 1990. Reprinted with permission of ASM International. All rights reserved.)

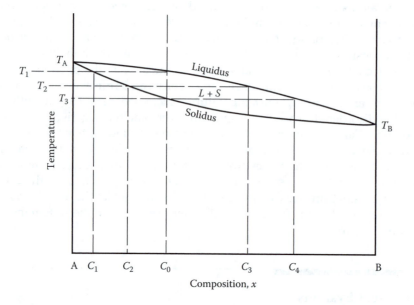

FIGURE 12.9

Equilibrium solidification process in a solid solution system. When a melt with composition C_0 is lowered to T_1, the first-to-freeze component will have composition C_1. The partition coefficient $k = C_1/C_0$. At T_2, solid with composition C_2 will be in equilibrium with liquid whose composition is C_3. Finally at T_3, solid with composition C_0 will be in equilibrium with the vanishingly small amount of the last-to-freeze liquid whose composition is C_4. But see Figure 13.3.

Instead of using the C_0 as the fulcrum, it is often convenient to use the end points of the tie line as a pivot point. For example, taking the pivot point at C_2 for T_2, the mole fraction of liquid in equilibrium with the solid can be found by balancing $(C_0 - C_2)$ times N against $(C_3 - C_2)$ times N_L resulting in

$$x_L = \frac{N_L}{N} = \frac{C_0 - C_2}{C_3 - C_2}. \tag{12.24}$$

It is assumed here that solidification is being carried out as a series of equilibrium states in which the solid continually adjusts through solid-state diffusion to the value corresponding to the intersection of the tie-line with the solidus. Eventually as the temperature falls to T_3, the final solid will have a uniform composition equal to the starting value.

12.5.4 Nonequilibrium Solidification—Coring

Solidification is rarely carried out slowly enough for the forming solid to continue to adjust to the equilibrium value. The first-to-freeze solid will retain its initial C_1 composition. The composition of the melt becomes depleted of the A component and moves down along the liquidus so that the forming solid becomes increasingly B-rich. The final solid will have an average composition of C_0, but the composition will range from C_1 in the first-to-freeze region to less than C_0 in the last-to-freeze region. This variation in composition between the first-to-freeze and last-to freeze is known as coring. More will be said about this in Chapter 13.

12.5.5 Order–Disorder Transitions

In a solid solution, A atoms may be freely substituted for B atoms and a random substitution of such atoms constitutes the disordered phase. However, an ordered phase may exist at low temperatures. For example, in bcc or β brass, an ordered phase exists below 468°C in which the Cu atoms sit on the corners of the lattice while the Zn atoms occupy the body-centered position. Other examples are A_3B systems in which the B atoms sit on the corners of an fcc lattice while the A atoms are on the faces in the ordered configuration. This ordering constitutes a lowering of entropy, which can be offset by the negative heat of mixing of the ordered phase. As the temperature is increased, the $-TS$ term with the higher entropy of the disordered phase will eventually overcome the negative heat of mixing of the ordered phase and the disordered phase will have the lowest free energy.

Order–disorder transitions are examples of a second-order transformation. An order-parameter can be assigned that goes from one for a perfectly ordered state to zero for a completely random state, i.e., a solid solution. Using a technique similar to the Curie–Weiss model for ferromagnetism (see Chapter 25), it can be shown that the order parameter goes from 1 at low temperatures to 0 at the transition temperature and the system then becomes a solid solution again.

12.6 Nonideal Systems

The phase diagram in Figure 12.8 will be altered in different ways depending on whether the excess heat of mixing is positive or negative. If there is a tendency for atoms in the solid to bond ionically or covalently, the solid phase will have a more negative heat of mixing than the liquid and a phase diagram similar to that shown in Figure 12.10 may result. These free energy curves were constructed from Equations 12.21 and 12.22 except an excess

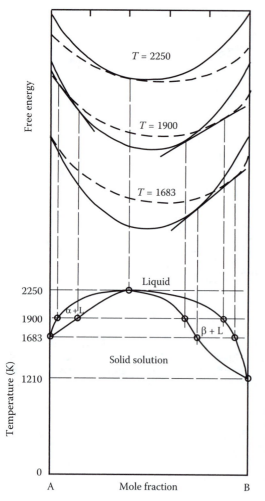

FIGURE 12.10

Hypothetical phase diagram for a system similar to Ge–Se but in which the solid excess heat of mixing was −30 kJ mol^{-1}. This high negative excess heat of mixing in the solid pushes the solidus line toward higher temperatures, which in this case allows congruent melting/solidification to occur at the point of highest temperature.

energy term $x(1-x)\varepsilon$ is added to the solid free energy and ε was set to −30 kJ mol^{-1}. Notice that the melting point was raised above the melting points of the pure components and that the two-phase region collapsed at one point where the system melts congruently. Only at the composition corresponding to a congruent melting point, can melting and freezing take place without a change in composition.

A similar but opposite effect will occur if the solid had a higher excess energy than the liquid. In this case the melting points will be below those of the pure components.

12.6.1 Intermediate Phases

If the heat of mixing of the solid is sufficiently negative, a new phase will form within some range of the composition x. This could be a new crystalline structure in which some substitution of A and B is permitted, either as a solid solution as in Al–Ni (Figure 12.11) or through its defect structure as discussed in Chapter 9. The formation of spinel ($MgAl_2O_4$) shown in Figure 12.12 is an example of the latter.

Or the new phase could be a line compound in which very limited deviations from stoichiometry are permitted as shown in Figure 12.13. Examples of such line compounds are the III–V and II–VI compound semiconductors such as GaAs, InP, CdTe, ZnSe, etc. When a

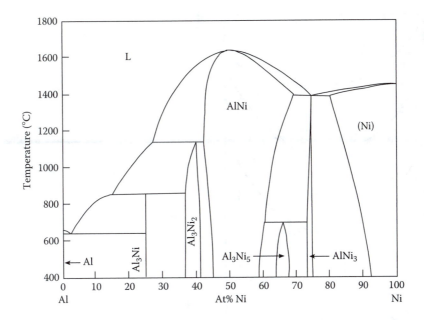

FIGURE 12.11

Al–Ni phase diagram. Intermetallic phases Al₃Ni, Al₃Ni₂, AlNi, Al₃Ni₅, and AlNi₃ are seen. Note that the intermetallic AlNi phase has a higher melting point than either Ni or Al. This stoichiometric Al-50 At % Ni phase has a broad range of existence (45%–58% Ni) at ambient temperature. (Reprinted from Massalski, T.B., *Handbook of Binary Alloy Phase Diagrams*, ASM, 1990. Reprinted with permission of ASM International. All rights reserved.)

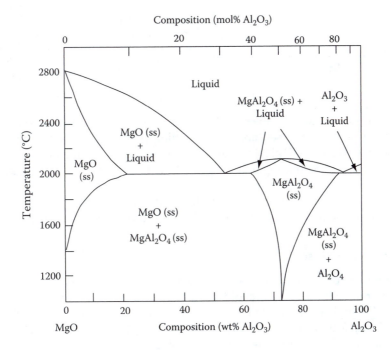

FIGURE 12.12

MgO–Al₂O₃ phase diagram. In this case the composition in wt% is along the bottom and At% is along the top. The compound spinel (MgAl₂O₃) forms at 50 At% and has a wide coexistence region near the melting temperature because of vacancy defect formation (see Problem 12.6). (From Halsedt, B., *J. Am. Ceramic Soc.*, 75/6, 1992. With permission.)

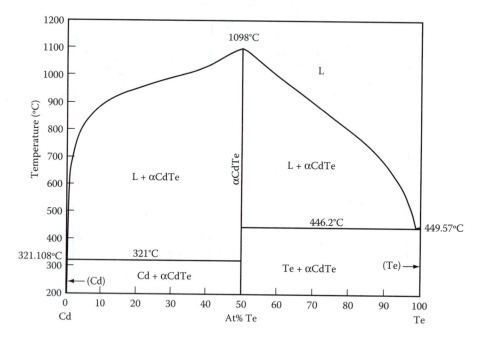

FIGURE 12.13

Cd–Te phase diagram. The line compound at 50 At% is CdTe. (Reprinted from Massalski, T.B., *Handbook of Binary Alloy Phase Diagrams*, ASM, 1990. Reprinted with permission of ASM International. All rights reserved.)

line compound is formed, the phase diagram is sometimes drawn with this compound as one of the endpoints on the composition axis as shown in Figure 13.2 in which the Fe–C phase diagram is shown for compositions ranging from pure Fe to the compound Fe_3C (cementite).

The free energy of an intermediate phase becomes much narrower in the vicinity of its stoichiometric composition, as illustrated in Figure 12.14a and b, because only then can the stronger ionic or covalent bonds really form. The free energy configuration for such a system must have a sharp wedge-like shape shown in Figure 12.14a so that when tangents are drawn from it to the liquid free energy, they meet at a point. The solid free energy curve for a system that produces a solid solution intermediate phase would have a more blunt or rounded nose so there is some space between the intersections of the tangents as illustrated in Figure 12.14b.

12.6.2 Eutectic and Eutectoid Systems

A positive heat of mixing in the solid phase tends to lower the solidus curve and will eventually result in solid immiscibility at low temperatures. If this heat of mixing is large enough for the solid–solid immiscibility to occur in the vicinity of the melting points of the pure components, a eutectic reaction may occur in which a melt at a certain composition and temperature may transform into two immiscible solids or symbolically $L \leftrightarrow S_1 + S_2$ or $L \leftrightarrow \alpha + \beta$. A similar reaction called a eutectoid reaction can also occur in the solid state in which $\gamma \leftrightarrow \alpha + \beta$.

A very simple model is now introduced to illustrate how a eutectic reaction may occur and to map out the phase diagram of a binary system in which a eutectic reaction occurs. We assume the freezing temperature of A $T_A > T_B$ and the eutectic temperature $T_E = 940\,K$ as shown in Figure 12.15.

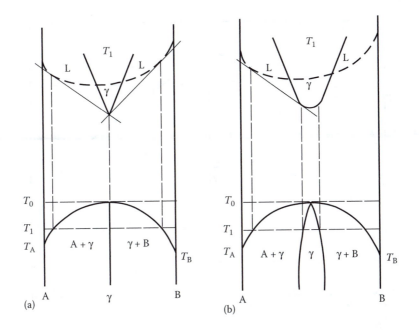

FIGURE 12.14
(a) Free energy diagram for a line compound. (b) Free energy for a solid solution intermediate phase.

At low temperatures ($T < T_E$), the liquid free energy (dashed line at $T = 700$ in Figure 12.15) is above the mutual tangent drawn between the two lowest points on the solid free energy curve and two segregated phases α and β exist in equilibrium as shown. The relative amounts of these two phases are determined from the overall composition using the lever rule.

As temperature is increased, the liquid free energy curve drops because of the $-T\Delta S$ term in the free energy, and encounters the mutual tangent line at T_E. At this temperature, three phases can coexist, a liquid plus the two solid phases. The Gibbs phase rule tells us that, at constant pressure, we have no additional degrees of freedom; therefore, there is only one composition and temperature for which this situation can occur. We call this particular combination of temperature and composition the eutectic point. This particular temperature is called the eutectic temperature and the composition is called the eutectic composition. Since temperature and composition are fixed at this particular point, it is called an invariant point. At this invariant point we say a eutectic reaction occurs. We will encounter other invariant points characterized by different invariant reactions.

As temperature is further increased, the liquid free energy curve descends below the mutual tangent line between the two solid phases so that now two tangent lines may be drawn, one from the solid free energy curve on the A-rich side to the liquid free energy curve, and the other from the liquid free energy curve to the solid free energy curve on the B-rich side. Mapping these compositions onto the temperature-composition plot produces the typical eutectic phase diagram shown at the bottom of Figure 12.15. Note that the temperature will have to be increased above T_E to push the liquid free energy curve to the point where it intersects the solid free energy curve at $x = 0$ or 1. Thus it may be seen that a eutectic system will have a lower melting point than either of its pure constituents.

This lowering of the melting temperature is exploited to form low melting point alloys such as Pb-61 wt% Sn which melts at 183°C; whereas, the melting point of pure Pb is 327.4°C and pure Sn is 231.9°C. A similar situation is seen in the Ag–Cu system as shown

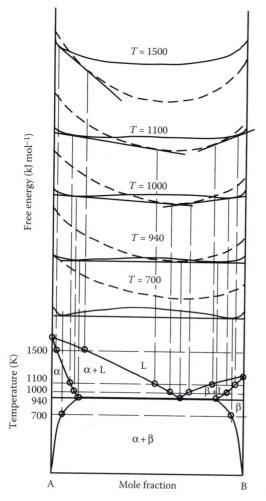

FIGURE 12.15

If the solid has a large enough positive heat of mixing, it will tend to separate into an A-rich α phase and a B-rich β phase when it solidifies. At the eutectic point, these two phases are in equilibrium with the melt. The eutectic temperature T_E will be lower than the melting point of either pure A or B.

in Figure 12.16. Note that Ag and Cu both crystallize into the fcc structure, but the atomic radius for Ag is 0.175 Å and Cu is 0.157 Å, which accounts for their relative insolubility. There is, however, some solid solubility on each side of the phase diagram denoted as the α phase on the A-rich side the β phase on the B-rich side. The line indicating the terminal solid solubility on either side is called the solvus. In the two-phase region, layers or lamellas of one phase will be interspersed between layers of the other phase. The relative amount and composition of these two phases can be determined from the lever rule.

Systems in which A and B crystallize into different structures have separate solid free energy curves that bend up sharply as the second component is added. If the two structures are vastly incompatible, such as fcc Au and diamond-like Si, the bend is extremely sharp, which accounts for the deep eutectic in this system shown in Figure 12.17. Pure Au melts at 1063°C and Si at 1404°C, while the eutectic composition Au 31 At% Si melts at 370°C.

This means a Si chip can be bonded to a Au-coated ceramic substrate by raising the temperature to 370°C at which point a melt will form with the eutectic composition. When cooled back down, even though the melt solidifies back into almost pure Au and pure Si because there is practically no solid solubility between the two phases, there will be a fusion weld in which there is intimate contact between the two components. Thus we have a practical, low temperature method for bonding Au interconnects to Si chips or Si chips to Au-coated ceramic substrates.

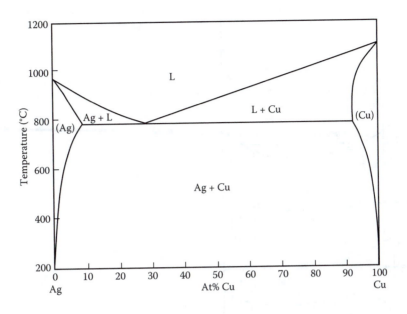

FIGURE 12.16
Ag–Cu phase diagram. (Reprinted from Massalski, T.B., *Handbook of Binary Alloy Phase Diagrams*, ASM, 1990. Reprinted with permission of ASM International. All rights reserved.)

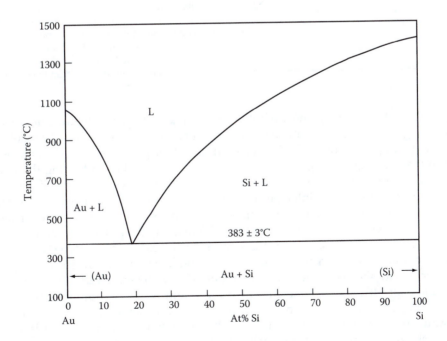

FIGURE 12.17
Au–Si phase diagram. Note that the Au–Si system has virtually no region of solid solution solubility between Au and Si because of their incompatible lattice structure. The large excess heat of mixing drops the eutectic temperature dramatically. (Reprinted from Massalski, T.B., *Handbook of Binary Alloy Phase Diagrams*, ASM, 1990. Reprinted with permission of ASM International. All rights reserved.)

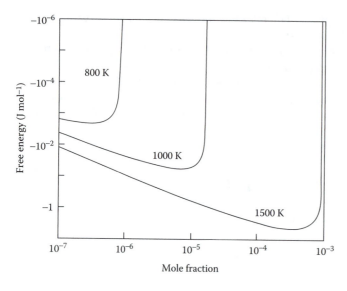

FIGURE 12.18
Contamination of a material by its crucible. It is assumed the binary melt-crucible has a 100 kJ/mol excess heat of mixing. The lowest point on the free energy curve is the mole fraction of crucible material dissolved in the melt at equilibrium at that particular temperature. At 1500 K, ~2 parts per thousand of the crucible material would be in equilibrium with the melt.

Recall that the derivative of the entropy of mixing becomes infinite at both 0 and 1, which means that the segregated free energy curve initially starts downward before going positive. This is demonstrated in Figure 12.18 in which the free energy of a segregated solid was assumed to have an excess heat of mixing of 100 kJ mol^{-1}. Assume that A component is the crucible in which the material B is being heated. There will be ~0.5 ppm of the crucible component in equilibrium at the interface with material B at 800 K and ~10 ppm at 1000 K. The choice of excess heat of mixing in this example may or may not be realistic, but the example serves to illustrate the difficulty in maintaining purity in materials during processing at high temperatures.

12.6.3 Formation of the Microstructure in Eutectics

Let us now examine the formation of the microstructure of eutectic systems of various compositions. First consider the case where the starting composition C_0 intersects the solvus line between the α phase and the $\alpha + \beta$ region in the generic eutectic phase diagram shown in Figure 12.19. Once the temperature falls below the solidus line between α and $\alpha + L$, the solid is α with average composition C_0. When the temperature falls below the solvus line, the α phase becomes supersaturated and small β particles will begin to

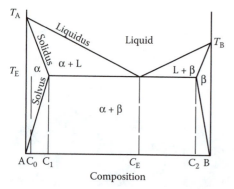

FIGURE 12.19
A generic eutectic phase diagram used as an example of the evolving microstructure when the starting compositing intersects the solvus line.

precipitate and grow within the α phase. The relative amounts and the compositions of the two phases are determined by the lever rule.

If the starting composition C_0 happens to be the eutectic composition C_E, the system will remain molten until the eutectic temperature T_E is reached. At this point the melt transforms into α with composition C_1 and β with composition C_2. The relative amount of the two phases is determined by the lever rule. Using one end of the tie-line as the pivot point, e.g., C_1 and balancing the torque from N atoms acting on lever arm $(C_E - C_1)$ against N_β acting on $(C_2 - C_1)$ we obtain

$$x_\beta = \frac{N_\beta}{N} = \frac{C_E - C_1}{C_2 - C_1}. \tag{12.25}$$

Taking the other end of the tie-line for the fulcrum,

$$x_\alpha = \frac{C_2 - C_E}{C_2 - C_1}. \tag{12.26}$$

12.6.4 The Hunt–Jackson Theory of Lamella Spacing

When a melt with the eutectic composition solidities, the two phases will form as thin layers or lamellas of alternating α and β phases as shown in Figure 12.20. Since the A atoms are being rejected from the β phase as it is forming and must diffuse over to the α phase and vice versa, and since the diffusion distance is proportional to the square root of the time, one might expect that the spacing between the lamella would be inversely related to the square root of the solidification rate. In other words, the faster the solidification rate R, the less time atoms would have to diffuse between alternating layers of materials and the closer the spacing λ. Indeed, a theory developed by Hunt and Jackson predicts that $\lambda^2 R$ is equal to a constant that depends on the material properties of the system.

The orientation of the lamella tends to be along the direction of heat flow. In polycrystalline grains solidified from a melt of eutectic composition, the orientation is more or less random but the spacing of the lamella is indicative of the rate the material was solidified. However, if a eutectic composition is directionally solidified, the lamella will be aligned along the solidification direction to form what is known as an in situ composite. If the eutectic system happens to be a low volume fraction eutectic (the eutectic composition is such that $A \gg B$ or vice versa), the minority phase will form a rod-like structure rather than the lamella structure described above. The spacing of the rods will still be governed by the

FIGURE 12.20
Schematic of the mechanism by which a melt of the eutectic composition solidifies into alternating α and β lamellas as described by the Jackson–Hunt theory.

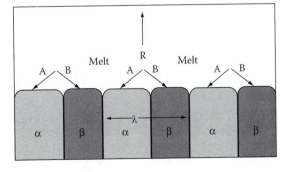

Hunt–Jackson relation described above. In either case, slightly more kinetic undercooling is required to drive the solidification of a eutectic because of the higher interfacial energy.

12.6.5 Solidification of Off-Eutectic Systems

If the starting composition is to the left of the eutectic composition as shown in Figure 12.21, the system is called hypoeutectic. If the starting composition is on the right of the eutectic composition, the system is said to be hypereutectic. Let us now consider the evolution of the microstructure of a hypoeutectic system with a starting composition C_0. When the melt is cooled through the $\alpha + L$ region, primary α phase will form. Just above the eutectic temperature T_E, primary α solid is in equilibrium with the melt at the eutectic composition. Using the lever rule, the mole fractions of primary α and of eutectic liquid are given by

$$x'_\alpha = \frac{C_E - C_0}{C_E - C_1}; \quad x_L = \frac{C_0 - C_1}{C_E - C_1}. \tag{12.27}$$

When the eutectic temperature is reached, the remaining melt solidifies with lamellas of α and β phases so that the final solid will be composed of grains of primary α (sometimes called proeutectic α) interspersed in lamellas of eutectic α and β. The total mole fraction of α is given by

$$x_\alpha = \frac{C_2 - C_0}{C_2 - C_1}. \tag{12.28}$$

Solidification of a hypereutectic system takes place in a similar manner except that primary β replaces the primary α that is interspersed throughout the eutectic matrix in the final solid.

It should be mentioned here that eutectic-like reactions can also take place in which a solid of composition α decomposes into two solids with compositions β and γ at a specific temperature. Such a reaction is called a eutectoid reaction and is symbolized as $\alpha \leftrightarrow \beta + \gamma$. The Pb–Bi system shown in Figure 12.23 contains a eutectoid reaction.

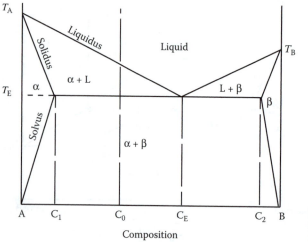

FIGURE 12.21
Solidification of a hypoeutectic melt described with a generic eutectic phase diagram.

12.6.6 Peritectic and Peritectoid Systems

Recall that in the eutectic system, the liquid free energy curve descended between the two lower portions of the solid free energy curve. However, if the free energy curves for a system with a positive solid heat of mixing are skewed so that the liquid free energy curve descends outside the two mutual tangent intersections with the solid free energy curve, then we have what is called a peritectic system. Factors that might cause such a skew in the relations between the solid and liquid free energy curve could be the formation of intermediate phases or large differences in the melting points of the pure materials.

The free energy curves shown in Figure 12.22 are shown for the temperature at which the mutual tangent intersects the liquid free energy curve as well as the lower extremities of the solid free energy curves. This temperature is the peritectic temperature and three phases may coexist at this temperature at the peritectic composition, x_p. Therefore, at this invariant point the peritectic reaction $\alpha + L \leftrightarrow \beta$ takes place. Figure 12.23 shows the phase diagram for the Pb-rich end of the Pb–Bi system which contains a simple peritectic reaction.

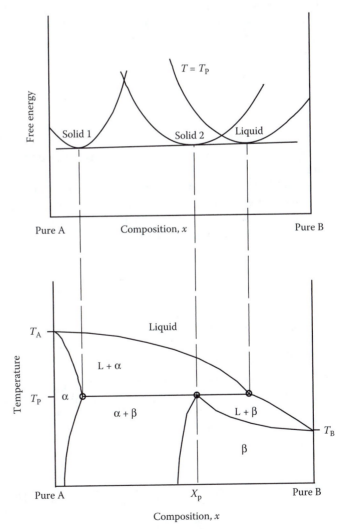

FIGURE 12.22
Schematic showing the construction of a peritectic phase diagram from free energy curves.

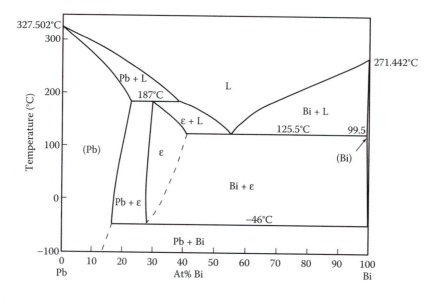

FIGURE 12.23

Pb–Bi system. There is a peritectic reaction at 187°C with composition 30 At% Bi. There is also a eutectoid reaction at −46°C with 27.5 At% Bi. (Reprinted from Massalski, T.B., *Handbook of Binary Alloy Phase Diagrams*, ASM, 1990. Reprinted with permission of ASM International. All rights reserved.)

As one might imagine, it is very difficult to achieve equilibrium solidification of a system with the peritectic composition. For the peritectic reaction to go to completion, grains of primary α must react with the B rich melt to form the β phase. This can only happen at the melt–solid interface. But once this occurs, the β phase is at the interface and the A component of the primary α must diffuse through the existing β phase to react with the melt in order to form new β. Since solid-state diffusion is slow, this can take a very long time. Therefore, practically speaking, any solidification at the peritectic composition will be highly segregated or cored. By this we mean that the individual grains will have cores that consist of the primary α composition surrounded by increasing B-rich solid. The situation will be compounded even more by the fact that the liquid and solid will invariably have different densities and will also tend to separate because of buoyancy effects.

Peritectic-like reactions can also occur in solids in which two components react at a specific temperature and composition to form a third component or α + β ↔ γ. Such reactions are called peritectoids and will be discussed in more detail in Chapter 13.

12.6.7 Monotectic and Monotectoid Systems

If the liquid phase has a positive heat of mixing and the consolute temperature is above the solidification temperature, a monotectic system will result. Generally speaking, if the melt has a positive heat of mixing, the solid will also. A typical free energy plot for this kind of system is shown in Figure 12.24. At the composition x_M, another kind of invariant reaction takes place, i.e., $L \leftrightarrow \alpha + L_2$, which is called a monotectic reaction. As the temperature is further decreased raising the liquid free energy curve still further, there will be a point at which a mutual tangent can be drawn between two points on the solid free energy curve to a point on the liquid free energy curve. This will result in either a eutectic point or a peritectic point, depending on the relative positions of the three points. Figure 12.24 shows a eutectic point. The monotonic system Na–Ca is shown in Figure 12.25.

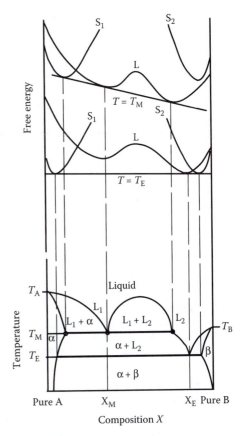

FIGURE 12.24
A schematic showing the construction of a monotectic phase diagram from free energy curves.

As a system with monotectic composition is solidified from the melt, the liquid transforms to α plus B-rich melt (L_2) at the monotectic temperature. Again the relative amounts are determined by the lever rule. This second phase liquid eventually transforms to a eutectic microconstituent at T_E. Since the B-rich phase is incorporated into the α phase as a liquid, it tries to reduce its interfacial energy by spherodizing. Therefore the final eutectic microconstiuents are generally found in the form of spherical particles.

Attempts to form alloys with compositions under the two-liquid dome will result in an almost completely segregated system because of the density differences between the two immiscible liquid phases. Even attempts to form such alloys under microgravity conditions led to similar results, but for a different reason. Instead of buoyancy effects, the phases became separated because of their relative interfacial energy differences with the container walls (see Cahn, J.W., *J. Chem. Phys.*).

A number of binary systems will form an AB solid solution at higher temperatures because of the entropy of mixing term, but will become immiscible at lower temperatures, segregating into α_1, an A-rich AB solid solution and α_2, a B-rich solid solution. In some cases a monotectoid reaction, $\alpha_2 \leftrightarrow \alpha_1 + \beta$, will also occur. Examples include Al–Zn, Hf–Ta, Nb–Zr, and Nb–U. The monotectoid reaction in the Al–Zn system is shown in Figure 12.26.

12.6.8 Mixed Valence Systems

One of the Hume–Rothery rules for isomorphic systems requires the components to have the same valence. However, when the components have different valences, some interesting phase diagrams can result. A classic example of a mixed valence system occurs

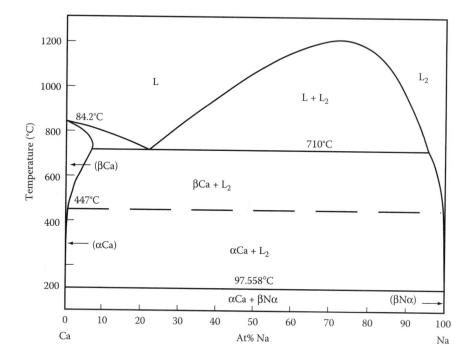

FIGURE 12.25

Ca–Na monotectic system. There is a solid phase transition from αCa to βCa at 447°C. The Na–αCa eutectic temperature is only 0.45°C below the melting point of Na. (Reprinted from Massalski, T.B., *Handbook of Binary Alloy Phase Diagrams*, ASM, 1990. Reprinted with permission of ASM International. All rights reserved.)

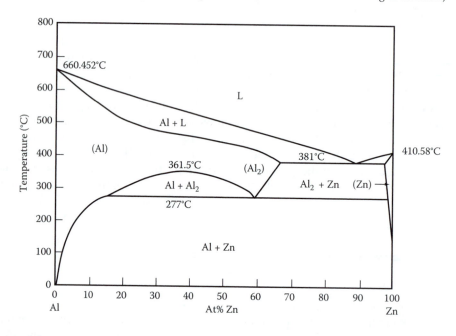

FIGURE 12.26

Al–Zn system. The monotectoid reaction occurs at 277°C with a composition of 59 At% Zn. The Al$_2$ is a solid solution of Al and Zn that is richer in Zn. These two components form a solid solution above 361.5°C up to ~60 At% Zn, but below that temperature, they will segregate into a Zn-rich phase and a Zn-poor phase. (Reprinted from Massalski, T.B., *Handbook of Binary Alloy Phase Diagrams*, ASM, 1990. Reprinted with permission of ASM International. All rights reserved.)

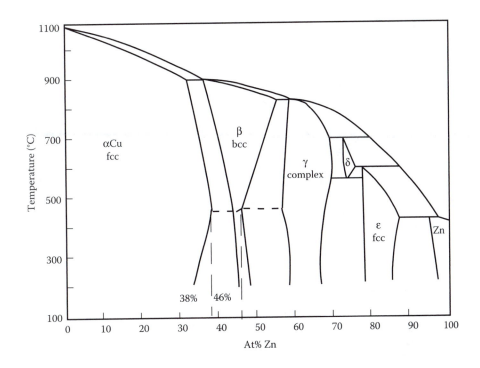

FIGURE 12.27
Cu–Zn phase diagram. Note the order–disorder transition in the β phase at ∼460°C. Also note the extent of the α phase to 38.27 At% Zn and the β phase to 48.2 At% indicated by the numbers on the diagram. (Reprinted from Massalski, T.B., *Handbook of Binary Alloy Phase Diagrams*, ASM, 1990. Reprinted with permission of ASM International. All rights reserved.)

when divalent Zn is added to monovalent Cu. Note in Figure 12.27 that α brass, which forms up to ∼38 At% Zn, is fcc. For higher concentrations of Zn, the system undergoes a phase transformation to β brass which is bcc up to ∼48 At%. At this point it becomes γ, a complex cubic cell containing 52 atoms. The addition of still more Zn produces the ε and η phases which are hcp with different *c* to *a* ratios. Cu–Al, Cu–Sn, Ag–Zn, Ag–Al, Ag–Cd and other mixed valence systems show similar behavior. As mentioned previously, this type of phase transformation can be understood in terms of the Fermi-sphere expanding through the Brillouin zone of the initial phase requiring a new phase with a larger Brillouin zone to accommodate the additional electrons. This subject will be treated in more detail in Chapter 19.

12.7 Summary

In two-component systems, the Gibbs phase rule allows one degree of freedom, usually in the form of a temperature vs. composition line. If there is little or no excess heat of mixing and the components are highly compatible, i.e., nearly same atomic radius and electronegativity, same crystal structure, and same valence (Hume–Rothery rules), the system can have complete solid and liquid solubility over the entire range of composition. Such systems are said to be isomorphous. However, even isomorphic systems do not solidify

congruently, i.e., do not go from melt to solid with the same composition. Instead, the phase diagram contains a lenticular region, bounded by an upper liquidus line and a lower solidus line, in which the liquid and solid can coexist. The segregation coefficient k is defined as the ratio of the solid to liquid composition.

The relative amounts of the two phases can be determined using the lever rule. A tie line is constructed across the two-phase region at the temperature of interest. Using the initial composition as the fulcrum, the amount of one phase times its lever arm (difference between its composition and the initial composition) must balance the amount of the second phase times its lever arm.

The first-to-freeze will have composition C_0/k, where C_0 is the initial composition of the melt. Equilibrium solidification in which the solid has the same composition as the melt is only possible if the solidification is carried out very slowly so that solid-state diffusion can take place. Otherwise, the grains are cored, meaning that centers of the first-to-freeze component are surrounded by the last-to-freeze composition.

Ordered phases can form at low temperatures in solid solutions. These will become disordered as the temperature is increased because of the $-T\Delta S$ term in the free energy where ΔS is higher entropy of the disordered phase. Order–disorder transitions are examples of second-order phase transitions that take place over a range of temperatures with no heat of transition.

If the solid and the liquid has a negative heat of formation, which can result if there is a large difference in electronegativity or if covalent bonds can form, the freezing points of all compositions become elevated and there is a tendency for intermetallic or other intermediate compound phases to form. Sometimes these intermediate phases form by congruent melting in which a melt goes directly into a solid with the same composition.

If the solid has a positive heat of mixing, which often occurs because of a lattice mismatch, the melting point is depressed for all composition and either a eutectic or peritectic reaction will occur. In a eutectic reaction, the liquid transforms into two solids, an A-rich α and a B-rich β phase. In a peritectic reaction, the liquid and the solid phase transform into a second solid phase. In either case, the solid will tend to segregate into two or more distinct phases. There is limited solid solubility in these two phases, and there will be a range of compositions between these phases where α and β will coexist in alternating layers or lamella. The spacing between these lamellas is governed by the Hunt–Jackson relation, which tells us that the spacing is inversely proportional to the square root of the cooling rate. The terminal solubility of these two phases is separated from the mixed $\alpha + \beta$ region by the solvus line.

Solidifying an off-eutectic composition from the melt will result in pro-eutectic grains with the first-to-freeze composition. The remaining melt then freezes with the eutectic composition. Attempts to solidify from the melt at the peritectic composition will result in highly cored grains because when the β first forms, it blocks the remaining α from reacting with the liquid to form new β phase.

Monotectic reactions, in which a liquid decomposes into a solid and a second liquid, can occur in systems that have excess heats of mixing in both the solid and liquid phase. Monotectic systems also have a region of liquid phase immiscibility. Unless done very rapidly, attempts to solidify through this two-phase liquid region will result in almost complete phase separation because of interfacial effects. Monotectoid reactions are also possible in which an AB solid solution separates into an A-rich solid solution and a B-rich β phase.

The various invariant points in which three phases can be present are described by the diagrams in Table 12.1.

TABLE 12.1

Summary of Invariant Reactions

Systems in which the components have different valences produce complex phase diagrams with a number of different solid phases and peritectic reactions as well as eutectoid and peritectoid reactions. These transitions can be explained in terms of interactions between the Fermi sphere and the Brillouin zones.

Bibliography

Askeland, D.R., *The Science and Engineering of Materials*, Books/Cole Engineering Division, Wadsworth, Bellmont, CA, 1984.

Barsoum, M.W., *Fundamentals of Ceramics*, Institute of Physic Publishing, Bristol and Philadelphia, 2003.

Cahn, J.W., Critical point wetting, *J. Chem. Phys.* 66, 3667, 1977.

Callister, W.D., *Materials Science and Engineering: An Introduction*, 7th edn., John Wiley & Sons, New York, 2007.

Gersten, J.I. and Smith, F.W., *The Physics and Chemistry of Materials*, John Wiley & Sons, New York, 2001.

Halsedt, B., Thermodynamic assessment of the system MgO–Al$_2$O$_3$, *J. Am. Ceramic Soc.*, 75/6, 1992.

Kingery, W.D., Bowen, H.K., and Uhlmann, D.R., *Introduction to Ceramics*, 2nd edn., John Wiley & Sons, New York, 1976.

Levin, E.M., Robins, C.R., and McMurdie, H.F., *Phase Diagrams for Ceramists*, The American Ceramic Society, Columbus, OH.

Massalski, T.B., Senior Editor, *Handbook of Binary Alloy Phase Diagrams*, Vols. 1, 2, and 3, ASM, Materials Park, OH, 1990.

Moffat, W.G., *Handbook of Binary Phase Diagrams*, Genium Publishing, New York, 1976.

Muller, O. and Roy, R., *The Major Ternary Structural Families and the Major Binary Structrual Families*, in *Crystal Chemistry of Non-Metallic Materials*, Vols. 3 and 4, Springer, Berlin, Heidelberg, New York, 1974 and 1977.

Ragone, D.V., *Thermodynamics of Materials*, Vol. II, John Wiley & Sons, New York, 1995.

Rosenberger, F.E., *Fundamentals of Crystal Growth*, 2nd edn., Springer-Verlag, New York, Heidelberg, Berlin, 1981.

Shunk, F.A., *Constitution of Binary Alloys*, McGraw-Hill, New York, 1969.

Problems

1. How would you know if a solid solution was ordered or not?

2. Si–Ge forms a solid solution over the entire range of compositions. But III–V compounds, which have similar diamond-like lattice structures, do not. Why do you suppose this is?

3. What is the wt% of Te in CdTe?

4. Identify all of the invariant points in the Cu–Zn phase diagram (Figure 12.27). Specify the reactants, the composition, the temperature, and the type of reaction that is taking place.

5. I am solidifying a 50 At% Ag–Cu alloy. Sketch what you would expect the microstructure to look like. With the help of Figure 12.16, describe the microstructure in terms of the amount and composition of the proeutectic and the amount and composition of the two eutectic phases.

6. In the MgO–Al_2O_3 phase diagram in Figure 12.12, it may be seen that the spinel ($MgAl_2O_4$) solid solution, which forms at 50 At% extends for nonstoichiometric compositions ranging from 40 At% Al_2O_3 to 80 At% Al_2O_3.

   (a) What is the type and fraction of vacancy defects at 80 At% Al_2O_3?

   (b) What is the type and fraction of vacancy defects at 40At% Al_2O_3?

Using the method of undetermined multipliers as before,

$$\sum \left[\ln \left(\frac{g_i - N_i}{N_i} \right) - \lambda - \beta E_i \right] \delta N_i = 0 \tag{15.24}$$

from which

$$\ln \left(\frac{g_i - N_i}{N_i} \right) = \lambda + \beta E_i$$

or

$$N_i = \frac{g_i}{e^{\lambda + \beta E_i} + 1}. \tag{15.25}$$

This is the F–D distribution function.

λ and β are determined as before giving

$$N_i = \frac{g_i}{e^{(E_i - \mu)/kT} + 1}. \tag{15.26}$$

Note that if $E_i > \mu$, $N_i/g_i = 0$ as $T \to 0$; but if $E_i < \mu$, $N_i/g_i = 1$ as $T \to 0$. Therefore, at $T = 0$, all states for which $E_i < \mu$ are filled and all states for which $E_i > \mu$ are empty. Thus we can identify μ as the Fermi energy.

The Fermi function $F(E)$ is defined as

$$F(E) = \frac{1}{e^{(E_i - \mu)/kT} + 1}. \tag{15.27}$$

which can be interpreted as the probability of a state with energy E being occupied at temperature T. A plot of the Fermi function at different temperatures is shown in Figure 15.1.

The density of states is found as before, except that we have made the cells in phase space half as small to assure no more than single occupancy. Thus Equation 15.18 becomes

$$g(E)dE = \frac{2\Delta p^3 \Delta x^3}{h^3} = \frac{8\pi 2^{1/2} m^{3/2} E^{1/2} V dE}{h^3}. \tag{15.28}$$

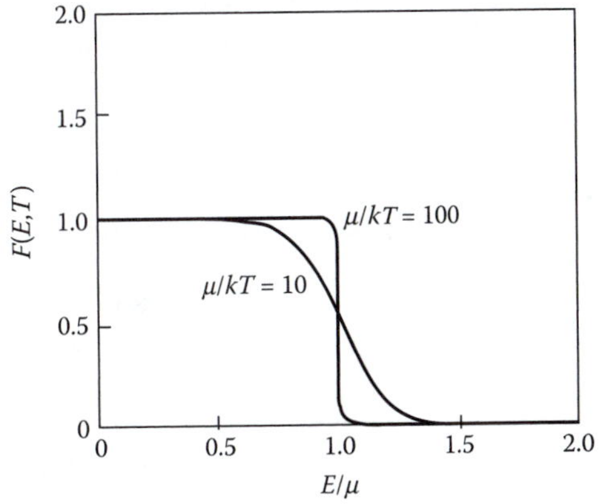

FIGURE 15.1

Fermi function as a function of E/μ for different values of μ/kT. At $T = 0$, $F(E,T) = 1$ for $E < \mu$, and 0 for $E > 0$.

However, if we put the numbers corresponding to electrons in a metal, say Al, the electron density $n = 1.8 \times 10^{23}$ electrons/cm^3 and $m = 9.11 \times 10^{-28}$ g into Equation 15.20, we see that the value for A computed in the classical manner is 1.6×10^4, which of course violates the assumption $A \ll 1$ that was made in taking the classical limit. Therefore, F–D statistics must be used in treating electrons in a metal.

15.4 Chemical Potential and Fermi Energy

The chemical potential μ can be found by evaluating the integral

$$N = \int_0^\infty \frac{g(E)\,\mathrm{d}E}{e^{(E-\mu)/kT} + 1} \tag{15.29}$$

and solving for μ. This cannot be done in closed form. However, at $T = 0$, the Fermi function is 1 for $E < \mu_0$ and 0 for $E > \mu_0$. Therefore, the integral reduces to

$$N = \int_0^{\mu_0} g(E)\mathrm{d}E = \frac{4\pi 2^{3/2} m^{3/2} V}{h^3} \int_0^{\mu_0} E^{1/2}\mathrm{d}E = \frac{4\pi 2^{3/2} m^{3/2} V}{h^3} \left(\frac{2}{3}\mu_0^{3/2}\right) \tag{15.30}$$

from which

$$\mu_0 = \frac{1}{2m}\left(\frac{3h^3 N/V}{8\pi}\right)^{2/3} = \frac{\hbar^2}{2m}\left(3\pi^2 N/V\right)^{2/3}. \tag{15.31}$$

The chemical potential at 0 K is the same as the Fermi energy which we obtained in Chapter 2 by summing over the states in the "electron in a box problem." The Fermi energy is a constant for a given material.

Sommerfeld obtained a series solution for Equation 15.29 which provides a good approximation of the chemical potential for $T_F \gg T$,

$$\mu = \mu_0\left[1 - \frac{\pi^2}{12}(T/T_F)^2 + \cdots\right], \tag{15.32}$$

where the Fermi temperature T_F is given by $kT_F = \mu_0$. Since $\mu_0 = O(5\text{ eV})$, which is equivalent to $T_F \sim 50{,}000$ K, this approximation will be good for most of our applications.

The chemical potential of electrons in a Fermi distribution is also called the Fermi level. The energy required to remove an electron from the Fermi level to infinity (the vacuum state) is the work function. Since the difference in chemical potential determines the flow of particles, when two materials with different Fermi levels are brought together as illustrated in Figure 15.2, electrons will flow from the material with the higher Fermi level (smallest work function) to the material with the lower Fermi level until equilibrium is reached. This transfer of charge results in the contact potential between the two materials.

In a metal at 0 K, the Fermi energy is the ground state energy of the sea of electrons. It is the energy level above which no states are occupied and below which all states are occupied. The Fermi level can also be defined as the energy at which the probability of higher states being occupied is the same as lower states being unoccupied. With this

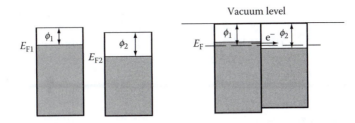

FIGURE 15.2
When material 1 with workfunction ϕ_1 and Fermi level E_{F1} is brought into contact with material 2 with work-function ϕ_2 and Fermi level E_{F2}, the vacuum levels line up and electrons flow from the material with the higher Fermi level (chemical potential) to the lower Fermi level (chemical potential) until equilibrium is reached and the Fermi levels in the two materials are the same.

definition, the Fermi level of a semiconductor will lie, not at the valence band, but in the bandgap between the valence and conduction bands (see Chapter 20).

15.5 Summary

Distribution functions are derived for various types of particles and waves. The distribution function specifies the number of particles that will that occupy a given energy state for a given temperature. Such distribution functions can be used for determining the average energy for an ensemble of particles (or waves) by multiplying the distribution function times the energy and summing (or integrating) over all energies. The distribution function contains two parts, the density of states, which is the number of states corresponding to a given energy, and the probability that those states are filled.

All particles can be classified into two categories, bosons and fermions. Bosons include protons, neutrons, most atoms, and molecules. They have integral units of spin and have no restriction on the number of particles that can occupy a quantum state. They obey B–E statistics and the B–E distribution function is given by Equation 15.13,

$$N_i = \frac{g_i}{e^{(E_i - \mu)/kT} - 1} = \frac{g_i}{A^{-1}e^{E_i/kT} - 1},\tag{15.33}$$

where
  N_i is the number in energy interval E_i
  g_i is the number of available states in interval E_i

$$A = e^{\mu/kT}$$

  μ is the chemical potential

In the classical limit, A becomes small so that the -1 in the denominator can be neglected and the distribution reduces to the M–B distribution given by

$$N_i = g_i A e^{-E_i/kT}.\tag{15.34}$$

For 3-D systems, it is important to remember that the density of states, $g_i \sim E_i^{1/2}$ so that the average energy of an ensemble of particles with three degrees of freedom can be obtained by

$$\langle E \rangle = \frac{E_{\text{total}}}{N} = \frac{\sum_i E_i g_i A e^{-E/kT}}{\sum_i g_i A e^{-E/kT}} = \frac{\int_0^\infty E^{3/2} e^{-E/kT} dE}{\int_0^\infty E^{1/2} e^{-E/kT} dE} = \frac{3}{2} kT. \quad (15.35)$$

In order to be within the classical limit under which M–B statistics apply, $A = nh^3/(2\pi mkT)^{3/2}$ must remain small. For an ideal gas such as N_2 at STP, $A \sim O(10^{-7})$. For condensed matter in which the number density, $n \sim 10^4$ larger than for an ideal gas, M–B statistics can still be used provided the temperature is not too low. At cryogenic temperatures, B–E statistics are required.

The number of wave-like particles, such as photons and phonons is not conserved since they can be created and destroyed. To lift the requirement that the number of particles be preserved, the chemical potential is set to zero and the B–E distribution reduces to the Planck distribution given by Equation 15.21,

$$N_i = \frac{g_i}{e^{E_i/kT} - 1} = \frac{g_i}{e^{\hbar\omega/kT} - 1}. \quad (15.36)$$

Fermions include electrons, protons, neutrons, and ^{3}He. They have half-integral units of spin and the Pauli exclusion principle prevents more than one particle from having the same quantum state. As a consequence, such particles must be treated using the F–D distribution function, given by

$$N_i = \frac{g_i}{e^{(E_i - \mu)/kT} + 1}. \quad (15.37)$$

Because the mass of the electron is so small, the $A = e^{\mu/kT}$ will generally be much too large to ignore the +1 in the denominator.

The Fermi function $F(E)$ defined as

$$F(E) = \frac{1}{e^{(E_i - \mu)/kT} + 1} \quad (15.38)$$

can be interpreted as the probability of a state with energy E being occupied at temperature T. As T approaches 0, the Fermi function is 1 for $E < \mu_0$ and 0 for $E > \mu_0$. In this limit

$$N = \int_0^{\mu_0} \frac{g(E) dE}{e^{(E-\mu)/kT} + 1} = \int_0^{\mu_0} g(E) dE. \quad (15.39)$$

Solving for μ_0,

$$\mu_0 = \frac{\hbar^2}{2m} (3\pi^2 N/V). \quad (15.40)$$

The chemical potential at 0 K is defined as the Fermi energy which we obtained in Chapter 2 by summing over the states in the "electron in a box problem." The Fermi energy is a constant for a given material. The chemical potential, which is a function of temperature and the number of free electrons in the material, is frequently called the Fermi level, especially when dealing with materials with a band gap (see Chapter 20).

Appendix

A.15.1 Derivation of the Identities

The derivation of the identities

$$\left.\frac{\partial S}{\partial U}\right)_{V,N} = \frac{1}{T} \quad \text{and} \quad \left.\frac{\partial S}{\partial N}\right)_{U,V} = -\frac{\mu}{T} \tag{A.15.1}$$

can be accomplished by first expanding the differential, dS

$$dS = \left.\frac{\partial S}{\partial U}\right)_{V,N} dU + \left.\frac{\partial S}{\partial V}\right)_{U,N} dV + \left.\frac{\partial S}{\partial N}\right)_{U,V} dN. \tag{A.15.2}$$

The combined first and second laws of thermodynamics may be written as

$$TdS = dU + pdV. \tag{A.15.3}$$

This is combined with the expansion of dS for constant N and V to give

$$\left.\frac{\partial S}{\partial U}\right)_{V,N} dU = \frac{dU + pdV}{T} \tag{A.15.4}$$

or

$$\left.\frac{\partial S}{\partial U}\right)_{V,N} = \frac{1}{T} \quad \text{Q.E.D.} \tag{A.15.5}$$

Now expand dG,

$$dG = \left.\frac{\partial G}{\partial U}\right)_{V,N} dU + \left.\frac{\partial G}{\partial V}\right)_{U,N} dV + \left.\frac{\partial G}{\partial N}\right)_{U,V} dN. \tag{A.15.6}$$

The definition of chemical potential is

$$\mu \equiv \left.\frac{\partial G}{\partial N}\right)_{U,V}. \tag{A.15.7}$$

For constant U and V, this may be combined with the expansion of dG to give

$$dG = \left.\frac{\partial G}{\partial N}\right)_{U,V} dN = \mu dN. \tag{A.15.8}$$

But, from the definition of Gibbs free energy,

$$dG = dU + pdV - TdS = dU + pdV \tag{A.15.9}$$

which, for constant U and V, may be combined with the above result to give

$$dG = \mu dN = -TdS. \tag{A.15.10}$$

Note that the expansion of dS for constant U and V reduces to

$$dS = \frac{\partial S}{\partial N}\bigg)_{U,V} \tag{A.15.11}$$

which, when combined with the above result, yields

$$\mu dN = -T\frac{\partial S}{\partial N}\bigg)_{U,V} dN \tag{A.15.12}$$

or

$$\frac{\partial S}{\partial N}\bigg)_{U,V} = -\frac{\mu}{T} \quad \text{Q.E.D.} \tag{A.15.13}$$

A.15.2 Entropy of an Ideal Gas

Having found the absolute activity of a M−B gas, it is a simple matter to obtain its entropy. The chemical potential was shown to be given by

$$\mu = kT \ln A = kT \ln\left(\frac{h^3 N/V}{(2m\pi kT)^{3/2}}\right), \tag{A.15.14}$$

which can also be written as

$$\mu = kT \ln\left(\frac{2\pi\hbar^2}{mkT}\right)^{3/2} + kT \ln(N/V). \tag{A.15.15}$$

The Gibbs energy for a single component system is

$$G = N\mu = NkT \ln\left(\frac{2\pi\hbar^2}{mkT}\right)^{3/2} + NkT \ln(N/V). \tag{A.15.16}$$

By definition, $G = U + pV - TS$. For an ideal gas, $U = (3/2)NkT$ and $pV = NkT$. Therefore

$$S = \frac{5}{2}NkT - \frac{G}{T}. \tag{A.15.17}$$

Inserting Equation A.15.17 for G,

$$S = \frac{5}{2}Nk + \frac{3}{2}Nk \ln\left(\frac{2\pi\hbar^2}{mkT}\right) - Nk \ln(N/V). \tag{A.15.18}$$

This is the Sakur−Tetrode equation for the entropy of an ideal gas.

A.15.3 Planck Theory of Black Body Radiation

Having developed the Planck distribution function for photons, we will now use it to obtain the Planck formula for the spectrum of black body radiation.

Recall from Chapter 2 when solving the Schrödinger wave equation for electrons in a box, we looked for wave functions with the form $\psi(x, y, z) = e^{i(k_x x + k_y y + k_z z)}$ that satisfied the boundary condition that the wavefunction vanished at the walls at 0 and L. This led to the requirement that $\sin(k_x L) \sin(k_y L) \sin(k_z L) = 0$ which requires $k_x L = n_x \pi$, $k_y L = n_y \pi$, and $k_z L = n_z \pi$, where n_x, n_y, and n_z are integers. Photons in a cavity that are in thermal equilibrium with conducting walls would likewise form standing waves whose amplitudes must vanish at the wall. The allowed momentum states are therefore

$$p = \sqrt{p_x^2 + p_y^2 + p_z^2} = \hbar\sqrt{k_x^2 + k_y^2 + k_z^2} = \frac{\hbar\pi}{L}\sqrt{n_x^2 + n_y^2 + n_z^2} = \frac{\hbar\pi n}{L}. \tag{A.15.19}$$

Since for photons the energy $E = cp$, the allowed energy states are

$$E = \frac{n\pi\hbar c}{L}, \quad n = \sqrt{n_x^2 + n_y^2 + n_z^2}. \tag{A.15.20}$$

The lowest values of n fall within a sector of a sphere of radius n in n_x, n_x, n_x space. The density of states in n-space is given by $2 \times 1/8 \times 4\pi n^2 dn$. The extra 2 accounts for the two possible polarization states for each wave. The density of states in energy is given by

$$g(E)dE = \pi\left(\frac{LE}{\pi\hbar c}\right)^2 \frac{dn}{dE} dE = \pi\left(\frac{L}{\pi\hbar c}\right)^3 E^2 dE. \tag{A.15.21}$$

Using Planck's distribution function, the radiation density in the cavity is given by

$$u(E)dE = \frac{Eg(E)dE}{V(\exp{(E/kT)} - 1)} = \pi\left(\frac{E}{\pi\hbar c}\right)^3 \frac{dE}{(\exp{(E/kT)} - 1)}. \tag{A.15.22}$$

Setting $E = \hbar\omega = h\nu$, we can write the distribution in terms of ω,

$$u(\omega)d\omega = \pi\left(\frac{\omega}{\pi c}\right)^3 \frac{\hbar d\omega}{(\exp{(\hbar\omega/kT)} - 1)} = \frac{8\pi h\nu^3}{c^3} \frac{1}{(\exp{(h\nu/kT)} - 1)} d\nu. \tag{A.15.23}$$

The spectral radiance is the amount of radiation emerging from a small hole in the cavity. Recall from kinetic theory, the number of molecules striking an area is given by $nv/4$, where n is the number density and v is the average velocity. Similarly, the photon flux emerging from a hole in the cavity is $nc/4$ and the energy flux will be $u(\nu)c/4$ and from Equation A.15.23

$$R(\nu)d\nu = \frac{cu(\nu)d\nu}{4} = \frac{2\pi h\nu^3}{c^2} \frac{1}{(\exp{(h\nu/kT)} - 1)} d\nu. \tag{A.15.24}$$

In terms of λ

$$u(\lambda)d\lambda = \frac{8\pi h}{\lambda^3} \frac{|d\nu/d\lambda|}{(\exp(hc/\lambda kT) - 1)} d\lambda = \frac{8\pi hc}{\lambda^5} \frac{1}{(\exp(hc/\lambda kT) - 1)} d\lambda. \tag{A.15.25}$$

$$R(\lambda)d\lambda = \frac{cu(\lambda)}{4} d\lambda = \frac{2\pi hc^2}{\lambda^5} \frac{1}{(\exp(hc/\lambda kT) - 1)} d\lambda. \tag{A.15.26}$$

We can integrate the spectral radiance in terms of frequency given by Equation A.15.24 by using the integral representation of a Riemann–Zeta function

$$\int_0^\infty \frac{x^3 dx}{e^x - 1} = 6 \sum_{n=1}^\infty \frac{1}{n^4} = \frac{\pi^4}{15} \tag{A.15.27}$$

to obtain

$$R = \int_0^\infty R(\nu)d\nu = \frac{2\pi^5 k^4 T^4}{15c^2 h^3} = \sigma T^4. \tag{A.15.28}$$

The Stefan–Boltzmann constant $\sigma = 2\pi^5 k^4/15c^2 h^3$ was computed in 1900 by Planck from the radiation data that was available which allowed him to determine the value for the Planck constant. His value of 6.56×10^{-34} J·s agreed very well with the value that Milliken determined in 1915 from the photoelectric effect and the presently accepted value of 6.626×10^{-34} J·s.

Bibliography

Constant, F.W., *Theoretical Physics*, Addison-Wesley, Reading, MA, 1958.
Kittel, C. and Kroemer, H., *Thermal Physics*, W.H. Freeman, San Francisco, CA, 1980.
Ragone, D.V., *Thermodynamics of Materials*, Vol. II, John Wiley & Sons, New York, 1995.
Sears, F.W., *Thermodynamics*, Addison-Wesley, Reading, MA, 1953.

Problems

1. Under what circumstances is it permissible to approximate the B–E distribution with the M–B distribution when dealing with solids?

2. Under what circumstances is it permissible to approximate the Fermi distribution with the M–B distribution when dealing with solids?

3. Under what circumstances is it permissible to approximate the Planck distribution with the M–B distribution when dealing with solids?

16

Lattice Vibrations and Phonons

The vibrations of the atoms in the crystalline lattice are important in understanding the thermal properties of both metallic and nonmetallic solids. The energy involved in these vibrations represents thermal energy; hence lattice vibrations are primarily responsible for the heat capacity of solids. Also, these vibrations are able to transport heat and are the dominant source of thermal conductivity in nonmetals. Therefore, in order to understand thermal properties of solids, it is necessary to start with a general understanding of the nature of lattice dynamics.

16.1 Vibrations in a Linear Homogeneous Medium

First consider an elastic longitudinal wave traversing a linear homogeneous medium of infinite length as shown in Figure 16.1. Pick an arbitrary element of length Δx located at some position x. Let $u(x)$ be the displacement from equilibrium. The strain, $e(x)$ is defined as $e(x) = \partial u / \partial x$, and the stiffness coefficient C_{11} is defined as the stress F/A required to produce a given strain, or $F/A = C_{11}e$.

Let the strain at x be $e(x)$, and the strain at $x + \Delta x$ be $e(x) + \partial e / \partial x \, \Delta x = e(x) + \partial^2 u / \partial x^2 \Delta x$. The force acting across the element Δx is $F/A = C_{11}[e(x + \Delta x) - e(x)]$ or $F/A = C_{11} \, \partial^2 u / \partial x^2 \Delta x$. The mass of the element Δx is $\rho A \Delta x$, where ρ is the density. Equating force on the element to mass times acceleration, we have

$$\rho A \Delta x \frac{\partial^2 u}{\partial t^2} = C_{11} \frac{\partial^2 u}{\partial x^2} A \Delta x$$

or

$$\frac{\partial^2 u}{\partial t^2} = \frac{C_{11}}{\rho} \frac{\partial^2 u}{\partial x^2}, \tag{16.1}$$

which we recognize as a wave equation in which the wave is propagating with velocity

$$v_0 = \sqrt{\frac{C_{11}}{\rho}}. \tag{16.2}$$

The solution can be written in the form of a traveling wave, $u(x, t) = \xi \exp i(\omega t \pm kx)$. Differentiating this twice with respect to t and to x and putting the derivatives back into

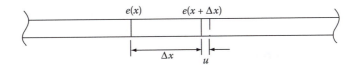

FIGURE 16.1
Schematic for analyzing waves in a linear homogeneous medium. The strain at x is $e(x) = \partial e/\partial x$ and at $x + \Delta x = e(x) + \partial e/\partial x \Delta x$.

the wave equation, we see that $-\omega^2 = -C_{11}/\rho k^2$ which gives us a dispersion relationship $\omega = \sqrt{C_{11}/\rho}\, k = v_0 k$ that plots as a straight line with slope v_0 on an ω vs. k plot. Recall that free electrons have a parabolic dispersion relationship ($\omega \sim k^2$). This is because for a particle with mass, $E = \hbar\omega = p^2/2m = \hbar^2 k^2/2\,m$, or $\omega = \hbar k^2/2m$. However, the momentum of a free massless particle such as a photon is $p = \hbar k = E/c = \hbar\omega/c$, so the dispersion will be a straight line with slope c, which for photons, represents the velocity of light. Therefore, a wave in a continuous elastic media would also be expected to have a straight-line dispersion relationship whose slope is the velocity of propagation.

16.2 Waves on a Chain of Like Atoms

To get a little more realistic with our model of lattice vibrations, we now consider the motions of a chain of discrete particles coupled together with little springs illustrated in Figure 16.2.

Considering only nearest neighbor interactions, the force on the nth atom in the chain can be written as

$$F_n = \beta(u_{n+1} - u_n) - \beta(u_n - u_{n-1}) = \beta(u_{n+1} + u_{n-1} - 2u_n), \tag{16.3}$$

where β is the spring constant. Note that the force is proportional to the relative displacement of the atoms. (If all were displaced the same amount, there would be no force.) Now the equation of motion becomes

$$m\ddot{u}_n = \beta(u_{n+1} + u_{n-1} - 2u_n) \tag{16.4}$$

and again we look for solutions in the form of traveling waves, or

$$u_n(x, t) = \xi e^{i(\omega t \pm kna)} \tag{16.5}$$

in which the x in the previous solution has been replaced by na, where n is the particle index and a is the equilibrium spacing between atoms. Substituting this back into the equation of motion,

$$-\omega^2 m \xi e^{i(\omega t \pm kna)} = \beta \xi e^{i\omega t} \left[e^{ik(n+1)a} + e^{\pm ik(n-1)a} - 2e^{\pm ikna} \right]. \tag{16.6}$$

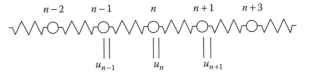

FIGURE 16.2
Schematic for analyzing waves on chain of like atoms.

Factoring out the $\xi \exp i(\omega t \pm kna)$, we get

$$-\omega^2 m = \beta(e^{ika} + e^{-ika} - 2)$$

The reader should recognize $\exp(ika) + \exp(-ika)$ as $2\cos(ka)$, and from the double angle formula in trigonometry, $\cos(2\theta) - 1 = -2\sin^2(\theta)$. Therefore, the solution requires

$$-\omega^2 m = -4\beta \sin^2(ka/2)$$

or

$$\omega = \sqrt{4\beta/m} \sin(ka/2). \tag{16.7}$$

This is quite a different dispersion relationship than we had before. For $ka \ll 1$ (wavelength long compared with atomic spacing), the $\sin(ka/2) \approx ka/2$ and $\omega \approx \sqrt{4\beta/m}\, ka/2$. Going from the microscopic to the macroscopic, we can replace m with ρAa. $F = \beta \delta u = C_{11} Ae = C_{11} A\, \delta u/a$, we can identify β as $C_{11}A/a$. Thus

$$\omega \approx \sqrt{\frac{4C_{11}A/a}{\rho Aa}}\, ka/2 = \sqrt{C_{11}/\rho}\, k, \quad ka \ll 1, \tag{16.8}$$

which is the same as the result from the linear homogeneous model. Thus if the wavelength is much longer than atomic dimensions, the discreteness of the atoms makes little difference, and the material can be treated as homogeneous.

However, note what happens as $ka \to \pi$. The $\sin(\pi/2) = 1$ and the $\omega \to \sqrt{4\beta/m}$ and no longer involves k. This means the group velocity $v_g = \partial\omega/\partial k = 0$, and this is the maximum frequency the lattice can propagate. Note also that as $k \to \pi/a$, the solution becomes $u_n(x, t) = \xi \exp i(\omega t \pm n\pi)$, which represents a standing wave. Physically, what is happening at the lower frequencies is the atoms are more or less moving in the same direction as the wave propagates through, but as $k \to \pi/a$, the atoms are vibrating in opposition to one another.

16.2.1 Bragg Reflections at the First Brillouin Zone

Note in Figure 16.3, that increasing k beyond π/a simply repeats the curve between $-\pi/a$ and 0, so that all relevant information is contained within the region from $-\pi/a$ and π/a. Recall that the reciprocal lattice vector for a simple cubic direct lattice has magnitude $2\pi/a$ and that the first Brillouin zone is formed by planes that are perpendicular bisectors

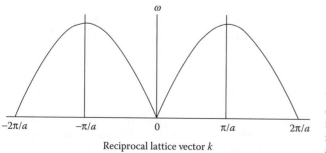

Reciprocal lattice vector k

FIGURE 16.3
Dispersion relationship for linear chain of like atoms. The $\omega(k)$ repeats over an interval of $2\pi/a$, so all of the pertinent information is contained between $-\pi/a$ and π/a.

of the nearest neighbor reciprocal lattice points. It can easily be shown that the Bragg condition $2d \sin \theta = \lambda$ is satisfied at $k = \pi/a$ by setting $\theta = \pi/2$, $d = a$, and $k = 2\pi/\lambda$.

At the lower frequencies, the velocity is given by $v_0 = a\sqrt{\beta/m}$, which is the velocity of sound in the medium, typically $\sim 3 \times 10^3$ m/s. The cutoff frequency, or the maximum frequency that can be propagated by the lattice is given by $\omega_{max} = 2\sqrt{\beta/m} = 2v_0/a \approx (2 \times 3 \times 10^5 \text{cm/s})/(3 \times 10^{-8} \text{ cm}) = 2 \times 10^{13}$ Hz, which corresponds to the far-infrared region of the spectrum.

16.2.2 Normal Modes in a Linear Chain of Atoms

Thus far, we have introduced no quantum effects into the discussion. However, we also considered the linear chain to be infinite in extent; thus we avoided applying any boundary conditions at the end points. However, now we must consider finite systems and apply appropriate boundary conditions that will allow waves to traverse the solid. Let the length of the chain be L and assume it contains $N+1$ atoms with spacing $a = L/N$. Recall from the treatment of electrons in a box, forcing the wave function to be zero at the boundaries of the crystal allowed only standing waves. To permit traveling waves, we chose periodic boundary conditions which, for the case of elastic waves, can be written as $u(x + L, t) = u(x, t)$. Thus the solution must be in the form $\exp(ikna) = \exp[ik(na + L)]$, which requires k to be given by

$$k = \pm \frac{2\pi}{L} j, \quad j = \pm 1, \pm 2, \pm 3, \ldots . \tag{16.9}$$

The only values of k that produce unique frequencies lie between $-\pi/a$ and π/a. If the allowed values of k are spaced $2\pi/L$ apart, the total number of states is

$$N = (2\pi/a)/(2\pi/L) = L/a. \tag{16.10}$$

Thus we arrive at the important conclusion that a finite chain having $N+1$ particles has N normal modes of longitudinal vibration. The same arguments used above can be generalized to three dimensions and it can be shown that two transverse modes of vibration are also possible. Therefore, a chain of $N+1$ atoms can have $3N$ discrete modes of vibration. Thus the k-vector, which was continuous for an infinite chain, is now quantized into $3N$ discrete states. The energy of each of these states is assigned to be $\hbar\omega$. Because of the close resemblance of these quantized vibrational states to photons in a cavity, these quantized waves are called phonons.

16.3 Motion of Atoms in a Diatomic Chain

Next consider a chain of two different atomic species in which the two species alternate. We can write the equations of motion for each species, which are similar to Equation 16.4.

$$m\ddot{u}_{2n} = \beta(u_{2n+1} + u_{2n-1} - 2u_{2n}) \tag{16.11}$$

$$M\ddot{u}_{2n+1} = \beta(u_{2n+2} + u_{2n} - 2u_{2n+1}). \tag{16.12}$$

Now we have a coupled set of second-order differential equations that must be solved simultaneously. Notice that coupling comes about through the odd nearest neighbors of the even atoms and vice versa. The solutions can be written as

$$u_{2n} = \xi e^{i(\omega t + 2nka)} \quad \text{and} \quad u_{2n+1} = \eta e^{i(\omega t + (2n+1)ka)}. \tag{16.13}$$

Putting these back into the differential equations, we get

$$-\omega^2 m\xi = \beta\eta\left(e^{ika} + e^{-ika}\right) - 2\beta\xi \tag{16.14}$$

$$-\omega^2 M\eta = \beta\xi\left(e^{ika} + e^{-ika}\right) - 2\beta\eta, \tag{16.15}$$

which have to be solved simultaneously. Identifying $(\exp(ika) + \exp(-ika)) = 2\cos(ka)$ and writing these equations in matrix form, we get

$$\begin{bmatrix} (2\beta - m\omega^2) & -2\beta\cos(ka) \\ -2\beta\cos(ka) & (2\beta - M\omega^2) \end{bmatrix} \begin{bmatrix} \xi \\ \eta \end{bmatrix} = \begin{bmatrix} 0 \\ 0 \end{bmatrix}. \tag{16.16}$$

If the inverse of the above matrix exists, the only solution possible is for ξ and $\eta = 0$, which is the trivial solution. The only other way to satisfy this equation is for the inverse matrix not to exist, which requires its determinant to vanish. Thus the solution to the simultaneous differential equations requires that

$$(2\beta - m\omega^2)(2\beta - M\omega^2) = 4\beta^2 \cos^2(ka) \tag{16.17}$$

or

$$\mu\omega^4 - 2\beta\omega^2 = -\frac{4\beta^2}{m + M}\sin^2(ka), \tag{16.18}$$

where the reduced mass $\mu = mM/(m + M)$. Solving for ω^2,

$$\omega^2 = \frac{\beta}{\mu}\left[1 \pm \sqrt{1 - \frac{4\mu}{m + M}\sin^2(ka)}\right]. \tag{16.19}$$

This solution is plotted in Figure 16.4.

If M had been set equal to m in Figure 16.4, the two branches would merge and the curve that starts at $-\pi/a$ and goes to zero at $k = 0$, would be identical to Figure 16.4 for a monatomic chain. The second solution is the same as the first solution, except that it is displaced by $\pm\pi/a$; thus all of the unique values for ω are contained in the interval $-\pi/2a$ to $\pi/2a$.

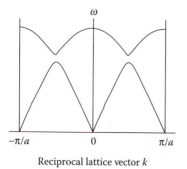

FIGURE 16.4
Dispersion relationship for chain of alternating atoms with different masses. For this plot, $M = 1.2\ m$. There is an energy gap at $\pm\pi/2a$ which closes as $m \to M$. The frequencies in this gap cannot be propagated by the chain. If $M = m$, the gap would close and the lower branch would be the same as Figure 16.3.

16.3.1 Limiting Cases

It is instructive to consider the limiting cases. For $\sin^2(ka) \ll 1$, which requires either $ka \ll 1$ or $ka \approx \pi$,

$$\omega^2 = \frac{2\beta}{\mu}; \quad \frac{2\beta}{m+M}(ka)^2. \tag{16.20}$$

The first solution represents the cutoff frequency, or the highest frequency the lattice can support. The second solution represents the low frequency limit where the ω is linear with k and the slope is the velocity of sound in the medium given by

$$v_0 = \sqrt{\frac{2\beta}{(m+M)}}\, a. \tag{16.21}$$

One can see that if $m = M$, these solutions reduce to those obtained for the chain of identical atoms, as they should.

Next consider the case where $ka \approx \pi/2$. Now, with some algebraic manipulation, the two solutions are

$$\omega^2 = 2\beta/m; \quad 2\beta/M. \tag{16.22}$$

16.3.2 Energy Gap at the First Brillouin Zone

Note that there is a gap between the two allowable frequencies (Equation 16.22) the diatomic chain can support. Also note that values of $k + \pi/a$ produce the same result as k, hence we can fold the dispersion curve into a reduced zone plot.

(Note: Some texts define a as the distance between a repeating pair of ions, whereas a is defined here as the distance between individual ions and would be half the distance between like ions. Thus the notation used in Figure 16.4 in which the Brillouin zone extends from $-\pi/2a$ to $\pi/2a$ would extend from $-\pi/a$ to π/a if a was taken as the distance between like ions. The choice of a as the distance between individual atoms, as used here, makes it easier to see how the two-atom model relates to the single-atom model as the masses of the ion pair become equal.)

As $m \rightarrow M$, the solution for the diatomic chain reduces to $\omega^2 = (\beta/\mu)[1 \pm \cos(ka)]$. The minus sign solution is identical to the solution for the chain of like atoms; however, the plus sign solution only exists for the case of the diatomic chain.

16.3.3 Physical Interpretation

Physically, what is happening can be described by the following: at small values of k (long wavelength) in the lower or acoustical branch, the atoms or ions in the chain move more or less together, as would be expected if a low frequency mechanical or pressure pulse was applied to the crystal, and this disturbance propagates with the velocity of sound as in the case of the monatomic chain, hence the term "acoustic" branch. As k approaches $\pi/2a$, (or $\lambda = 4a$), the lighter atoms (mass m) will remain stationary while the heavier atoms (mass M) vibrate back-and-forth with a frequency $(2\beta/M)^{1/2}$ as a standing wave, or the heavier atoms can remain stationary while the lighter atoms vibrate back-and-forth with a frequency $(2\beta/m)^{1/2}$ as a standing wave. The lattice cannot propagate frequencies between $(2\beta/M)^{1/2}$ and $(2\beta/m)^{1/2}$. The atoms in the upper or optical branch start to vibrate in opposition to one another until at $ka = \pi$ (0 in the reduced zone), they form standing waves.

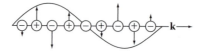

FIGURE 16.5

Electromagnetic wave propagating along a chain of oppositely charged ions can excite a transverse optical (TO) mode. To excite a longitudinal optical (LO) mode, the k-vector must make an angle with the chain in order to project a component of its E-vector along the chain (Berreman effect).

In ionically bonded systems, the heavy and light ions will have different charges, which means they can be excited by an electromagnetic wave as illustrated in Figure 16.5. A wave propagating along a chain of crystals can excite a transverse-polarized mode, but not a longitudinal-polarized mode because its **E**-vector would have no component along the chain. In order to excite a longitudinal mode in a crystal, the incident radiation must enter obliquely to one of the (100) faces. Only the portion of the wave polarized normal to the face can excite the longitudinal mode.

A photon–phonon interaction requires the conservation of both energy and momentum. Recall the momentum of a photon is $\hbar\omega/c$ and the momentum of a phonon is $\hbar\omega/v$, where v is the velocity of sound. Because of the much larger momentum/energy of the acoustical phonons, exchange between photons and acoustical phonons is limited to inelastic collisions in the form of radiant heating. However, in the vicinity of $k=0$ in the optical branch, the energy/momentum of phonons can more easily match that of photons and there will be a strong absorption peak near the top of the optical branch in the vicinity of $k=0$ where the frequency is $(2\beta/\mu)^{1/2}$. This phenomenon is discussed in more detail in Chapters 23 and 24.

It should be noted that sp^3-bonded materials such as Si and Ge also exhibit both an optical and an acoustic branch even though they contain only one type of atom. The covalent sp^3 bond places much of the electron density between the atoms so that the charge distribution is similar to that of an ionic system.

The optical and acoustic modes for NaCl were computed from the elastic coefficients in Table 16.1 and displayed in the reduced zone scheme in Figure 16.6. As k increases beyond $\pi/2a$, the dispersion curve is the same as for $-k\pi/a$; therefore, all the pertinent information is contained in the reduced zone from $0 \geq k\pi/2a$. These optical and acoustic modes can be mapped by neutron inelastic scattering.

16.3.4 Normal Modes in a Diatomic Chain

Applying periodic boundary conditions to the wave equations in Equation 16.13, we see that $k = 2\pi j/L$ as in Equation 16.9. But now the first Brillouin zone extends from $-\pi/2a$ to $\pi/2a$

TABLE 16.1

Comparison of Calculated and Measured Optical Cut Off Frequencies

	C_{11} (GPa)	C_{44} (GPa)	a (nm)	μ (amu)	ω_T (cal)	ω_T (act)	ω_L (cal)	ω_L (act)
LiF	119	53	0.201	5.08	5.02	5.80	7.53	12.00
LiCl	49.4	24.6	0.249	5.80	3.57	3.60	5.05	7.50
NaCl	48.6	12.8	0.283	13.95	1.77	3.10	3.45	5.00
NaBr	33	13	0.298	17.85	1.62	2.50	2.58	3.90
KCl	40	6.2	0.319	18.59	1.13	2.70	2.88	4.00
KI	27	4.2	0.358	29.89	0.78	1.90	1.97	2.60

Source: From Kittel, C., *Introduction to Solid State Physics*, 7th edn., John Wiley & Sons, New York, 1966.

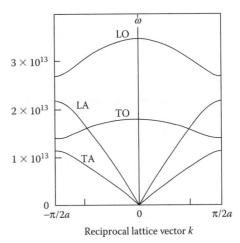

FIGURE 16.6
Dispersion relations for NaCl in the reduced zone scheme for the various modes of oscillation: longitudinal optical (LO), transverse optical (TO), longitudinal acoustic, and transverse acoustic. The curves were computed from Equation 16.19 using the elastic coefficients for NaCl given in Table 16.1. It should be noted that the *a* in this figure is the distance between the individual ions. Usually this dispersion relation is shown plotted between $-\pi/a$ and π/a where *a* is taken as the distance between ion pairs.

instead of $-\pi/a$ to π/a as before as seen in Figure 16.6. So now the extent of *k*-space is π/a and the spacing between each node is $2\pi/L$, which means the number of modes is $L/2a$. Since *a* is the nearest neighbor distance, L/a is the number of ions and $L/2a$ is the number of ion pairs. The number of modes is now 3*N* (two transverse and one longitudinal) in the acoustic branch and 3*N* (two transverse and one longitudinal) in the optical branch, where *N* is the number of ion pairs. (The result obtained here would be the same if we had defined *a* as the distance between repeating ions, which would be the length of an ion pair. The *k*-space would extend over $2\pi/a$ and the number of modes would be $L/a = N$ ion pairs.)

16.4 Tests of the Model

16.4.1 Relating the Force Constant to the Elastic Coefficients

It was stated previously that a strong absorption of photons can occur in the vicinity of $k = 0$ at $\omega = \sqrt{2\beta/\mu}$. In order to correlate the observed absorption frequency with predictions, it is necessary to relate the spring constant β to known or measurable quantities. We can obtain the velocity of sound in the acoustic branch from Equation 16.20:

$$v_s = \left(\frac{\partial \omega}{\partial k}\right)_{k \to 0} = \frac{\partial}{\partial k}\sqrt{\frac{2\beta}{m + M}}\, ka = \sqrt{\frac{2\beta}{m + M}}\, a, \qquad (16.23)$$

where *a* is the nearest neighbor distance. Recall from Equation 16.2 that $v_L = \sqrt{C_{11}/\rho}$ and $v_T = \sqrt{C_{44}/\rho}$. Therefore, we can relate the transverse and longitudinal force constants to the elastic coefficients by

$$\beta_{11} = \frac{C_{11}(m + M)}{2\rho a^2} \quad \text{and} \quad \beta_{44} = \frac{C_{44}(m + M)}{2\rho a^2}, \qquad (16.24)$$

where *a* is the nearest neighbor distance. For the NaCl lattice, $\rho = 4(M + m)/(2a)^3$, so Equation 16.24 simplifies to

$$\beta_{11} = C_{11}a \quad \text{and} \quad \beta_{44} = C_{44}a. \qquad (16.25)$$

16.4.2 Comparison with Observed Data

Assuming the elastic coefficients for the acoustic branch are the same as for the optical branch, we can now compute the ω_L and ω_T values from the lattice constants using Equation 16.20 and compare them against experimental values in Table 16.1.

The elastic coefficients C_{11} and C_{44} are obtained from velocity of sound measurements, and the values in Table 16.1 were taken from Kittel, *Introduction to Solid State Physics*, 2nd edition. The values for ω_T(act), taken from Kittel, *Solid State Physics*, 7th edition, were obtained by observing the onset of strong reflectance in the infrared. In Chapter 24, it is shown that ionic materials reflect strongly as well as absorb at frequencies slightly above their transverse resonant frequency. The values for ω_L(act) were obtained from the ratio of the static to optical dielectric constants using the Lyddane, Sachs, Teller relation (see Chapter 23) and were also taken from Kittel, *Solid State Physics*, 7th edition. All frequencies are in units of $10^{13}/s$.

We see that the values for ω_T and ω_L predicted by the model are in general agreement but are somewhat below the observed values. Considering the simplifying assumptions made in the model, i.e., using only nearest neighbor interactions, perhaps this is all that can be hoped for.

16.5 Applications

One of the more useful predictions from the model is that the absorption peak from Equation 16.20 is proportional to the square root of the lattice stiffness divided by the reduced mass of the ion pairs. In the design of glasses and other optical components that must operate in the far-infrared, one wants to operate at frequencies well above the absorption at $\omega = \sqrt{2\beta/\mu}$. Therefore, one looks for materials with heavy ion pairs and low sound velocities. Single crystalline NaCl is used for infrared windows out to about 15 µm while KBr is good out to about 25 µm. KRS-5, (thallium bromoiodide) is used for infrared applications out to ~35 µm. The development of heavy metal glasses is discussed further in Chapters 14 and 24.

16.6 Summary

For continuous media, the propagation velocities for longitudinal and transverse waves is given by $v_L = \sqrt{C_{11}/\rho}$ and $v_T = \sqrt{C_{44}/\rho}$ and the dispersion relation is $\omega = v_0 k$. For a monatomic chain of atoms, the dispersion relation is given by Equation 16.7, $\omega = \sqrt{4\beta/m}\ \sin(ka/2)$. The ratio of β/m can be related to the elastic stiffness coefficient C_{11} (or C_{44}) by $\beta/m = C_{11}/\rho a^2$, where a is the spacing between the atoms in the chain. There is a cutoff frequency at $k = \pi/a$ given by $\omega_L = \sqrt{4\beta/m} = \sqrt{4C_{11}/\rho a^2} = 2v_L/a$.

The dispersion relation simply repeats itself between $-\pi/a > k > \pi/a$, which is the first Brillouin zone for a simple cubic lattice, so that all of the relevant information is contained in this zone. When a finite length L of the chain of N atoms is specified, the dispersion relation is quantized into N modes that are spaced at $2\pi/L$ apart throughout the interval $-\pi/a > k > \pi/a$. Each state has energy $\hbar\omega$ where the ω is related to k by the dispersion relation. Traveling waves with these discrete values of k are called phonons. Since a wave can have three polarization states, two transverse and one longitudinal, there are $3N$ normal modes.

If the wavelength is long compared to the spacing between particles ($ka \ll 1$) the dispersion relation reduces to $\omega = \sqrt{\beta a^2/mk} = \sqrt{C_{11}/\rho}k = v_0 k$, which is the same as the dispersion relation for continuous media.

For a chain of diatomic atoms (or ions) with different masses, the dispersion relation separates into two branches: an acoustic branch and an optical branch. In the vicinity of $k = 0$, the dispersion relations are given for the optical branch and the acoustic branch, respectively, by $\omega^2 = 2\beta/\mu$; $2\beta(ka)^2/(M+m)$, where a is the spacing between the individual atoms and μ is the reduced mass. In the optical branch, the atoms pairs are moving back-and-forth in opposition to each other; whereas, in the acoustical branch they move collectively, as in the case of a monatomic chain of atoms. It can be seen that the dispersion relation reduces to that of a monatomic chain when $m = M$. The dispersion relation repeats itself as $k \geq \pi/2a$. The entire information is contained in the first Brillouin zone that extends from $-\pi/2a \leq k \leq \pi/2a$ ($-\pi/a \leq k \leq \pi/a$ if a is taken as the spacing between the repeating ion pairs). The number of normal modes is now $3N$ for both the acoustic branch and the optical branch, where N is the number of ion pairs.

At $k = \pi/2a$, an energy gap develops. The frequencies of the optical branch and acoustic branch are given, respectively by $\omega^2 = 2\beta/m$; $2\beta/M$ and the lattice cannot support any frequency between these values. In the acoustic branch, the lighter atom remains stationary and the heavier atom oscillates back-and-forth with frequency $\sqrt{2\beta/M}$; in the optical branch the heavier atom remains stationary and the lighter atom oscillates back-and-forth with frequency $\sqrt{2\beta/m}$.

A very efficient coupling between photons and transverse phonons can occur in the vicinity of $k = 0$ in the optical branch leading to strong absorption at frequencies given by

$$\omega_L = \sqrt{\frac{2\beta_{11}}{\mu}} = \sqrt{\frac{2C_{11}a}{\mu}} \quad \text{and} \quad \omega_T = \sqrt{\frac{2\beta_{44}}{\mu}} = \sqrt{\frac{2C_{44}a}{\mu}}$$

for the NaCl lattice. Comparison of this result with actual measurements for several alkali halides shows only fair agreement.

Materials for infrared applications must have their absorption peak well below the frequency they must transmit. This requires heavy atoms with small elastic coefficients.

Bibliography

Ashcroft, N.W. and Mermin, N.D., *Solid State Physics*, Brooks Cole, Philadelphia, 1976.

Kittel, C., *Introduction to Solid State Physics*, 7th edn., John Wiley & Sons, New York, 1966.

Problem

1. For a diatomic chain of atoms, use Equation 16.19 to obtain Equation 16.22 and find the energy gap.

17

Thermal Properties of Solids

Now that we know something about how the molecules in a solid vibrate, we are in a position to connect these vibrations to the thermal properties such as heat capacity, thermal conduction, and thermal expansion.

17.1 Lattice Heat Capacity

The measurement of heat capacity of solids has been of great theoretical as well as practical interest because the departure of the observed heat capacity from the predictions based on classical concepts was one of the early hints that something was quite wrong with the classical models and that new models based on quantum concepts were necessary to understand what was going on.

17.1.1 Classical Approach

Let us use our statistical approach to obtain the average energy of a molecule of a classical monatomic gas using Maxwell–Boltzmann (M–B) statistics. The average energy per molecule is given by

$$\langle E \rangle = \frac{\int_0^\infty E N(E) \mathrm{d}E}{N} = \frac{\int_0^\infty E N(E) \mathrm{d}E}{\int_0^\infty N(E) \mathrm{d}E}. \tag{17.1}$$

For M–B statistics, $N(E) = A e^{-E/kT} g(E) \mathrm{d}E$ and for a three-dimensional (3-D) system, recall that $g(E) \mathrm{d}E = C E^{1/2}\, \mathrm{d}E$ where C is a constant. Putting this in the above,

$$\langle E \rangle = \frac{\int_0^\infty E^{3/2} e^{-E/kT}\, \mathrm{d}E}{\int_0^\infty E^{1/2} e^{-E/kT}\, \mathrm{d}E} = \frac{1}{2m} \frac{\int_0^\infty p^4 e^{-p^2/2mkT}\, \mathrm{d}p}{\int_0^\infty p^2 e^{-p^2/2mkT}\, \mathrm{d}p}. \tag{17.2}$$

The kinetic energy of a molecule with a momentum component p is $p^2/2m$ and $\mathrm{d}E = p\,\mathrm{d}p/m$. Carrying out the integration,

$$\langle E \rangle = \frac{1}{2m} \frac{3/8\sqrt{32\pi(mkT)^5}}{1/4\sqrt{8\pi(mkT)^3}} = \frac{3}{2} kT. \tag{17.3}$$

Now consider a 1-D gas confined to a tube of length, L. In this case the volume in phase space, $\Delta x \Delta p_x = h$. The number of states between p_x and $p_x + \Delta p_x$ is given by $L \Delta p_x / h$ or

$$g(E)dE = \frac{m^{1/2} L dE}{h(2E)^{1/2}}.$$ (17.4)

It is useful to remember that a distribution function such as $g(E)dE$ is the number of states between E and $E + dE$. As such it must be dimensionless. It is also useful to remember that if there is a relationship between two variables such as $p^2 = 2mE$ so that dE and dp can be related, then their distribution functions can be written as $g(E)dE = g(p)dp$.

Now the average energy per molecule is

$$\langle E \rangle = \frac{\int_0^\infty E^{1/2} e^{-E/kT} dE}{\int_0^\infty E^{-1/2} e^{-E/kT} dE} = \frac{1}{2m} \frac{\int_0^\infty p_x^2 e^{-p_x^2/2mkT} dp_x}{\int_0^\infty e^{-p_x^2/2mkT} dp_x}.$$

$$= \frac{1}{2m} \frac{1/4\sqrt{8\pi(mkT)^{3/2}}}{1/2\sqrt{2\pi(mkT)^{1/2}}} = \frac{kT}{2}.$$ (17.5)

Similarly one can show that a molecule of gas with two degrees of freedom has average energy of kT. This example illustrates the classical theorem of equipartition of energy, i.e., thermal energy equal to $kT/2$ is assigned to each degree of freedom. (It can also be shown that the same partition of energy per degree of freedom can be assigned to each rotational and vibrational mode provided the temperature is high enough to excite that particular mode.)

The total energy of an oscillator such as an atom in a crystal lattice is the sum of the kinetic and potential energies. The kinetic energy is just $p_x^2/2m$. The potential energy is

$$\text{Potential energy} = \int_0^x F \cdot dx = \int_0^x kx \, dx = \frac{kx^2}{2}.$$ (17.6)

Since the natural frequency $\omega = \sqrt{k/m}$, the potential energy is given by $m\omega^2 x^2/2$, and the total energy may be written as $E = p_x^2/2m + m\omega^2 x^2/2$.

In thermal equilibrium, the average energy can be found by

$$\langle E \rangle = \frac{1}{2m} \frac{\int_0^\infty p_x^2 e^{-p_x^2/2mkT} dp_x}{\int_0^\infty e^{-p_x^2/2mkT} dp_x} + \frac{m\omega^2}{2} \frac{\int_0^\infty x^2 e^{-m\omega^2 x^2/2kT} dx}{\int_0^\infty e^{-m\omega^2 x^2/2kT} dx}.$$ (17.7)

We previously showed that the first term was $kT/2$. By a similar process, it can be shown that the potential term also becomes $kT/2$. Therefore, the average energy for the x-component of an oscillator is kT. The y- and z-component also each contribute kT, so the total average energy is $3kT$. Therefore, the equipartition theorem assigns $kT/2$ to each of the six degrees of freedom, three kinetic and three potential, of an atom in a crystal lattice.

17.1.2 Dulong–Petit Classical Limit

The heat capacity, C_V, is defined as

$$C_V = \frac{\partial \langle E \rangle}{\partial T} = 3k \text{ per atom or } 3R \text{ per mole,} \tag{17.8}$$

where R is the gas constant, which is Avogadro's number times the Boltzmann constant. Since $R \approx 2$ kcal/mol, one would expect the heat capacity of solids to be 6 kcal/mol. The observed heat capacity of most solids approaches this value at moderate to high temperatures and this value is called the Dulong–Petit or classical limit of heat capacity. However, the observed heat capacity is much lower at low temperatures and, for non-metals, approaches 0 with a T^3 dependence.

Metals provided additional problems for the classical theory of heat capacity. Metals are generally much better conductors of heat than nonmetals because most of the heat is carried by the free electrons. According to the classical theory, this electron gas should contribute an additional $(3/2)R$ to the heat capacity. But the measured heat capacity of metals approached nearly the same $3R$ Dulong–Petit limit as the nonmetals. How can the electrons be a major contributor to the thermal conductivity and not provide significant additional heat capacity?

Careful measurements at low temperatures indicated that metals do have a small electronic contribution to heat capacity that approaches 0 with a first power dependence on T, as will be shown later. In fact, at very low temperatures, the electronic contribution can be greater than the lattice contribution. In order to explain these departures from classical theory, we must reformulate the problem quantum mechanically.

17.2 Debye Model

Using the Planck distribution, the energy of an ensemble of oscillators can be found by

$$\langle E \rangle = \int_0^{\omega_0} \frac{\hbar \omega C(\omega)}{e^{\hbar \omega / kT} - 1} d\omega, \tag{17.9}$$

where
  $C(\omega)$ is the number of modes between ω and $\omega + d\omega$ (the density of states)
  ω_0 is the cutoff frequency

Now the trick is to find the appropriate $C(\omega)$. We start by finding the distribution in k-space and use the relation $C(\omega)d\omega = W(k)dk$ to obtain $C(\omega)$. To get $W(k)$, first take the number of states on the surface of a sphere in k-space, which can be found by dividing the volume of a shell in k-space, $4\pi k^2 dk$, by the volume of an individual state, which we found from Equation 16.9 to be $(2\pi/L)^3$, and multiply by three polarization states to get

$$W(k)dk = 3\frac{4\pi k^2 dk}{(2\pi/L)^3} = \frac{3k^2 L^3 dk}{2\pi^2}. \tag{17.10}$$

To compute $dk/d\omega$, we should use the dispersion relation developed in Chapter 16 for the chain of atoms. From Equation 16.7

$$k = \pm \frac{2}{a} \sin^{-1}(\omega/\omega_0) \qquad \frac{dk}{d\omega} = \frac{2}{a(\omega_0^2 - \omega^2)^{1/2}}$$

$$C(\omega) = 3\frac{k^2 L^3}{2\pi^2} \frac{dk}{d\omega} = \frac{3L^3}{2\pi^2} \frac{2}{a(\omega_0^2 - \omega^2)^{1/2}} \left[\frac{2}{a} \sin^{-1}(\omega/\omega_0)\right]^2 = \frac{12N}{\pi^2} \frac{\left[\sin^{-1}(\omega/\omega_0)\right]^2}{(\omega_0^2 - \omega^2)^{1/2}}.$$

Here is where we run into a problem. As we saw before, it is necessary for

$$\int_0^{\omega_c} C(\omega)d\omega = 3N. \text{ But } \frac{12N}{\pi^2} \int_0^{\omega_0} \frac{\left[\sin^{-1}(\omega/\omega_0)\right]^2 d\omega}{(\omega_0^2 - \omega^2)^{1/2}} = \frac{12N}{\pi^2} \frac{\pi^3}{24} = \frac{N\pi}{2} \neq 3N.$$

Debye dodges this little problem by using the dispersion relation for a homogeneous solid which has no cutoff frequency and in which $k = \omega/v_0$ and $dk/d\omega = 1/v_0$.

$$C(\omega) = W(k)\frac{dk}{d\omega} = \frac{3k^2}{2\pi^2} \frac{L^3}{v_0} = \frac{3\omega^2 L^3}{2\pi^2 v_0^3}. \tag{17.11}$$

To be more precise, $3/v_0^3$ is sometimes written as $1/v_\ell^3 + 2/v_t^3$ to distinguish between the different velocities associated with the longitudinal and transverse modes. Debye then proceeds to determine a cutoff frequency by integrating the density of states (Equation 17.11) to a cutoff frequency ω_D that produces the required number of modes, i.e.,

$$3N = \int_0^{\omega_D} C(\omega)d\omega = \int_0^{\omega_D} \frac{3\omega^2 L^3}{2\pi^2 v_0^3}d\omega = \frac{\omega_D^3 V}{2\pi^2 v_0^3}. \tag{17.12}$$

Solving for the cutoff frequency, we obtain

$$\omega_D = v_0(6\pi^2 N/V)^{1/3}. \tag{17.13}$$

We can now solve for the energy by putting Equation 17.11 for the $C(\omega)$ into Equation 17.9 and integrating the cutoff or Debye frequency, ω_D.

$$\langle E \rangle = \frac{3\hbar L^3}{2\pi^2 v_0^3} \int_0^{\omega_m} \frac{\omega^3}{e^{\hbar\omega/kT} - 1}d\omega. \tag{17.14}$$

It is convenient to set $x = \hbar\omega/kT$. The integral in Equation 17.14 can then be written as

$$\langle E \rangle = \frac{3k^4 T^4 V}{2\pi^2 \hbar^3 v_0^3} \int_0^{x_D} \frac{x^3}{e^x - 1}dx, \tag{17.15}$$

where $x_D = \hbar\omega_D/kT$. It is also convenient to define a Debye temperature $\Theta_D = \hbar\omega_D/k$ so that $x_D = \Theta_D/T$.

In the high temperature limit $(T \gg \Theta_D)$, $x \ll 1$ and the integral may be approximated as

$$\langle E \rangle \approx \frac{3k^4 T^4 V}{2\pi^2 \hbar^3 v_0^3} \int_0^{x_D} x^2 dx = \frac{3k^4 T^4 V}{2\pi^2 \hbar^3 v_0^3} \frac{x_D^3}{3} = 3NkT, \tag{17.16}$$

which is the classical limit.

In the low temperature limit, x_D can be allowed to $\rightarrow$ infinity and the integral may be recognized by the more mathematically astute as the integral representation of a Riemann–Zeta function (the rest of us look it up in tables) and has the value

$$\int_0^\infty \frac{x^3 dx}{e^x - 1} = 6 \sum_{n=1}^\infty \frac{1}{n^4} = \frac{\pi^4}{15}. \tag{17.17}$$

Therefore, the energy of the ensemble of oscillators in the low temperature limit becomes

$$\langle E \rangle \approx \frac{3k^4 T^4 V}{2\pi^2 \hbar^3 v_0^3} \frac{\pi^4}{15} = \frac{3\pi^4 NkT}{5} \left(\frac{T}{\Theta_D} \right)^3, \tag{17.18}$$

from which the heat capacity is obtained by

$$C_v = \frac{\partial \langle E \rangle}{\partial T} \bigg)_V = \frac{12\pi^4 Nk}{5} \left(\frac{T}{\Theta_D} \right)^3, \tag{17.19}$$

which describes the observed T^3 behavior at very low temperatures.

17.2.1 Critique of the Debye Model

Let us examine the validity of the primary assumption of the Debye model, i.e., that the dispersion relation of a chain of atoms can be represented by a linear relationship corresponding to that of a homogeneous medium. Clearly this assumption is valid in the region where the wavelength is long compared to the atomic spacing, as was shown in Chapter 16, but what about the behavior near the cutoff frequency? The results are shown in Figure 17.1.

As can be seen from Figure 17.1, there is significant departure from the chain of atoms dispersion relation before the Debye cutoff frequency is reached. The primary justification for the Debye model seems to be that it gives reasonably good agreement with the measured heat capacity data as shown in Figure 17.2, which suggests that the heat capacity is relatively insensitive to the dispersion relation. The deviations between the observed heat capacity and Debye's model are discussed by Dekker (see Bibliography).

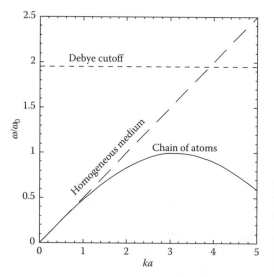

FIGURE 17.1

Dispersion relation for a continuous medium assumed by Debye (dashed line) compared with that of a chain of identical atoms (solid line). The frequency ω is normalized by the cutoff frequency for the chain of atoms $\omega_0 = (4\beta/m)^{1/2}$ which occurs at $k = \pi/a$.

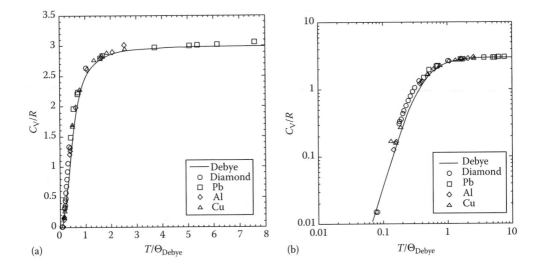

FIGURE 17.2
(a) Heat capacity predicted from Debye theory compared with measured values. The Debye model appears to slightly under-predict the actual data in the region where it departs most from the theoretical dispersion relation. (b) Log–log plot of the heat capacity predicted from Debye theory. The third power dependence at low temperatures is more evident in this plot.

17.3 Electronic Heat Capacity

We can now straighten out the problems we encountered in computing the heat capacity and thermal conductivity of electrons in a metal. The average energy of the electrons in a Fermi distribution is found from

$$\langle U \rangle = \int_0^\infty \frac{g(E)E\,dE}{e^{(E-\mu)/kT}+1}. \tag{17.20}$$

This may also be expanded in a series to give

$$\langle U \rangle = \frac{3}{5}n\varepsilon_F + \frac{\pi^2}{4}\frac{nk^2T^2}{\varepsilon_F} + \cdots, \tag{17.21}$$

from which we can find the electronic heat capacity

$$C_e = \frac{\partial U}{\partial T} = \frac{\pi^2}{2}\frac{nk^2T}{\varepsilon_F} = \frac{\pi^2}{2}\frac{nkT}{T_F} = \frac{\pi^2}{2}\frac{RT}{T_F}. \tag{17.22}$$

Note that the electronic contribution to the heat capacity is linear in T, but for $T \ll T_F$, this contribution is much less than the $3R$ contribution from the ion cores. This explains why at normal temperatures the electronic contribution is negligible. Only near absolute zero, where the ionic contribution approaches zero as T^3, does the electronic contribution dominate as shown in Figure 17.3.

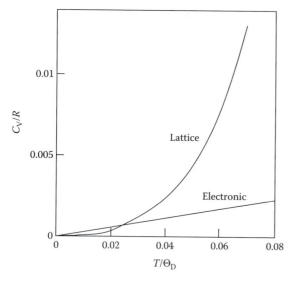

FIGURE 17.3
Lattice heat capacity and electronic heat capacity at very low temperatures. The electronic heat capacity is drawn for a material whose $\Theta_D/T_F = 0.006$. At very low temperature, the electronic heat capacity can exceed the lattice heat capacity.

17.4 Thermal Conductivity

The conduction of heat is related to the thermal gradient ∇T by Fourier's first law of heat conduction (similar to Fick's first law of diffusion),

$$\dot{Q} = -K\Delta T, \tag{17.23}$$

where
   $\dot{Q}$ is the heat flux (W/m^2)
   K is the thermal conductivity (W/m · deg)

17.4.1 Ideal Gas Model

We can understand the transport of heat in solids by phonons as well as electrons by considering the transport of heat in an ideal gas. Consider a unit volume dV located at some radial distance r from a point on a surface dA in which r makes angle θ with the normal to the surface as shown in Figure 17.4.

The number of particles in dV is $n\,dV$, where n is the number density. Let ν be the collision frequency so that the number of colliding particles emerging from all directions from dV in time dt will be $n\nu\,dV\,dt$. (Actually the number of collisions per unit time will be $1/2\,n\nu\,dV\,dt$, but each collision produces two particle paths.) If the collisions are random,

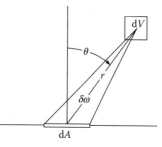

FIGURE 17.4
Geometry for obtaining the thermal conductivity of an ideal gas.

then the number of paths in solid angle $d\omega$ is $d\omega/4\pi$. Since the element of area dA subtends a solid angle $d\omega$ given by $dA \cos\theta/r^2$, the number crossing dA per increment of time is

$$\frac{dA \cos\theta}{4\pi r^2} v n dV dt. \qquad (17.24)$$

Now we ask how many molecules emerge from dV and encounter dA without making another collision? From elementary Poisson statistics, the probability of an event not occurring in time Δt when τ is the average time between events is given by $\exp(-\Delta t/\tau)$. It follows therefore that the probability of a particle traveling a distance r without making a collision is given by $\exp(-r/\lambda)$ where λ is the mean free path. Now we replace dV with $r^2 dr \sin\theta d\theta d\phi$ and integrate over the hemisphere above dA,

$$\frac{dAdtvn}{4\pi} \int\limits_0^{2\pi} d\pi \int\limits_0^{\pi/2} d\theta \sin\theta \cos\theta \int\limits_0^\infty dr e^{-r/\lambda} = \frac{1}{4} n v \lambda dA dt. \qquad (17.25)$$

Note that the mean free path λ is just the average velocity $\bar{v}$ times the average time between collisions, which is just the reciprocal of the collision frequency v. Thus the above result is identical to the well-known equation for the number of particles incident on a surface per unit time, which is given by $1/4n\bar{v}$.

However, the reason for going through all this trouble is that we need the average height above the surface dA at which the molecules make their last collision. To get this, we put $h = r\cos\theta$ into the above integral and normalize by dividing by the integral without the $r\cos\theta$.

$$\langle h \rangle = \frac{\dfrac{dAdtvn}{4\pi} \int\limits_0^{2\pi} d\pi \int\limits_0^{\pi/2} d\theta \sin\theta \cos^2\theta \int\limits_0^\infty r dr e^{-r/\lambda}}{\dfrac{1}{4} n v \lambda dA dt} = \frac{2}{3}\lambda. \qquad (17.26)$$

Now assume there is a uniform vertical temperature gradient, dT/dy in the system. Molecules crossing the surface from above will have a temperature given by $T + \langle h \rangle dT/dy$, while molecules crossing the surface from below will have temperature given by $T - \langle h \rangle dT/dy$. Therefore, the net heat transferred to surface dA in time dt will be given by

$$\frac{dQ}{dAdt} = -\frac{1}{4} n\bar{v}c_V \times 2\langle h \rangle \frac{dT}{dy} = -\frac{1}{3} n\bar{v}c_V \lambda \frac{dT}{dy}, \qquad (17.27)$$

where c_V is the molecular heat capacity. (The negative sign is introduced because the heat is being transferred in the $-y$ direction.) Since the thermal conductivity K is defined by Fourier's first law, $\dot{Q} = -KAdT/dy$, the thermal conductivity becomes

$$K = \frac{nc_V\bar{v}\lambda}{3} = \frac{C_V\bar{v}\lambda}{3}, \qquad (17.28)$$

where C_V is the volume specific heat of the material.

This important result should be committed to memory for it applies not only to the thermal conductivity of gases, but also to thermal transport from phonons and, with some modification, to thermal transport from electrons.

17.4.2 Lattice Thermal Conductivity

Since we know that C_V approaches 0 with a T^3 dependence at low temperatures, one would expect the thermal conductivity to do likewise and this accounts for the low thermal conductivity of nonmetals at low temperature. At higher temperatures, thermal conductivity is limited by the mean free path λ. Phonons are scattered inelastically from defects, impurities, and grain boundaries. Phonons may also interact with other phonons through the anharmonic terms in the lattice potential. For example, two phonons may combine to produce a third phonon. The conservation of momentum requires

$$\mathbf{k_1} + \mathbf{k_2} = \mathbf{k_3} \text{ provided } |\mathbf{k_3}| \text{ inside first B-zone } (N\text{-process}) \tag{17.29}$$

$$\mathbf{k_1} + \mathbf{k_2} = \mathbf{k_3} + \mathbf{G} \text{ if } |\mathbf{k_1} + \mathbf{k_2}| \text{ outside first B-zone } (U\text{-process}), \tag{17.30}$$

where $\mathbf{G}$ is the reciprocal lattice vector ($2\pi/a$ for a 1-D lattice). The collision is elastic in the normal or N-process and the thermal energy is carried by phonon $\mathbf{k_3}$. However, if the vector sum of the two original phonons is greater than half reciprocal lattice vector $\mathbf{G}/2$ such that the resultant vector is outside the first Brillouin zone, the resulting momentum is Bragg reflected by $-\mathbf{G}$ so that the phonon momentum $\mathbf{k_3}$ remains inside the first Brillouin zone as shown in Figure 17.5. This event is referred to as an *Umklapp* (German for "turn over") or a U-process and the $\mathbf{G}$ portion of momentum goes into what is called the crystal momentum as an inelastic collision which results in heating the crystal rather than conducting heat.

If both k_1 and $k_2 < \pi/2a$, the resultant vector will always remain in the first Brillouin zone and the collision will be an N-process. However, as they become larger than $\pi/2a$, a U-process becomes more likely. Using the Debye model, the frequency associated with a phonon whose $k = \pi/2a$ is $\omega = v_0\pi/2a$. The ratio of this frequency to the Debye frequency is

$$\frac{\omega}{\omega_D} = \frac{v_0\pi}{2a} \frac{1}{v_0(6\pi^2 N/V)^{1/3}} = 0.403. \tag{17.31}$$

The probability of exciting a mode with frequency ω is given by

$$e^{-\hbar\omega/kT} = e^{-0.403\hbar\omega_D/kT} = e^{-0.403\Theta_D/T}. \tag{17.32}$$

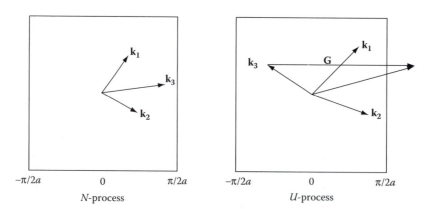

FIGURE 17.5

Illustration of an N-process in which the collision is elastic and a U-process in which the collision is inelastic. If the resulting vector from a phonon–phonon collision falls outside the first Brillouin zone (the square box), the momentum is Bragg reflected back into the first Brillouin zone with a transfer of $\mathbf{G}\hbar$ momentum to the lattice in the form of crystal momentum.

Thus the *U*-processes can be expected to become important when the temperature approaches ~$0.4\Theta_D$ and will limit the conductivity as the temperature increases.

Since thermal conduction of nonmetals is limited at low temperatures by the falloff of the heat capacity and at high temperatures by increased Umklapp scattering, a peak in thermal conductivity would be expected near $0.4\theta_D$.

Ceramics are generally poor thermal conductors because of the scattering from impurities in the form of ions of different sizes and from grain boundaries as well as other defects. Room temperature thermal conductivities are on the order of 10–50 W/m · K. However, pure single crystals can have very high thermal conductivities, especially those systems such as diamond that have large elastic constants which result in high sound velocities as shown in Figure 17.6. Room temperature conductivities for diamond can range from 2000 to 2500 W/m · K, several times higher than metals. Graphite also has a very high thermal conductivity in its basal plane. Commercially available annealed pyrolytic graphite has a conductivity of 1700 W/m · K and is used to remove heat in laptop computers as well as in other thermal management applications. High-conductivity graphite fibers can be aligned in a polymeric matrix to make a strong, highly conductive sheet useful in space radiator panels. Such graphite fibers can have conductivities up to 1100 W/m · K.

Glasses, on the other hand, are very poor thermal conductors because of their lack of a periodic structure, which makes for very efficient phonon scattering. Their conductivities are on the order of 1 W/m · K. Polymers are even better thermal insulators with conductivities ranging from 0.1 to 0.5 W/m · K. Woods are also good insulators with conductivities ranging from 0.1 to 0.4 depending on the type and moisture content. Still air is an extremely poor conductor so porous ceramics are very good insulators. The best insulator is silica aerogel. An aerogel is a gel in which the liquid has been replaced by a gas. Silica aerogels have been made with densities as low as 0.001 g/cm^3 and with thermal conductivities as low as 0.003 W/m · K.

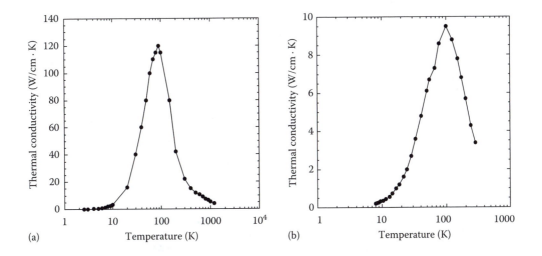

FIGURE 17.6
Thermal conductivities of (a) diamond and (b) cubic SiC. The conductivity of diamond reaches a peak of 12,000 W/m · K at 100 K, 30 times greater than Cu. The thermal conductivities are limited by heat capacity below 100 K and by phonon collisions above 100 K. (Data taken from Adachi, *Handbook of Physical Properties of Semiconductors.*)

17.4.3 Electron Thermal Conductivity

One can calculate the contribution to the thermal conductivity from the electrons classically and almost obtain the correct result. Classically, the heat capacity of an electron gas with n electrons per unit volume, $C = 3nk/2$. The mean free path is given by $v\tau$, where v is the average thermal velocity and τ is the time between collisions. Therefore

$$K = \frac{1}{3}n(3/2)kv^{-2}\tau = \frac{4nk^2T}{m\pi}\tau, \tag{17.33}$$

since $\bar{v} = \sqrt{(8kT)/(\pi m)}$ for an ideal classical gas and τ is the average time between collisions.

Now if the correct value for the electronic heat capacity given by Equation 17.22 is put into Equation 17.28 along with $v = v_F$ and $\lambda = v_F\tau$, where $v_F^2 = 2\varepsilon_F/m$, the thermal conductivity of the electrons becomes

$$K = \frac{1}{3}Cv\lambda = \frac{1}{3}\frac{\pi^2}{2}\frac{nk^2T}{\varepsilon_F}v_F^2\tau = \frac{\pi^2}{3}\frac{nk^2T}{m}\tau. \tag{17.34}$$

Note that the smaller heat capacity of the electron gas is compensated for by the higher Fermi velocity and an expression close to the classical result is obtained but for totally different reasons.

17.5 Thermal Expansion

Most solids expand when heated. Their length increases according to

$$L(T) = L_0(1 + \alpha T), \tag{17.35}$$

where α is the coefficient of linear thermal expansion (CTE) which is $O(10^{-6}\,\text{K}^{-1})$. A similar expression can be written for volume expansion and it is easy to show that since α is small, the coefficient of volumetric expansion is 3α.

Thermal expansion comes about through the asymmetry of the bond energy function. This asymmetry produces anharmonic oscillations of the atoms in the lattice; i.e., instead of oscillating with equal amplitude on both sides of their equilibrium position, they can swing farther to the soft side of the bond energy function than toward the steep side. As a result, the average positions of the atoms are displaced from the equilibrium position at the bottom of the bond–energy curve, which corresponds to the $0\,\text{K}$ location. The average displacement of the lattice atoms can be found from

$$\langle x \rangle = \frac{\displaystyle\int_{-\infty}^{\infty} xe^{-\beta U(x)}\,dx}{\displaystyle\int_{-\infty}^{\infty} e^{-\beta U(x)}\,dx}, \tag{17.36}$$

where $U(x)$ is the bond energy function as a function of the displacement x from the equilibrium and $\beta = 1/kT$.

The bond–energy curve can be approximated by $U(x) = Cx^2 - Dx^3$ in which C represents the harmonic part of the potential and D is the anharmonic part. It is assumed that

$\beta D x^3 \ll 1$ so that $\exp(-\beta D x^3)$ can be expanded as $1 + \beta D x^3 + \cdots$. Keeping only terms through second order, the integrals can then be written as

$$\langle x \rangle = \frac{\displaystyle\int_{-\infty}^{\infty} e^{-\beta C x^2} (x + \beta D x^4 + \cdots)\, dx}{\displaystyle\int_{-\infty}^{\infty} e^{-\beta C x^2} (1 + \beta D x^3 + \cdots)\, dx} = \frac{3 D k T}{4 C^2}. \tag{17.37}$$

This model was used with the bond–energy relations introduced in Chapter 7 to estimate the thermal expansion of ionic compounds.

17.6 Coupled Transport Effects

Fourier's law, Fick's law, and Ohm's law all relate a flux to a transport coefficient times a gradient of some field. These three laws can be written in matrix form as

$$\begin{pmatrix} J_Q \\ J_M \\ J_E \end{pmatrix} = \begin{pmatrix} c_{11} & c_{12} & c_{13} \\ c_{21} & c_{22} & c_{23} \\ c_{31} & c_{32} & c_{33} \end{pmatrix} \begin{pmatrix} -\nabla T \\ -\nabla C \\ -\nabla V \end{pmatrix}, \tag{17.38}$$

where J_Q, J_M, and J_E are the heat, mass, and electric current fluxes, respectively, that respond to the temperature, composition, and electric potential gradient. The diagonal terms can be immediately identified as the thermal conductivity K, the chemical diffusivity D, and the electrical conductivity, σ. In addition, the off-diagonal terms represent coupling that occurs because of the interaction between electrons, phonons, and the ion cores. This coupling produces some interesting and useful effects.

17.6.1 Dufour and Soret–Ludwig Effects

Consider a homogeneous bar with a thermal gradient. From the matrix equation (Equation 17.38) we see

$$J_M = -c_{21} \nabla T = \frac{c_{21}}{K} J_Q; \quad \nabla V = 0. \tag{17.39}$$

Equation 17.39 predicts a mass flux driven by a thermal gradient, a phenomenon known as the Soret–Ludwig effect or sometimes as thermomigration or thermal diffusion. One can think of the mass transport arising from the interactions of the phonons carrying the heat flux (J_Q) with the different constituents of the material. Components with positive heat of transport migrate to the hot end while those with negative heat of transport migrate to the cold end. The effect is more pronounced in liquids and gases where the mobility of the particles is higher. Compositions of the melt have been altered in directional solidification experiments by Soret diffusion, although such effects are often masked by convective transport. Thermal diffusion has also been used in Clausius–Dickel columns for isotope separation. The Soret effect for solids is generally small but can become significant when there are high temperature gradients, such as those found in fuel pellets and in the cladding on fuel rods in nuclear reactors.

The complement of the Soret effect is the Dufour effect in which a concentration gradient causes a heat flow according to

$$J_Q = -c_{12}\nabla C = \frac{c_{12}}{D}J_M; \quad \nabla V = 0 \tag{17.40}$$

by interactions between the diffusion current (J_M) and phonons in the solid. Given the fact that diffusion is a very slow process in solids, there is little diffusion current to interact with the phonons, which makes the Dufour effect virtually undetectable.

17.6.2 Electrotransport

The c_{23} term is associated with electrotransport in which the electrons in the current flow interact with impurities and tend to drag them along. From the matrix Equation 17.38,

$$J_M = -c_{23}\nabla V = \frac{c_{23}}{\sigma}J_E; \quad \nabla T = 0. \tag{17.41}$$

Under most circumstances, electrotransport (or electromigration as it is commonly called) is a small effect. However, in integrated circuits, as the interconnects become smaller and smaller, the current densities J_E can become large enough to cause structural changes in the Al interconnects, which caused reliability problems in some of the early VLSI circuits. Naturally, this problem prompted a great deal of research into electrotransport, which resulted in methods for preventing the degradation of interconnects from its effects.

The complement of electrotransport predicts a current flow resulting from a concentration gradient as the diffusion current of atoms or ions drag electrons along

$$J_E = -c_{32}\nabla C = \frac{c_{32}}{D}J_M. \tag{17.42}$$

Again because solid state diffusion is so slow, the effect is negligible.

17.6.3 Seebeck and Peltier Effects

Probably the most important of the cross-coupling effects are those that couple heat flow with current flow. Consider the coefficient c_{31}, which connects an electric current to a temperature gradient. Using the matrix Equation 17.38 with Ohm's law,

$$J_E = -c_{31}\nabla T = -\sigma\nabla V; \quad \nabla C = 0. \tag{17.43}$$

The Seebeck coefficient is defined as $S(T) = dV/dT$ and we can identify $c_{31} = S(T)\sigma$. The Seebeck effect arises partly from the small temperature dependence of the Fermi level, which causes electrons to migrate from the region of higher Fermi level (higher chemical potential) to the lower. But the dominant effect comes from the interaction of the phonons with the electrons as they move through the solid, dragging the electrons along with them.

Next consider the case of a homogenous conductor with an applied electric field. From the matrix Equation 17.38 we see that

$$J_Q = -c_{13}\nabla V; \quad \nabla C = 0. \tag{17.44}$$

From Ohm's law, $\nabla V = -J_E/\sigma$, which combined with Equation 17.44 gives

$$J_Q = c_{13}\frac{J_E}{\sigma} = \Pi J_E, \tag{17.45}$$

where Π is the Peltier coefficient. The electron flow interacts with phonons to move heat along with the current flow. The Peltier effect has a number of practical applications, primarily as a solid-state electronic heat pump which is the heart of a variety of thermoelectric devices (TEDs), which will be discussed later.

17.7 Applications

17.7.1 Thermocouples

The Seebeck voltage can only be measured in terms of the difference in Seebeck coefficients between two conductors as illustrated in Figure 17.7. Junctions are formed by spot welding wires with compositions A and B at each end. Wire B may be broken and connected to a voltmeter with leads of a different composition as long as these two connections are held at the same temperature. The voltage read by the voltmeter is given by

$$V_{AB} = \int_{T1}^{T2} (S_A(T) - S_B(T))dT, \tag{17.46}$$

where S_A and S_B are the Seebeck coefficients of the two wires. These coefficients are generally dependent on temperature and can be approximated in the form $a + bx + \cdots$. Integrating Equation 17.46, we see the Seebeck voltage will be a cubic (or higher order polynomial) of the difference in temperature between the two junctions. There will also be a contact potential at the junctions between the wire B and the meter leads, but if these two junctions are at the same temperature, there will be no potential difference and the effect from these junctions will cancel.

Before electronic compensated reference junctions became available, the reference junction T_1 was immersed in ice water and the temperature of the other junction was determined by reading the microvoltmeter and either looking up the temperature corresponding to that voltage in tables or evaluating a polynomial representing the calibration data. Nowadays, the two thermocouple leads can be connected directly to a A/D converter whose contacts are compensated for contact potential and resistance electronically and the digital voltage is converted to temperature using a built-in algorithm.

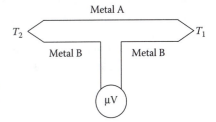

FIGURE 17.7
Typical thermocouple arrangement with junctions held at T_1 and T_2. The Seebeck voltage is read by a microvolt meter in series with one of the wires.

TABLE 17.1

Thermocouple Types and Their Properties

Types	Materials	Range (°C)	EMF/deg. (μV/°C)
B	Pt-30%Rh–Pt	0 to 1800	−0.2[a]
E	Chromel (Ni–Cr)-constantin (Cu–Ni)	−270 to 1000	59
J	Fe-constantin	−210 to 1200	50
K	Chromel–alumel (Ni–Al)	−270 to 1372	39
N	Ni-14%Cr-1.4%Si – Ni-4.4%Si-0.1%Mg	−200 to 1300	0.038
R	Pt-10%Rh–Pt	−50 to 1768	5.2
S	Pt-13%Rh–Pt	−50 to 1768	5.4
T	Cu–Constantin	−270 to 400	39

Source: Data taken from *Omega Temperature Handbook.*

Note: EMF/deg. is taken at 0°C.

[a] Type B thermocouples show a negative slope that becomes positive after a few degrees above 0°C.

Thermocouples are widely used in measuring temperatures and are particularly useful in the processing of materials because of their high temperature range. The most commonly used thermocouples are listed in Table 17.1.

Thermocouples are also used in the safety shut-off valves in gas appliances. The current developed by a thermocouple inserted in the pilot flame flows though a solenoid that holds the gas supply open. If the pilot flame were to go out, the valve would automatically close, preventing the release of gas. It may seem surprising that such a little voltage could produce enough current to activate a solenoid valve, but there is practically no internal resistance in the thermocouple junction. This means that the current that flows is the Seebeck voltage divided by the resistance in the wires of the solenoid, which can be a small fraction of an ohm.

The Seebeck voltage generated at a moving solidification interface in a directionally solidified alloy has been used to measure the kinetic undercooling and to observe the incipient breakdown of a plane interface to dendritic growth (see Section 13.2.6).

17.7.2 Thermoelectric Generators

The small voltage output of a thermocouple can be increased by wiring a large number of thermocouples in series to make a thermopile as illustrated in Figure 17.8. Thermoelectric generators made from thermopiles are used to power spacecraft that must operate in deep

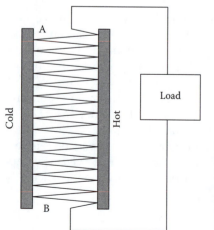

FIGURE 17.8

Schematic of a thermopile or thermoelectric generator. Alternating pairs of thermocouple wires are joined so that the hot junctions are embedded in the hot plate and the cold junctions are embedded in a cold plate capable of rejecting heat. Current will flow through the load as long as a temperature difference is maintained between the hot and the cold plates.

space where solar cells do not receive enough sunlight. The hot plate is heated by the radioactive decay of Pu-238 and the cold plate radiates to space. Such a device is called a radioisotope thermoelectric generator or RTG. Similar devices are used on earth to make electricity from low-level process heat that would otherwise be wasted.

17.7.3 Thermoelectric Cooling

If the voltmeter in Figure 17.7 is replaced by a battery or other current source, heat can be made to flow from one junction to the other. The heat flux is given by $J_Q = (\Pi_B - \Pi_A)J_E$. To obtain a useable amount of cooling, the thermocouple wires are connected in series as shown in Figure 17.8 except the load is replaced by a current source and heat is pumped from cold to hot just as in a mechanical heat pump, except here there are no moving parts. Such an arrangement is called a TED. By thermally insulating the cold plate and removing heat from the hot plate convectively using fins or a fan, one can make a Peltier refrigerator. By reversing the current flow one has a Peltier oven although there are easier ways to heat electrically.

Peltier coolers can be stacked so that the hot plate of the first cooler is cooled by the second cooler and so on. This type of stack is useful for making a small area, such as the focal plane of an infrared imager, very cold in order to reduce thermal noise in the image.

Peltier pulsing can also be used to mark a liquid–solid interface in a directionally solidified alloy. Sending a current pulse through the specimen quickly cools and advances the solidification interface to capture some of the rejected solute that is being pushed ahead of the advancing interface (see the section on macrosegregation in Section 13.2.4). The sudden change in composition can be seen in the specimen after it has been cut and polished. By sending a series of timed Peltier pulses, the growth rate as well as the shape of the interface at different times can be determined.

17.8 Summary

Classical thermodynamics using the equipartition of energy principle predicts that the lattice molar heat capacity will be given by $1/2R$ for each of the six degrees of freedom in a solid (three kinetic and three potential energy) for a total of $3R$. As the temperature is increased, the observed heat capacity of materials approaches this value, which is known as the Dulong–Petit limit. However, at low temperatures, the observed heat capacity approaches zero as T^3.

Using the Planck distribution, Debye constructed a model in which the lattice heat capacity goes to zero as T^3. The Debye model assumes a linear dispersion relation characteristic of a continuous medium rather than a chain of discrete atoms and cuts off the distribution at a frequency such that the number of normal modes is equal to 3 × the number of atoms. This frequency, known as the Debye frequency, is given by $\omega_D = v_0(6\pi^2 N/V)^{1/3}$, where v_0 is the velocity of sound in the medium and N/V is the atoms per unit volume. A Debye temperature Θ_D is defined in terms of the Debye frequency $\Theta_D = \hbar\omega_D/k$. For $T \gg \Theta_D$, the Debye model approaches the classical Dulong–Petit limit.

Even though the Debye model uses an unrealistic assumption for the dispersion relation, its primary justification comes from its excellent agreement with the observed heat capacity, especially at the lower temperatures. Apparently, the heat capacity is not particularly sensitive to the exact form of the dispersion relationship.

The small electronic contribution to the heat capacity of metals can be understood by applying Fermi–Dirac statistics to the free electron gas. Since the ground state energy is much higher than ambient thermal energy, only those electrons at the top of the Fermi sea can be thermally excited. As a result the electronic heat capacity is given by $C = (\pi^2/2)/(RT/T_F)$ instead of the classical value of $3/2R$. At very low temperatures, the electronic contribution, which has a linear temperature dependence, can exceed the lattice contribution to heat capacity, which has a T^3 temperature dependence.

The thermal conductivity can be calculated for phonons as well as electrons using the ideal gas model derived from kinetic theory, $K = C_v \bar{v} \lambda/3$, where λ is the mean free path of electrons or phonons. The thermal conductivity of the lattice is low at low temperatures because the lattice heat capacity goes to zero as the temperature approaches 0 K. The lattice thermal conductivity is limited by scattering from lattices with different sized ions and from defects such as grain boundaries, which is why ceramics are generally poor thermal conductors. Glasses are especially poor conductors because of their lack of a periodic structure. However, single crystal ceramics can be very good thermal conductors, especially diamond and sapphire whose stiff lattices give them high sound velocities. Aligned pyrolytic graphite has thermal conductivity seven times higher than aluminum. Thermal conductivity tends to decrease beyond 0.4 times the Debye temperature because of inelastic phonon–phonon collisions (*U*-processes) that become more prevalent at higher temperatures.

Metals generally have higher thermal conductivity than nonmetals because of the presence of free electrons, but the electrons do not contribute to the heat capacity as they would be expected to form classical considerations. The low heat capacity is compensated by the high Fermi velocity, so the conductivity calculated using quantum mechanics is almost the same as the classical result.

Thermal expansion is a result of the asymmetry of the energy–bond curve, which produces anharmonic oscillations on the part of the atoms in the lattice. Thus the time–average position of the atoms is displaced from the equilibrium position at the bottom of the bond–energy well. This displacement increases with temperature, giving rise to the coefficient of linear expansion.

Interactions between electrons, phonons, and ion cores can produce a number of interesting coupling effects between heat flow, mass flow, and current flow. The most important of these coupling effects are the Seebeck effect, the production of current flow in conjunction with heat flow, and the Peltier effect, the flow of heat in conjunction with current flow. These two effects make possible thermocouples for temperature measurement, thermoelectric generators and refrigerators that have no moving parts.

Bibliography

Adachi's *Handbook of Physical Properties of Semiconductors*, Springer, New York, 2004.

Ashcroft, N.W. and Mermin, N.D., *Solid State Physics*, Brooks Cole, Philadelphia, 1976.

Barsoum, M.W., *Fundamentals of Ceramics*, Institute of Physics Publishing, Bristol and Philadelphia, 2003.

Dekker, A.J., *Solid State Physics*, Prentice Hall, Engelwood Cliffs, New Jersey, 1962.

Kittel, C., *Introduction to Solid State Physics*, 7th edn., John Wiley & Sons, New York, 1966.

Omega Temperature Handbook, available online at www.omegaeng.cz/temperature/Z/zsection.asp.

Problems

1. Show that the density of states for a 2-D ideal gas is a constant that does not contain E.

2. Show that the energy of a 2-D ideal gas is kT.

3. Adjust the density of states we developed,

$$C(\omega) = \frac{12N}{\pi^2} \frac{\left[\sin^{-1}(\omega/\omega_0)\right]^2}{\left(\omega_0^2 - \omega^2\right)^{1/2}}$$

by multiplying it by a constant so that it gives the proper number of modes when integrated. Then use it to obtain the heat capacity. How does this model compare with the Debye model?

4. In the high temperature limit, at what temperature will the electron heat capacity equal the lattice heat capacity for Al?

18

Free Electrons in Metals

In order to understand the properties of metals, it is necessary to understand the role of electrons in metals. We already know that the metallic bond has to do with electrons that become delocalized from their parent ion cores and are more or less free to move around in the metal. To the first approximation, treating the electrons as an ideal Fermi gas does a fairly good job of explaining most of the properties of metals, but there is more to it as we shall see.

18.1 Drude Theory of Free Electrons in Metals

The properties that characterize metals (good electrical conductivity, good thermal conductivity, bright, shiny, highly reflective in the visible spectrum) are attributed to the presence of free electrons. Shortly after the discovery of the electron by J.J. Thompson in 1897, P.K.L. Drude formulated a theory in 1900 to explain these properties of metals in terms of an ideal electron gas that permeates the spaces between the positive ion cores that form the metal lattice. These electrons were supposed to be the carriers of heat and electricity that made metals much better conductors than ceramic or polymeric systems. When light was incident on a metal, the electrons in the metal responded to the oscillating E-field of the incoming light wave, thus reradiating a wave of the same frequency which explained the high reflectivity exhibited by metals. Even the metal bond could be explained qualitatively by this electron gas, which played essentially the same role as the anions in the rock salt face-centered cubic structure. This simple theory was successful in predicting a number of properties of metals as will be seen in the following sections.

18.1.1 Electrical Conduction

Metals, for the most part, obey a simple linear relationship between the applied voltage V and the current I that flows through it. This empirical relationship was discovered by Ohm and is expressed as

$$V = IR. \tag{18.1}$$

(Actually this not really Ohm's law, but the definition of the resistance R.) The constant of proportionality R is called the resistance and is measured in ohms (1 $\Omega = 1$ V/A) and depends not only on the metal but also on its size and configuration. This is what is known

as an extensive property since it depends of the size or extent of the material. R, however, can be shown to be directly proportional to the length of the conductor and inversely proportional to the cross-sectional area; thus, for a particular metal

$$R = \frac{\rho l}{A} = \frac{l}{\sigma A}, \tag{18.2}$$

where
ρ is the resistivity, Ω m
σ is the conductivity, $(\Omega\text{ m})^{-1}$ or mho/m

Both resistivity and conductivity are intensive properties that depend only on the metal itself. (The International unit of conductivity is now siemens per meter (S/m), where S is defined as 1 A/V.) If we rewrite Equation 18.1 and substitute the above for R,

$$I = \frac{V}{R} = \frac{V\sigma A}{l} = E\sigma A \tag{18.3}$$

and then divide by the area (A), we get the general form of Ohm's law:

$$\mathbf{J} = \sigma\mathbf{E}, \tag{18.4}$$

where
$\mathbf{J}$ is the current density
$\mathbf{E}$ is the applied electric field

The essential feature of Ohm's law is that $\mathbf{J}$ is directly proportional to the applied field $\mathbf{E}$ and σ, being a property of the material, is independent of the field. Note that $\mathbf{J}$ and $\mathbf{E}$ are vector quantities while σ is a scalar (tensor of rank zero) for an isotropic media; however, it will be a tensor of rank 2 for an anisotropic material such as a single crystal.

Treating the electrons as a classical ideal gas, the average speed would be

$$v_{\text{th}} = \sqrt{\frac{8kT}{\pi m}}. \tag{18.5}$$

This velocity is random, so there are just as many electrons moving to the left as to the right resulting in no net current flow. To cause a current to flow, this thermal velocity distribution must be biased by applying an electric field to produce a drift velocity that carries the current.

The current density $\mathbf{J}$ can be expressed in terms of the drift velocity as

$$\mathbf{J} = -ne\mathbf{v}_{\text{d}}, \tag{18.6}$$

where
n is the electron density
e is the magnitude of the electronic charge (the drift velocity for electrons will be negative as well as the charge carried by the electrons so the current density will be positive).

Another intensive property of the material is the mobility of the electrons μ, which is defined for electrons by the relation

$$\mu \equiv \frac{v_d}{E}, \tag{18.7}$$

where v_d is the average or drift velocity of the electron when an electric field E is applied. Like the conductivity, mobility can be a tensor of rank 2 and is a scalar only for an isotropic media. (The sign of the mobility is always taken to be positive for charge carriers of either sign.)

Since

$$\mathbf{J} = \sigma\mathbf{E} = -ne\mathbf{v_d} = ne\mu\mathbf{E}, \tag{18.8}$$

σ is related to μ by

$$\sigma = ne\mu \tag{18.9}$$

18.1.2 Classical Electron Dynamics

Consider the equation of motion for an electron at rest when an electric field is suddenly switched on

$$m\ddot{\mathbf{x}} = -e\mathbf{E}, \tag{18.10}$$

where m is the electron mass.

The solution is

$$\dot{\mathbf{x}}(t) = -(e/m)\mathbf{E}t \quad \text{and} \quad \mathbf{x}(t) = \frac{-(e/m)\mathbf{E}t^2}{2}. \tag{18.11}$$

Assume a collision occurs at time τ. The average drift velocity can be found from

$$\mathbf{v_d} = \langle \dot{\mathbf{x}}(\boldsymbol{\tau}) \rangle = \frac{\mathbf{x}(\boldsymbol{\tau})}{\tau} = -\frac{(e/m)\mathbf{E}\boldsymbol{\tau}}{2}. \tag{18.12}$$

However, this model assumes that the electrons come to a complete stop after each collision. A better model can be obtained by writing Newton's law with a drag term,

$$\dot{\mathbf{v}}_d - \frac{1}{\boldsymbol{\tau}}\mathbf{v_d} = -\frac{e}{m}\mathbf{E}. \tag{18.13}$$

The solution is

$$\mathbf{v_d} = -(1 - e^{-t/\tau})\frac{e\boldsymbol{\tau}}{m}\mathbf{E}. \tag{18.14}$$

Here it may be seen that τ is the e-folding response time to a change in the $\mathbf{E}$ field and that the steady-state velocity is the terminal velocity at which the drag force, v_d/τ is balanced by the force exerted by the field. The τ can also be interpreted as the average time between collisions. The electrons give up some of their energy in these collisions which create phonons, adding to the thermal energy of the lattice. The result is known as ohmic heating.

Putting this v_d into Equation 18.7, the mobility and conductivity can be written as

$$\mu = \frac{e}{m}\tau \quad \text{and} \quad \sigma = \frac{ne^2\tau}{m}. \tag{18.15}$$

Note that since mobility and conductivity, which are material properties, depend on the collision time, then this collision time must be independent of the applied field if Ohm's law is to be obeyed. Why should the collision time be a material property? The mean free path for electrons to travel before collisions would seem to be a much more reasonable material property since it is determined by the size and number density of atoms in the structure. But the mean free path λ is the velocity times the collision time τ and since v_d is directly proportional to τE, the collision time would be inversely proportional to the square root of the field, in violation of Ohm's law. This paradox can be resolved (at least temporarily) by assuming that the thermal velocities of the electrons are much higher than v_d so that $\lambda = \tau(v_{th} \pm v_d) \approx \tau v_{th}$. Let us now check to see if this assumption is valid.

The thermal velocity of electrons at 300 K from Equation 18.5 is 1.07×10^5 m/s. For Cu, the measured conductivity is 6×10^7 s/m. Assuming each atom contributes one free electron to the electron gas, the number density is

$$n = \frac{8.96 \text{ g/cm}^3}{63.55 \text{ g/mol}} 6.023 \times 10^{23} \text{ atoms/mol} = 8.5 \times 10^{22} \text{ electrons/cm}^3 = 8.5 \times 10^{28}/\text{m}^3.$$

A #12 Cu wire is approximately 1 mm in diameter and is rated to carry 20 A without overheating. The current density is

$$J = \frac{4I}{\pi d^2} = \frac{4(20)}{\pi(0.001)^2} = 2.5 \times 10^7 \text{ A/m}^2$$

and the drift velocity is found from Equation 18.14 to be

$$v_d = \frac{J}{ne} = \frac{2.5 \times 10^7}{(8.5 \times 10^{28})(1.6 \times 10^{19})} = 0.00184 \text{ m/s}.$$

Clearly the drift velocity is much smaller than the thermal velocity of the electrons, so we can take τ as λ/v_{th} and Ohm's law is obeyed. Of course we know that it does not take this long for the current to run through the wire. The fact that a light turns on the instant the switch is thrown can be explained by the marble-in-the-tube analogy. If the tube is full of marbles and you insert a new marble in one end, a marble will come out of the other end almost instantly, even though the average velocity of the marbles in the tube is very slow.

Using the measured value for the conductivity of Cu, its mobility must be

$$\mu = \frac{\sigma}{ne} = \frac{6 \times 10^7}{8.5 \times 10^{28} \times 1.6 \times 10^{-19}} = 4.4 \times 10^{-3} \text{ m}^2/\text{V s}.$$

The collision time must be given by

$$\tau = \frac{m\mu}{e} = \frac{(9.11 \times 10^{-31})(4.4 \times 10^{-3})}{1.6 \times 10^{-19}} = 2.5 \times 10^{-14} \text{ s}.$$

The collision time, which is also the response time to a changing field, is much shorter than the period of electronic frequencies, so that Ohm's law can be expected to be obeyed for alternating currents as well as for direct currents. However, we will find in later chapters that deal with the interaction of light with metals that modifications will have to be made to Ohm's law.

18.2 Matthiessen's Rule

The typical behavior of the resistivity of a metal with temperature is shown in Figure 18.1 and can be expressed by Matthiessen's rule, which can be stated as

$$\rho(T) = \rho_i + \rho_d + \rho_{th} = \rho_0 + \alpha T; \quad \text{for } T > \Theta_{Debye}, \tag{18.16}$$

where ρ_i, ρ_d, and ρ_{th} are respectively the impurity, defect, and thermal contributions to the resistivity. The impurity and defect contributions can be combined into ρ_0, which is the extrapolated value at 0 K.

18.2.1 Effect of Impurities and Defects

At very low temperatures, electrons can move through a uniform lattice of ions with little scattering so the resistivity at 0 K (ρ_0) for pure metals can be quite small. The presence of interstitial or substitutional impurity atoms disrupts the periodicity of the lattice and can add a major contribution to the resistivity of a metal since the mean free path can be very short due to the random distribution of the impurity atoms. Hence solid solution hardening is not a good method for strengthening electrical conductors as can be seen in Figure 18.1. Adding 2% Be to Cu to form a solid solution produces a much greater change in resistivity than adding 35% Zn to form α-brass because of the 12% size difference between the Be and Cu versus only 4% between Zn and Cu. Age-hardening Be-Cu

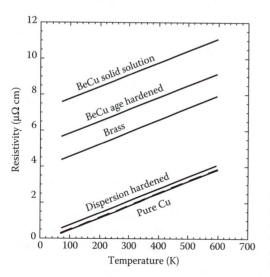

FIGURE 18.1
Resistivity of Cu as a function of temperature. The dashed line above pure Cu represents work-hardened Cu. (From Askeland, D.R. and Phule, P.P., *The Science and Engineering of Materials*, 5th edn., Thompson, Canada, 2006.)

decreases the resistivity of solid solution Be-Cu because the Be atoms coalesce to form dispersed Be clusters rather than a random distribution of Be atoms. The added resistivity contributed by a mole fraction x of a solid solution impurity atom can be described by Nordheim's rule:

$$\rho_0(x) = \rho_a(1 - x) + Ax(1 - x) + \rho_b x, \tag{18.17}$$

where A is the defect resistivity coefficient that is a function of the host and impurity atoms and can be quite large. For example, Cu-60 At% Ni has a resistivity that is 30 times higher than pure Cu. In systems that exhibit order–disorder transitions, the resistivity of the ordered phase is less than the disordered phase, as might be expected.

Dispersion or age hardening has less effect than solid solution hardening as seen in Figure 18.1 because the atoms are coalesced as a second phase. Also note that dispersion hardening with 7% Al_2O_3 particles produces only a modest increase in resistivity. The presence of a second phase follows the rule of mixtures, i.e.,

$$\rho_0(V_b) = \rho_a V_a + \rho_b V_b = \rho_a(1 - V_b) + \rho_b V_b, \tag{18.18}$$

where the Vs are the volume fractions of the two phases, respectively. This relation also applies to the conductivity of eutectics.

Dislocations and grain boundaries are also much less effective in scattering electrons because they leave the vast bulk of atoms in the lattice undisturbed; therefore, work hardening and grain refining have relatively little effect on resistivity. This is fortunate because Cu house wiring tends to work harden from bending as electrical outlets are being installed. If this resulted in a large resistance change, hot spots would develop and fires could result.

18.2.2 Temperature Dependence of Resistivity

For temperatures above 20% of the Debye temperature, the resistivity increases linearly with temperature as suggested by Matthiessen's rule. The slope α is pretty much the same for a given metal, regardless of the defects and impurities. The first-order dependence of resistivity on T can be understood from a simple model. The collision cross section between an electron and the core ions in a metal is proportional to the square of the amplitude of its vibration. The square of the amplitude of an oscillator is proportional to its thermal energy kT (assuming $T > \Theta_D$, so all of the modes are active). The collision time τ is directly proportional to the mean free path λ, which is inversely proportional to the collision cross section. Therefore, conductivity is inversely proportional to T.

18.2.3 Grüneisen Model of Resistivity

As stated previously, the Matthiessen formulation for resistivity holds only for temperatures greater than about 20% of the Debye temperature because all of the vibrational modes that can scatter electrons are not active at low temperatures. Grüneisen used the Debye model to formulate an exact theory of the temperature dependence of resistivity for metals that extends the temperature dependence of resistivity to low temperatures.

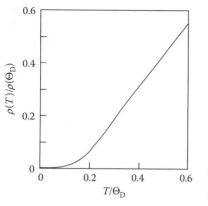

FIGURE 18.2
Universal reduced temperature plot for metals. It may be seen that the resistivity becomes linear with temperature for $T \geq 0.20\,\Theta_D$.

$$\rho(T) = A\left(\frac{T}{\Theta_D}\right)^5 \int\limits_0^{\Theta_D/T} \frac{x^5 dx}{(e^x - 1)(1 - e^{-x})}, \tag{18.19}$$

where A is a constant for a particular metal. However, if we divide $\rho(T)$ by $\rho(\Theta_D)$ and plot this reduced resistivity against the reduced temperature T/θ_D, we obtain a universal plot that is valid for all metals as shown in Figure 18.2.

18.3 Problems with the Classical Free Electron Gas Theory

18.3.1 Temperature Dependence

The resistivity can be written in terms of collision times as

$$\rho = \frac{1}{\sigma} = \frac{m}{ne^2}\left(\frac{1}{\tau_0} + \frac{1}{\tau_{th}}\right) = \frac{mv_{th}}{ne^2}\left(\frac{1}{\lambda_0} + \frac{1}{\lambda_{th}}\right), \tag{18.20}$$

where the $1/\tau_0 = v_{th}/\lambda_0$ is the contribution from impurities and defects and the $1/\tau_{th} = v_{th}/\lambda_{th}$ is the temperature-dependent contribution from the ion cores. Since the collision cross section is dependent on the square of the amplitude of the vibrating ion cores and the vibrational amplitude is proportional to the square root of the temperature (assuming $T >$ the 20% of the Debye temperature), λ_{th} must be inversely proportional to T as we argued before. Therefore, in order for the resistivity to increase linearly with temperature, v_{th} would have to be independent of temperature. But v_{th} is proportional to $T^{1/2}$, so assuming that the electrons in a metal behave an ideal gas in thermal equilibrium would lead to a temperature dependence of resistivity in the form of $\rho(T) \propto AT^{1/2} + BT^{3/2}$, which is contradictory to the observed thermal behavior of the conductivities of metals.

18.3.2 Mean Free Path Considerations

In addition to the failure of the free electron gas model to properly account for the observed temperature behavior of resistivity of metals, there are serious problems when other properties are considered. First, let us make a few simple calculations to illustrate one of its major difficulties. Multiplying the collision time of 2.5×10^{-14} s by the thermal velocity 1×10^5 m/s, one gets a mean free path of 2.5 nm, which is on the order of a few lattice spacing. This seems incredibly small, especially considering that Cu is one of the

best conductors. What about the mean free path of poor conductors whose resistivity can be several orders of magnitude higher than Cu? Their mean free paths would have to be much shorter than the distance between the atoms. What could cause this scattering?

18.3.3 Electronic Heat Capacity and Paramagnetism

Recall the other serious difficulty discussed in Chapter 17 that arises from the fact that the classically predicted heat capacity of the electrons is not observed even though they are the major contributor to both the thermal and electrical conductivity of metals. We will find yet another problem with the classical theory when we take up the topic of paramagnetism and find that the electronic contribution expected from classical theory is not observed. Despite the success of the classical Drude theory of the free electron gas in being able to describe many of the observed properties of metals, it was these discrepancies between the classical theory and observation that prompted theorists to reexamine the classical theory of the electron and to apply the quantum mechanical treatment that had been developed to explain the electronic structure of atoms and molecules to describe the behavior of electrons in metals.

18.4 Quantum Theory of Free Electrons

Now that we are better equipped to handle some of the quantum theory necessary to understand the behavior of electrons in metals, let us review some of the problems we encountered in the simple Drude theory.

First, the problem of relating the collision time τ to an identifiable material property, such as electron mean free path, required a velocity of the electron gas that was large compared to the drift velocity imposed by the applied electric field and was essentially independent of both the applied field and of the temperature. This problem can be resolved by considering the electrons-in-a-box model in which it was shown that the ground state energy or the Fermi energy ε_F (the energy state of the last electron added) is very large (typically $\sim$5 eV). It is useful to define a Fermi momentum whose magnitude is given by

$$|\mathbf{p}_F| = \sqrt{2m\varepsilon_F} \tag{18.21}$$

and a Fermi velocity given by

$$|\mathbf{v}_F| = \left|\frac{\mathbf{p}_F}{m}\right| = \sqrt{\frac{2\varepsilon_F}{m}}. \tag{18.22}$$

For Cu, we had found $n = 8.5 \times 10^{28}$ electrons/m^3, which yields a Fermi energy of 1.131×10^{-18} J (or 7.058 eV) and a Fermi velocity $v_F = 1.57 \times 10^6$ m/s, which is an order of magnitude larger than the thermal velocity. Since the Fermi velocity is independent of the applied field and only a very weak function of temperature, substituting v_F for v_{th} resolves the problems with the temperature dependence of resistivity described in Section 18.3.1. Now the mean free path for Cu is on the order of 10^{-8} m, which corresponds to $\sim$100 lattice parameters. (We will find a more sophisticated way of calculating resistance in Chapter 19, which explains the large variations of resistivity among the transition metals.)

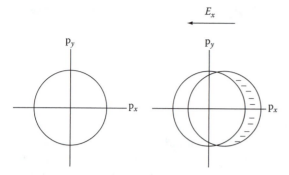

FIGURE 18.3
Fermi sphere in the absence of an applied field (left). When an electric field is applied in the $-x$ direction, each electron on the surface of the Fermi sphere gains momentum in the $+x$ direction and moves to a higher energy state. This frees up energy states just below the Fermi surface allowing the other electrons to move to higher energy states, the result being that the entire Fermi sphere is shifted to the right.

One can visualize the distribution of momenta in momentum space in the absence of an applied electric field as a sphere with radius p_F centered about the origin as shown in Figure 18.3. This we call the Fermi sphere and its surface is called the Fermi surface. Later, we show that the Fermi surface is distorted by Bragg reflections of electrons near boundaries of the Brillouin zones. The electrons below the Fermi energy are in filled states and cannot respond directly to the applied electric field. Only those electrons at the Fermi energy can be accelerated by the applied field to a velocity $v_d + v_F$. However, this creates new states at the Fermi surface in which electrons just below the Fermi level can move into, and so on. The net result is that each electron is able to increase its velocity by v_d.

The drift velocity when an electric field is applied is still given by $\mathbf{v}_d = -e\tau\mathbf{E}/m$, so the entire Fermi sphere is shifted by an amount $\delta\mathbf{p} = -e\tau\mathbf{E}$. In the Drude model, all of the electrons were assumed to have random velocities with an average speed v_{th} but the average velocity would be zero in the absence of an applied field. An applied field would increase each electron's average velocity by $\mathbf{v}_d$ so the current is carried by all of the electrons with a velocity $\mathbf{v}_d$.

However, the electrons in a Fermi gas do not have random velocities. Those on the leading edge of the Fermi sphere have velocities v_{Fx} and those on the back edge have velocities $-v_{Fx}$. While it is true that the applied field increases each electron's velocity by v_d, it is not true that the current is carried by each electron moving at v_d. To illustrate this, consider a simple one-dimensional Fermi model with 11 particles in which the Fermi velocity is 5 m/s and quantum states are separated by 1 m/s as shown below.

| -5 | -4 | -3 | -2 | -1 | 0 | 1 | 2 | 3 | 4 | 5 | |

Now let an applied field increase each particle's velocity by 1 m/s.

| | -4 | -3 | -2 | -1 | 0 | 1 | 2 | 3 | 4 | 5 | 6 |

Note that the particle with velocities from 4 to -4 m/s cancel each other and carry no net charge. All the charge is carried by the two particles with velocities 5 and 6 m/s. So instead of 11 particles moving at 1 m/s, as would be the case with the Drude model, we have two particles moving with an average speed of 5.5 m/s to carry the current. The sum of the products of the charge carriers and their velocities is the same in either model.

A rough estimate of the number of electrons in the Fermi sphere that are able to participate in carrying charge when an electric field is applied is nv_d/v_F. So instead of n electrons moving at v_d, we have $n(v_d/v_F)$ electrons moving at v_F. Therefore for simple metals in which the Fermi surface is essentially spherical, the conductivity and mobility given by Equation 18.15, $\mu = (e/m)\tau$ and $\sigma = ne^2\tau/m$ are still valid, but the transition metals

in which the *d*-electrons play a larger role required a more rigorous analysis, as discussed in Chapter 19.

18.5 Hall Effect

In 1879, the American physicist Edwin Herbert Hall reasoned that since charges moving through a magnetic field are deflected to one side, it should be possible to measure a voltage produced by this deflection. This phenomenon is known as the Hall effect and the voltage resulting from this deflection is known as the Hall voltage. (This is rather curious since the electron was not discovered until 1897. Come to think of it, Maxwell formulated his famous equations in 1865 before the electron was discovered. The concept of positive and negative charge was well known at the time even though the electron itself had not been identified.)

Since the force on an electron moving through a magnetic field is acted on by the Lorentz force given by $\mathbf{F} = -e(\mathbf{E} + \mathbf{v} \times \mathbf{B})$, the equation of motion can be written as

$$\dot{\mathbf{v}} + \frac{\mathbf{v}}{\tau} = -\frac{e}{m}(\mathbf{E} + \mathbf{v} \times \mathbf{B}). \tag{18.23}$$

For a direct current in the $x-y$ plane in the presence of a magnetic field B in the z-direction, the steady-state solution is

$$
\begin{aligned}
v_x &= -\frac{e\tau}{m}\left(E_x + B_z v_y\right) \\
v_y &= -\frac{e\tau}{m}\left(E_y - B_z v_x\right).
\end{aligned}
\tag{18.24}
$$

Assume the applied field is in the x-direction causing a current j_x to flow. The presence of the field will deflect electrons in the y-direction to the walls of the conductor. Since the current cannot flow out through the walls, a field $E_y = B_z v_x$ will develop such that $v_y = 0$. Putting this into the first of the above equations gives

$$E_y = -\frac{eB_z\tau}{m}E_x. \tag{18.25}$$

The Hall coefficient is defined as $R_H = E_y/j_x B_z$. Putting $j_x = \sigma E_x$ into Equation 18.24, we find

$$R_H = \frac{E_y}{j_x B_z} = -\frac{eB_z \tau E_x}{m j_x B_z} = -\frac{e\tau E_x}{m \sigma E_x} = -\frac{e\tau}{m}\frac{m}{ne^2\tau} = -\frac{1}{ne} \tag{18.26}$$

for electrons. Note that measurement of the Hall coefficient determines not only the carrier concentration n but the sign of the carriers—in this case, electrons. (Note: This n is the total number of charge carriers, not just those at the Fermi level. This relationship holds for both the Drude and the Fermi model as long as the conductivity can be written as $\sigma = ne^2\tau/m$).

We shall see in later chapters that it is possible in semiconductors (and in some metals) for holes in the electrons band structure to act as positively charged carriers, Had the carriers been holes, the signs in the equation of motion would have been reversed resulting in a positive Hall coefficient. (Note: This simplified derivation of the Hall effect is only

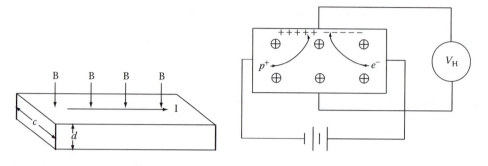

FIGURE 18.4

Schematic of a Hall measurement. Both positive and negative carriers will be swept to the same side. The magnitude and polarity of the Hall voltage measures the number density and sign of the majority carriers.

correct if there is a single carrier. If there are both positive and negative carriers, a more complicated derivation is required as discussed in Chapter 20.)

The Hall coefficient is measured by flowing a known current I through a rectangular bar of the material of interest. Let the width of the bar perpendicular to the B-field be c and its height d as shown in Figure 18.4. The E_y field produces a potential difference V_H across the bar, which is given by $-E_y c$. The current density $j_x = I/cd$. Therefore,

$$R_H = \frac{E_y}{j_x B_z} = -\frac{V_H d}{I B_z}. \tag{18.27}$$

Had the carriers been holes, the v_x would have the opposite sign, they would be deflected to the same side of the conductor as the electrons, but would have produced an opposite E-field and opposite Hall voltage V_H. Thus the measurement of the Hall coefficient is a powerful tool that gives both the sign and the number density of the majority carriers in metals as well as in semiconductors.

Recall that $\sigma = ne\mu$; therefore, $\mu = R_H \sigma$. Thus if σ is known or measured separately, the Hall coefficient also gives a measure of the majority carrier mobility.

The measured Hall coefficients for various metals are given in Table 18.1 below together with the carrier density inferred from the Hall coefficient and the valence electron density. The agreement is close but not exact for reasons we will get into when we study the band theory and introduce the effective mass of the electrons. But look what happens with Zn. The Hall coefficient has a positive sign. Does this mean that the carriers in Zn are holes? or electrons with negative mass? To make it even stranger, at high magnetic fields, Al has a positive Hall coefficient (Ashcroft and Mermin, 1976). We will examine these phenomena in the later chapters.

TABLE 18.1

Hall Coefficients of Simple Metals

Metal	R_H (SI Units)	$n_{carriers}$ (m^{-3})	$n_{valence}$ (m^{-3})
Na	-2.07×10^{-10}	3.02×10^{28}	2.5×10^{28}
Cu	-0.54×10^{-10}	1.16×10^{29}	8.5×10^{28}
Al	-0.39×10^{-10}	1.6×10^{29}	1.8×10^{29}
Zn	$+0.33 \times 10^{-10}$	1.89×10^{29}	1.31×10^{29}

Source: From Srivastava, C.M. and Srinivasan, C., *Science of Engineering Materials*, Wiley Eastern, New Delhi, India, 1987.

Small Hall-effect sensors are widely used in conjunction with small magnets as proximity detectors in door and window alarms in security systems and have replaced the old breaker points in automotive ignition systems. They can also be used as ammeters to measure current flow in wires.

18.6 Wiedemann–Franz Ratio

The fact that the ratio of the thermal conductivity to the electrical conductivity of any metal is a constant times the absolute temperature was observed by Wiedemann and Franz and this relationship is known as the Wiedemann–Franz ratio. This relationship works because the collision time τ for the electron carriers is the same in both models and cancels out when taking the ratio of the two conductivities. From Chapter 17, the classical electronic thermal conductivity was found in Equation 17.33 to be $K = (4nk^2T/m\pi)\tau$. The classical electrical conductivity from the Drude model is given by Equation 18.15 and the Wiedemann–Franz ratio becomes

$$\frac{K}{\sigma} = \frac{4nk^2T}{\pi m}\frac{m}{ne^2} = \frac{4}{\pi}\left(\frac{k}{e}\right)^2 T. \tag{18.28}$$

The constant of proportionality $L = k/\sigma T$ is called the Lorenz number. Putting in the appropriate values for the Boltzmann constant and the electronic charge, we get for the Lorenz number

$$L = \frac{K}{\sigma T} = \frac{4}{\pi}\left(\frac{k}{e}\right)^2 = \frac{4}{\pi}\left(\frac{1.38 \times 10^{-23}}{1.602 \times 10^{-19}}\right)^2 = 0.94 \times 10^{-8} \text{ W}\Omega.$$

Observed values for the Lorenz number range from 2.06×10^{-8} for Cd to 3.13×10^{-8} for W.

However, when the corrected electronic conductivity $K = (\pi^2/3)(nk^2T/m)\tau$ (Equation 17.36) is used to compute the Lorenz number,

$$L = \frac{K}{\sigma T} = \frac{\pi^2}{3}\left(\frac{k}{e}\right)^2 = 2.445 \times 10^{-8} \text{ W}\Omega, \tag{18.29}$$

which is much closer to the observed values.

Note that the temperature dependence in the Lorenz number has nothing to do with the temperature dependence of resistivity (Matthiessen's rule) that affects the collision time τ because this effect is eliminated when taking the ratio of the conductivities. Instead this T originates with the first power dependence of the electronic heat capacity with temperature.

Also note that the Wiedemann–Franz law assumes that all of the heat is carried by the electrons and therefore it only applies to metals that are good thermal conductors (or where the heat conduction from the electrons is much greater than the conduction by phonons).

18.7 Conductive Polymers

Until a few decades ago, it was believed that metals in which electrons were detached from the ion cores and were free to respond to electric fields, were the only materials capable of

being good conductors of electricity. In fact, polymers were among the best insulators known with conductivities $<10^{-10}$ S/cm. Then came the discovery of ceramic superconductors by Muller and Bednortz in the late 1960s and the discovery of conductive polymers by Heegar, MacDiarmid, and Shirakawa in the 1970s. The later discovery is bound to have far more technological impact because, unlike ceramics, polymers are more versatile and are conductive at ambient rather than at cryogenic temperatures. More recently, researchers have found ways to engineer semiconducting properties into polymers resulting in materials with controllable bandgaps in which the charge carriers can be either positive or negative. Such polymers can now fill the full range of electronic and photonic applications like their silicon and compound semiconductor cousins. Because of their low cost, ease of fabrication, and flexible nature, photonic polymers stand to revolutionize flat panel display technology.

18.7.1 Charge Conjugation

As discussed in Chapter 3, carbon has a unique capability of being able to form an sp^2 hybrid bond in which one s-electron is promoted to a p-electron and two of the three p-electrons, the p_x and p_y electrons, combine their wavefunctions with the remaining s-electron to form a three-lobed planar pattern with the three lobes 120° apart. These three-lobed orbitals form σ bonds with similar three-lobed orbitals resulting in a hexagonal ring molecule. If hydrogen atoms attach to the outer bonds of the six carbons making up the ring, a benzene molecule is formed. The six remaining p_z electrons have their figure-8 wavefunctions perpendicular to the plane of the ring and would like to form π bonds with each other, but they can only form three such bonds, giving rise to alternating single and double bonds around the ring. Since the carbon atoms are equivalent, the double bonds can be thought of as alternating between different pairs of carbon atoms in the ring. Quantum mechanically, the state may be described as a weighted average between the two equally possible states, a phenomenon described by Linus Pauling as resonance bonding.

Instead of the outside bonds of the hexagonal rings bonding with hydrogen atoms, the hexagonal rings can bond with each other to form a planar array of hexagonal rings, much like chicken wire fencing, which is the structure of graphite, the stable form of carbon under ambient conditions. As we know from Chapter 4, the planes in graphite are held together by very weak induced dipolar bonds called van der Walls bonding, which accounts for the slippery feel of graphite as the planar arrays of hexagons slide over one another.

Now instead of the double bonds alternating between the six atoms in a ring, they alternate throughout the entire plane. Since the electrons making up the π bonds do not belong to any particular carbon atom, we say they are delocalized and can move in response to an electric field. Graphite, of course, has a very high electrical (and thermal) conductivity in its basal plane. Consequently, any molecule that has alternating single and double bonds has delocalized or conjugated electrons, which can be treated with the free electron approximation just as they are in a metal.

Graphite is an example of a zero-bandgap polymer whose behavior is metal-like. Most polymers behave more like semiconductors and require doping to achieve high conductivity (see Chapter 20 for more details).

18.8 Summary

The Drude theory treats the electrons in a metal as a classical monatomic gas. This simple theory is able to explain many observed properties of metals such as why they are good

electrical and thermal conductors and reflectors of light. The electrical conduction obeys Ohm's law $\mathbf{J} = \sigma \mathbf{E}$, where $\mathbf{J}$ is the current density and $\mathbf{E}$ is the applied electric field. The conductivity σ is a property of the material. Another material property is the mobility μ of the charge carriers defined as their average velocity when subjected to an applied electric field. The mobility is related to the conductivity by $\sigma = ne\mu$, where n is the electron density and e is the electronic charge. Using Newton's law we find the drift velocity of electrons being accelerated by the electric field to be $\mathbf{v}_d = -e\tau\mathbf{E}/m$ and the conductivity to be given by $\sigma = ne^2\tau/m$, where τ is the relaxation time or the time between collisions.

To satisfy Ohm's law, this collision time must be independent of the field. If we assume that these collisions are between the electrons traveling over some mean free path λ that is associated with the lattice, the average speed of the electrons would have to be much larger than the drift speed produced by the applied electric field. This would be the case if the electrons behaved as an ideal gas in thermal equilibrium with the lattice.

The resistivity $\rho = 1/\sigma$ is found to have a linear relationship with temperature in the form $\rho(T) = \rho_0 + \alpha T$ (Matthiessen's rule), where ρ_0 is due to collisions with structural and impurity imperfections and the temperature dependence comes about from collisions with the lattice ions whose cross sections increase linearly with temperature. Impurity atoms, such as found in solid solution alloys, produce a much larger increase in resistivity than structural defects such as dislocations and grain boundaries or condensed second phases because they are more widely dispersed. Also there is a departure from the linear temperature dependence of the resistivity at low temperatures because all of the phonon modes are active. Grüneisen used the Debye theory to develop a universal relationship between reduced resistivity and reduced temperature that holds for all metals.

The observed first power relation between resistivity and temperature implies that the collision velocity must not only be independent of the applied field and be much greater than the drift velocity but must also be independent of temperature, which rules out the thermal velocity as the motion responsible for the observed collision rate. With the discovery of quantum mechanics, the difficulties encountered by the classical electron gas model are resolved by having the electrons travel at the Fermi velocity.

But only the electrons at the surface of the Fermi sphere can move with the Fermi velocity. So instead of all of the electrons moving at the drift speed, we now have the current carried by much fewer but much faster electrons at the Fermi surface that can respond to the applied field. This new Fermi model requires a reformulation of the conductivity and it will be shown later that the Drude formula for conductivity still holds for simple metals but for wrong reason.

Charges moving through a conductor are deflected to the sides of the conductor by a magnetic field causing a voltage difference called the Hall voltage to appear across the lateral surfaces. Measurement of the Hall voltage for a given magnetic field and conductor geometry determines the sign and number density of the charge carriers in the conductor. If the conductivity is also known, the mobility of the charge carriers can be determined. Again, for simple metals, the number density of charge carriers measured from the Hall coefficient is the total number of free electrons, not just those actually carrying the current.

The ratio of the thermal conductivity to the electrical conductivity times the absolute temperature is known as the Wiedemann and Franz ratio and involves only universal physical constants. Therefore, this ratio should be the same for any metal (provided that the heat is predominately carried by the electrons). This relationship depends on the fact

that the collision times that inhibit the flow of the electrons is the same whether they are driven by an applied field of a thermal gradient.

Bibliography

Ashcroft, N.W. and Mermin, N.D., *Solid State Physics*, Brooks Cole, Philadelphia, 1976.

Askeland, D.R. and Phule, P.P., *The Science and Engineering of Materials*, 5th edn., Thompson, Canada, 2006.

Hush, N.S., An overview of the first half-century of molecular electronics, *Ann. N.Y. Acad. Sci.*, 100, 2003, 1–20.

Ibach, H. and Lüth, H., *Solid State Physics*, 3rd edn., Springer-Verlag, New York, 1990.

Kittel, C., *Introduction to Solid State Physics*, 7th edn., John & and Sons, New York, 1966.

Srivastava, C.M. and Srinivasan, C., *Science of Engineering Materials*, Wiley Eastern, New Delhi, India, 1987.

Problem

1. The Earth's magnetic field is ~0.5 G (1 G = 10^{-4} T). Do a feasibility analysis on the possibility of building a magnetometer using the Hall effect. Assume you can detect a Hall voltage as small as 1 μV. Also you want to minimize the power consumption. What material would you want to use? What dimensions would be required? What voltages and currents would be required?

19

Band Theory of Metals

In spite of the success of the free electron theory after the introduction of the Fermi energy and Fermi velocity to straighten out some of the inconsistencies in the original Drude theory, the free electron theory still does not account for all that is known. It is necessary to introduce a modification to the free electron theory to account for the interaction of electrons with the charged ion cores. There are two ways to go about this. One could start with the empty lattice (free electrons) and add the interactions of the core ions as a perturbation to the energy of the electrons (nearly free electron model). Or one could start with the core atoms and describe the energy bands that form as the electron wavefunctions overlap (tight-binding approximation). These models give a rough approximation to the actual band structure and serve only to give the reader some insight into the importance of the band structure and how it originates. The actual calculation of the energy band diagrams is enormously complicated and is beyond the scope of this chapter.

19.1 Nearly Free Electron Model

19.1.1 Wavefunctions for Traveling Electrons

In the previous derivation of the Fermi energy using the electrons in a box, the boundary conditions required that the wavefunction vanish at the ends of the crystal. This allows only standing waves and is useful for determining the static energy levels of electrons in a metal (or a molecule with conjugated electrons such as heme or chlorophyll). However, if we are to describe phenomena such as current flowing through a conductor, it is necessary to allow for traveling electron waves through the solid as we did for phonons in Chapter 16. To accommodate traveling waves, we apply "periodic boundary conditions" that require the wavefunction to be the same at the end of the solid as it was at the beginning, i.e., $\psi(x+L, y, z) = \psi(x, y, z)$, where L is the length of the solid.

The solution to the Schrödinger wave equation is

$$\psi_n = (1/L)^{3/2} e^{2\pi i \left(n_x x + n_y y + n_z z\right)/L},$$ (19.1)

where
$n_x = 0, \pm 1, \pm 2, \ldots$
$n_y = 0, \pm 1, \pm 2, \ldots$
$n_z = 0, \pm 1, \pm 2, \ldots$

One can easily see that this solution satisfies the boundary conditions

$$\psi(x+L,y,z) = (1/L)^{3/2} e^{2\pi i\left(n_x(x+L)+n_y y+n_z z\right)/L} = \left(\frac{1}{L}\right)^{3/2} e^{2\pi i n_x} e^{2\pi i\left(n_x x+n_y y+n_z z\right)/L} = \psi(x,y,z).$$

Putting this back into the Schrödinger wave equation,

$$E_n = \frac{\hbar^2}{2m}\left(\frac{2\pi}{L}\right)^2\left(n_x^2 + n_y^2 + n_z^2\right) = \frac{\hbar^2}{2m}\left(\frac{2\pi}{L}\right)^2 n^2, \quad \text{where } \vec{n} = \hat{i}n_x + \hat{j}n_y + \hat{k}n_z. \qquad (19.2)$$

Recall in the previous treatment where we allowed only standing waves, we found energy eigenvalues given by

$$E_n = \frac{\hbar^2}{2m}\left(\frac{\pi}{L}\right)^2 n^2 \quad \text{where } n_x = 1,\,2,\,3,\ldots;\quad n_y = 1,\,2,\,3,\ldots;\quad n_z = 1,\,2,\,3,\ldots. \qquad (19.3)$$

Clearly, employing periodic boundary conditions produces a different set of energy levels; however, it will also produce a different density of states. Since $E = \hbar^2 k^2/2m$, $\vec{k} = (2\pi/L)\,\vec{n}$. Thus the volume associated with each energy state in k-space is $(2\pi/L)^3$. The minimum energy configuration is a Fermi sphere with radius k_{max}. The volume of this sphere in k-space is $(4\pi/3)k_{max}^3$. The number of electrons that can be accommodated is

$$N_{electrons} = 2\frac{(4\pi/3)k_{max}^3}{(2\pi/L)^3} = \frac{L^3}{3\pi^2}\left(\frac{2m\,E_F}{\hbar^2}\right)^{3/2} \qquad (19.4)$$

and the Fermi energy is

$$E_F = \frac{\hbar^2}{2m}\left(3\pi^2 n_e\right)^{2/3}, \qquad (19.5)$$

where n_e is the number of electrons per unit volume. This is the same as we calculated from the previous model that admitted only standing waves. The reason for this agreement is that in the previous model, the volume per state in k-space was $(\pi/L)^3$ which is only $1/8$ as large as in the present model. But since the present model allows both plus and minus values of n, we now integrate over the entire sphere in n-space (or k-space) rather than over just one octant. Thus in either case, the relationship between the minimum or Fermi energy and number of states or electron density is the same.

If one considers a one-dimensional (1-D) line of atoms in the limit of no electron interaction (free electron approximation), the energy dispersion curve (E vs. k) is parabolic since $E \sim k^2$ as shown in Figure 19.1. Actually, the curve is not continuous, but a locus of

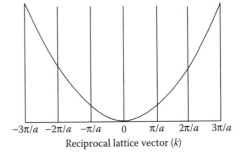

FIGURE 19.1
Energy versus k for a 1-D line of atoms with spacing a assuming no interaction with the ion cores (free electron approximation). These curves are actually the locus of the energy states spaced at $2\pi/L$.

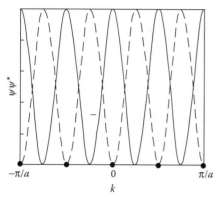

FIGURE 19.2

Two standing wave solutions of the Schrödinger equation. Note that the $\sin^2$ solution (dashed curve) places the electrons between the ion cores, which gives it a higher energy than the $\cos^2$ solution (solid curve) that puts the electrons on the lattice ions.

points spaced at $2\pi/L$ along the k-axis which represent the energy states. As waves, the electrons are subject to Bragg reflections from the lattice planes just as x-rays. Since $k = 2\pi/\lambda$, the Bragg criteria, $2a \sin \theta = n\lambda$, is satisfied for $\theta = 90°$ when $k = n\pi/a$. For these values of k, the electrons are reflected constructively to form standing waves. The first such reflection occurs at $n = 1$ or $k = \pm\pi/a$, which corresponds to the first Brillouin zone for a simple cubic lattice.

Recall that a standing wave can be decomposed into two traveling waves moving in opposite direction, or $k = \pm\pi/a$. There are two possible reflection modes

$$\psi_a \sim e^{i\pi x/a} - e^{-i\pi x/a} \sim \sin(\pi x/a) \text{ (nodes at lattice points)} \tag{19.6}$$

$$\psi_b \sim e^{i\pi x/a} + e^{-i\pi x/a} \sim \cos(\pi x/a) \text{ (antinodes at lattice points).} \tag{19.7}$$

19.1.2 Energy Gap at the Brillouin Zone

The electron density represented by $\psi_b \psi_b^*$ is concentrated mainly near the ion cores, whereas $\psi_a \psi_a^*$ is concentrated mainly in the space between the lattice ions as shown in Figure 19.2. A traveling wave would spend more-or-less equal time near and away from the ion cores. Therefore, $\psi_b \psi_b^*$ would have lower energy (more negative) than $\hbar^2 k^2/2m$ and $\psi_a \psi_a^*$ would have greater energy than $\hbar^2 k^2/2m$. Thus at the first Brillouin zone, two solutions to the Schrödinger equation are possible that yield different energies, but no solution exists with energies between these two values. This gives rise to an energy gap at the first Brillouin zone boundaries. Similar gaps will arise at other Brillouin zone boundaries where higher order reflections occur as seen in Figure 19.3.

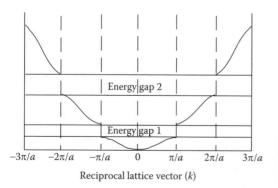

Reciprocal lattice vector (k)

FIGURE 19.3

Energy versus k for a 1-D line of atoms in which interactions with ion cores are considered. The electron wavefunctions are Bragg-reflected at the Brillouin zone boundaries ($n\pi/a$). As seen in Figure 19.2, the resulting standing waves then have two possible energies, depending on whether the maximum $\Psi\Psi^*$ sits on the ion cores (lowest energy solution) or in between the ion cores (highest energy solution). This difference in energy causes energy gaps Eg_1 and Eg_2 to appear at the Brillouin zone boundaries.

19.2 Binary Phase Diagrams for Mixed Valency Metals

Now that we have shown that there exists an energy gap at the Brillouin zones, we are in a position to understand the succession of phases from face-centered cubic (fcc) to body-centered cubic (bcc) to various types of hexagonal close-packed (hcp) phases seen in the Cu–Zn (Figure 19.4) and other systems with mixed valences as discussed in Section 12.6.8.

Recall the conditions suggested by Hume-Rothery that must be satisfied in order for two elements to combine in all proportions to form a substitutional solid solution:

1. Less than 15% difference in size

2. Near equal electronegativity

3. Identical crystal structure

4. Same valency.

If these conditions are not met, the two elements will have limited solid solubility or may form either compounds or new phases. The ratio of Zn/Cu atomic radii is 1.04, so the first condition is satisfied. Cu crystallizes in the fcc configuration; whereas, Zn is hcp. Also, Cu is univalent and Zn is divalent, so the last two conditions are not met.

Note that α-brass, that forms up to ∼38 At% Zn, is fcc. For higher concentration of Zn, the system undergoes a solid-phase transformation to β-brass which is bcc up to ∼48 At%. At this point it becomes γ (a complex cubic unit cell containing 52 atoms). The addition of still more Zn produces the ε and η phases which are hcp with different c/a ratios. Cu–Al, Cu–Sn, Ag–Zn, Ag–Al, Ag–Cd and other mixed valence systems show similar phase behavior.

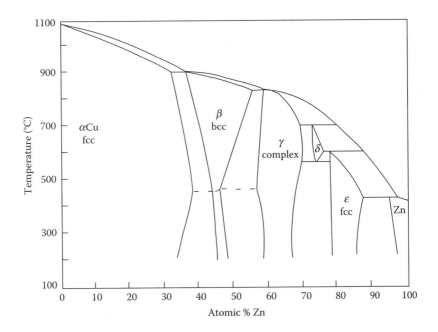

FIGURE 19.4
Cu–Zn phase diagram. The α phase is fcc and extends to ∼36 At% Zn. The β phase is bcc and extends to ∼48 At%. (From Massalski, T.B., Senior Editor, *Handbook of Binary Alloy Phase Diagrams*, Vols. 1–3, American Society for Metals, 1990. Reprinted with permission of ASM International. All rights reserved.)

This behavior can be understood from Brillouin zone theory. The number of electrons contained in a Fermi sphere is

$$N_e = \frac{2(4\pi/3)k_F^3}{(2\pi/L)^3} = \frac{k_F^3 L^3}{3\pi^2} \tag{19.8}$$

or

$$k_F^3 = 3\pi^2 n_e. \tag{19.9}$$

where n_e is the electron density.

From Chapter 6, the largest Fermi sphere that can just fit inside the first Brillouin zone for a fcc direct lattice has a radius $k_F = \pi\sqrt{3}/a$ (proof left to the student). A fcc unit cell contains four atoms and contains 4ν electrons, where ν is the average number of electrons per atom. Therefore, the electron number density is $4\nu/a^3$. Putting this into Equation 19.9 and equating the resulting k_F to the largest k_F that can fit inside the first Brillouin zone,

$$k_F^3 = 3\pi^2 \frac{4\nu}{a^3} = \frac{\pi^3 3\sqrt{3}}{a^3}. \tag{19.10}$$

Solving for ν, we find $\nu = 1.36$, which is the largest number of electrons per atom that can fit in a Fermi sphere that touches the first Brillouin zone.

Now if we form an alloy with $(1 - x)$ Cu atoms, each of which contributes one electron per atom, and x Zn atoms, each of which contributes two electrons per atom, the average number of electrons per atom will be $(1 - x) + 2x = 1 + x = \nu = 1.36$. Therefore, the Fermi sphere will just touch the first Brillouin zone when $x = 0.36$. What happens as we add more Zn? There is still unfilled volume in the first Brillouin zone as seen in Figure 19.5 between the Fermi sphere and the polyhedral zone boundaries, but these are higher energy states and it costs energy to fill these. (Recall that the lowest energy states are obtained for nearly equal values of n_x, n_y, and n_z, or states within the Fermi sphere.)

The Fermi sphere could expand across the zone boundary, but remember there is an energy gap at the boundary and this also would cost energy. Is there another configuration that the system could go into that would allow more electrons per atom before the Fermi sphere encounters the first Brillouin zone?

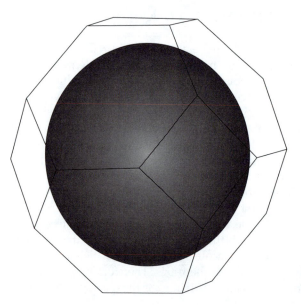

FIGURE 19.5
Fermi surfaces for the noble metals (Cu, Ag, and Au) fit inside the first Brillouin zone (however, see Figure 19.15). The volume inside the Fermi surface represents the filled minimum energy states. The space between the Fermi surface and the polygon are the unfilled higher energy states.

It turns out that there is. A bcc direct lattice has a fcc reciprocal lattice and it can be shown that the largest k_F that will just fit inside is $k_F = \pi\sqrt{2}/a$ (again proof left to the student). This is shorter than the maximum radius for the previous case, but the bcc lattice has only two atoms per unit cell. Therefore Equation 19.10 can be written as

$$k_F^3 = 3\pi^2 \frac{2\nu}{a^3} = \frac{\pi^3 2\sqrt{2}}{a^3} \tag{19.11}$$

and $\nu = \pi\sqrt{2}/3 = 1.48$. which implies that when $x > 0.48$, the system will try to lower its energy by seeking other structures that can accommodate more electrons in the first Brillouin zone. As it turns, out the γ phase is a complex cubic phase with 52 atoms in the unit cell and the ε and η phases are hcp with different c/a ratios.

19.3 Band Structure in Metals

19.3.1 The Bloch Theorem and the Reduced Band Scheme

Bloch obtained a one-electron solution to the Schrödinger equation for a periodic potential and was able to show that

$$\psi_k(r) = u_k(r)e^{ik \cdot r}, \tag{19.12}$$

where $u_k(r)$ has the periodicity of the lattice, i.e., $u_k(r + a) = u_k(r)$ (see a standard solid-state physics text for proof). The theorem simply states that the wavefunction is modulated by the lattice periodicity. One of the consequences is that a wavefunction is not changed after being displaced by $\mathbf{G}$, i.e.,

$$\psi_{K+G}(r) = \psi_K(r). \tag{19.13}$$

(see Appendix). Putting these Bloch wavefunctions into the Schrödinger equation,

$$H\psi_K(r) = H\psi_{K+G}(r) = E(K)\psi_K(r) = E(\mathbf{K} + \mathbf{G})\psi_{K+G}(r) \tag{19.14}$$

with the result that $E(\mathbf{k}) = E(\mathbf{k} + \mathbf{G})$, which may be generalized to $E(\mathbf{k}) = E(\mathbf{k} + n\mathbf{G})$. This important result says that possible energy states are not restricted to a single parabola in k-space, but can be found equally well on parabolas shifted by the integral values of the reciprocal lattice vector $\mathbf{G}$. Therefore, it is not necessary to plot the E versus $\mathbf{k}$ curves for $\mathbf{k}$ beyond the first Brillouin zone; the higher energy states can be represented by simply shifting $\mathbf{k}$ back into the first Brillouin zone by adding (or subtracting) multiples of $\mathbf{G}$. Another way of saying this is, for a given value of $\mathbf{k}$, there are multiple solutions of the Schrödinger equation in which the energy increases by $(\mathbf{k} + n\mathbf{G})^2$. The result is shown in Figure 19.6 for the 1-D line of atoms. Recall that the $\mathbf{G}$ for a simple cubic lattice is $2\pi/a$, so all of the energy states can be shown by shifting the energy values in the expanded zone scheme we saw in Figure 19.6 by integral values of $2\pi/a$ so they appear in the first Brillouin zone. This is called the reduced band scheme. Figure 19.7a shows the bands for electrons in the absence of core interactions; Figure 19.7b shows the effects of interactions with the core electrons. Recall that we saw something like this in Chapter 16 in dealing with the acoustic dispersion relations.

Recall that the spacing between states is $2\pi/L$ along the k-axis. Therefore, the number of states between $-\pi/a$ and π/a is just the number of ion cores since a is the spacing between the

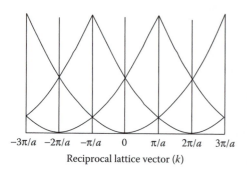

−3π/a −2π/a −π/a   0   π/a   2π/a   3π/a

Reciprocal lattice vector (*k*)

FIGURE 19.6

Bloch solution for *E* versus *k* for periodic lattice in the limit of vanishing interaction between electrons and ion cores. The different parabolas represent the energy values for **k** shifted by integral values of **G** which, for a simple cubic lattice, is 2π/*a*.

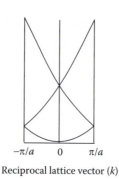

−π/a   0   π/a

(a)   Reciprocal lattice vector (*k*)

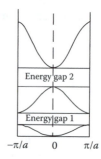

Energy gap 2

Energy gap 1

−π/a   0   π/a

(b)   Reciprocal lattice vector (*k*)

FIGURE 19.7

Reduced zone representation for a 1-D line of atoms in the empty lattice model (a) and with core interactions (b) which give rise to energy gaps E_{g_1} and E_{g_2}.

atoms in a simple cubic lattice. If each atom contributes one electron, the first zone will be half full; two electrons per atom would fill the band. Similarly, four electrons per atom would fill the second band. Thus group IV semiconductors (as well as III–V and II–VI compounds), have filled valence bands. The empty band above this is the conduction band. However, as we shall see later, there is more to this story. Given this picture, one would expect elements with even valences would be semiconductors or insulators. Indeed, this would be the case if it were not for band overlap, which allows group II elements to be conductive.

19.3.2 Effective Electron Mass

When an electric field is applied to a conductor, the electrons feel not only the external field, but they are also acted upon by the lattice. The equation of motion can be written

$$m \frac{\partial v_g}{\partial t} = F_{ext} + F_{lattice},$$ (19.15)

where $F_{lattice}$ is generally not known. The v_g is the group velocity which, by definition,

$$v_g = \frac{\partial \omega}{\partial k} = \frac{1}{\hbar} \frac{\partial E}{\partial k}.$$ (19.16)

Therefore

$$\frac{\partial v_g}{\partial t} = \frac{1}{\hbar} \frac{\partial}{\partial t} \left(\frac{\partial E}{\partial k} \right) = \frac{1}{\hbar} \left(\frac{\partial^2 E}{\partial k^2} \right) \left(\frac{\partial k}{\partial t} \right).$$ (19.17)

The external force should equal the time rate change in momentum,

$$F_{ext} = \frac{\partial p}{\partial t} = \hbar \frac{\partial k}{\partial t}, \tag{19.18}$$

which, using the previous result, can be written

$$F_{ext} = \hbar \frac{\partial k}{\partial t} = \frac{\hbar^2}{\partial^2 E / \partial k^2} \frac{\partial v_g}{\partial t}. \tag{19.19}$$

We can put this in the form of Newton's law and include the effects of $F_{lattice}$ by assigning the coefficient of the time derivative of the velocity term as an effective mass, i.e.,

$$m^* \equiv \frac{\hbar^2}{\partial^2 E / \partial k^2}. \tag{19.20}$$

For a free electron,

$$E = \frac{\hbar^2 k^2}{2m}; \quad \frac{\partial^2 E}{\partial k^2} = \frac{\hbar}{m}; \quad \text{and} \quad m^* = m. \tag{19.21}$$

Notice that the effective mass varies as the curvature of the band. In the conduction band in the vicinity of $\mathbf{k} = 0$ where the band is curved upward, the effective mass of an electron is positive. However, near an inflection point the effective mass becomes infinite, and then becomes negative in the regions where the band curves downward. It follows that electrons in the second or valence band near $\mathbf{k} = 0$, have negative mass.

19.3.3 Holes

If an electron is removed from the valence band, it leaves a hole. A hole can be thought of as a particle with positive charge, but with negative electron mass. Therefore, holes near the top of a valence band with negative curvature will have positive mass and positive charge and will move in the direction of an applied electric field. An electron in the conduction band with positive curvature will move opposite to the field because of its negative charge and positive mass.

19.3.4 Cyclotron Resonance

The effective masses of electrons and holes can be measured by irradiating the material with circularly polarized radio frequency (microwave) radiation in the presence of a magnetic field. The equation of motion is

$$\frac{d\mathbf{v}}{dt} + \frac{\mathbf{v}}{\tau} = \frac{e}{m^*} (\mathbf{E} + \mathbf{v} \times \mathbf{B}). \tag{19.22}$$

Let the magnetic field be in the z-direction and the right-hand circularly polarized radiation be given by $E_x = E_0 e^{i\omega t}$ and $E_y = -iE_0 e^{i\omega t}$. The solution to Equation 19.22 can be written as $v_x = v_0 e^{i\omega t}$ and $v_y = -iv_0 e^{i\omega t}$. Inserting these quantities into Equation 19.22 and identifying the cyclotron frequency as $\omega_c = eB_z / m^*$, the v_0 can be written as

$$v_0 = -\frac{e\tau}{m^*} \frac{E_0[1 - i(\omega - \omega_c)\tau]}{1 + (\omega - \omega_c)^2 \tau^2}.$$

(19.23)

The current density is then

$$J_x = -nev_x = \frac{ne^2\tau}{m^*} \frac{E_0[1 - i(\omega - \omega_c)\tau]}{1 + (\omega - \omega_c)^2 \tau^2} e^{i\omega t}$$

(19.24)

and

$$J_y = -nev_y = -i\frac{ne^2\tau}{m^*} \frac{E_0[1 - i(\omega - \omega_c)\tau]}{1 + (\omega - \omega_c)^2 \tau^2} e^{i\omega t}.$$

(19.25)

The power density is $P = \oint_{\text{cycle}} Re(\mathbf{J}) \cdot Re(\mathbf{E}) d(\omega t)$ which becomes

$$P = \frac{\sigma E_0^2}{1 + (\omega - \omega_c)^2 \tau^2}.$$

(19.26)

The resonant or cyclotron frequency is detected by absorption of power from the microwave beam which becomes a maximum when $\omega = \omega_c$, as can be seen from Equation 19.26. The effective electron mass can then be determined from $m^* = eB_z/\omega_c$. The effective mass of holes can be found in a similar manner using left-hand polarized microwave radiation.

In conductive metals, it is difficult to penetrate to much depth with microwave radiation. The magnetic field and the microwave beam are oriented along the surface of the conductor. The circulating electrons will only feel the field when they are within the skin depth of the radiation, so when the applied frequency is some integer times the resonance frequency, absorption will occur. This technique is known as the Azbel–Kramer cyclotron resonance (AKCR) method.

19.3.5 Band Structure in Real Systems

As you might suspect, the band structure in real systems can get considerably more complicated than shown in these simple models. Figure 19.8 shows the band structure for Al. The band structure is little affected by the lattice interactions. The energy gaps are small and the electrons behave almost as though they were free. This type of behavior is typical of simple metals, particularly of the alkali metals. The complexity of the diagram is due primarily to the folding of the bands along different **k**-directions back into the first Brillouin zone.

Things get even more complicated in the transition metals because of the interactions of the d-electrons. Figure 19.9 shows the band structure for Cu. The parabolic bands originate from the s-electrons. The flat bands are a result of the localization of the d-electrons. The flatness of these bands indicates these electrons are quite immobile. The steepness and curvature of the s-bands near the Fermi level indicate high mobility for these electrons, hence the high conductivity of Cu.

There is a high electron density in the $3d$ band a few electron volts below the Fermi level. Cu preferentially absorbs blue light as these $3d$ electrons make transitions to $4s$ states giving Cu its characteristic reddish color. Alloying Cu with Zn shifts the absorption edge giving brass its yellowish color. Gold's yellow color is the result of similar interband transitions.

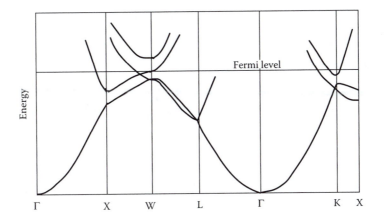

FIGURE 19.8
Band structure for Al along directions of high symmetry designated by X, W, L, and K as shown in Figure 6.7(a). (From Segal, B., *Phys. Rev.*, 125, 115, 1962. With permission.)

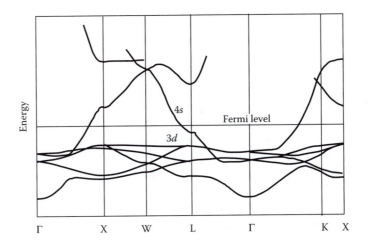

FIGURE 19.9
Band structure of Cu. Transitions of the 3*d* electron states (flat lines just below the Fermi level to 4*s* states at the Fermi level absorb blue light giving Cu is reddish color. (From Segal, B., *Phys. Rev.*, 125, 114, 1962. With permission.)

19.4 Conductivity and the Fermi Surface

It was stated previously that the expressions for conductivity and mobility obtained from the simple Drude theory also held in the quantum mechanical treatment of electrons in simple metals. Since only the electrons near the Fermi surface are able to respond to an applied electric field, we must integrate the available states over the Fermi surface. We can write the current density as

$$\mathbf{J} = e \iint_{E_F} \mathbf{v}(\mathbf{k}) S(\mathbf{k}) d\mathbf{k}, \tag{19.27}$$

where

    $\mathbf{v}(\mathbf{k})$ is the Fermi velocity

    $S(\mathbf{k})$ is the density of states near the Fermi surface.

When an electric field, E_{ap}, is applied in the x-direction, the symmetry is disturbed and we must expand the integrand

$$\mathbf{v}(\mathbf{k})S(\mathbf{k}) = \mathbf{v}(\mathbf{k})S_0(\mathbf{k}) - v_x \frac{\partial S(\mathbf{k})}{\partial k_x} \Delta k_x. \tag{19.28}$$

The first term of the expansion integrates to zero by symmetry. We can write the incremental change in momentum in the x-direction resulting from the applied field as $\Delta k_x = mv_d/\hbar = -e\tau E_{ap}/\hbar$. We can also write $\partial S(k)/\partial k_x = \hbar v_x \, dS(k)/dE$ and $d|\mathbf{k}| = dE/\hbar|\mathbf{v}|$. Putting all of this back into Equation 19.27, we can write the conductivity as

$$\sigma = \frac{J_x}{E_{ap}} = \frac{e^2}{\hbar} \iint\limits_{E_F} \frac{v_x^2}{|\mathbf{v}|} \tau dS(\mathbf{k}). \tag{19.29}$$

The square of the Fermi velocity $\mathbf{v}^2 = v_x^2 + v_y^2 + v_z^2$, but since $v_x^2 = v_y^2 = v_z^2$, $\mathbf{v}^2 = 3v_x^2$, and $v_x^2/|\mathbf{v}| = (v/3) = \hbar k_F/3m^*$ where $\hbar k_F$ is the Fermi momentum. (Here we use the effective mass to be more general.) Assuming τ is not a function of k, Equation 19.29 can now be written as

$$\sigma = \frac{e^2 \tau}{3m^*} \iint\limits_{E_F} k_F dS(\mathbf{k}). \tag{19.30}$$

For a simple metal, such as Cu where the s-electrons are the primary carries, the Fermi surface is nearly spherical and the integral over its surface is $4\pi k_F^2$. The total volume in k-space included in the integral is $4\pi k_F^3$. If the system has volume L^3, the number of states is therefore $4\pi k^3/(2\pi/L)^3$ which can accommodate $8\pi k^3/(2\pi/L)^3$ electrons or $8\pi k^3/(2\pi)^3$ electrons per unit volume. Remembering from Equation 19.4 $k_F^3/3\pi^2 = n_e$, the electron density of the Fermi sphere, Equation 19.30 can be written for simple metals as

$$\sigma = \frac{e^2 \tau}{3m^*} \frac{(8\pi k_F^3)}{(2\pi)^3} = \frac{e^2 \tau}{m^*} \frac{k_F^3}{3\pi^2} = \frac{e^2 \tau}{m^*} n_e, \tag{19.31}$$

which is identical to the Drude model except that the effective mass of the electrons replaces the normal mass. What is happening here is the smaller number of electrons available to carry the current is compensated for by multiplying their number by the Fermi momentum in Equation 19.31. This gives the same volume in k-space as would be obtained by multiplying all of the electrons in the Fermi sphere by the drift velocity, as was illustrated in the 1-D model in Chapter 18.

Being able to compute the conductivity from band theory gives new insight into the widely differing conductivities that are found among the various metals. For example, Cu, which is a good conductor, has its Fermi level in the region where the s-bands are steep (see Figure 19.12), meaning the electrons have high velocity ($v_g = (1/\hbar)\partial E/\partial \kappa$). Fe, on the other hand, has its Fermi level lower in the bands populated by the d-electrons where, despite the larger density of states, the bands are flat making the electrons less mobile because they are partially involved in forming covalent bonds. Consequently, Fe is not as good a conductor of electricity than Cu.

19.5 Tight Binding Approximation

Previously we had started with free electrons and then added interactions with the ion cores to modify the E versus $\mathbf{k}$ relationships which provide the basis for the bands and energy gaps in conductors. Now we start with the bound states of the core electrons in the atoms making up the lattice and add the interactions that lead to similar bands in insulators and semiconductors.

The potential well of an individual atom with bound states A, B, C can be represented in Figure 19.10. When assembled into a crystal, the potentials of the adjacent atoms overlap to form a periodic potential as illustrated in Figure 19.11. In metals, the outer *s*-electrons represented as C become delocalized and form the Fermi band. The electronic states represented as B are only partially localized. Recall from quantum mechanics that an electron wavefunction does not go to zero at a finite potential barrier and a portion of its wavefunction exists outside the barrier; the lower and thinner the barrier, the more wavefunction is outside of the potential barrier. For this reason, the B electrons are only partially localized.

The tight-binding approximation uses the method of linear combination of atomic orbitals (LCAO) to approximate the wavefunction of one electron under the influence of the other atoms in the crystal.

$$\psi_{\mathbf{k}}(\mathbf{r}) = \sum_j C_{kj}\phi(\mathbf{r} - \mathbf{r}_j), \qquad (19.32)$$

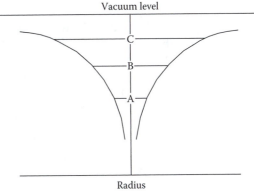

FIGURE 19.10
Energy well of an individual atom in which A, B, and C are bound states.

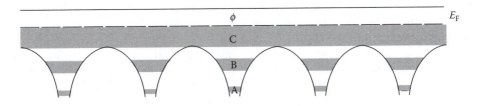

FIGURE 19.11
Illustration of a periodic potential with a band structure of bound electrons (A and B) and nearly free electrons in the Fermi region (C).

where $\phi(\mathbf{r} - \mathbf{r}_j)$ is the wavefunction of the jth isolated atom at distance $\mathbf{r}_j$. Recall we used this technique to describe the binding process when the wavefunctions of two H atoms began to overlap as they were brought together. The coefficients C_{kj} are assumed to be given by $N^{-1/2} \exp(i\mathbf{k} \cdot \mathbf{r}_j)$ to put the function in the Bloch form since it is periodic in $\mathbf{r}_j$. Putting these coefficients into the original wavefunction gives

$$\psi_{\mathbf{k}}(\mathbf{r}) = N^{-1/2} \sum_j \exp(i\mathbf{k} \cdot \mathbf{r}_j) \phi(\mathbf{r} - \mathbf{r}_j). \tag{19.33}$$

We now compute the energy of the electrons under the influence of the other atoms in the crystal.

$$E = \langle k|H|k \rangle = N^{-1} \sum_j \sum_m \exp(i\mathbf{k} \cdot (\mathbf{r}_j - \mathbf{r}_m)) \langle \phi_m|H|\phi_j \rangle. \tag{19.34}$$

Since all j atoms are equivalent, we pick the $j=0$ atom, set $\mathbf{r}_j = 0$, and multiply by the N atoms in the ensemble.

$$E = \sum_{m=0} \exp(-i\mathbf{k} \cdot \mathbf{r}_m) \langle \phi_m|H|\phi_0 \rangle = \langle \phi_0|H|\phi_0 \rangle + \sum_{m=1} \exp(-i\mathbf{k} \cdot \mathbf{r}_m) \langle \phi_m|H|\phi_0 \rangle. \tag{19.35}$$

Restricting the sum to nearest neighbor atoms, we can write

$$E = \langle k|H|k \rangle = -E_0 - E_1 \sum_{m=1} \exp(-i\mathbf{k} \cdot \mathbf{r}_m), \tag{19.36}$$

where $\langle \phi_0|H|\phi_0 \rangle = -E_0$ is the binding energy of the electron in the $j=0$ atom due to the potential from its nucleus, and $\langle \phi_m|H|\phi_0 \rangle = -E_1$ is the influence of the nearest neighbor atoms on the electron. The binding energy $-E_0$ will be slightly lower than that of an isolated atom due to a perturbation on the potential from its nucleus from the overlap of nearest neighbor wavefunctions.

For a simple cubic lattice, $\mathbf{r}_m = (\pm a, 0, 0), (0, \pm a, 0), (0, 0, \pm a)$. The dispersion relations are given by

$$E = -E_0 - 2E_1 \left[\cos(k_x a) + \cos(k_y a) + \cos(k_z a) \right]. \tag{19.37}$$

A bcc structure has nearest neighbors at $\mathbf{r}_m = (\pm a/2, \pm a/2, \pm a/2)$. The dispersion relations are given by

$$E = -E_0 - 8E_1 \left[\cos(k_x a/2) \cos(k_y a/2) \cos(k_z a/2) \right]. \tag{19.38}$$

An fcc structure has nearest neighbors at $\mathbf{r}_m = (\pm a/2, \pm a/2, 0), (\pm a/2, 0, \pm a/2), (0 \pm a/2, \pm a/2)$. The aspersion relations are given by

$$E = -E_0 - 4E_1 \left[\cos(k_y a/2) \cos(k_z a/2) + \cos(k_z a/2) \cos(k_x a/2) \right.$$
$$\left. + \cos(k_x a/2) \cos(k_y a/2) \right]. \tag{19.39}$$

The band structure for a simple cubic structure computed in the k_{100} and the k_{110} directions is shown in Figure 19.12. The Fermi surface can be determined from Equations 19.37, 19.38, or 19.39 by setting $E(k) = E_F$ and solving for the values of k_x, k_y, and k_z. An example of

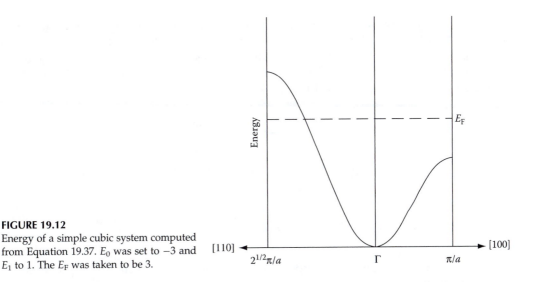

FIGURE 19.12
Energy of a simple cubic system computed from Equation 19.37. E_0 was set to -3 and E_1 to 1. The E_F was taken to be 3.

such a Fermi surface is for a simple cubic lattice shown as a contour plot in Figure 19.13 and in three dimension in Figure 19.14.

The approximations used in the tight-bonding approach are only valid for non-degenerate s-electron (spherically symmetric) wavefunctions. It is introduced here only

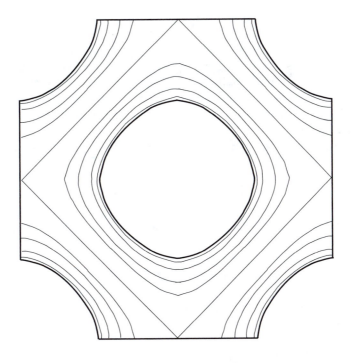

FIGURE 19.13
A contour map of the Fermi surface of the simple cubic system in the first Brillouin zone that was described in Figure 19.12. The contours go in steps of 0.1 k_z with the outer contour corresponding to $k_z = 0$ and the inner to $k_z = 1$.

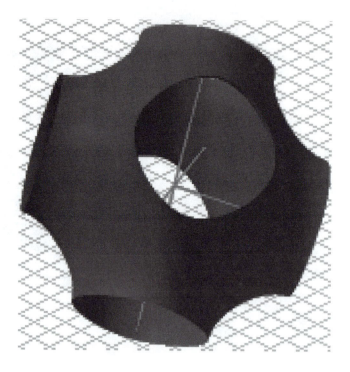

FIGURE 19.14
3-D rendition of the Fermi surface of a simple cubic system described in Figures 19.12 and 19.13.

to illustrate that energy bands do indeed exist even in systems with no free electrons. Actual band calculations require much more sophisticated analytical and numerical methods (Ashcroft and Mermin, 1976).

The effect of the energy gap in the vicinity of the Brillouin zone boundary is to distort the Fermi surface such that a face will be formed at the zone interface as we saw for the case of the simple cubic system. This schematic in Figure 19.15 illustrates a better approximation of the actual Fermi surface of the noble metals as calculated from Equation 19.39 than that shown in Figure 19.5.

19.5.1 Conductivity in Divalent Metals

The anomalous sign of the Hall coefficient for zinc can be understood in terms of a simplified 2-D model of the band structure illustrated in Figure 19.16 which is similar to the system described in Figure 19.12. If the band at one Brillouin zone boundary overlaps the energy gap at another zone boundary, there will be an overlap in the density of states that can be filled by the electron, hence there will be no energy gap in the system. If the Fermi energy happens to lie in the overlap region, as was depicted in Figure 19.16, there will be hole states below the Fermi energy and electron states above. Both hole and electrons contribute to conduction and in the case of Zn, the holes apparently dominate giving a positive Hall coefficient.

This mechanism of band overlap also offers another possible mechanism for the conductivity of divalent metals in addition to the possibility of s- and p-band overlap described in Chapter 3.

FIGURE 19.15
Sketch of the actual Fermi surface of the noble metals.

FIGURE 19.16
Simple 2-D model to explain conductivity in divalent systems. If the band in the [110] direction overlaps the energy gap in the [100] direction, there will be no effective gap in energy. If there are two electrons per atom, the Fermi energy for this band will lie in the region of overlap and electrons in the region of overlap will leave holes near the top of the band in the [100] direction.

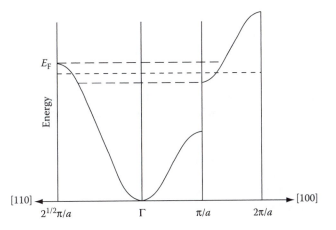

19.6   Experimental Methods

19.6.1   Photoelectric Effect

When a photon strikes a metal with sufficient energy to overcome the work function of the metal, it can eject a photoelectron. Einstein observed that there was a minimum frequency of the incident light for which photoelectron emission was possible and used this fact to argue that light had a particle-like nature and that these light particles (photons) had energy given by Planck's constant times the frequency. Of course, we now know that the minimum energy required for a photon to eject a photoelectron is equal to the work function and that the electron came from the top of the Fermi distribution.

19.6.2 Photoelectron Spectroscopy

The minimum energy required to eject a photoelectron is the work function of the metal and the ejected photoelectrons come from the top of the Fermi surface. By increasing the photon energy, electrons that lie deeper in the Fermi well can be ejected and the numbers and energies of the photoelectrons coming off can be measured. This technique is called photoelectron spectroscopy. If an ultraviolet lamp is used as the photon source, the technique is called ultraviolet photoelectron spectroscopy or UPS. Similarly, if x-rays are used as the photons, the technique is called x-ray photoelectron spectroscopy or XPS.

In either case, for an incident photon of given energy, the ejected photoelectrons with the smallest energy have emerged from deeper in the Fermi distribution and those with the highest energies come from the top of the Fermi distribution. By plotting the number of photoelectrons as a function of their energies, not only can the work function and the width of the Fermi distribution be determined, but the density of states within the Fermi distribution can also be obtained. With very high resolution analysis of the electron energy distribution, chemical shifts can be detected and the technique can be used for detailed chemical analysis. This technique goes by the name electron spectroscopy for chemical analysis (ESCA).

19.6.3 X-Ray Spectroscopy

When high-energy electrons bombard a metal target, they knock the inner shell electrons out of their energy levels and characteristic x-rays are emitted as electrons in higher states fall to fill the vacant level. For example, if a p-electron in an atom is knocked out, an s-electron in a higher shell will fall and emit its characteristic x-ray. However, in a metal, when p-electrons are knocked out, s-electrons from various levels within the Fermi distribution will fall to take their places. The distribution of x-ray energies and their relative intensities offers another method for mapping out the density of states within the Fermi distribution.

19.6.4 Thermionic Emission

Heated cathodes in television picture tubes, x-ray tubes, and the vacuum tubes of yesterday are used to produce large quantities of electrons which are then accelerated to the anode by a high voltage. The process by which these electrons are produced is called thermionic emission. The number of electrons produced by heating the cathode material can be obtained by integrating over the high-energy portion of the F–D distribution

$$N = \int_{E_\mathrm{F}+\phi}^{\infty} \frac{g(E)\mathrm{d}E}{e^{(E-\mu)/k_\mathrm{B}T}+1} \approx \int_{E_\mathrm{F}+\phi}^{\infty} CE^{1/2}e^{-(E-E_\mathrm{F})/kT}. \tag{19.40}$$

Accounting for the momenta of the electrons traveling in the direction of the surface, the thermionic current is given by the Richardson equation,

$$J = \frac{4\pi em}{h^3}(kT)^2 e^{-\phi/kT}. \tag{19.41}$$

The equation has the Arrhenius form so plotting $\log J$ against $1/kT$ offers an alternative method for obtaining the work function. Also, it is clear that efficient cathode materials must have a small work function.

19.7 Summary

Despite the success with the free electron theory, especially when quantum effects are taken into consideration, it must be recognized that electrons are not entirely free; they still interact with the core ions. We introduce periodic boundary conditions to describe traveling waves of electrons moving through a conductor. These periodic boundary conditions produce a different set of quantum states than the stationary waves we determined by requiring the wavefunctions to vanish at the edge of the conductor. The density of states found for traveling waves is also different, but when the new energy states are weighted by the new density of states and summed over the occupied states, the same value for the Fermi energy is obtained as was found by summing over the standing wave states.

These traveling waves are diffracted at the Brillouin zones to form two sets of standing waves. One set of standing waves places the electrons between the ion cores, which produces a higher energy (less negative) than the wave that places the electrons on top of the ion cores. There are no energy states between the two diffracted waves at the Brillouin zone boundary, hence an energy gap is formed at the Brillouin zone boundary.

The Fermi sphere for atoms with one valence electron per atom can easily fit in a bcc reciprocal lattice (fcc direct lattice). If a divalent element is added, the additional electrons will increase the diameter of the Fermi sphere. When the Fermi sphere begins to touch the Brillouin zone, the additional electrons must either go to the unfilled states in the corners of the first Brillouin zone, which are higher energy states, go across the Brillion zone, which costs energy because of the energy gap, or undergo a solid-phase transition to a new reciprocal lattice with a Brillouin that can accommodate more electrons. This model explains the solid-phase transformations of many mixed-valency binary alloys in which the phases progress from fcc to bcc to hcp as more divalent component is added.

The Bloch theorem states that the wavefunction of a single electron moving in a periodic potential is modulated by the potential according to $\psi_k(r) = u_k(r)e^{ik \cdot r}$. As a consequence, $\psi_{K+G}(r) = \psi_K(r)$, which means wavefunctions are not changed after being displaced by the reciprocal lattice vector G. It therefore follows that $E(k) = E(k + nG)$, which means that energy bands on an E versus k plot are the same if displaced along the k-axis by G. It also follows that all of the bands can be shifted back into the first Brillouin zone by adding or subtracting G.

As the electrons move through the lattice under the influence of an applied field, they interact with the lattice and transfer momentum to the lattice (or vice versa). This momentum transfer is accounted for by assigning the electron an effective mass that is defined as $m^* \equiv \hbar^2(\partial^2 E/\partial k^2)^{-1}$. Since the effective mass is inversely proportional to the curvature of the energy band, bands that curve upward have positive effective mass and those that curve downward have negative effective mass. Electrons occupying sharply curved bands have small effective mass and tend to be very mobile, while those occupying flat bands are heavy and sluggish.

The band structure in real systems is more complex than the simple models we have developed in trying to understand the electronic properties of metals. This is especially true in the transition metals where the d-electrons contribute significantly to the band structure. We see that conductivity is a function of the Fermi surface and the density of states in the vicinity of the Fermi level. By examining the calculated or measured band structure we can obtain much more insight into interband transitions, which account for the color of metals, and why some metals are much better conductors than others.

One method for approximating the band structure that is more suitable for the description of the bands in semiconductors and insulators is the tight-binding method. This method treats the perturbations on the bound electrons due the overlap of the nearest

neighbor electrons which give rise to the partial delocalization of these bound electrons and the formation of electronic bands similar to those formed by the nearly free electrons. The top of these bands is called the valence band. If these bands are completely full, no conduction can take place. The antibonding states form a conduction band and the separation between the highest state in the valence band and the lowest antibonding state is the bandgap energy. Materials with bandgap energies $\lesssim 3.1$ eV are considered semiconductors and those with bandgap energies $\gtrsim 3.1$ eV are considered insulators (3.1 eV is the top of the visible spectrum).

The presence of energy gaps near the Brillouin zone boundaries distort the Fermi surface causing it to penetrate into the zone interface. The tight-binding approximation allows a better approximation of the actual shape of the Fermi for simple cubic, bcc, and fcc structures for simple metals with only s-electrons.

Even though energy gaps appear in bands at the Brillouin zone, the Brillouin zones occur at different values of **k**, depending on the direction taken. Since the bands are symmetrical about **k** = 0, it is common practice to plot the bands in one k-direction on one side of the origin and use the other half of the plot to show the bands in a different k-direction. If the band in one direction overlaps an energy gap in another direction, electrons have new energy states to move into and no actual gap exists. This explains the conduction in divalent metals whose two electrons per atom would fill the first energy band. The Fermi energy in these circumstances lies in the overlap region causing holes to form in the first band. Conduction takes place by both electrons and holes, but if the holes dominate, the metal will exhibit a positive Hall coefficient.

Appendix

A.19.1   Proof That a Wavefunction Displaced by G Is Unchanged

Show that $\psi_{\mathbf{k}}(r) = \psi_{\mathbf{k}+\mathbf{G}}(r)$.

From the Bloch theorem,

$$\psi_{\mathbf{k}+\mathbf{G}}(r) = e^{i(\mathbf{k}+\mathbf{G})\cdot r} u_{\mathbf{k}+\mathbf{G}}(r). \tag{A.19.1}$$

Since $u_{\mathbf{k}}$ has the periodicity of the lattice, it can be expressed in a Fourier series as

$$u_{\mathbf{k}} = \sum_n A_n e^{-in\mathbf{k}\cdot\mathbf{r}}, \tag{A.19.2}$$

where $k = \pi/a$ for a simple cubic lattice. Then

$$u_{\mathbf{k}+\mathbf{G}} = \sum_n A_n e^{-in(\mathbf{k}+\mathbf{G})\cdot\mathbf{r}} = \sum_n A_n e^{-in\mathbf{k}\cdot\mathbf{r}} e^{-in\mathbf{G}\cdot\mathbf{r}}. \tag{A.19.3}$$

Since $G = 2\pi/a$ for a simple cubic lattice, $\exp(-in\mathbf{G}\cdot\mathbf{r}) = \exp(-i\mathbf{G}\cdot\mathbf{r})$ and Equation A19.3 can be written as

$$u_{\mathbf{k}+\mathbf{G}} = e^{-i\mathbf{G}\cdot\mathbf{r}} \sum_n A_n e^{-in\mathbf{k}\cdot\mathbf{r}} = e^{-i\mathbf{G}\cdot\mathbf{r}} u_{\mathbf{k}}. \tag{A.19.4}$$

Putting this back into Equation A.19.1 and using the Bloch theorem,

$$\psi_{k+G}(r) = e^{i(k+G)\cdot r}u_k(r)e^{-iG\cdot r} = u_k(r)e^{-ik\cdot r} = \psi_k(r) \cdot \text{QED}. \qquad (A.19.5)$$

Bibliography

Ashcroft, N.W. and Mermin, N.D., *Solid State Physics*, Brooks Cole, Belmont, MA, 1976.

Hummel, R.E., *Electronic Properties of Materials*, 3rd edn. Springer-Verlag, New York, 2000.

Ibach, H. and Lüth, H., *Solid State Physics*, 3rd edn. Springer-Verlag, New York, 1990.

Kittel, C., *Introduction to Solid State Physics*, 7th edn. John Wiley & Sons, New York, 1966.

Levy, R.A., *Principles of Solid State Physics*, Academic Press, New York, 1968.

Livingston, J.D., *Electronic Properties of Engineering Materials*, Wiley, MIT series in Materials Science and Engineering, John Wiley & Sons, New York, 1999.

Massalski, T.B., Senior Editor, *Handbook of Binary Alloy Phase Diagrams*, Vols. 1–3, American Society for Metals, Materials Park, Ohio, 1990.

Omar, M.A., *Elementary Solid State Physics*, Addison-Wesley, Reading, MA, 1975.

Segal, B., Fermi surface and energy bands of copper, *Phys. Rev.*, 125(1),115, 1962.

Problems

1. Show that the largest Fermi sphere that can just fit inside the first Brillouin zone for a fcc direct lattice has a radius $k_F = \pi\sqrt{3}/a$.

2. Show that the largest Fermi sphere that can just fit inside the first Brillouin zone for a bcc direct lattice has a radius $k_F = \pi\sqrt{2}/a$.

3. Make a scale drawing of the first two Brillouin zones for a fcc direct (bcc reciprocal) lattice in the (110) plane shown in Figure 6.7a. (Hint: plot the lattice points on graph paper and construct perpendicular bisectors.) Now draw a set of circles representing the Fermi spheres for simple metals containing 1, 2, and 3 electrons per atom.

4. Repeat question 3 for a bcc direct (fcc reciprocal) lattice shown in Figure 6.7.

20

Semiconductors

Semiconductors form the core of modern electronic and photonic devices (photonics refers to the generation, manipulation, and detection of photons). Unlike metals, the conductivity of pure (intrinsic) semiconductors increases with temperature, which cannot be explained by the Drude theory. The formal definition of a semiconductor is a material whose conductivity lies between that of a metal and an insulator. Perhaps a better definition would be a material with a bandgap greater than zero and less than some arbitrary value such as ~3 eV, which is the highest energy of a visible photon. At least this definition alludes to the photonic applications of semiconductors. What makes semiconductors useful is the ability to control both the sign and the number of charge carriers by adding impurity atoms to form extrinsic semiconductors.

20.1   The Group IV Systems

Most of the semiconductors used in devices are members of the group IV elements or III–V or II–VI compounds, which behave similarly to group IV elements.

20.1.1   The Chemical Picture

Consider the group IV elements, C, Si, Ge, Sn, and Pb. The higher atomic number atoms (Sn and Pb) have larger radii, hence their s- and p-bands are sufficiently narrow so that they do not overlap as illustrated in Figure 20.1. The s-band holds two electrons and its bond is saturated; the p-band can hold six electrons, but only contains two. Hence, there are plenty of unfilled states in the p-band and the system is an electrical conductor and behaves as a metal. However, at a critical radius close to that of Sn, the bands begin to overlap and it becomes energetically favorable for the system to form an sp^3 hybrid bond. One of the s-electrons is promoted to a p-state in order to form the lower energy sp^3 covalent bond as described in Chapter 3. The tetrahedral nature of this directed bond is responsible for the diamond structure seen in C, Si, and Ge. When this sp^3 bond is formed, each of the four atoms involved forms two orbitals: one bonding and one antibonding. The four electrons fill the bonding orbitals, leaving the antibonding orbitals empty. The gap between the bonding and antibonding orbitals is the bandgap of the system. As mentioned previously, Sn is right at this critical radius; consequently, two forms of Sn exist, gray tin or α-Sn which is a semimetal (a semiconductor with a very small bandgap) and white tin which is a metal.

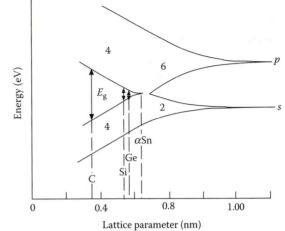

FIGURE 20.1

Band scheme illustrating the transition from metal-like behavior for the larger atoms, to semi-conductor and insulator behavior in the smaller diameter atoms that are closer together. The band-gaps for C, Si, Ge, and α-Sn as well as their lattice parameters are shown to scale. The number of electrons that can be accommodated in each band is indicated by the numeral in the band.

20.1.2 The Bandgap Energy

Once the sp^3 hybrid is formed, all the electrons are tied up as valence electrons forming saturated covalent bonds. Since these tend to be strong bonds, especially in the case of C, there is a significant bandgap between the valence band and the conduction band because one or more of the covalent bonds must be broken to provide a conduction electron. Thus, diamond with a bandgap of ~5 eV is a good insulator. As may be seen in Figure 20.1, the bandgap decreases with Si, which has $E_g = 1.107$ eV and lattice parameter 0.54307 nm; decreases further with Ge, which has $E_g = 0.67$ eV and lattice parameter $= 0.5657$ nm; and essentially vanishes with a-Sn, which has $E_g = 0.08$ eV and lattice parameter $= 0.649$ nm. Similar behavior may be seen in the group III–V and II–VI partially covalently bonded compound semiconductors (Table 20.2). This description of the origin of the bandgap energy is known as the chemical picture.

Since the bandgap energy increases as the lattice parameter decreases, one might expect that bandgap energy would increase with pressure. Conversely, since the lattice parameter increases with temperature due to thermal expansion, one would expect the bandgap energy to decrease with temperature.

It is interesting to note the relation between the bandgap energy and the cohesive energy of these group IV elements as shown in Table 20.1. There appears to be an approximate linear relation between the cohesive energy and the bandgap energy given by $E_C(eV) = 5.47 + 1.78E_g(eV)$.

TABLE 20.1

Cohesive Energies and Bandgap Energies for Group IV Semiconductors

	C	Si	Ge	Sn
$E_C(eV)$	14.7	7.75	6.52	5.5
$E_g(eV)$	5.2	1.107	0.67	0.08

Source: Data from Sapoval, B. and Hermann, C., *Physics of Semiconductors*, Springer-Verlag, New York, 1995.

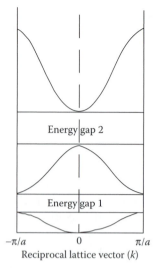

-π/a   0   π/a
Reciprocal lattice vector (k)

FIGURE 20.2
Energy band structure obtained from the nearly free electron model discussed in Chapter 19. For systems with four electrons per atom, the two lower bands are full and the energy gap 2 is the bandgap energy between the valence band and the conduction band.

20.1.3 The Energy Band Picture

In Chapter 19, we showed, using the nearly free electron model, that systems with four free electrons per atom fill the first two bands and that there was an energy gap between the two filled bands and the next band as shown in Figure 20.2. Even though these four valence electrons are tied up forming the sp^3 bonds, these electrons exhibit an effective mass as measured by cyclotron resonance that is consistent with a band structure similar to that shown in Figure 20.2. Since the band is full, there is no electrical conduction unless the electrons are given enough energy to cross the bandgap into the conduction band. How can these electrons behave in one sense as they are delocalized and yet be tied up in sp^3 bonds?

We saw in the covalent bond model (Chapter 3) that when we brought atoms together to form a crystal, both bonding and antibonding states were formed due to the overlap of the electronic wavefunctions. One explanation is that the electrons are only partially localized when forming the sp^3 bond and that the delocalized portion has band structure similar to the electrons in the nearly free electron model. This is similar to the d-electrons in the transition metals. These d-electrons are involved in forming covalent bonds, which give added strength to these metals, yet are still involved in conduction.

The bandgap energy can be measured by optical absorption (or reflection) by determining the wavelength where the material begins to absorb radiation and is no longer transparent. The effective masses of the electrons and holes can be measured directly by cyclotron resonance as discussed in Chapter 19. A very pure single crystal at cryogenic temperatures is placed in a magnetic field and illuminated with optical radiation above the bandgap energy to create electron–hole pairs that can respond to the microwave radiation.

20.1.4 Amorphous Semiconductors

The energy gaps in the nearly free electron model were the results of a periodic lattice with long-range order. How do we explain the bandgaps in amorphous solids such as fused quartz and amorphous silicon, which have energy gaps similar to their crystalline counterparts? Here it helps to resort back to the chemical picture. In amorphous Si, the same sp^3 tetrahedral bonds form to produce short-range order, but these bonds are distorted and some are left dangling because of the amorphous structure. These dangling bonds produce

energy states that lie in the bandgap making the system electronically impure. These dangling bonds can be tied up by exposing Si to H_2 at high temperatures to form Si–H bonds. This allows amorphous Si to replace single crystal Si in large-scale applications such as solar cells, which is desirable because amorphous Si can be produced in large scale at a small fraction of the cost of single crystal Si.

20.2 Intrinsic Semiconductors

Intrinsic semiconductors are extremely pure group IV elements or stoichiometric III–V or II–VI compounds that are insulators at absolute zero with bandgaps $\sim$1–3 eV (Si, Ge), (GaAs, InP, CdTe, CdS, etc.). Intrinsic means no imparities are present that could create ionized states in energy gap. The only conduction mechanism then is the formation of holes by electrons leaving the valence band and jumping to the conduction band. This can happen either by thermal excitation or by the absorption of a photon with energy equal or greater than the bandgap energy. Thus in intrinsic material, the number of conduction electrons equals the number of holes. A list of the common semiconductors with their bandgap energies is presented in Table 20.2.

20.2.1 Fermi Level in Intrinsic Semiconductors

In metals, we think of the Fermi level as the energy of the highest electron state, which at $T=0\,\mathrm{K}$ is the same as the Fermi energy. However, in a semiconductor, which has a bandgap between the last filled state and the next available state, the Fermi level must lie somewhere between the two states. A more general definition of the Fermi level is—the Fermi level is that energy for which the probability of the state just above the Fermi level being occupied is the same as the probability of the state just below the Fermi level being unoccupied.

The probability of an electron being thermally excited to a conduction band is given by the Fermi function times the density of states at E_C

$$P_e(E_C,T) = \frac{g(E_C)}{\exp[(E_C - E_F)/kT] + 1} \qquad (20.1)$$

and the probability of a hole being formed in the valence band is $1 -$ probability of its remaining in the valence band or

$$P_p(E_V,T) = 1 - \frac{g(E_V)}{\exp[(E_V - E_F)/kT] + 1}. \qquad (20.2)$$

Since these probabilities must be equal,

$$\frac{g(E_C)}{\exp[(E_C - E_F)/kT] + 1} = 1 - \frac{g(E_V)}{\exp[(E_V - E_F)/kT] + 1}. \qquad (20.3)$$

If the density of states below the valence band $g(E_V)$ is the same as the density of states above the conduction band $g(E_C)$, it may be seen that the Fermi level is $E_F = (E_C + E_V)/2$

TABLE 20.2

Lattice Parameters and Bandgap Energies for Selected Semiconductors

System	Lattice Parameter (Å)	Bandgap (eV)
C	3.567	5.3
Cubic-SiC	4.3596	6.27
Si	5.4307	1.107
Ge	5.657	0.67
α-Sn	6.4912	0.08
Zn-S	5.409	3.54
ZnSe	5.667	2.58
ZnTe	6.101	2.26
CdS	5.5818	2.42
CdSe	6.05	1.74
CdTe	6.477	1.44
HgSe	6.084	0.30
HgTe	6.460	0.15
Cubic-AlN	4.38	6.28
AlP	5.467	2.48
AlAs	5.6622	2.2
AlSb	6.1355	1.6
α-GaN	3.189/5.184	3.42
Cubic-GaN	4.52	3.23
GaP	5.4505	2.24
GaAs	5.6315	1.35
GaSb	6.0854	0.67
InN	3.548/5.760	1.95
InP	5.8687	1.27
InAs	6.0583	0.36
InSb	6.4788	0.165

Sources: Data taken from Gersten, J.I. and Smith, F.W. *The Physics and Chemistry of Materials,* John Wiley & Sons, New York, 2001; Kittel, C., *Introduction to Solid State Physics,* 7th edn. John Wiley & Sons, New York, 1966; Navon, D.H., *Electronic Materials and Devices,* Houghton Mifflin, Boston, MA, 1975.

and lies in the middle of the energy gap. (Actually the density of states in the conduction band is not necessarily the same as in the valence band, but this will be addressed in Section 20.2.2.)

20.2.2 Electron–Hole Product

Next, we need to find the number of electrons and holes present as a function of T. At thermal equilibrium, the number of electrons occupying the conduction band will be given by

$$N_e = \int_{E_c}^{\infty} g(E_C)F(E,T)dE. \tag{20.4}$$

The density of states $g(E_C)$ is the same as our previous result (Equation 15.28) with the effective electron mass m^* substituted for m and $E - E_C$ substituted for E, or

$$g(E_C) = \frac{V}{2\pi^2}\left(\frac{2m_e^*}{\hbar^2}\right)^{3/2}(E - E_C)^{1/2}. \tag{20.5}$$

Since $kT \sim 0.026$ eV and $(E - E_F) \geq \sim 1$ eV, the Fermi function may be approximated as

$$F(E,T) = \frac{1}{e^{(E-E_F)/kT} + 1} \approx \frac{1}{e^{(E-E_F)/kT}} = e^{(E_F-E)/kT}. \tag{20.6}$$

The number of electrons per volume residing in the conduction band is therefore,

$$n = \frac{N_e}{V} = \frac{V}{2\pi^2}\left(\frac{2m_e^*}{\hbar^2}\right)^{3/2} e^{E_F/kT} \int_{E_C}^{\infty} (E - E_C)^{1/2} e^{-E/kT} dE. \tag{20.7}$$

Integrating Equation 20.7, we get

$$n = 2\left(\frac{2\pi m_e^* kT}{h^2}\right)^{3/2} e^{-(E_C-E_F)/kT} = N_{eff}\, e^{-(E_C-E_F)/kT}, \tag{20.8}$$

where N_{eff} is called the effective density of states for electrons.

Similarly, the number of holes can be found from

$$N_p = \int_{-\infty}^{E_V} g(E_V)F'(E,T)dE, \tag{20.9}$$

where $g(E_V)$ is the density of states below the valence band, i.e., it will contain the factor $(E_V - E)^{1/2}$ and m will be replaced by the effective hole mass m^*. The $F'(E,T)$ is given by

$$F'(E,T) = 1 - \frac{1}{e^{(E-E_F)/kT} + 1} = \frac{e^{(E-E_F)/kT}}{e^{(E-E_F)/kT} + 1} = \frac{1}{e^{(E_F-E)/kT} + 1} \approx e^{-(E_F-E)/kT}, \tag{20.10}$$

since $(E_F - E_V) \gg kT$.

Integrating as before, we get

$$p = \frac{N_p}{V} = 2\left(\frac{2\pi m_p^* kT}{h^2}\right)^{3/2} e^{-(E_F-E_V)/kT} = P_{eff}\, e^{-(E_F-E_V)/kT}, \tag{20.11}$$

where P_{eff} is the effective density of states for holes. A useful number to remember is

$$2\left(\frac{2\pi m_e\, kT}{h^2}\right)^{3/2} = 2.476 \times 10^{25} \text{ electrons (or holes)/m}^3 \quad \text{at } T = 300°\text{K}. \tag{20.12}$$

Thus the $N_{eff} = 2.476 \times 10^{25}\, (m_e^*/m_e)^{3/2}(T/300)^{3/2}$ and $P_{eff} = 2.476 \times 10^{25}\, (m_p^*/m_p)^{3/2}(T/300)^{3/2}$.

The electron–hole product is an important quantity that is given by

$$np = N_{eff}P_{eff}e^{-(E_C-E_V)/kT} = N_{eff}P_{eff}e^{-(E_g/kT)}. \tag{20.13}$$

This is sometimes referred to as the law of mass action. Note that the number of charge carriers depends only on the effective masses, the temperature, and the energy gap. Also note that the Fermi level does not appear in this product, which means that no assumptions have been made concerning the origin of the electrons and holes. Therefore Equation 20.13 applies to electrons that have been extrinsically generated (from donor or collector states) as well as to intrinsic electrons and holes. As more electrons are added, some will go to annihilating holes until the equilibrium (Equation 20.13) is reached.

If the material is intrinsic (no electrical active impurities), at equilibrium there must be a hole for every electron promoted to the conduction band. Therefore, $n_i = p_i$ and

$$n_i = p_i = \sqrt{N_{eff}P_{eff}}\,e^{-E_g/2kT} \tag{20.14}$$

and the Fermi level is given by setting $n = p$ and solving Equations 20.8 and 20.11 for E_F,

$$E_F = \frac{E_C + E_V}{2} + \frac{kT}{2}\ln\left(\frac{P_{eff}}{N_{eff}}\right). \tag{20.15}$$

20.2.3 Energy Band Structure

The energy band structures for Si and GaAs are shown in Figures 20.3 and 20.4, respectively. The lowest curve that starts at Γ and curves upward is the bonding orbital for the s-electrons. The curves starting at Γ and $E = 0$ that curve downward are the bonding orbitals for the three p-electrons. (Two of these curves are degenerate in some directions from Γ.) The antibonding orbitals lie above the energy gap in the conduction band.

As may be seen in Figure 20.3, the minimum in the conduction band does not coincide with maximum in the valence band for Si. The same is true for Ge and some compound semiconductors, which are called indirect bandgap systems. Compare this energy band structure with that of GaAs in Figure 20.4.

GaAs is an example of a direct bandgap system. Recall that the momentum of a photon is E/c while the momentum of a phonon is E/v_0. Since the velocity of sound v_0 is very much less than the velocity of light, photons have energy but virtually no momentum. So, direct transitions from a valence band to a conduction band must be represented by a vertical line in the band diagram. Since the minimum in the conduction band is directly above the peak

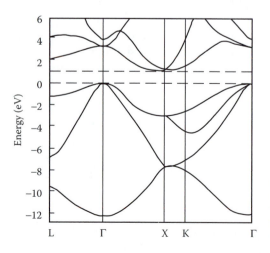

FIGURE 20.3

Band structure of Si. Note that the minimum in the conduction band does not coincide with maximum in the valence band. (From Cheilkowski, J.R. and Cohen, M.L., *Phys. Rev.*, B 14/2, 559, 1976. With permission.)

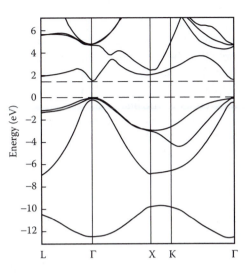

FIGURE 20.4
Band structure of GaAs. Here the minimum in the conduction band does coincide with maximum in the valence band. (From Herman, F. and Spicer, W.E., *Phys. Rev.*, 174/3, 906, 1968. With permission.)

in the valence band in the GaAs band diagram, a direct transition is possible at the bandgap energy.

In an indirect bandgap material such as Si and Ge, a transition at the bandgap energy must be accompanied by a phonon to supply the needed momentum to reach the conduction band at its lowest point, as illustrated in Figure 20.5. Such transitions are possible but are not as likely as the direct transitions possible in a direct bandgap material. This difference is reflected in the absorbance at the band edge as seen in Section 20.6.

Photons may induce direct or vertical transitions from any occupied band to a higher energy band that is not completely full. Such transitions are called interband transitions and contribute to the absorption spectra. So even in an indirect bandgap material, valence electrons can be promoted vertically to the conduction band by adsorbing photons of sufficient energy. Once in the conduction band, the hot electrons become thermalized through collisions and will eventually move to the lowest point in the band.

Electrons may also undergo intraband transitions in which they are promoted to a higher energy within their own band by absorbing a photon. But since such transitions must be assisted by a phonon, they, like other indirect transitions, are not as likely to occur as interband transitions.

In a direct bandgap system such as GaAs, an electron may also make a transition from the conduction band directly to the valence band by giving off a photon with energy E_g. However, such a transition is not likely in an indirect bandgap system because the minimum in the conduction band, where the electrons are likely to reside, does not occur at $k = 0$ (at Γ). In this case the transition, called a nonradiative transition, must involve a phonon and the energy E_g eventually goes into lattice vibrations or heat. Thus either

FIGURE 20.5
Schematic of direct and indirect bandgap transition. The vertical transitions are allowed direct transitions. The indirect transition from the $k = 0$ to the minimum of the conduction band requires the photon to combine with a phonon in order to conserve momentum.

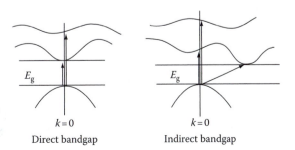

system may be used as a photodetector, but only systems with direct bandgaps are useful for light-emitting diodes (LEDs) or solid-state lasers.

20.2.4 Effective Electron and Hole Masses

We showed previously that the effective electron mass depends on the curvature of the band. In the conduction band in the vicinity of $k=0$, the band is curved upward, so the effective mass of a conduction electron is positive. However, near the top of the valence band, the bands curve downward, giving the particles near the top of the valence band negative mass.

If an electron is removed from the valence band, it leaves a hole. A hole can be thought of as a particle with positive charge, but with negative electron mass. Therefore, holes in the valence band will have positive charge and positive mass. If an electric field is applied in the positive direction, an electron in the conduction band will move opposite to the field because of its negative charge, and a hole will move of the field because of its positive charge and positive effective mass. Thus the current becomes the sum of electrons carrying negative charge in one direction and holes carrying positive charge in the other.

Also note that the greater the curvature of the band, the lighter the electron becomes. Recall from Equation 18.15 that the mobility of the electron is inversely proportional to the effective mass. Compare the sharpness of the conduction band in the vicinity of Γ for GaAs to that of Si. The higher mobility inherent in GaAs, along with its direct bandgap makes it more desirable than Si for specialized applications. But Si, because of its native oxide for making insulating regions, its low cost for high quality single crystal material, and because of the vast processing technology that has been developed, will always be used for less demanding applications. Thus the axiom—if you can make it using Si, then do it.

20.3 Extrinsic Semiconductors

We have shown that the carrier concentration in intrinsic semiconductors is a strong function of temperature. This would allow us to make useful devices such as thermistors or other temperature-sensitive devices, but the property that makes semiconductors so indispensable for modern electronic applications is the ability to drastically alter the electronic properties of the host material by the addition of trace quantities of electrically active impurities called dopants. For example, the addition of a pentavalent (group V) element such as Sb, P, or As to Si results in an additional electron that is loosely bound to the impurity atom. You can think of the orbital for this impurity atom as being similar to that of a hydrogen atom. Recall that the first ionization energy of a hydrogen atom is given by $me^4/(4\pi\varepsilon_0)^2\,\hbar^2$. If we replace the mass of a free electron with the effective mass ($m^* \sim 0.5\,m$) and replace ε_0 by $12\,\varepsilon_0$ (since the dielectric constant of Si is $\sim$12), we find the ionization potential for this electron is $\sim$0.05 eV. Since this is on the order of kT (0.026 eV at ambient temperature), one would expect a reasonable fraction of these impurity atoms to be ionized at ambient temperature, thus providing enough charge carriers to make the material a fairly good conductor of electricity. Conversely, the addition of trivalent (group III) dopants such as In, Ga, B, or Al creates a situation in which an electron is missing from the sp^3 bond (or one could say that the outer shell needs one electron in order to be full). Thus it becomes easy for an electron from an adjacent Si atom to fill this need, which creates a hole in the valence band.

TABLE 20.3

Ionization Energies (eV) for Donors
and Acceptors

Dopant	Si	Ge
P	0.045	0.012
As	0.049	0.0127
Sb	0.039	0.010
B	0.045	0.0104
Al	0.057	0.0102
Ga	0.065	0.0108
In	0.16	0.0112

Source: Data from Navon, D.H., *Electronic Materials and Devices*, Houghton Mifflin, Boston, MA, 1975.

Therefore, we see that adding group V impurities creates states in the energy gap close to the conduction band which are called donor states because they can donate electrons to the conduction band. In this case, the carriers are electrons and we call the material n-type because the carriers have a negative charge. Adding group III impurities creates states in the energy gap close to the valence band which are called acceptor states because they can accept electrons from the valence band. In this case, the carriers are holes and we call the material p-type because the carriers have a positive charge. It is important to remember that in either case the material does not have a net charge; the extra electrons (or holes) are balanced by the extra (or deficit) positive charge in the impurity ion cores. If equal number of donor and acceptor impurities are added, the extra electrons from the donor states occupy the acceptor sites (or you could say that the holes annihilated the electrons by recombining with them) and the material becomes semi-insulating again, similar to the intrinsic material. This is called compensation doping.

The ionization energies of the donor and acceptor states for Si and for Ge are given in Table 20.3. The energies are referred to the nearest band edge (conduction band for donors; valence band for acceptors).

20.3.1 Carrier Concentrations

Let N_D be the number of donor atoms. Assume that the main contribution to the majority carriers are the electrons from ionized donor states so that the number of ionized donors $N_D^+ \gg$ number of intrinsic electrons, n_i. The number of electrons from the ionized donors is given by $N_D - N_D^0$, where N_D^0 are the number of neutral or nonionized donors, or

$$N_D^+ \approx N_D\left(1 - \frac{1}{1 + e^{(E_D - E_F)/kT}}\right) = N_D\left(\frac{1}{1 + e^{(E_F - E_D)/kT}}\right), \tag{20.16}$$

where E_D is the energy of the donor state. Recall from Equation 20.8

$$n = N_{eff}e^{-(E_C - E_F)/kT}. \tag{20.17}$$

Solving for E_F/kT,

$$e^{E_F/kT} = e^{E_C/kT}n/N_{eff}. \tag{20.18}$$

Using Equation 20.18 into Equation 20.16,

$$N_D^+ \approx N_D \left(\frac{1}{1 + e^{(E_C - E_D)/kT}(n/N_{eff})} \right). \tag{20.19}$$

Generally in extrinsic semiconductors, $N_D^+ \gg n_i$ so $n = N_D^+ + n_i \approx N_D^+$. Solving Equation 20.19 for n,

$$n + \frac{n^2}{N_{eff}} e^{\Delta E/kT} = N_D \tag{20.20}$$

or

$$n = 2N_D \left(1 + \sqrt{1 + 4\frac{N_D}{N_{eff}} e^{\Delta E/kT}} \right)^{-1}, \tag{20.21}$$

where $\Delta E = E_C - E_D$.

Let us examine some limiting cases:

At very low temperatures, such that $4 (N_D/N_{eff}) \exp\Delta E/kT \gg 1$,

$$n \approx \sqrt{N_D N_{eff}} e^{-\Delta E/2kT}. \tag{20.22}$$

When $\ln(n)$ is plotted against $1/T$, the slope is $-\Delta E/2k$. This is characteristic of what is known as the freeze-out region and the observed slope can be used to determine the ionization energy, ΔE.

At high temperatures, where $4 (N_D/N_{eff}) \exp(\Delta E/kT) \ll 1$,

$$n = N_D \left(1 - \frac{N_D}{N_{eff}} e^{\Delta E/kT} + \cdots \right) \approx N_D \text{ provided } \frac{N_D}{N_{eff}} e^{\Delta E/kT} \ll 1. \tag{20.23}$$

The region in which $n \approx N_D$ is called the saturation region.

At still higher temperatures, the thermal energy becomes large enough to promote more electrons across the E_g than provided by the dopant and the material behaves as if it were intrinsic once again.

Figure 20.6 shows the carrier concentration in n-doped Si calculated from the above analysis plotted against $1/T$. The intrinsic, saturation, and freeze-out regions can easily be identified.

Remember that throughout this process, the law of mass action holds which says that the product of hole and electron carrier concentration is a function only of the ratio of the effective masses of the electrons, the E_g, and the absolute temperature. The dopant concentration does not enter into the np product. Increasing the number of donor impurities raises n, but at the expense of p. Conversely, increasing the number of acceptor states increases p, but at the expense of n. In other words, for a given material at a given temperature, increasing the majority carrier concentration is done at the expense of the minority carrier concentration such that their product remains the same.

20.3.2 Fermi Levels in Extrinsic Semiconductors

For reasons that will become clear in Chapter 21, it is very important to understand the influence of doping on the Fermi level. From Equation 20.18,

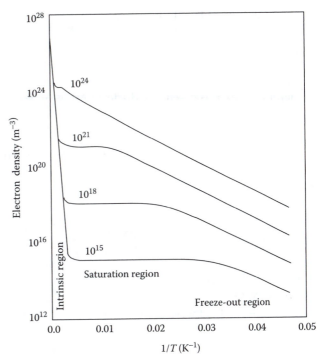

FIGURE 20.6
Computed carrier concentration in *n*-doped Si as a function of reciprocal temperature at various doping levels.

$$E_F = E_C + kT \ln\left(\frac{n}{N_{eff}}\right), \tag{20.24}$$

where *n* is the total number of electrons. Now consider the limiting cases for the number of ionized donors obtained from Equation 20.21. First consider the freeze-out region, where $n^+ \approx \sqrt{N_D N_{eff}} \exp\left(-(\Delta E/2kT)\right)$. Putting this into Equation 20.24,

$$E_F = E_C - \frac{\Delta E}{2} + \frac{kT}{2} \ln\left(\frac{N_D}{N_{eff}}\right). \tag{20.25}$$

Notice that at low temperatures, the Fermi level moves to between E_C and E_D which allows a large number of donors to be ionized even if $kT \ll \Delta E$.

Next consider the saturated region $n \approx n^+ = N_D$. Using Equation 20.23, Equation 20.24 becomes

$$E_F = E_C + kT \ln\left[\frac{N_D\left(1 - (N_D/N_{eff})e^{\Delta E/kT} + \cdots\right)}{N_{eff}}\right] \approx E_C + kT \ln\left(\frac{N_D}{N_{eff}}\right). \tag{20.26}$$

One can see that adding donors raises the Fermi level. We see from Equation 20.24 that it is possible to raise the E_F above the conduction band in order to make the material ohmic (metal-like) by making $n > N_{eff}$. One is tempted to think that this could be accomplished by setting $N_D = N_{eff}$. However, $\exp(\Delta E/kT) = 6.66$ at 300 K for Si. So as $N_D \rightarrow N_{eff}$, the approximation used in Equation 20.26 is no longer valid and the Fermi level must be found by using Equation 20.21 with Equation 20.24.

What happens when acceptors are added? Here we have to go back to Equation 20.11, $E_F = E_V - kT \ln (p/P_{eff})$. For p_i, recall that E_F was between the conduction and the valence

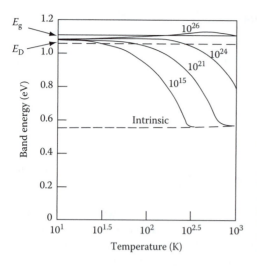

FIGURE 20.7
Fermi level as a function of temperature for various *n*-doping levels. At low temperatures, the Fermi level lies between the donor level and the conduction band energy which allows the donor states to be ionized at lower temperatures. At higher temperatures, the Fermi level falls to the intrinsic level. At 10^{26} electrons/m^3, the Fermi level is above the conduction band and the material becomes ohmic.

band. We see that increasing the hole concentration pushes the E_F toward the valence band and that by sufficient doping, it is also possible to make *p*-type material ohmic.

As the temperature is increased, the n_i and p_i increases while the number of donors (or acceptors) remains constant (assuming they are all ionized), which drives E_F back toward E_F^i as seen in Figure 20.7.

20.4   Hall Coefficient for Both Electrons and Holes

The determination of the Hall coefficient in the presence of multiple carriers is a little more involved than the derivation in Chapter 19. The current density is given by

$$\mathbf{J} = n\mu_n e(\mathbf{E} + \mathbf{v}_n \times \mathbf{B}) + p\mu_p e(\mathbf{E} + \mathbf{v}_p \times \mathbf{B}), \tag{20.27}$$

where the subscripts n and p refer to the electrons and holes, respectively. Setting $v_y = 0$,

$$J_x = eE_x\left(n\mu_n + p\mu_p\right) \tag{20.28}$$

and

$$J_y = eE_y\left(n\mu_n + p\mu_p\right) - eB_z\left(n\mu_n v_{nx} + p\mu_p v_{px}\right). \tag{20.29}$$

The velocities of the electrons are $v_{nx} = -\mu_n E_x$ and the holes are $v_{px} = \mu_p E_x$. Putting these into Equation 20.29,

$$J_y = eE_y\left(n\mu_n + p\mu_p\right) - eB_z E_x\left(p\mu_p^2 - n\mu_n^2\right). \tag{20.30}$$

Setting $J_y = 0$ and substituting Equation 20.28 for E_x,

$$E_y = \frac{B_z J_x \left(p\mu_p^2 - n\mu_n^2 \right)}{e\left(n\mu_n + p\mu_p \right)^2}. \tag{20.31}$$

The Hall coefficient is given by

$$R_H = \frac{E_y}{J_x B_z} = -\frac{1}{e} \left[\frac{n\mu_n^2 - p\mu_p^2}{\left(n\mu_n + p\mu_p \right)^2} \right]. \tag{20.32}$$

One can easily see that if $n \gg p$, R_H reduces to $-1/ne$. Hall measurements are easiest to interpret in doped materials when either $n \gg p$ or $n \ll p$. Otherwise one is faced with four unknowns, which require other measurements to resolve. For example, except for the difference between electron and hole mobilities, the Hall effect would be zero for intrinsic materials. One can also see that doing Hall measurements as a function of temperature offers a means of determining the occupancy number and energy levels of the various impurity states in the freeze-out region through Equation 20.22.

20.5 Conductivity of Semiconductors

In a semiconductor, the current is carried by both electrons and holes. Thus the current density may be written as

$$J = J_n + J_p = ne\mu_n E + pe\mu_p E = \sigma E \tag{20.33}$$

or

$$\sigma = ne\mu_n + pe\mu_p, \tag{20.34}$$

where e is the magnitude of the charge of an electron. The mobilities of the electrons and holes are inhibited by scattering from lattice vibrations (phonon scattering) as well as from impurity atoms and from lattice defects.

Recall that for metals, we argued that $v_{drift} \ll v_{th} < v_F$, and since the Fermi velocity was independent of the electric field and temperature, v was therefore independent of temperature. But in metals, $v_F > v_{th}$ because of the fact that every atom contributed at least free one electron to the Fermi distribution. (We do not count the bound electrons in the inner core of the atoms.) Similarly, the electrons in the valence band of a semiconductor are bound and do not contribute to the Fermi level as long as they are in this bound state. Recall also that we placed the Fermi level between the valence and the conduction bands of an intrinsic semiconductor so that the probability of promoting an electron to the conduction band would equal the probability of creating a hole in the valence band. So for semiconductors, the Fermi velocity will be substantially less than the thermal velocity since the number of free electrons will be much less than for a metal. Therefore, we can assume for semiconductors that the average electron velocity $v \sim T^{1/2}$.

At very low temperatures ($<10\,K$) scattering from lattice defects dominates in semiconductors, just as in metals. However, in the case of charged defects such as ionized donor

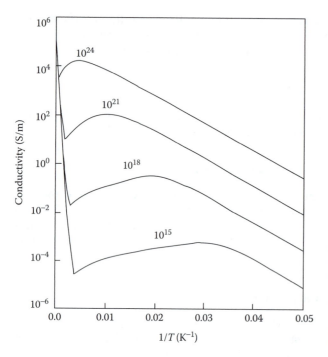

FIGURE 20.8

Calculated conductivity for *n*-doped Si as a function of reciprocal temperature. The electron and hole mobilities were taken to be 0.14 and 0.048 $m^2/V \cdot s$, respectively at 300 K and were assumed to fall as the $-3/2$ power of temperature. In the intrinsic region, $\sigma \sim \exp - (E_g/2kT)$.

or acceptor atoms, there is an additional Coulomb scattering (Rutherford scattering) whose cross section $\sim \langle v \rangle^{-4}$. Since the number of scattering events per unit time $\sim$ cross section $\times$ average velocity, $1/\tau \sim \langle v \rangle^{-3}$ and the mobility $\mu \sim \tau \sim \langle v \rangle^{3} \sim T^{3/2}$.

At higher temperatures, thermal scattering from collisions with the lattice ion cores dominates. We argued in the case of metals that the scattering cross section is proportional to $\sim T$. This would also be true for semiconductors, except now the average electron velocity $\sim T^{1/2}$. Hence the number of scattering events per unit time $\sim T^{3/2}$ and the mobility $\mu \sim T^{-3/2}$.

Electrons and holes have different mobilities but, at moderate temperatures, the mobilities of both would be expected to go as $T^{-3/2}$. Therefore, when we compute the conductivity of a semiconductor, this $T^{-3/2}$ term in the mobilities cancels the $T^{3/2}$ in the pre-exponential term for the carrier concentration, so that plotting $\ln(\sigma)$ versus $1/T$ for intrinsic material yields a straight line with a slope $= -E_g/2k$, which provides one method for determining the bandgap energy.

As can be seen in Figure 20.8, conductivity for doped material drops off at low temperatures because not all of the impurity sites are ionized. In the saturation region, the conductivity drops with increasing temperature because of the increased thermal scattering until the intrinsic curve is encountered, at which point the carrier density increases rapidly with increasing temperature.

20.6 Optical Properties

Semiconductors are transparent at wavelengths longer than their cut-on wavelength, $\lambda_C \geq hc/E_g$ or at photon energies less than the bandgap energy E_g. At shorter wavelengths, the incident photons have sufficient energy to promote electrons from the valence to the conduction band and the material begins to absorb strongly. Much information can be

gleaned from the analysis of the absorption edge in which the absorption coefficient α is measured as a function of ω (or λ). The α can be obtained from a transmission measurement using Beer's law, $I/I_0 = \exp-(\alpha x)$ where the transmitted intensity I/I_0 through a thin slab of material as a function of frequency (or λ). For very large values of α, it becomes no longer feasible to measure the transmitted intensity and similar information can be obtained from reflectivity measurements. Materials that absorb strongly also reflect strongly for reasons that are discussed in Chapter 24. The structure of the absorption edge will depend on whether the semiconductor has a direct or indirect bandgap.

20.6.1 Absorption Edge for Direct Bandgap Semiconductors

The absorption coefficients for Si and GaAs are shown in Figure 20.9. Note that the absorption coefficient for GaAs, which has a direct bandgap, rises very quickly at the bandgap energy, 1.34 eV. There is a secondary peak at ~4.6 eV, which corresponds to the direct transition from the valence band to the second conduction band along Γ in Figure 20.4.

20.6.2 Absorption Edge for Indirect Bandgap Semiconductors

The absorption coefficient for Si seen in Figure 20.9 is more interesting. It begins to rise at the indirect bandgap energy but then starts to taper off until the bandgap energy of the direct transitions, 4.3 eV, where it increases rapidly. This behavior reflects the fact that indirect transitions are not as likely as direct transitions because they require phonon assistance. A second peak occurs at 4.2 eV, which corresponds to the second direct transition as shown seen in Figure 20.3. These direct transitions can be seen more easily in reflective measurements shown in Figure 20.10.

20.6.3 Excitons

Although we tend to think of semiconductors being transparent to frequencies below the bandgap energy, it is possible for a photon with less than the bandgap energy to lift an electron from the valence band, but instead of going all the way to the conduction band, the electron becomes attracted to the hole it created and forms a bound pair. Such a bound electron–hole pair is called an exciton and can be described by hydrogen-like wavefunctions. Since excitons can have several energy states, they are observed as small absorption

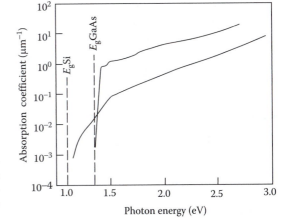

FIGURE 20.9
Absorption coefficients for Si and GaAs. The bandgap edges are located by vertical lines. (Absorption coefficients were computed from Palik, E.D., Ed., *Handbook of Optical Constants of Solids*, Vols. 1–3, Academic Press, 1997.)

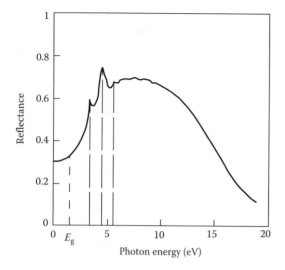

FIGURE 20.10
Reflectance spectrum of Si. Peaks at 3.4, 4.5, and 5.4 eV are direct interband transitions. (Reflectivity was computed from Palik, E.D., Ed., *Handbook of Optical Constants of Solids*, Vols. 1–3, Academic Press, 1997.)

lines at energies just short of the absorption edge (or at wavelengths just beyond the cut-on wavelength). Excitons can move through the crystal carrying energy (but not charge) and eventually decay giving up their energy as a photon.

20.6.4 Photoconductive Detectors

The electron–hole pairs produced by the absorption of photons can be detected by applying a potential across the semiconducting material and measuring the current. The resulting photocurrent is directly proportional to the rate of electron–hole production, which is proportional to the photon flux. Thus a semiconductor crystal provides a very simple device for measuring the intensity of light.

Such crystals are also useful for measuring penetrating radiation such as x-ray or gamma radiation. An energetic photon passing through the crystal will produce many photoelectrons and holes, which can be collected by applying an electric field across the crystal. Since the number of photoelectrons produced is proportional to the energy of the incident photon, analyzing the total charge produced by each incident photon with a multichannel pulse height analyzer can determine the gamma or x-ray spectrum and is used in x-ray fluorescence for elemental analysis. Such a detector is called an energy dispersive spectrometer.

20.7 Semiconducting Polymers

Recall from Chapter 3 that when two atoms are brought together and a covalent molecular bond is formed in a conjugated polymer, two quantum states are created: a bonding π orbital and an antibonding π^* orbital. If the energy of the bonding state is negative, a stable bond is formed. The bonding states comprise the valence band and the antibonding states the conduction band.

Conjugated polymers tend to have bandgaps ∼1–2 eV and are considered to be semiconductors. Intrinsic semiconductors, we know, are generally poor conductors and dopants are added to the traditional semiconductors such as Ge and Si to create donor or acceptor states within the bandgap in order to make them more conductive. Similarly,

dopants, such as the iodine that Heeger, MacDiarmid, and Shirakawa added to polyace-tylene, are required to increase the conductivity of polymers with conjugated electrons. Dopants such as Na, K, Li, Ca, and tetrabutyl ammonium act as donors for *n*-type polymers while I_2, PH_6, BF_4, Cl, AsH_4 act as acceptors for *p*-type polymers. Dopants are added by exposing the polymer (or its precursor) to vapors or solutions containing the dopant or by electrochemical methods. Unlike semiconductor in which the dopants take their place as substitutional defects, dopants in polymers take their places interstitially among the polymer chains and donate to or accept charges from the polymer backbone thus forming new three-dimensional structures. The original I_2-doped polyacetylene developed by Heeger et al. had a conductivity of ~100 S/cm but now polymer conductivities have exceeded 10^5 S/cm, which exceeds many metals.

20.7.1 Energy Levels of Confined Electrons

Free electrons that are confined in a potential well must satisfy the Schrödinger wave equation with boundary conditions that require the potential energy to be infinite at the boundaries of the molecule as explained in Chapter 2. This requires the wavefunction to vanish at the boundaries and to form standing waves within the molecule, i.e., the classic electron-in-a-box problem. The Pauli exclusion principle requires that only one electron may occupy a particular quantum state, so only two electrons (one with spin up and the other with spin down) can occupy each energy level. As more delocalized electrons are added, they must go into higher energy states. The energy state of the last delocalized electron is called the highest occupied molecular orbital (HOMO). The next available energy level is the lowest unoccupied molecular orbital (LUMO). Electrons occupying the energy levels below the HOMO are in the valence band and the energy levels above the LUMO are in the conduction band. The energy difference between the HOMO and LUMO is the bandgap. Typically in molecules with conjugated electrons, this bandgap is ~2.0 eV. The Fe-heme molecule, for example, has 26 conjugated electrons confined to a box that is ~1 nm^2. The bandgap is 1.882 eV, which corresponds to a photon with wavelength of 695 nm, which is just above the red part of the spectrum. As a result, photons with shorter wavelengths can excite HOMO to LUMO transitions and are absorbed, while longer wavelengths are reflected, thus accounting for the red color of blood. Similar arguments account for the green color of the chlorophyll molecule or the yellow color of β-carotene molecule.

Also recall from Chapter 2 that the difference between energy levels, i.e., the bandgap, for the electron-in-a-box analysis, goes inversely as the square of the length of the box. Thus very large molecules, such as a sheet of graphite, will have a very small bandgap so that thermal energy, typically 0.025 eV, is sufficient to populate the conduction band with electrons so that no dopants are required to obtain high electrical conductivity.

20.8 Summary

Most semiconductors used in electronic and photonic devices are group IV elements or compounds formed by III–V or II–VI elements. They are characterized by an sp^3 hybrid bond in which an *s*-electron is promoted to a *p*-state in order to form a tetrahedral directed diamond-like bond. The four valence electrons occupy the bonding states from which the antibonding states, which form the conduction band, are separated by the energy gap E_g.

The width of the energy gap decreases as the lattice parameter increases, or as the atoms become larger and farther apart. Thus, diamond has the highest bandgap energy and the bandgap decreases with Si, Ge, and disappears with Sn and Pb, which are metals. The compound systems follow the same trend.

Even though the valence electrons are tied up in covalent bonds, they are not completely localized and have similar band structure and dispersion relationships (E vs. $\mathbf{k}$) as nearly free electrons. Thus the band structure becomes a powerful tool for understanding the properties and behavior of semiconductors.

When an electron is promoted to the conduction band, it leaves a hole in the valence band, which frees up states in the valence band so that the valence electrons can also respond to an electrical field. The hole in the valence band can be treated as a particle with positive mass and positive charge, while the electron in the conduction band behaves as a particle with positive mass and negative charge. Thus in response to an applied electrical field, electrons carry the charge in one direction while holes carry the opposite charge in the opposite direction and both contribute to the conduction of current.

Systems in which the minimum of the conduction band coincides with the maximum in the valence band at $\mathbf{k} = 0$ are called direct bandgap materials. In such materials, electrons may go directly from the valence band to the conduction by absorbing a photon with energy equal or greater than E_g. GaAs is an example of a direct bandgap material. Si and Ge are indirect bandgap materials in which the minimum energy gap is at some distance $\mathbf{k}$ from 0. In such systems, a photon with energy E_g must interact with a phonon with momentum $\hbar\mathbf{k}$ to promote an electron to the conduction band, which has a lower probability than a direct transition. Electrons can be promoted directly to a higher conduction band at $\mathbf{k} = 0$ if they have sufficient energy.

In direct bandgap materials, electrons may go directly from the conduction band to the valence band with a photon emission, which makes them suitable as LEDs or solid-state lasers. Indirect transitions from the conduction to the valence band must be accompanied by the emission of a phonon and are nonradiative transitions, which results in heating the material.

Pure materials, or at least materials with no impurity states in the bandgap region, are called intrinsic semiconductors. The creation of an electron leaves a hole; therefore, the number of holes must equal the number of electrons in an intrinsic material. The electron–hole product is directly proportional to the Boltzmann factor, $\exp -(E_g/kT)$. The Fermi level, the energy for which the probability of creating a conduction electron is the same as creating a hole is shown to be somewhere in the bandgap.

Pentavalent atoms may be added to group IV semiconductors to form donor states close to the conduction band. Electrons are loosely attached to these donor atoms and are thermally ionized to produce conduction electron without producing holes. The number of conduction electrons, hence the conductivity of the materials, can be controlled by the doping level or the number of donor atoms added. Material in which the majority carriers are electrons are referred to as n-type materials. Adding donor states moves the Fermi level toward the conduction band.

Similarly, trivalent atoms may be added to group IV semiconductors to form acceptor states close to the valence band. Atoms from the valence band may be easily thermally ionized to replace the missing electrons in the covalent bond, leaving holes in the valence band. Materials in which the majority carriers are holes are referred to as p-type materials. Adding acceptor states moves the Fermi level toward the valence band.

Neither n- or p-type materials have a net charge. The excess electrons (or holes) are balanced by the charge of the ion cores. Also the np product remains the same as that of

intrinsic materials regardless of the dopants added. Adding electrons reduces the number of holes and vice versa.

At low temperatures, not all of the donors (or acceptors) are ionized and the number of carriers $n(\text{or } p) \sim \exp{-(\Delta E/2kT)}$, where ΔE is the ionization energy of the donor (or acceptor). This energy can be determined by making Hall measurements in this freeze-out region.

At ambient temperatures, most of the donors or acceptors will be ionized and the carrier concentration will essentially be the number of donors or acceptors in the system. Most devices are designed to operate in this plateau or saturation region so their performance will not be affected by modest changes in temperature.

At still higher temperatures where $kT \sim E_g$, the intrinsic electron concentration becomes larger than the donor (or acceptor) concentration and the material becomes intrinsic again.

The Hall coefficient is more complicated when both electrons and holes are present because they tend to cancel each other out in the Hall current. The Hall measurement is the easiest to interpret in doped materials when the minority carriers can be ignored.

The conductivity of a semiconductor is given by $\sigma = ne\mu_n + pe\mu_p$. Except for very low temperatures, the mobilities of the charge carriers in a semiconductor $\sim T^{-3/2}$; therefore in the plateau region, where the carrier concentration is more or less constant, $\sigma \sim T^{-3/2}$. However, in the freeze-out region and in the intrinsic region, the carrier concentration is increasing with temperature faster than the mobility is decreasing and the conductivity increases with temperature.

Bibliography

Adachi, S., Ed., *Handbook on Physical Properties of Semiconductors*, Vols. 1–3, Springer-Verlag, New York, 2004.

Cheilkowski, J.R. and Cohen, M.L., Nonlocal pseudo calculations for the electronic structure of eleven diamond and zinc-blende semiconductors, *Phys. Rev.*, B 14/2, 559, 1976.

Gersten, J.I. and Smith, F.W., *The Physics and Chemistry of Materials*, John Wiley & Sons, New York, 2001.

Herman, F. and Spicer, W.E., Spectral analysis of photoemissive yields in GaAs and related crystals, *Phys. Rev.*, 174/3, 906, 1986.

Hummel, R.E., *Electronic Properties of Materials*, 3rd edn. Springer, New York, 2000.

Ibach, H. and Lüth, H., *Solid State Physics*, 3rd edn. Springer-Verlag, New York, 1990.

Kittel, C., *Introduction to Solid State Physics*, 7th edn. John Wiley & Sons, New York, 1966.

Livingston, J.D., *Electronic Properties of Engineering Materials*, Wiley, MIT series in Materials Science and Engineering, John Wiley & Sons, New York, 1999.

Navon, D.H., *Electronic Materials and Devices*, Houghton Mifflin, Boston, MA, 1975.

Palik, E.D., Ed., *Handbook of Optical Constants of Solids*, Vols. 1–3, Academic Press, San Diego, CA, 1997.

Sapoval, B. and Hermann, C., *Physics of Semiconductors*, Springer-Verlag, New York, 1995.

Problems

You are given a piece of unknown semiconductor material and are asked to characterize it electrically. Using cyclotron resonance you have determined the effective

masses of the holes to be $0.5m_e$ and the electrons to be $0.4m_e$. You have made a series of Hall measurements and conductivity measurements with the results given below:

T (K)	R_H (m³/C)	σ (S/m)
20	−470,900	1.727×10^{-5}
30	−3,992	1.109×10^{-3}
40	−1,723	1.669×10^{-3}
50	−1,693	1.214×10^{-3}
100	−1,693	4.297×10^{-4}
200	−1,693	1.519×10^{-4}
300	−815.6	1.796×10^{-4}
400	−4.10	0.024
500	−0.118	0.584
600	−0.011	4.971

1. Determine the bandgap energy E_g.
2. Determine the dopant concentration.
3. Determine the ionization energy of the dopant.
4. Determine the mobility of the electrons at 300 K (assume mobility varies as $T^{-3/2}$).
5. Determine the mobility of the holes at 300 K (assume mobility varies as $T^{-3/2}$).
6. What donor level would be required to make this material ohmic?
7. Construct carrier concentration versus $1/T$ curves similar to Figure 20.6 for In-doped Si.

21

Theory and Applications of Junctions

Now that we have laid the groundwork for understanding how we can engineer various carrier concentrations and Fermi levels into semiconducting materials, we can explore some of the numerous devices that have become possible using these novel materials.

21.1 The p–n Junction

Most elementary texts give a very simple and easy to understand description of the operation of a p–n junction (Figure 21.1).

Although this description is true in some respects, it only tells part of the story. There is much more to the operation of a p–n junction that must be understood in order to appreciate the operation of the more sophisticated devices that have evolved in recent years. Now let us understand what really happens at a p–n junction.

When a p-type material is brought into intimate contact with n-type material, there will be more electrons on the n-side than on the p-side. In Chapter 20, we showed that the addition of donor states shifts the Fermi level toward the conduction band and the addition of acceptor states shifts the Fermi level toward the valence band. Thus the Fermi level will be higher in the n-type material than in the p-type material. Recall that the Fermi level is also the chemical potential and that particles will move from the higher chemical potential to the lower chemical potential until equilibrium is reached and the Fermi level is the same in both materials as seen in Figure 21.2. The electrons flowing from the n-material to the p-material will produce a negative charge in the p-material, which decreases its electron affinity χ_p (the energy required to lift an electron from the conduction band to the vacuum state). Similarly, the n-material becomes positively charged and its electron affinity is increased. The difference between the Fermi levels is the contact potential, eV_0.

The transfer of charge from the n- to the p-material creates an electric field $-E_x$ in the depletion zone that starts at x_n, peaks at $x = 0$, and diminishes at x_p as illustrated in Figure 21.3

Another way to look at what is happening is to consider that there will be a sharp gradient in electron density at the junction. Electrons will diffuse from the n-type to the p-type material driven by this gradient. Similarly, holes will diffuse from the p-type to the n-type material. (Or you can think of the electrons filling the holes in the p-type material leaving holes in the n-type material.) However, both the n- and p-type materials were electrically neutral at the beginning. Electrons migrating across the junction create an excess negative charge on the p-side of the junction, and the deficiency of electrons (or excess of holes) creates an excess positive charge on the n-side of the junction. The electric field created by this charge separation limits the diffusion process so that equilibrium is reached in which the diffusion current is balanced by the field it created. The region over

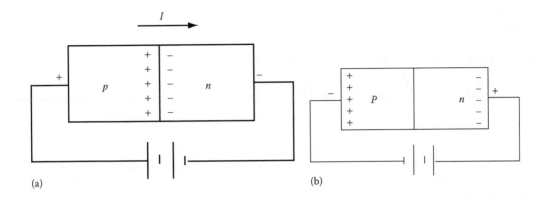

FIGURE 21.1
(a) Forward-biased *p–n* junction. The applied field pushes the electrons and holes together at the junction where they combine causing current *I* to flow. (b) Reverse-biased *p–n* junction. The applied field draws the charge carriers away from the junction thus preventing their recombination and no current flows.

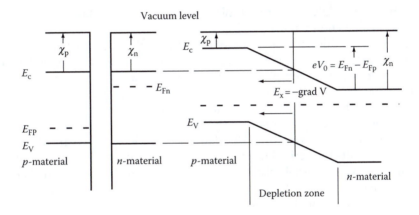

FIGURE 21.2
When a *p*-type material is brought into intimate contact with *n*-type material, electrons flow from the *n*- to the *p*-material until their Fermi energies (chemical potentials) become equal, causing a shift in the conduction and valence bands in both the *p*-side and the *n*-side, which creates a contact potential difference V_0. The charge separation results in a strong electric field $E = -\nabla V$ (right-to-left).

which this field acts is called the depletion region or the space charge region. In this region, the electrons and holes have recombined; so there are no free carriers. Even though an electron combines with a hole and is no longer a free carrier, it still deposits a charge. The built-in electric field caused by the diffusion current is really the feature that makes *p–n* junctions useful as we shall see.

21.1.1 Analysis

Now let us see if we can describe this process analytically. From Equation 20.18 we can see that adding N_D donor states will raise the Fermi level above the intrinsic Fermi level in the *n*-material by

$$E_F^n - E_F^i = kT \ln\left(\frac{n_n}{n_i}\right) = kT \ln\left(\frac{N_D + n_i}{n_i}\right) \text{ (if all donors ionized).} \quad (21.1)$$

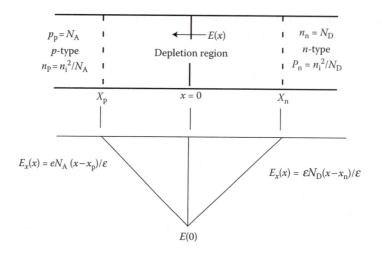

FIGURE 21.3

Schematic of the electron and hole concentration and their associated electric field near an unbiased *p–n* junction. The electric field as a function of *x* in the depletion zone is depicted underneath the junction.

Similarly, adding acceptor states lowers the Fermi level below the intrinsic Fermi level in the *p*-material by

$$E_F^i - E_F^p = kT \ln\left(\frac{p_p}{n_i}\right) = kT \ln\left(\frac{N_A + n_i}{n_i}\right) \text{ (if all acceptors ionized)}. \qquad (21.2)$$

Subtracting Equation 21.2 from Equation 21.1, the contact potential V_0 is obtained as

$$eV_0 = E_F^n - E_F^p = kT \ln\left(\frac{n_n p_p}{n_i^2}\right). \qquad (21.3)$$

If all donor and acceptor states are ionized and if $N_A \gg n_i$ and $N_D \gg n_i$, the number of electrons in the *n*-type material, $n_n = N_D$, and the number of holes in the *p*-type material, $p_p = N_A$, and the contact potential V_0 is given by

$$eV_0 = E_F^n - E_F^p = kT \ln\left(\frac{N_D N_A}{n_i^2}\right). \qquad (21.4)$$

Recall that $n_i^2 = n_i\,p_i = n_n\,p_n = n_p\,p_p$, where p_n and n_p are the minority carriers in the *p*-type and *n*-type materials, respectively.

$$eV_0 = kT \ln\left(\frac{n_n p_p}{n_p p_n}\right) = kT \ln\left(\frac{p_p}{p_n}\right) = kT \ln\left(\frac{n_n}{n_p}\right). \qquad (21.5)$$

For $N_A = N_D = 10^{24}$ electrons/m^3 and $n_i\,n_p = 10^{32}$/m^3 at 300 K, the contact potential will be

$$V_0 = 0.026 \ln\left(\frac{10^{48}}{10^{32}}\right) = 0.958 \text{ V}.$$

An alternative derivation of the above result can be obtained by writing the hole current as the sum of the current produced by the electric field and the diffusion current,

$$J_p = e\left(p\mu_p E_x - D_p \frac{\partial p}{\partial x}\right). \tag{21.6}$$

At equilibrium, the total current must vanish; hence,

$$D_p \frac{\partial p}{\partial x} = p\mu_p E_x = -p\mu_p \frac{\partial V}{\partial x}. \tag{21.7}$$

Since the hole gradient is negative, hole flow is positive or from the *p*- to *n*-type material. This means the E_x must be negative or in the direction of the *p*-type material.

Integrating this expression with the boundary condition that $V = V_p$ and $p = p_p$ in the *p*-type material far removed from the junction, we get

$$D_p \ln\left(\frac{p_p}{p}\right) = \mu_p(V - V_p). \tag{21.8}$$

Similarly, if we take the boundary condition that $V = V_n$ and $p = p_n$ in the *n*-type material far away from the junction,

$$D_p \ln\left(\frac{p_n}{p}\right) = \mu_p(V - V_n). \tag{21.9}$$

Subtracting Equation 21.9 from Equation 21.8, and using the Einstein relationship, $D_p = kT\mu_p/e$, we get

$$e(V_n - V_p) = eV_0 = kT \ln\left(\frac{p_p}{p_n}\right), \tag{21.10}$$

which is the same result as Equation 21.5.

21.1.2 Width of the Depletion Layer

The width of the depletion layer can be estimated from the following rather crude model. Assume the depletion layer extends from 0 to $-x_p$ into the *p*-type material and x_n into the *n*-type material as in Figure 21.3. Assume uniform charge distribution within these regions. Charge conservation requires that $N_A x_p = N_D x_n$.

The electric field may be found from one of the Maxwell's equation, $\nabla \cdot D = \nabla \cdot \varepsilon E = \rho$,

$$\frac{dE_x}{dx} = -\frac{eN_D}{\varepsilon} \text{ from 0 to } x_n, \tag{21.11}$$

where ε is the dielectric constant. Integrating with the boundary condition that requires the electric (E) field to vanish at x_n, we get

$$E_x = \frac{eN_D}{\varepsilon}(x - x_n). \tag{21.12}$$

Integrating again with the boundary condition that requires the potential to vanish at x_n,

$$V = \frac{eN_D}{2\varepsilon}(x_n^2 - x^2).$$ (21.13)

Setting $V = V_0$ and using Equation 21.4,

$$x_n^2 = \frac{2\varepsilon V_0}{eN_D} = \frac{2\varepsilon kT}{eN_D} \ln\left(\frac{N_D N_A}{n_i^2}\right).$$ (21.14)

A similar argument can be made for the width of the depletion zone in the p-side. Note that the width of the depletion zone varies as the inverse square root of the doping level. Using the same doping level as above and taking the dielectric constant to be 12, the width of the depletion zone into the n-side is

$$x_n = \sqrt{\frac{2 \times 12 \times 8.85 \times 10^{-12} \times 0.985}{1.6 \times 10^{-19} \times 10^{24}}} = 3.6 \times 10^{-8} \text{ m}.$$

Thus we see that the depletion layers are extremely thin, on the order of tens of nanometers. Putting this width into Equation 21.12, we obtain an electric field of 5.54×10^7 V/m.

21.1.3 Biasing the Junction

Consider Equation 21.6 in which the diffusion of holes from the p- to the n-type material is balanced by the gradient of the contact potential. If the p-side is connected to the positive terminal of a battery (holes flow from $+$ to $-$, electrons from $-$ to $+$), some of the electrons that had diffused into the p-side are taken up by the battery. This applied voltage V reduces the contact potential as seen in Figure 21.4 as well as the electric field across the depletion zones as seen in Figure 21.5 and more electrons can diffuse across the junction. The positive terminal of the battery continues to supply holes to the p-material by taking up excess electrons while the electrons flowing from the negative terminal of the battery continue to annihilate the excess holes that were injected into the n-type material.

With a positive voltage V is applied to the p-type material, Equation 21.10 may be written as

$$\frac{p_p)_{x_p}}{p_n)_{x_n}} = e^{e(V_0 - V)/kT}[e(V_0 - V)/kT],$$ (21.15)

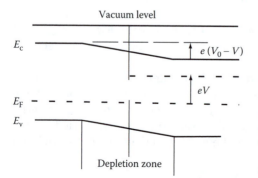

FIGURE 21.4
Forward-biased p–n junction. Applying a plus voltage to the p-material reduces the contact potential and raises the effective Fermi level allowing more electrons to diffuse from the n-side to the p-side.

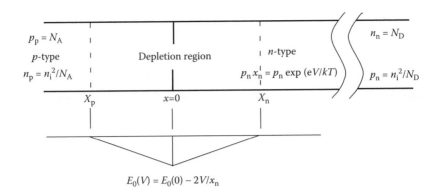

FIGURE 21.5
Carrier distribution and related electric field in a forward-biased *p–n* junction. The number of holes that were injected at the edge of the depletion zone on the *n*-side is given by $p_n \exp(eV/kT)$ where p_n is the hole density in the *n*-material away from the junction region. Note also that the electric field has been reduced by $2V/x_n$.

where
$p_p)_{x_n}$ is the hole concentration in the *p*-type material at the edge of the space charge region

$p_n)_{x_n}$ is the hole or minority carrier concentration in the *n*-type material at the edge of the space charge region

But since $p_p)_{x_p} \approx p_p$ and from Equation 20.3,

$$\frac{p_p}{p_n} = e^{e(V_0)/kT}, \tag{21.16}$$

then

$$p_n)_{x_n} = p_n e^{eV/kT}, \tag{21.17}$$

This is known as the junction law and says that the minority carrier density at the edge of the space charge region is enhanced by $\exp(eV/kT)$, where V is the applied forward bias. The excess hole density is

$$\Delta p_n = (p_n)_{x_p} - p_n = p_n e^{eV/kT} - 1. \tag{21.18}$$

This process of enhancing the minority carrier concentration at the edge of the space charge region is called minority carrier injection. The holes that were injected into the *n*-type material by the applied voltage will migrate toward the negative electrode and will be met by electrons drifting from the negative electrode. The holes are annihilated by the electrons so that $J = e(n\mu_n + p\mu_p)E$ remains constant.

When the bias is reversed as shown in Figure 21.6, the equilibrium is disturbed in the other direction. The increased potential forces minority holes from the *n*-side of the depletion zone into the space charge region where they are swept into the *p*-material resulting in a reverse current flow. However, in this case, the current is carried by the minority carriers so the reverse-biased current will be much less than the forward-biased current.

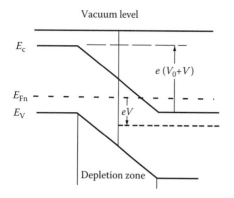

Vacuum level

E_c

$e\,(V_0 + V)$

E_{Fn}

E_V

eV

Depletion zone

FIGURE 21.6
Reversing the bias increases the electric field, which shuts down the flow of majority carriers, but now allows minority carriers to diffuse through the depletion zone carrying a small amount of leakage current in the direction of the reverse bias.

21.1.4 Current Flow

The net current flow through a *p–n* junction is directly proportional to minority carriers injected at the x_n edge of the depletion region. Therefore, from Equation 21.18, the response of junction device can be expressed as

$$I = I_0 \left(e^{eV/kT} - 1 \right), \tag{21.19}$$

where I_0 is the limit of the leakage current due to the minority carriers flowing down the potential gradient under reverse bias. However, because minority carriers have a limited lifetime τ before they are annihilated by a majority carrier, their velocities are restricted to their diffusion speeds given by D/L, where D is their diffusion coefficient and L is the diffusion length which is given by $\sqrt{D\tau}$.

The leakage current I_0 can be expressed as

$$I_0 = e \left(\sqrt{\frac{D_P^n}{\tau_P^n}} n_p + \sqrt{\frac{D_n^p}{\tau_n^p}} p_n \right) = e n_i^2 \left(\frac{1}{N_A} \sqrt{\frac{D_P^n}{\tau_P^n}} + \frac{1}{N_D} \sqrt{\frac{D_n^p}{\tau_n^p}} p_n \right), \tag{21.20}$$

where
   D_P^n is the diffusion coefficient of holes in the *n*-material
   D_n^p is the diffusion coefficient of electrons in the *p*-material
   τ_P^n and τ_n^p are the lifetimes of the respective minority carriers

These lifetimes are extremely short because there are so many majority carriers to combine with. Neither the diffusion coefficients nor the lifetimes are particularly temperature dependent, but $n_i^2 \sim \exp\left(-E_g/kT\right)$. The leakage current can be minimized by selecting a high bandgap material and by heavy doping. Typical values for the leakage current density J_0 are $O(10^{-6}\ \mathrm{A/m^2})$ at ambient temperatures.

Equation 21.20 is known as the rectifier equation because of the fact that a large current can flow if V is positive (*p*-side made positive or forward biased), but only a small current results as V is made negative (reverse biased) as shown in Figure 21.7. Circuit elements that utilize *p–n* junctions are known as diodes. They find use as rectifiers, blocking devices for preventing unwanted reverse current flow, and even as logic devices.

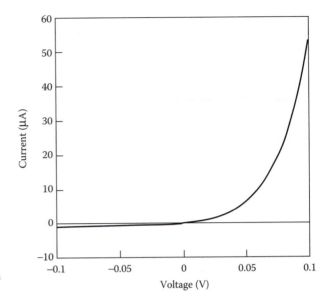

FIGURE 21.7
Typical *I–V* response curve of a diode with $I_0 = 10^{-6}$ A.

21.1.5 Zener Effect

As the reverse bias is increased, the reverse current approaches $-I_0$ asymptotically. However, if the reverse voltage exceeds a certain value, a breakdown occurs and a large reverse current will result. This can be due to two possible effects. One, if the electric field in the space charge region becomes high enough so that the leakage current carriers are accelerated to the point where they cause secondary ionization, an avalanche effect occurs in which the number of carriers are greatly multiplied. In the second case, the field in the space charge region can become high enough to directly excite valence electrons to the conduction band, thus producing a large number of carriers. In either case, this breakdown is known as the Zener effect. In designing blocking diodes for isolation or rectification, this should be avoided. However, the breakdown is predictable and can be controlled by device design; hence, it can serve as a voltage reference in power supplies or other systems. Diodes used in this manner are known as Zener diodes.

21.1.6 Schottky Diodes

When a metal is brought into intimate contact with a *n*-type semiconductor with a higher Fermi level (or more precisely, $\phi_m > \phi_n$) electrons from the conduction band in the semiconductor will diffuse into the metal giving it a negative charge and leaving behind positively charged holes in the depletion region. The charge separation creates a field that inhibits additional electron flow after equilibrium is reached and the Fermi levels coincide as illustrated in Figure 21.8.

This type of junction is known as a Schottky barrier diode or a Schottky diode. The contact potential is the difference between the two Fermi levels or $eV_0 = \phi_m - \phi_s$. The field created by the charge separation is represented by the upturn of the conduction band at the junction as seen in Figure 21.8. The potential barrier the electrons in the metal will have to overcome in order to flow back into the semiconductor is $\phi_m - \chi_n$ where χ_n is the electron affinity or the potential required to lift an electron from the conduction to the vacuum state. If the equilibrium is disturbed by forward biasing with a voltage V, the current flow will be

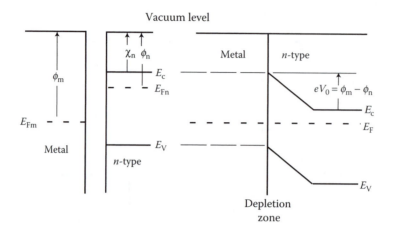

FIGURE 21.8

Band structure at the interface of an n-type semiconductor and a metal with a lower Fermi level. Electrons will flow from the conduction band into the metal giving it a negative charge, which results in a contact potential V_0. The resulting electric field limits the electron diffusion just as in a p–n junction.

proportional to $J \sim \exp -[e(V_0 - V)/kT]$. Since $J = 0$ when $V = 0$, the current flow can be written as

$$J \sim e^{-e(V_0 - V)/kT} - e^{-eV_0/kT} = J_0 \left(e^{eV/kT} - 1 \right), \qquad (21.21)$$

which is similar to Equation 21.19 for a p–n diode.

Thus an n-metal junction operates much the same as a p–n junction except when the metal is forward biased, charge carriers enter the metal directly from the conduction band of the n-type material rather than by minority carrier injection as occurs in a p–n junction, which means the current is conducted by majority carriers rather than by injected minority carriers.

A Schottky diode can also be made using a p-type semiconductor with a metal interface if $\phi_m < \phi_p$. In this case, the holes flow from the valence band of the semiconductor into the metal (or the electrons flow from the metal into the valence band filling the holes) leaving the metal with a positive charge. The operation of the p-metal diode is much the same as the n-metal diode except it conducts in the opposite direction. The negative terminal of the battery supplies electrons to the metal where they flow into the p-semiconductor to be annihilated by the hole flow supplied by the positive terminal of the battery. Again the current is carried by majority carriers.

Since the switching time of a diode is limited by the lifetime of the minority carriers, the switching time for Schottky diodes is limited only by the capacitance of the depletion region, making them more suitable for high frequency switching operations. (Schottky diodes were used as detectors in radar and microwave receivers until the development of the GaAs transistor in the 1980s.) Another advantage of Schottky diodes is that one of the active elements is a metal surface that aids in removing heat in high power applications.

21.1.7 Ohmic Contacts

In order to make useful devices, it is necessary to be able to make metal contacts with semiconductors without forming Schottky barriers so the contacts would not act as diodes. Rectifying junctions can be avoided in n-metal contacts by selecting metals with Fermi levels higher than those of the n-semiconductor ($\phi_m < \phi_n$). When these materials are

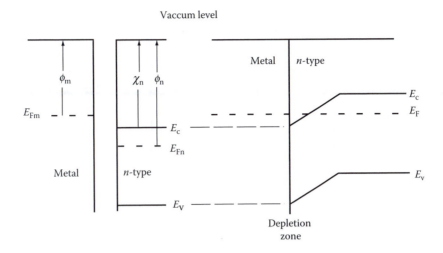

FIGURE 21.9
Ohmic contact at an *n*-metal interface in which $\phi_m < \phi_n$. Here electrons can flow directly from the metal into the *n*-semiconductor and vice versa and no rectification takes place.

brought together, electrons will flow from the metal to the semiconductor creating a layer of negative charge at the semiconductor interface. The resulting field bends the conduction band down allowing the electrons from the metal to flow directly into the conduction band as illustrated in Figure 21.9. Similarly, ohmic junctions can be made in *p*-type semiconductors by selecting metals in which $\phi_m > \phi_p$.

Of course it is not always possible to select metals for the interconnects that have the desired Fermi levels. As an alternative, heavy doping moves the Fermi level closer to the conduction band in *n*-type semiconductors as we saw in Chapter 20. This reduces the Schottky barrier as well as the width of the depletion region so that electrons can tunnel through the barrier. An even better alternative, if possible, is to dope heavily enough to move the Fermi level into the conduction band. Then the semiconductor becomes metal-like and the barrier is eliminated. Similar arguments apply to doping *p*-type semiconductors so that the Fermi level moves into the valance band.

21.2 Applications of Diodes

The nonlinear response of a diode makes analyzing the current flow in a circuit containing a diode somewhat more involved than a circuit with pure resistive elements because the effective resistance of the diode depends on the voltage. For small voltages, the effective resistance of a diode at ambient temperature $(kT/e = 0.025 \text{ V})$ can be written as

$$R(V) = \frac{V}{I_0(e^{eV/kT} - 1)} \approx \frac{kT}{eI_0} - \frac{V}{2I_0} = \frac{1}{I_0}(0.025 - V/2), \quad -\frac{kT}{e} < V < \frac{kT}{e}. \tag{21.22}$$

For large negative voltages, $R(V) \sim -V/I_0$. For $V \gg -0.025$,

$$R(V) = \frac{Ve^{-V/0.025}}{I_0}, \tag{21.23}$$

which effectively becomes 0 for $V > 0.5$ V.

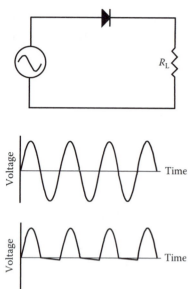

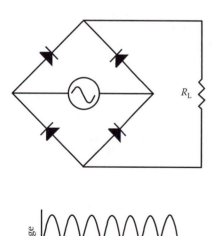

FIGURE 21.10
Simple half-wave rectifier circuit (upper). The symbol for a diode is an isosceles triangle with a line at its vertex. The alternating voltage supplied by the source (middle) is rectified by the diode, so that the voltage at the load R_L (lower) is just the positive half of the source waveform with a small negative voltage due to the leakage current when the applied voltage is reversed.

21.2.1 Diode as a Rectifier

A circuit element that has a large resistance with a negative applied voltage and virtually zero resistance with a positive applied voltage makes an ideal rectifier for converting AC to DC. Figure 21.10 shows a simple half-wave rectifier circuit.

A full-wave rectifier can be constructed by using a bridge circuit as in Figure 21.11. For a sine wave input, the average voltage delivered to the load is $1/\sqrt{2}$ of the peak voltage. The diode plate in the alternator in your automobile is a bridge rectifier to convert the AC from the alternator to DC in order to charge the battery.

Historically, rectifiers using stacks of Cu-Cu_2O plates were in use long before the mechanism of the Schottky barrier was understood.

FIGURE 21.11
Full-wave rectifier using a bridge circuit. The bridge circuit provides positive voltage to the load regardless of the polarity of the source voltage.

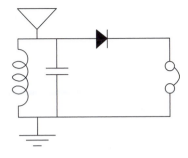

FIGURE 21.12
Diagram of a simple crystal receiver for AM broadcasts. The carrier frequency is picked up at the antenna and appears across the *L–C* tank circuit. The diode clips the negative half of the carrier waveform and the headphones respond to the amplitude of the audio waveform that was impressed on the carrier waveform.

21.2.2 Diodes as Demodulators in AM Radios

Diodes can be used as demodulators or detectors in AM (amplitude modulated) radios. Figure 21.12 is a diagram of a simple crystal receiver that kids used to build. AM radio modulates the amplitude of a carrier frequency (550–1650 kHz) with the audio signal. The carrier wave is picked up through the antenna and goes to ground through the *L–C* tank circuit. The *L* and *C* in this circuit are chosen to resonate at the carrier frequency of the particular station of choice, which causes the oscillating voltage of this frequency to appear across the tank circuit. The diode detects the audio signal by cutting off the negative portion of the carrier waveform. The inductance of the headphones is such that they cannot respond to the higher frequency carrier, so one hears just the audio signal.

This type of radio is called a crystal set because, before you could buy good Si *p–n* diodes, one would go to a radio shop and buy a natural occurring galena (ionic form of PbS) crystal mounted in a conducting base and a cat's whisker, which was a piece of spring wire to make contact with the crystal. We were actually making a point contact Schottky diode, but we did not know it at the time. Point contact Schottky diodes are still used as detectors and mixers in microwave deices because of their high switching speed.

21.2.3 Applications in Logic Circuits and ROMs

Diodes and resistors can be used to represent simple Boolean logic in computer circuits. Figure 21.13 shows how an OR and an AND circuit can be constructed.

Diode matrices were used as ROMs (read-only memories) in early computers. The rows and columns represented address lines and the intersections represented bits. A simple lookup table to convert decimal to binary is shown in Figure 21.14.

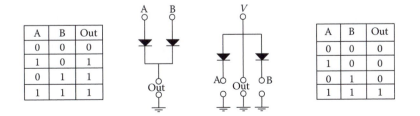

A	B	Out
0	0	0
1	0	1
0	1	1
1	1	1

A	B	Out
0	0	0
1	0	0
0	1	0
1	1	1

FIGURE 21.13
Logical OR circuit (left) and AND circuit (right) with their corresponding truth tables. The applied voltage *V* is typically 5 V and a logical "1" is 5 V and a logical "0" is ground.

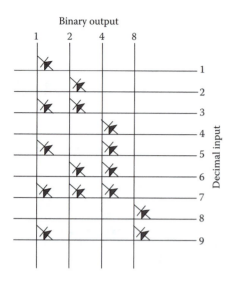

Binary output

FIGURE 21.14
Decimal to binary converter matrix using diodes as a ROM. For example, an input on the "6" line will produce "0110" on the output lines.

21.3 Tunnel Diode and Negative Resistance

When both n- and p-regions are heavily doped so that the Fermi level is below the valence band in the p-side and above the conduction band on the n-side, the depletion region is very thin and the internal fields are very high. When a small forward bias is applied, there are unoccupied states in the valence band that the conduction electrons in the n-type material can reach by tunneling through the depletion region, thus producing a current flow (Figure 21.15). A reverse bias causes a reverse tunneling current to flow as seen in Figure 21.16. However, when the forward bias raises the conduction band in the n-type material above the valence band in the p-type material, tunneling is no longer possible and the junction behaves as a regular diode. This produces a region in the I–V curve in which the current decreases with increasing voltage, as shown in Figure 21.16. Such a behavior is the same as if the resistance had become negative. As the current decreases, the voltage across the device decreases causing the current to increase resulting in an oscillating current and voltage. Instead of electrons giving energy to the lattice as heat, the lattice is giving energy to the electrons in this region and a device operation in this region can be used as a microwave amplifier or oscillator.

21.3.1 The Gunn Effect

While J.B. Gunn was studying the electrical properties of n-GaAs, at IBM in 1963, he observed that high frequency oscillations broke out when the applied voltage reached a threshold of $\sim$25 V, signifying a region of negative resistance similar to the I–V curve in Figure 21.16. Recall that GaAs is a direct bandgap material and in addition the direct bandgap at 1.35 eV, there are secondary valleys that are 0.36 eV above the bottom of the central valley as seen in Figure 21.17. At low fields most of the conduction electrons are in the central valley, but as the field is increased, some of these electrons will spill over into the satellite valleys due to intraband transitions. The effective mass is much greater in the secondary valleys; hence, the mobility is decreased. The decrease in mobility causes less current to flow as the voltage is increased, giving rise to the regions of negative resistance responsible for the oscillations observed by Gunn. GaAs Gunn oscillators can operate up to $\sim$200 GHz while GaN Gunn oscillators can reach terahertz frequencies.

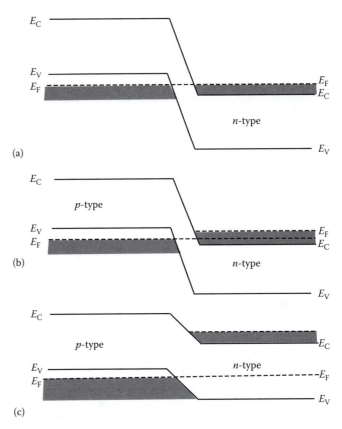

FIGURE 21.15
Unbiased tunnel diode (a). Small forward bias (b) causes tunneling current flow. When forward bias raises E_C in the n-material above E_V in the p-material (c), tunneling stops and the device becomes an ordinary diode.

Gunn-effect oscillators (or Gunn diodes as they are frequently called, although technically, they are not diodes) in a resonant cavity can be used to make very simple low power microwave oscillators. Handheld radar speed detectors use Gunn-effect oscillators as their microwave source.

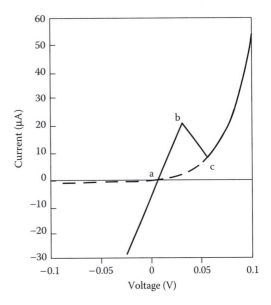

FIGURE 21.16
I–V curve for a tunnel diode. Tunneling current increases with forward bias from a to b. When tunneling stops, the current drops to c. The region b–c exhibits negative resistance.

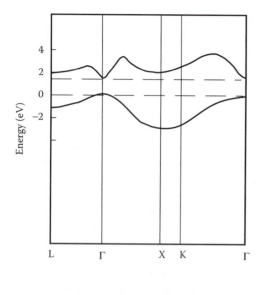

FIGURE 21.17

Band structure of GaAs in the vicinity of the energy gap (adapted from Figure 20.4). Note the shallow secondary valley (X) to the right of the steep valley centered at the direct bandgap (Γ). Increasing the applied voltage moves high mobility electrons through intraband transitions from Γ to X where their mobilites are lower. Thus, increase in the voltage results in decreased current flow and negative effective resistance.

21.4 Light-Emitting Diodes

We saw previously that forward biasing a *p–n* junction causes electrons and holes to flow into the junction region where they recombine. In a direct bandgap material, this recombination will produce a photon of light, which is the basis for light-emitting diodes (LEDs) and solid-state lasers. In indirect bandgap materials, such as Si and Ge, phonons are required for recombination and the recombination energy goes into heating the crystal.

21.4.1 Development of Light-Emitting Diodes

The earliest LEDs were made from GaAs, which emits in the infrared. In 1962, Nick Holonyak Jr. invented the first practical visible-spectrum LED by alloying P with GaAs to raise the bandgap into the visible region (see Figure 22.12). Later yellow and green LEDs became available as researchers tinkered with alloy-type compound semiconductors. However, a blue LED remained a difficulty until recently. SiC has the required bandgap, but unfortunately SiC is an indirect bandgap material. The breakthrough came in the 1990s when Shuji Nakamura developed the two-flow metallorganic chemical vapor deposition (MOCVD) reactor, which allowed him to grow InGaN and AlGaN epitaxial layers with precise chemical control. InGaN is a direct bandgap material, which accounts for the brightness of the new blue–white LEDs.

Nakamura had to overcome a number of problems in order to develop this technology. The LEDs are grown on sapphire substrates, which have a 15% lattice mismatch with the InGaN material. The resulting lattice strain produced severe dislocations and even cracking when the material cooled from the growth temperature. Nakamura solved this problem by growing a buffer layer of AlGaN between the sapphire and the GaInN active layer. (It would be more desirable to grow the devices on GaN substrates, but high quality substrates of this material are not available.) However, it was the two-flow MOCVD growth reactor that provided the real breakthrough for producing these brilliant blue-white LEDs.

Because of their high brightness, efficiency, and long life, LEDs are rapidly replacing incandescent bulbs in many applications. Use of LEDs in traffic lights results in 50%–70% savings, primarily in bulb replacement costs. LEDs need to be changed in 5–10 years, as opposed to 1 year for conventional bulbs. In another application, the use of blue–white

lasers for CD and DVD will allow four times the storage of information because of the smaller diffraction-limited spot size.

21.4.2 OLED and PLEDs

We saw in Chapter 20 that some organic molecules and polymers can be doped to make them either p-type or n-type semiconductors. Therefore, it should be possible to make a p–n junction and, if there is a direct bandgap, it should also be possible to make a small molecule organic light-emitting diode (OLED) and a polymer light-emitting diode (PLED).

In 1987, Tang and VanSlyke fabricated the first OLED using a sandwich of indium tin oxide (ITO)-coated glass, diamine, Alq_3 (tris 8-hydroxyquinoline aluminum(III)), and a Mg-10%Ag cathode. The polymer is dip-coated or spin-coated onto the ITO-coated glass to form a 20–40 nm layer. A 50–70 nm layer of Alq_3 along with 200 nm of the cathode material is then vacuum deposited onto the polymer layer. The Alq_3 serves as the emitter and electron transport material and the diamine is the hole transport material. The holes and electrons combine at the interface to produce photons of visible light. The device had a quantum efficiency of 1% and a brightness of 1000 cd/m^2 with a bias of 10 V. Other charge transport polymers such as polyvinylcarbazole (PVK) can be used as the emitter layer.

Later a Cambridge group made a single-layer OLED from polyphenylene vinylene with a Ca cathode that emits yellow light with a brightness of 500 cd/m^2 with an 8 V bias.

The potential advantages offered by light-emitting organic and polymers for flat screen displays and large electroluminescent (EL) panels for area lighting have spurred intensive research in this area in recent years. Unlike liquid crystal displays, OLEDs do not require backlighting. Also, they are relatively inexpensive to produce.

Small molecule OLEDs, are fabricated by vacuum-depositing molecules with conjugated electronic structures onto suitable substrates using effusion cells. PLEDs may be fabricated by dip-coating or spin-coating the polymer or its precursor onto a substrate. OLEDs seem to have the advantage of being able to produce brighter and more vivid colors, while PLEDs have the advantage of simpler fabrication, particularly where large surfaces are required.

The heart of the device can either be a single layer or a bilayer of organic molecules or polymers. In the bilayer, one layer transports electrons and serves as the emitter while the other layer conducts the holes. The electron–hole recombination takes place at the interface between the two layers. In single-layer devices, the metal cathode acts as a Schottky diode, injecting electrons into the organic material. The holes migrate from opposite sides of the film and recombine near the cathode. Bilayer devices are generally brighter and more efficient. One of the electrodes must be transparent and conductive. Glass coated with ITO is the material of choice for this purpose. It is used as the substrate for the fabrication of the device and serves as the anode. The cathode is usually in the form of a deposited metallic film using a metal with a low work function to facilitate electron injection.

One of the difficulties that had to be overcome in the development of light-emitting organic molecules was the low quantum efficiency. If a polymer that contains conjugated electrons is placed between two conducting plates and a potential is applied, electrons will be injected at the cathode plate and removed at the anode plate, leaving a hole. The electrons and holes are attracted to each other to form a bound electron–hole pair called an exciton. Since electrons have spin of 1/2, depending on how the spins of the electron–hole pair are aligned, the exciton may have net spin of zero ($e{\uparrow}h{\downarrow}-h{\uparrow}e{\downarrow}$), which is called a singlet state, or 1 ($e{\uparrow}h{\uparrow}$, $e{\downarrow}h{\downarrow}$, and $e{\uparrow}h{\downarrow}+h{\uparrow}e{\downarrow}$), which is called a triplet state. Singlet excitons have an allowed radiative transition with lifetime of ~1 ns, which is responsible for the luminance. Radiative decay of triplet excitons is forbidden; consequently, they dissipate their energy through phonon interactions in the form of heat. Unfortunately,

the formation of triplet excitons is favored by 3:1, so the quantum efficiency of an OLED material appeared to be inherently limited to less than 25%.

An important breakthrough came with the discovery that with the addition of phosphorescent dopant emitters such as Pt, Ir, Os, etc., the energy could be transferred from the triplet states to the dopant molecules which could decay radiatively. As a result, the internal quantum efficiency (the number of photons generated inside the device per electron–hole pair injected) of these phosphorescent organic light-emitting diodes (PHOLEDs) (PHOLED is a trademark of the Universal Display Corporation, Ewing, New Jersey) can approach a quantum efficiency 100% at a luminescence of 100 cd/sr.

Despite the gains in internal quantum efficiency, the overall efficiency is still less than 19%. As much as 80% of the photons produced cannot get out because of the high index of refraction of the polymer material. Approaches such as texturing the surface of the glass, index matching or using low index substrate materials, and the use of arrays of microlenses have made increases in the extraction efficiency.

OLED flat and flexible panel display technology is advancing rapidly and full color displays are currently being used in cell phones. Sony recently announced a 2.5 in. flexible screen TV that is only 0.3 mm thick and now has an 11 in. OLED TV in production. Samsung also announced a prototype 17 in. high definition (1600×1200 pixels) active matrix OLED display panel.

Flexible PLED electro luminescent (EL) panels are now available from several manufacturers. Research is still underway to improve the color, durability, and efficiency of the EL panels. A replacement for the ITO-coated glass or plastic, which presently is the only practical hole injecting material, is being sought. ITO is brittle and can be easily cracked. The ITO glass has a transmission of only 0.85, so 15% of the light is lost at the window. However, the biggest problem with ITO is that its conductivity is too low to deliver power to large areas.

21.4.3 Solid-State Lasers

A laser is a little more complicated than a LED in which an inverted population of excited states must be created so that stimulated emission can occur. One way of accomplishing this is by doping the Fermi level on the *n*-side above the conduction band and the *p*-side below the valence band as shown in Figure 21.18a. Now when the junction is biased, the filled states in the *n*-materials are lifted above the empty states in the *p*-material and the population is inverted as seen in Figure 21.18b. The battery continues to supply electrons to the *n*-material to replace those that fell into the unfilled state by emitting photons with energy equal to the difference between the conduction band and the unfilled states near the valence band.

In order to get stimulated emission and lasing action, we still need a resonant cavity. The GaAs single crystal is carefully cleaved on each end to form the cavity. The high dielectric constant of the GaAs causes a fairly large reflection of the radiation that is trying to be emitted, thus assuring sufficient photons remain in the cavity to stimulate the emission of other photons to produce the coherent laser radiation.

21.5 Photodiode

Photodiodes convert photons into electrical current and are used as power sources as well as photodetectors. Recall that the internal fields setup in the depletion region sweep holes toward the *p*-region and electrons toward the *n*-region. Photons above the energy gap absorbed in or near the depletion region produce the electron–hole pairs, which are swept

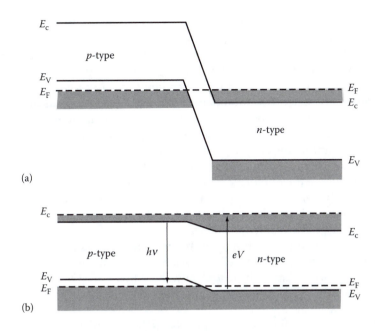

(a)

(b)

FIGURE 21.18

Schematic showing the operation of a solid state laser. In the unbiased state (a) the Fermi level lies above the conduction band in the n-side and below the valence band on the p-side. (b) Biasing the junction lifts the filled states in the conduction band in the n-region above the above empty states below the valence band in the p-side, thus inverting the electron population.

to their respective regions where they may be collected by an external circuit. As opposed to photoconductive detectors whose conductance depends on the photon flux, photovoltaic (p.v.) detectors produce a voltage that is proportional to photon flux.

21.5.1 Solar Cells

Solar cells that convert sunlight directly into electricity work on the same principle as shown in Figure 21.19. The top electrode can be a web of thin wires to collect the charge or a transparent conductor such as ITO.

The equivalent circuit of a solar cell consists of current source from the incident photons in parallel with a forward-biased diode as shown in Figure 21.20. The circuit current produced by the cell can be expressed as

$$I_{circuit} = -I_0\left(e^{eV/kT} - 1\right) + eAG(L_p + L_n) = -I_0\left(e^{eV/kT} - 1\right) + I_{photo}, \tag{21.24}$$

FIGURE 21.19

Schematic of the structure of a typical solar cell. Electrons and holes created by the photons are swept by the internal field in the depletion region toward the negative and positive electrodes respectively, thus creating a source of current flow.

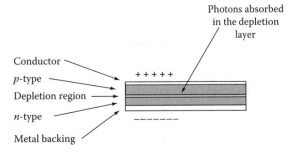

Photons absorbed in the depletion layer

Conductor

p-type

Depletion region

n-type

Metal backing

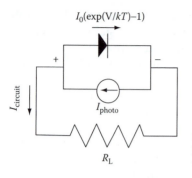

$$I_0(\exp(V/kT)-1)$$

FIGURE 21.20
Equivalent circuit of a solar cell as a current source in parallel with a forward-biased diode.

where

    I_0 is the leakage current

    A is the exposed area of the cell

    G is the rate per volume at which photons produce electron–hole pairs

    L_p and L_h are the diffusion lengths of the minority carriers.

Some texts define a photodiode current I_{pd} that flows in the direction of the current in the diode or $I_{pd} = -I_{circuit}$ as shown in Figure 21.21. The operating point is in the lower right hand corner and the negative *I–V* product signifies power being generated.

In the absence of a load, the photocurrent increases the voltage across the cell until it is balanced by the current flowing through the forward-biased diode. The open circuit voltage

$$V_{oc} = \frac{kT}{e} \ln\left(\frac{I_{photo}}{I_0} + 1\right). \tag{21.25}$$

The short-circuit current is obtained by setting $V = 0$ in Equation 21.22, which gives $I_{sc} = I_{photo}$.

High quality single crystal Si solar cells have a theoretical efficiency of ~24%, but the losses drop the actual efficiency to ~14%–17%. For terrestrial use, the noonday sun is considered to deliver 1000 W/m^2, so a 14% efficient cell would deliver 140 W into a

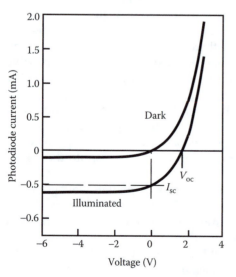

FIGURE 21.21
I–V curve for a solar cell in the dark and when illuminated. The operating point of a solar battery is in the fourth quadrant where the negative *I–V* product indicator power being generated.

matched load. Single crystal Si is expensive to produce since wafers must be sawed from Czochralski-grown ingots. It is much more cost effective to use polycrystalline or amorphous Si and accept the reduction in efficiency due to the shortened carrier lifetimes in these cheaper materials. Polycrystalline cells have efficiencies that range from 13% to 15% and amorphous cells are only 5%–7% efficient.

The primary factor that reduces the efficiency of a solar cell is the fact that photons with energies less than the bandgap cannot produce electron–hole pairs. We saw in Figure 20.9 that even though the absorption edge for Si is 1.1 eV, the absorptivity increases fairly slowly until around 3–4 eV because of the indirect bandgap, thus ~30%–50% of the available spectrum is lost to begin with. Materials with lower bandgaps can be used, but they are not as efficient in capturing the shorter wavelength radiation.

One solution, although by no means a cheap alternative, is to make compound cells. A stack of cells starts with a high bandgap cell on the sun side to tap the higher energy photons. Photons with energies below the bandgap of the high bandgap cell are not absorbed and go on the second and third layers where the lower energy photons get absorbed. For example, Spectrolab, Inc. manufactures a triple junction cell for use in spacecraft that sandwiches p–n $GaInP_2$/GaAs/Ge cells in a stack. The $I_{SC} = 170$ A/m^2 and the $V_{OC} = 2.665$ V. The power delivered to a matched load is 392 W/m^2 for an efficiency of 28% (assuming the full unattenuated solar input of 1400 W/m^2) is incident on the cell. As one might imagine, such cells are extremely expensive to manufacture and are generally used only in space applications where weight and transportation costs can justify the manufacturing cost of such cells.

Another alternative is to use mirrors or to concentrate the sun on the cells. Recently a team led by Allen Barnett (University of Delaware) announced an efficiency of 42.8% obtained by concentrating 20 suns onto a beam splitter that divided the solar input into three spectral bands and imaged each of these bands onto a cell optimized for that band.

21.6 Summary

Since the Fermi level is moved closer to the conduction band in n-type semiconductors and closer to the valence band in p-type semiconductors, bringing these materials into intimate contact will cause electrons to flow from the n-type into the p-type material in order to equalize their Fermi levels. The electrons that diffuse into the p-type material create a depletion zone in the vicinity of the junction where there are no free carriers. The annihilation of an electron–hole pair leaves a negative charge in the p-material and a positive charge in the n-material. This charge separation creates a contact potential and an electric field in the depletion zone.

Equilibrium is reached at the contact potential V_0 when the field in the depletion zone balances the diffusion current. The field is in the direction of the p-type material, so any hole that happens to be in the depletion region is swept toward the p-type material. Likewise any electron that drifts into the depletion region will be swept into the n-type material.

Reducing the contact potential by applying a forward bias or positive voltage to the p-material upsets the equilibrium and increases the flow of holes, resulting in current flow through the circuit. The current flow requires electrons to leave the n-type material and flow through the junction, leaving excess holes in the vicinity of the depletion zone. These holes will migrate toward the negative electrode until they recombine with the electrons provided by the current source, which also replenishes the holes in the p-type material.

This process is called minority carrier injection and the time before these minority carriers recombine is the minority carrier lifetime.

When the bias is reversed, the minority carriers are driven into the depletion region by the applied potential resulting in a reverse current, but since the reverse current consists of the minority carriers, it will be much smaller than the forward-biased current. The circuit response is described by the rectifier equation, $I = I_0[\exp(eV/kT) - 1]$, where I_0 is the leakage current from the minority carriers driven by the reverse bias potential. This non-linear response is useful for a number of circuit applications such as rectifiers to convert AC to DC.

Diodes may be constructed to break down at a given reverse potential allowing them to be used as voltage regulators. This type of diode is known as a Zener diode.

Diodes can also be formed by a metal–semiconductor junction. These diodes are called Schottky diodes. In order to make metal to semiconductor contacts that do not act as diodes, it is necessary to make the semiconducting material ohmic or metal-like in the vicinity of the contact by raising the doping level to put the Fermi energy in the conduction (or valence) band.

Using the negative resistance characteristics of tunnel diodes and Gunn diodes, simple low-cost oscillators are possible that have found their way into handheld radar units and other microwave applications.

LEDs can produce infrared or visible light from the recombination of carriers in the junction regions. By inverting the electron population and providing a resonant cavity, stimulated emission may be produced making solid-state lasers possible.

It is also possible for diodes to capture photons near the junction and produce voltage and current, thus making p.v. detectors and solar cells possible.

Bibliography

Hummel, R.E., *Electronic Properties of Materials*, 3rd edn., Springer-Verlag, New York, 2000.

Navon, D.H., *Electronic Materials and Devices*, Houghton Mifflin, Boston, MA, 1975.

Kwong, R.C. et al., High efficiency light emitters, *Org. Electron.*, 4, 2003, 155–164.

Livingston, J.D., *Electronic Properties of Engineering Materials*, Wiley, MIT series in Materials Science and Engineering, John Wiley & Sons, 1999.

Sapoval, B. and Hermann, C., *Physics of Semiconductors*, Springer-Verlag, New York, 1995.

Tang, C.W. and Van Slyke, S.A., Organic electroluminescent diode, *Appl. Phys. Lett.*, 51, 1987, 914.

Problems

Spectrolab makes a GaAs/Ge single junction solar cell that has an open circuit voltage $V_{OC} = 1.025$ V and a short-circuit current $J_{SC} = 30.5$ mA/cm^2 in full sun (0 air mass).

1. Find J_0.
2. What is the maximum power a 1 m^2 array of these cells could deliver to a resistive load?
3. What would the load resistance need to be in order to obtain maximum power?
4. Solar constant at zero air mass is 1354 W/m^2. What is the efficiency of this cell?

22

Transistors, Quantum Wells, and Superlattices

Before the invention of the transistor in 1947 at Bell Laboratories by John Bardeen, Walter H. Brattain, and William B. Shockley, vacuum tubes were the workhorse of electronics. Vacuum tubes were bulky and fragile, required high voltage (>45 V) and a heater current to operate, and had lifetimes comparable to electric light bulbs. It is certainly no exaggeration to say that the transistor made possible modern solid-state electronics. Transistors serve much the same function as the old vacuum tubes but are much smaller, rugged, operate on low voltages, and have virtually infinite lifetimes.

Once the purification and processing technology required to make transistors was developed, a virtual explosion of new concepts became possible resulting in a vast array of new electronic and photonic devices, a few of which will be described in this chapter.

22.1 Transistor Theory and Applications

22.1.1 Bipolar Junction Transistor (BJT)

A bipolar junction transistor (BJT) can be thought of as back-to-back n–p or p–n diodes to form an *npn* or *pnp* configuration as shown in Figure 22.1.

The symbols for *pnp* and *npn* transistors in circuit diagrams along with the current flow in their active state are shown in Figure 22.2.

The emitter and collector are basically the same, although, generally the emitter is more heavily doped than the collector in order to make the depletion zone between the emitter and base thinner than the depletion zone between the base and the collector. Also, the base–collector junction cross section is often made larger than the emitter junction since it is usually required to carry more current.

There are four basic modes of transistor operation:

1. Active mode; emitter forward biased, collector reverse biased
2. Saturated mode; both emitter and collector forward biased
3. Cutoff mode; both emitter and collector reverse biased
4. Inverse mode; emitter reverse biased, collector forward biased

The active mode is of most interest if the transistor is to be used as an amplifier. The saturated and cutoff modes are of interest if the transistor is to be used as a switch. The inverse mode is used only in very special applications and will not be considered here.

In the active mode, minority carriers are injected into the base region by forward biasing the emitter, just as in a forward-biased diode. Because of the narrow depletion zone

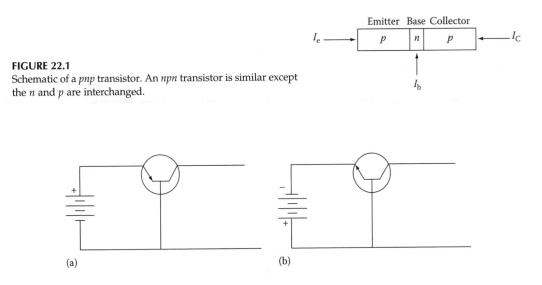

FIGURE 22.1
Schematic of a *pnp* transistor. An *npn* transistor is similar except the *n* and *p* are interchanged.

FIGURE 22.2
Standard symbols for (a) *pnp* and (b) *npn*. BJTs biased in the active or "on" state. The arrow is in the direction of the forward bias of the emitter–base portion.

between the emitter and base and aided by the reverse bias between the base and collector, most of the carriers are swept into the collector before they can recombine and contribute to the base current.

22.1.2 Common Base Amplifier

In the common base circuit shown in Figure 22.3, the current in the collector circuit is essentially the same as the current in the emitter circuit, which is controlled by the forward bias voltage. Let V_e be the forward bias voltage. The current in the emitter circuit is given by the junction equation:

$$I_e = I_{0e}\left(e^{eV_e/kT} - 1\right) \approx I_{0e}e^{eV_e/kT}, \quad eV_e \gg kT. \tag{22.1}$$

The current in the collector circuit is given by

$$I_c = I_{0c}\left(e^{eV_c/kT} - 1\right) + \alpha I_e \approx \alpha I_e, \tag{22.2}$$

where α is the fraction of the minority carriers that are swept to the collector. Since V_c is negative and $I_{0c} \ll \alpha I_e$, $I_c \approx \alpha I_e$.

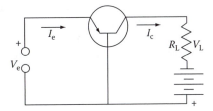

FIGURE 22.3
Common base amplifier circuit using a *pnp* transistor. An equivalent circuit can be constructed using an *npn* transistor by reversing the above polarities.

If a load is placed in series with the source providing the reverse bias to the collector circuit, the voltage across the load is

$$V_L = \alpha I_e R_L, \tag{22.3}$$

while the emitter voltage from Equation 22.1 is

$$V_e = \frac{kT}{e} \ln\left(\frac{I_e}{I_{0e}}\right). \tag{22.4}$$

The voltage gain is given by

$$\text{Voltage gain} = \frac{dV_L}{dV_e} = \frac{\alpha R_L dI_e}{(kT/e)(dI_e/I_e)} = \frac{\alpha R_L I_e}{(kT/e)}. \tag{22.5}$$

Let $R_L = 5$ KΩ, $I_e = 5$ mA, $kT/e = 0.026$ V, $\alpha \sim 1$, the voltage gain is 1000. In this case the voltage gain is determined by the load resistor and can be virtually any value.

22.1.3 Common Emitter Amplifier

There are varieties of ways to use junction transistors in circuits as shown in the following examples. The common emitter circuit shown in Figure 22.4 is widely used for amplifier as well as switching applications.

When the base is made positive relative to the emitter, the junction is forward biased and the transistor is turned on. The collector current is given by the αI_e and the current in the base circuit is $(1 - \alpha)I_e$. The ratio of collector to base current is

$$\frac{I_c}{I_b} = \frac{\alpha I_e}{(1 - \alpha)I_e} = \beta, \tag{22.6}$$

where the short circuit current gain β is defined as $\alpha/(1 - \alpha)$. Recall that α is the fraction of the injected minority carriers that make it to the collector and is a function of the design of the transistor, i.e., the width of the base region and the dopant levels that determine the width of the depletion zones. Therefore, the β is also a function of the transistor design and is used to characterize the transistor.

22.1.4 Basic Flip-Flop Circuit

Biopolar junction transistors are often used as switches and logic devices. Figure 22.5 is an example of a bistable or flip-flop circuit. When voltage is applied, either T_1 or T_2 will conduct slightly more than the other. If T_1 is conducting, the base or T_2 is driven to ground,

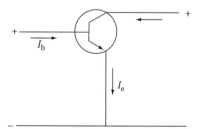

FIGURE 22.4
Common emitter current amplifier using an *npn* transistor.

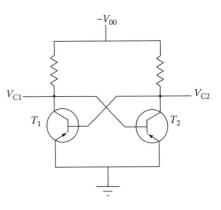

FIGURE 22.5
Basic flip-flop circuit using pnp BJTs. If T_1 is conducting, T_2 is turned off and vice versa. A negative pulse at V_{C1} will turn on T_2 and turn off T_1. This type of circuit is the basic element in counters and serves as the memory element in a SRAM.

shutting it off. Conversely, if T_2 is conducting, T_1 is shut off. So the circuit is bistable: either one or the other transistor is "on" and the other is "off."

If T_1 happens to be on, the state may be changed by applying a negative pulse to V_{C1}. This will cause T_2 to start conducting, shutting off T_1.

Flip-flop circuits can be used as data storage by assigning the state in which one transistor is conducting a logical 1 and the state when the other transistor is conducting a logical 0. They are also useful as binary counters. A series of flip-flops is wired in sequence so that when the first is cycled, it changes the state of the second flip-flop so that it counts every second pulse. The second flip-flop, in turn, changes the state of the next flip-flop so that it counts every fourth pulse, and so on. Thus n flip-flops can count 2^{n-1} pulses.

22.1.5 Differences between BJTs and Vacuum Tubes

BJTs are basically current-controlled conductors or current amplifiers in which large currents can be controlled by small currents. This feature is quite different from the vacuum tubes that the transistors began replacing in the 1950s. Vacuum tubes controlled the current flowing from the plate to the cathode by changing the voltage imposed on the grid of the tube as illustrated in Figure 22.6. As a result, virtually no current flows from the grid to cathode. Thus the vacuum tube is said to have high input impedance since the source voltage is sensed without drawing much current, just as though a resistor with a high resistance was connected across the source. The equivalent resistance of such a resistor is the input impedance of the system.

BJTs, on the other hand, have low input impedances which makes them unsuitable for applications such as charge-sensitive amplifiers in which it is necessary to sense a voltage with little or no disturbance to the system, or in systems such as digital watches or other applications where the flow of current must be held to a minimum.

FIGURE 22.6
Typical vacuum tube circuit. A high-voltage source ($B_+ = {\sim}100$ V) is connected to the plate of the tube through a load resistor, R_L. Electrons boil off of the cathode when heated by the heater filament and are attracted to the plate by the high plate voltage. The plate current, I_p, can be regulated by the grid voltage. A negative bias of a few volts on the grid, V_g, repels the electrons trying to get to the anode and shuts off the plate current. The changes in plate current resulting from a small signal imposed on the grid is seen as a much larger voltage appearing across the load resistor.

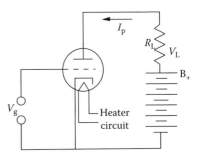

22.2 Field Effect Transistors

Field effect transistors (FETs) work on an entirely different principle than junction transistors. These devices are sometimes called unipolar devices since only one type of carrier is involved. Although they are much simpler and faster than junction transistors, their development did not begin as early because specialized methods and controls had to be invented for their manufacture.

The term field effect is derived from the fact that the current flow between the source and the drain is modulated by a field imposed by a voltage applied to the gate, very much as the electron flow from the cathode to anode of a triode vacuum tube is modulated by a voltage applied to the grid. The gate can either be a junction or it can be insulated with a metal oxide. Since virtually no current is drawn from an insulated gate circuit, such devices have extremely high input impedances (in excess of 10^{14} Ω) and can be used as electrometers to measure the amount of charge without significant draining of the charge.

FETs can be categorized as p-channel or n-channel depending on which carriers are used and as enhancement-type, which are nonconductive in their normal state, and depletion type, which are conductive in their normal state. We will consider the enhancement-type first.

22.2.1 Enhancement-Type FETs

A typical p-channel enhancement-type metal oxide semiconducting FET (MOSFET) is shown in Figure 22.7.

With no voltage applied to the gate, the p^+n junctions at the source and the drain act as back-to-back diodes and no conduction takes place. Applying a negative bias to the gate attracts minority holes in the n-type material in the vicinity of the gate. When the negative bias is high enough, the material in that region becomes p-type and a conductive path between the source and drain is established.

22.2.2 Depletion-Type FETs

A depletion-type MOSFET, shown in Figure 22.8, has an existing p-channel connecting the source and drain so that it conducts with no voltage applied at the gate. Also note that the gate region is shorter than for the enhancement-type MOSFET. Now a positive voltage applied to the gate will repel the holes from the p-channel and reduce the conductive path.

Because of the insulated gate, MOSFETs have extremely high impedance $\sim 10^{14}$ Ω. This makes them useful for electrometers and other charge-sensitive circuits. However, they are also highly susceptible to static electricity that can punch through the thin insulated gate and ruin the device. Once installed, they are protected, but extreme care must be exercised to prevent static discharges when handling them.

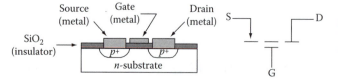

FIGURE 22.7
Schematic of a p-channel enhancement-type insulated gate MOSFET. The circuit symbol is shown to the right. The arrow is reversed for n-channel.

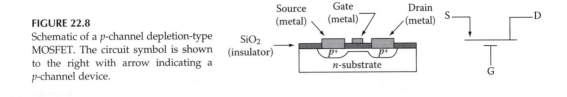

FIGURE 22.8
Schematic of a *p*-channel depletion-type MOSFET. The circuit symbol is shown to the right with arrow indicating a *p*-channel device.

22.2.3 Complimentary Metal Oxide Semiconductors

Of course it is possible to reserve the polarities of these field-effect devices by exchanging the *n*- and *p*-materials. For many applications it is desirable to operate *n*- and *p*-devices side-by-side on the same substrate. This is accomplished by *p*-doping a region in an *n*-type substrate large enough to accommodate an *n*-channel device so that it may work in parallel with an adjacent *p*-channel device. Such a combination is called complimentary metal oxide semiconductors (CMOS).

The flip-flop circuit described previously is the basic digital counter and makes wide use of CMOS technology in applications such as digital watches where current drain must be held to a minimum. A quartz oscillator provides the basic clock whose pulses are repeatedly divided by two by a series of flip-flop circuits in order to count seconds, minutes, and hours.

CMOS flip-flop circuits also form the basis of static random access memories (SRAMs) which store each bit as the state of one of the two transistors in a flip-flop circuit.

22.2.4 Junction Field Effect Transistors

Another type of FET is the junction field effect transistor (JFET) shown in Figure 22.9. The source and drain are connected by an *n*-channel, which is sandwiched between two p^+ regions. The gate is not insulated, but contacts the p^+ material as does the metal back.

The gate and metal back are connected together. With no bias on the gate, conduction takes place through the *n*-channel. Applying a negative bias to the gate reverse biases the np^+ junctions, which widens the depletion region at the junction and narrows the conductive path and eventually "pinches off" the *n*-channel when the negative bias is high enough. The circuit symbol is shown to the right.

FETs lend themselves to very large scale integration (VLSI) circuits and with their low power requirements, they have found a wide variety of applications, especially in computers and related devices.

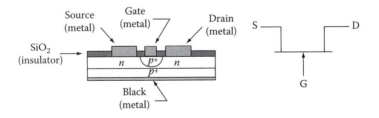

FIGURE 22.9
Schematic of an *n*-channel JFET. The circuit symbol to the right indicates a *p*-type gate for this *n*-channel device. The arrow is reversed for a *p*-channel JFET.

22.3 Random Access Memory

Random access memory (RAM) allows a computer to store and retrieve data instantly anywhere in its storage system (as opposed to disc memory which must wait for the disc to rotate to where the desired information is stored). The advent of very large quantities of very fast RAM has made today's personal computers possible.

22.3.1 Static Random Access Memory

There are two general types of RAM, static RAM (SRAM), and dynamic RAM (DRAM). SRAM is based on the flip-flop circuit discussed previously. Two cross-coupled transistors comprise the basic storage unit, a logical "1" if one transistor is "on" and a logical "0," if the other is "on." Two more transistors sense the state, and two more transistors control the access to the storage location, so six transistors are required to store one bit of information. Access time ranges from 10 to 30 ns. The term "static" refers to the fact that the information remains stored as long as the system is powered. When power is removed, the state of the flip-flops becomes indeterminate; hence, data is lost unless stored on the hard drive. Such a memory is called a volatile memory.

22.3.2 Dynamic Random Access Memory

DRAM stores information as charge on a capacitor. Since the charge can leak out of the capacitor, the memory has to be periodically refreshed, generally once every 64 μs, which is the reason it is called dynamic. Like SRAM, DRAM is volatile, meaning that data will be lost if power is removed. DRAM is slower than SRAM and requires more complicated circuitry because of the refresh requirement, but the memory is much smaller and cheaper than SRAM because only one transistor and one capacitor is required to store a bit of information. A schematic of a typical DRAM memory is shown in Figure 22.10.

22.3.3 Flash Memory

The recent advent of large capacity nonvolatile flash memory cards or sticks has made digital cameras and portable computer memories possible. A flash memory is essentially an electronic, erasable, programmable read-only memory (EEPROM). The basis of operation of EEPROMs is a process known as Fowler–Nordheim tunneling. A second gate, called a control gate, is situated above and separated from the regular gate of a MOSFET by a thin oxide layer. Thus the regular gate is not directly connected to anything and acts as a floating gate. The control gate is connected to the bit line. A logical "1" is written into the floating gate by applying a negative high voltage to the control gate which forces electrons to tunnel through the oxide layer. When the voltage is removed, the electrons are not able to escape so they simply stay there, causing the FET to either turn "on" or "off," depending on whether it is an enhancement or depletion type. The charge of the floating gate may be electronically erased by reversing the bias on the control gate.

22.4 Charge Coupled Devices

The heart of simple CCD video cameras as well as digital cameras that are now available is the metal–insulator–semiconductor (MIS) structure. This is similar to the MOS structure

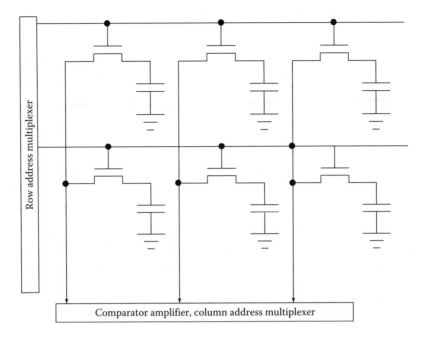

FIGURE 22.10
Schematic of a DRAM. Information is stored in the form of charge on the capacitors. Information is stored as charges on the capacitors.

discussed previously. Metal contacts grown on a thin oxide (or other insulator) back of an n-type silicon crystal are held at a negative potential, which attracts minority holes to the region next to the metal contact, forming a depletion layer when equilibrium is reached. Each metal contact represents a pixel of information. Light incident on the silicon will produce additional electron–hole pairs and the holes will be swept by the applied field into the depletion layer where they are stored as excess positive charge, the magnitude of the charge being proportional to the integrated number of photons incident since the last time the system was read out.

The readout of the system consists of manipulating the voltage on the metal contacts in order to shift the accumulated charges at the silicon–insulator interface along each row to the edge of the focal plane array where they are collected by a charge-sensitive amplifier. The mechanism for performing this operation is called a shift register. A schematic shift register readout is shown in Figure 22.11.

22.5 Moore's Law

In 1965 Gordon Moore, cofounder of Intel, made the observation that the number of transistors per area of an integrated circuit had doubled every year since the discovery of the transistor. This trend has been known as Moore's law and it has continued until the present, although the doubling time has slowed to ~18 months. Experts in the field expect the trend to continue through the next two decades.

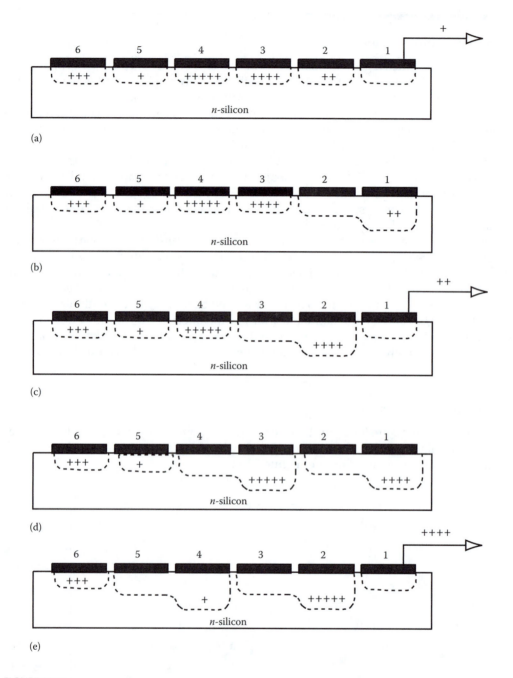

FIGURE 22.11

Schematic of a shift register readout of charges gathered under the pixels of an Si CCD video camera. The amount of charge is proportional to the light received. In (a) the charge under electrode 1 has been read. Making electrode 1 more negative transfers the charge under electrode 2 to electrode 1 (b). Charge from electrode 2 is read out in (c) and charge from electrode 3 is transferred to electrode 2. The charge from electrode 4 is transferred to electrode 3 and the charge from electrode 2 is transferred to electrode 1 in (d) where it gets read out in (e). In this way each of the charges collected can be clocked out serially to form an analog video stream for each row of pixels.

22.6 Heterojunctions

With the use of epitaxial growth methods such as molecular beam epitaxy (MBE) or metal organic vapor deposition (MOCVD), it is possible to grow single crystalline ternary alloy systems such as $Al_xGa_{1-x}As$ or quaternary systems such as $Ga_xIn_{1-x}As_yP_{1-y}$ with controlled composition as well as to form heterostructures by growing one compound semiconductor epitaxially on top of another compound semiconductor. (Epitaxy means that the lattice periodicity is maintained across the growth interface.)

By varying x in $Al_xGa_{1-x}As$, one can adjust the bandgap from that of AlAs (2.2 eV) to GaAs (1.35 eV). At $x = 0.45$, the alloy changes from a direct bandgap system like GaAs to an indirect bandgap system like AlAs. Figure 22.12 plots bandgap energy against lattice constant. By choosing systems with a minimum lattice mismatch, the lattice strain is reduced and defect density is lowered.

It is now possible to form heterostructures in which the transition from one system to the other takes place over one atomic layer. When two such semiconductors are brought into intimate contact, the electron affinities (energy required to remove an electron from the bottom of the conduction band to vacuum state) must line up, and at thermal equilibrium, the Fermi levels must also line up. The result is band bending with discontinuities in both the valence (ΔE_V) and conduction band (ΔE_C) at the interface as shown in Figure 22.13. (Recall that in homojunctions between n- and p-type material, the vacuum levels already lined up because both sides of the junction were the same material so no such discontinuities appeared.)

22.6.1 Modulation Doping

Of particular interest are junctions between heavily n-doped wide bandgap alloys such as AlGaAs and lightly n-doped or intrinsic smaller bandgap systems such as GaAs, which produces a band structure shown in Figure 22.14.

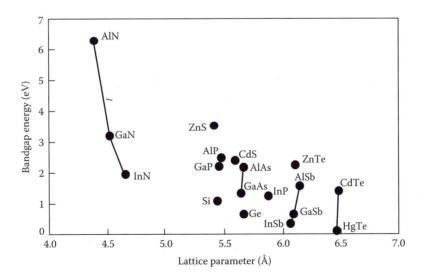

FIGURE 22.12
Bandgap energy versus lattice constant for various elemental and compound semiconductors. The lines connecting the different compounds indicate the range of energy gaps that can be obtained by alloying the different components with similar lattice constants.

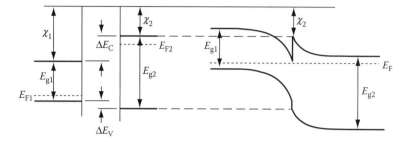

FIGURE 22.13

Two different semiconductor materials with different energy gaps before forming a junction (left). The differences between the two conduction bands and between the two valence bands relative to the vacuum energy must be preserved at the junction. This condition causes discontinuities in the bands when a junction is formed and the Fermi levels are matched (right).

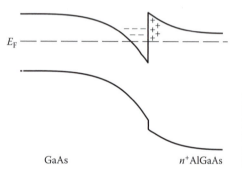

FIGURE 22.14

Heterojunction between heavily n-doped AlGaAs and lightly n-doped GaAs. The higher concentration of electrons in the AlGaAs causes them to diffuse across the junction into the nearly intrinsic GaAs where they congregate in the well between the conduction band and the Fermi level.

The ionized donors in the heavily doped AlGaAs system produce a high concentration of electrons which diffuse across the junction and wind up above the conduction band of the GaAs, where they remain as free carriers. Thus these electrons are removed from the material containing the donors that created them and are transported to a material with few ionized donors or acceptors. Rutherford scattering from ionized impurities (donors) is a mobility-limiting mechanism at low temperatures. This separation of free electrons from the donors that produced them is call modulation doping and results in a dramatic increase in mobility at low temperatures. It should be understood that the high mobility electrons are confined to plane parallel to the heterojunction.

22.6.2 High Electron Mobility Transistor

So what use can a two-dimensional (2-D) very high mobility electron gas be put to? In 1985, Takashi Mimura, invented the high electron mobility transistor (HEMT). The concept is illustrated in Figure 22.15.

The heavily n-doped AlGaAs donates the electrons that are forced into the semi-insulating (near intrinsic) GaAs to form an n-channel depletion FET. Since the ionized donors stay in the AlGaAs layer, they do not scatter the electrons in the n-channel which permits the conducting electron to have very high mobilities. The high electron mobility allows the transistor to operate at frequencies in excess of 300 GHz.

The first applications of HEMTs were in radio astronomy where they improved the performance of the existing radio telescopes. Later they became widely used in the receivers for the direct broadcast satellites and in cell phones. Recently HEMT technology has been extended to GaN and InP to make high power, high frequency transmitters.

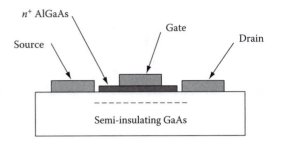

FIGURE 22.15
Schematic of a HEMT. The highly doped AlGaAs layer forces its electrons into the semi-insulating GaAs where they have very high mobility. Changing the bias on the gate alters the number of electrons in the *n*-channel.

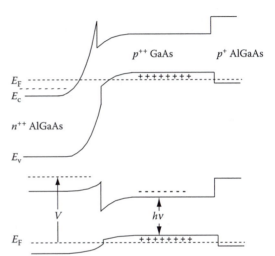

FIGURE 22.16
Schematic of the energy bands of a double heterojunction laser. The Fermi level is above the conduction band edge in the *n*-AlGaAs and below the valence band in the *p*-GaAs. Forward biasing the junction (below) causes the electrons to spill over into the active *p*-GaAs region where they are confined by the higher bandgap AlGaAs. The confined inverted population in the active region promotes stimulated emission.

22.6.3 Double Heterojunction Lasers

Another application for the use of heterojunctions is the double heterojunction laser, which has greatly improved efficiency over the simple junction laser. Heavily doped *p*-GaAs is sandwiched between *n*-AlGaAs and *p*-AlGaAs as shown in Figure 22.16. The doping level is chosen to place the Fermi level below the valence band in the *p*-GaAs and above the conduction in the *n*-AlGaAs. This places a heavy population of holes in the valence band of the *p*-GaAs and a heavy population of electrons in the conduction band of the *n*-AlGaAs.

When forward biased, the quasi-Fermi level exceeds the conduction band edge in the *p*-GaAs material, as shown in the lower part of Figure 22.16, electrons flow from the *n*-AlGaAs into the *p*-GaAs active layer where they are confined by the wider bandgap *p*-AlGaAs material at the second heterojunction. Here they combine radiatively with the holes in the *p*-GaAs region. Cleaving the faces of the crystals to form a Fabry-Perot cavity partially confines the photons to produce stimulated emission.

22.7 Superlattices

With the development of molecular beam technology (MBE, MOCVD), it is possible to not only make heterojunctions, but epitaxial layers can now be grown with thicknesses that can be controlled to atomic dimensions. By sandwiching a lower bandgap material

between two layers of higher bandgap materials, electrons in the lower bandgap material are confined in 1-D very much as in the electron-in-a-box model. Instead of being in a conduction or valence band, these electrons are in discrete energy states whose levels can be controlled by the width of the lower bandgap material. Such a structure is called a single quantum well. A series of periodic quantum wells can also be constructed to form multiple quantum wells (MQWs) which constitute a compositional superlattice. Such structures can lead to some very novel electronic and photonic devices as illustrated by a few examples in the following sections.

22.7.1   Resonance Tunneling Diodes

By placing intrinsic GaAs between two thin (5 nm) layers of AlGaAs as shown in Figure 22.17, the electrons in the well have only a limited number of states by analogy with the electron-in-a-box problem. Suppose one of the bound energy levels of the electrons in the well is above the Fermi level of the electrons in the outer layers of n^+ GaAs, in which the Fermi level is above the conduction band. When a voltage is applied, some tunneling through the AlGaAs can take place from the electrons that have energies equal to the bound state of the well, but this tunneling current will increase rapidly when enough voltage is applied to align the bound state energy of the well with the Fermi level of the n^+ GaAs to the left of the well. This is called resonance tunneling. However, a further increase in voltage brings the bound state in the well below the Fermi level of the GaAs on the left and since there are no allowable states in the well the current decreases, thus producing a negative resistance. Such a device is called a resonance tunneling diode (RTD).

Oscillators based on RTDs are able to operate in the hundreds of GHz and are beginning to close the gap at the terahertz portion of the electromagnetic spectrum, the region between the far-infrared and submillimeter radio waves.

22.7.2   *nipi* Structures

Another type of superlattice, called a doping superlattice, is created by alternating the doping of GaAs from n- to p- and back to n-type. Since an intrinsic region must necessarily occur as the doping is alternated, such structures are sometimes called "*nipi* structures." As seen in Figure 22.18, the periodicity of the doping leads to a periodic band structure as the Fermi levels in each region line up.

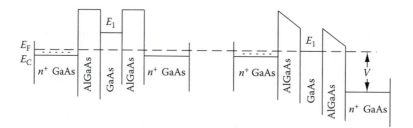

FIGURE 22.17

RTA: A thin layer of intrinsic GaAs is sandwiched between two thin layers of intrinsic AlGaAs, which has a higher bandgap than GaAs, to form a quantum well. The two AlGaAs layers are sandwiched between two layers of heavily doped n^+ GaAs. In the unbiased state (left), a bound state energy of the quantum well E_1 is above the Fermi energy and only a few of the electrons from the n^+ GaAs can tunnel through. As the bias is increased, the tunnel current increases until it reaches its peak (right), when the bound state is in resonance with the Fermi level. A further increase in voltage produces decreasing current because there are no available states in the well.

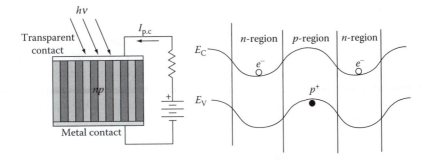

FIGURE 22.18
Schematic of a *nipi* structure (left). Very thin alternating layers of *n*- and *p*-doped GaAs form a sandwich-like structure. When an incident photon creates an electron–hole pair, the holes are swept into the *p*-layer and the electrons are swept into the *n*-layer by the fields created by the junctions. Since there are few minority carriers to recombine with, a larger fraction of these charges can be collected.

An interesting feature of this structure is that the free electrons will be found in the minima of the conduction bands while the holes reside in the maxima of the valence band, thus are physically separated from each other. This provides extremely long recombination times for photo-induced electron–hole pairs. Thus a photocurrent can exist much longer in a nipi structure than in a homogeneously doped semiconductor.

Increasing the minority carrier lifetime is very important in photonic devices. In a heavily doped material, the majority carriers far out number the minority carriers under equilibrium conditions because of the law of mass action that requires the *np* product to be constant. However, absorption of photons produces equal numbers of electron–hole pairs. Under this nonequilibrium situation, the majority carriers are increased by only a small fraction; whereas, the minority carriers are increased by a very large fraction. Since there are a very large number of majority carriers to annihilate them as the system returns to equilibrium, it is necessary for the minority carrier lifetime to be long enough for them to reach an electrode in order to be counted. By using a *nipi* structure to capture and separate the electrons and holes, highly efficient photoconductive devices can be fabricated.

22.7.3 Quantum Well Cascade Lasers

Of all the far-infrared sources, the quantum cascade (QC) laser is the most advanced. Instead of utilizing the radiative recombination of electron–hole pairs, which is the basis for conventional solid-state lasers, the QC laser operates with only electrons. Quantum wells are created by forming thin (few atomic layers) sandwiches of AlInAs and GaInAs, materials with different bandgaps. Electrons moving perpendicularly through these quantum wells are quantized into discrete energy levels that can be tailored by the thicknesses of the sandwiches. Photons are generated by the electrons hopping from higher to lower energy levels in the conduction band as they move through the series of quantum wells. Unlike conventional lasers in which the electron–hole pair is annihilated when they recombine, the electron remains in the conduction band where it can be re-injected into the next stage to repeat the process. As seen in Figure 22.19, the applied voltage bias lowers the level of each of the quantum wells the electron moves through, so the electron literally falls down a series of steps of an energy staircase, emitting a photon at each step, sort of an

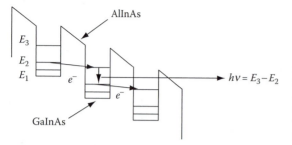

FIGURE 22.19

Schematic of a QC laser. A single electron stays in the conduction band of the two semiconductors as it traverses the superlattice. It loses energy when it is in the lower bandgap material by dropping from a higher quantum level to a lower level, emitting photons as it drops. Thus a single electron can produce as many photons as there are stages in the lattice.

"electronic water fall" as it was described by its inventor, Frederico Capasso, which is the basis for the term QC describing the process (Faist et al., 1996).

QC lasers are unique in many ways. Rather than requiring a specific bandgap material to laze at a particular wavelength, the wavelength may be selected by the dimensions of the InGaAs layers that determine the energy levels in that region. Since they can be tuned to look for specific molecular bands in the 3.4–17 μm, this makes such devices very useful for molecular spectroscopy in the mid- and far IR region. The fact that a single electron can produce up to 75 photons makes them very efficient and powerful (as solid-state lasers go), up to 0.5 W at room temperature in a pulse mode, and 0.2 W at LN_2 temperature in a CW mode. As a result, these devices have found use in trace gas detection, air pollution monitoring, industrial process control, and breath analysis for early detection of disease.

22.8 Quantum Wires and Quantum Dots

Whereas quantum wells provide 1-D confinement of electrons to dimensions such that their individual quantum states become significant, it is possible to form thin (<100 nm) bars or strips of semiconductors called quantum wires that provide 2-D confinement, and nano-sized particles called quantum dots (QD) that provide 3-D confinement.

The interest in quantum wires until recently has been primarily academic. Considerable work has been done to understand electron transport in such structures as well as their optical properties and applications are beginning to emerge. For example, Kouklin et al., have found that quantum wires have a very large resistance change when illuminated with a nanowatt of infrared radiation, thus are possible candidates for very sensitive IR detectors (Kouklin et al., Appl. Phys. Lett., Vol. 76, (2000) 460). Several groups have reported micro-cavity quantum wire lasers (e.g.,). Other applications being pursued are quantum wire HEMTs.

QDs can be thought of as artificial atoms whose energy levels can be tailored by the size of the structure. They can be designed to interact strongly with visible light by adjusting the energy levels between the highest occupied molecular orbital (HOMO) and the lowest unoccupied molecular orbital (LUMO) to the wavelength of interest. One application of QDs is the luminescent tagging of active regions of biological macromolecules. The use of QDs in lasers has several advantages over quantum well lasers: less temperature sensitivity, lower threshold current, and much narrower line width. The Zia Corp., Albuquerque, New Mexico, and other groups are developing QD lasers for commercial use. Another potential application under consideration is the possibility of filling or emptying states in QDs with a single electron, which would be the heart of the single electron (SET) transistor (see Kastner, 2000).

22.9 Summary

The first transistors, which were BJTs, were essentially back-to-back diodes with a control gate between them. In the active mode, the transistor could be turned on by forward biasing the emitter–gate diode. Most of the minority holes that were injected into the gate region are swept by the internal fields into the collector. The α of the transistor is the fraction of the injected minority carriers that leave the emitter and get to the collector. The fraction of the minority carries that go to the gate is $1 - \alpha$, so the current gain, the ratio of the collector current to the gate current, is called the β and is given by $\beta = \alpha/(1 - \alpha)$. BJTs are low impedance devices, meaning they draw current, and are primarily used as current amplifiers, using the small gate current to control a larger collector current.

FETs control the flow of current from their source to drain by a voltage applied to their gate. The current that flows in a FET can either be electrons (n-channel) or holes (p-channel) and they can either be normally off when unbiased (enhancement type) or normally on (depletion type). Metal–insulator–semiconductor FETs (MISFETs or MOSFETs) have insulated gates and draw practically no current. They are used in applications such as flip-flop counters in digital watches and in SRAMs where current flow must be kept to a minimum. DRAMs store digital information as a charge or no-charge on a capacitor that is controlled by a single transistor. Since a DRAM only requires one transistor per bit of information compared to six in a SRAM, DRAM memory is smaller and cheaper than SRAM, although DRAM memory has to be continually refreshed as the charge leaks off of the capacitor.

Both the SRAM and DRAM memories are volatile, meaning information is lost if the power is removed. EEPROMs utilize a floating insulated gate on a FET. The floating gate can be charged or discharged by tunneling from a control electrode and will remain in that state until changed by a voltage pulse on the control electrode. EEPROMs are used in today's memory sticks and digital cameras.

Photoelectrons or holes can be collected under charged metal electrodes that form the pixels in a CCD video camera. By manipulating the voltages on the electrodes, these charges may be transferred serially in a shift register to a charge-sensitive amplifier to produce an analog video output.

In 1965 Gordon Moore observed that the number of transistors per area of an integrated circuit had doubled every year since its invention. This trend has continued and has become known as Moore's law.

With the development of sophisticated growth techniques that can lay down epitaxial layers of different materials with atomic precision and form heterojunctions of different bandgap materials, new dimensions in device technology have opened up. Thus one can engineer discrete energy levels of the electrons confined in a low bandgap material between two higher bandgap materials. The 1-D quantum wells can be used to make highly efficient double heterojunction lasers and RTDs that can operate in the THz region. By repeating these layered structures to make superlattices, cascade quantum well lasers can be made in which a single electron can generate many photons. Carrying the quantum confinement to 2-D and 3-D, quantum wire and quantum dots have been produced in which their electrical and optical properties can be designed into their structure.

Bibliography

Bhattacharya, P., *Properties of III-V Quantum Wells and Superlattices*, Institution of Electrical Engineers, now the Institution of Engineering and Technology, U.K., 1996.

Faist, Capasso, et al., *Appl. Phys. Lett.*, 68, 1996, 3680–3682.

Faist, J., Capasso, F., Sivco, D.L., et al., Quantum cascade laser, *Science*, 264, 1994, 553.

Hummel, R.E., *Electronic Properties of Materials*, 3rd edn., Springer, New York, 2000.

Ibach, H. and Lüth, H., *Solid State Physics*, 3rd edn., Springer-Verlag, New York, 1990.

Kastner, M.A., The single electron transistor and artificial atoms, *Ann. Phys. (Leipzig)*, 9(11–12), 2000, 885–894.

Kittel, C., *Introduction to Solid State Physics*, 7th edn., John Wiley & Sons, New York, 1966.

Kouklin, et al., *Appl. Phys. Lett.*, 76, 2000, 406.

Livingston, J.D., *Electronic Properties of Engineering Materials*, Wiley, MIT series in Materials Science and Engineering, John Wiley & Sons, New York, 1999.

Navon, D.H., *Electronic Materials and Devices*, Houghton Mifflin, Boston, 1975.

Percival, et al., GaAs quantum wire lasers grown on v-grooved substrates isolated by self-aligned ion implantation, *IEEE Trans. Electron Devices*, 47, 2000, 1769–1772.

Sapoval, B. and Hermann, C., *Physics of Semiconductors*, Springer-Verlag, New York, 1995.

23

Dielectrics and the Dielectric Function

Dielectric materials in the form of ceramics or polymers are characterized by the fact that they contain virtually no free charges and therefore are electrical insulators. The lack of free charge implies a bandgap energy that is at least large enough to prevent thermal ionization of electrons and is generally high enough for them to be transparent in the visible spectrum. However, just because their conductivity is small, the electrical and optical properties of dielectrics are by no means uninteresting. The presence of an electric field distorts the atoms and molecules so that the center of their positive charge is displaced from the center of their negative charge and electric dipoles are formed. The electrical and optical properties are determined primarily by the dielectric material's ability to form electric dipoles in the presence of an electric field. But, since one of the dielectric's primary functions is that of an electrical insulator, we shall consider this attribute first.

23.1 Conductivity of Dielectrics

Even though the conductivity in dielectrics is small, it is not zero. Both free electrons and the migration of ions can contribute to conductivity in ceramics. We shall consider the electronic contribution first.

23.1.1 Electronic Conduction

We saw in Chapter 20 that the electron–hole product in an intrinsic semiconductor is given by $np \sim (2.5 \times 10^{25}/\mathrm{m}^3)^2 \exp(-E_g/kT)$ so the number of charged carriers will be $\sim(5 \times 10^{25}/\mathrm{m}^3)\exp(-E_g/2kT)$. Even with a bandgap as low as 3 eV, the number of carriers present at ambient temperature would be less than 1 electron or hole/m^3. However, if electrically active impurities are present that could act as donors or acceptors, the electron or hole population could increase dramatically. For example, 1 ppm of a donor impurity, if fully ionized, would produce a carrier concentration of 10^{22} electrons/m^3.

23.1.2 Ionic Conduction

Ions can move through an ionic compound by vacancy diffusion driven by an electric field rather than by a concentration gradient. The current density is $\mathbf{j}_i = n_i z_i e \mu_i \mathbf{E}$, where n_i is the ion concentration, $z_i e$ is the charge on the ion, and μ_i is the ion mobility. The ion mobility is

given by the Einstein relationship $D_i = kT\,\mu_i/e$ (see Chapter 21). Therefore, the ionic current is given by

$$\mathbf{J}_i = \frac{n_i z_i^2 e^2 D_i}{kT}\mathbf{E}. \tag{23.1}$$

For $N_i = 10^{-28}/m^3$ and $D_i \sim 10^{-12}\ m^2/s$, the conductivity $\sigma = n_i z_i^2 e^2 D_i/kT \sim 10^{-5}$ S/m.

23.1.3 Fast Ion Conductors

Ion conductivity is necessary in any battery to provide an internal conduction path to combine with the electrons flowing through the external load. In most batteries a liquid electrolyte such as KOH provides this ionic conductivity. However, there is great interest in β-alumina solid electrolyte (BASE) for use in high power Na–S batteries being considered for electric automotive systems. Na^+ ions can be complexed with Al_2O_3 to form β-alumina which acts as a substrate with channels through which Na^+ ions can easily flow giving conductivities $\sim$1 S/m. Other fast ion conductors include α-AgI and nonstoichiometric oxides with the fluorite structure that have large numbers of oxygen vacancies.

23.1.4 Dielectric Breakdown

Since one of the primary uses of dielectrics is in the form of electrical insulators, another electrical property of interest is the dielectric strength or breakdown potential. Since the conductivity of a dielectric is not zero, some current will flow when a strong field is applied. If there are weak points in the dielectric, more leakage current will flow at these points. This leakage current causes local resistive heating, which increases the electrons in the conduction band, leading to still more current flow. At high enough fields, thermal runaway can cause catastrophic breakdown.

Small inclusions or cavities in the dielectric can intensify local fields because of their higher curvature. High electric fields can cause arcing across such a cavity; ions and electrons in the cavity can be accelerated by these fields to further erode the walls of the cavity causing it to grow and eventually break down the dielectric.

Ionic conductivity, although small, can transport conductive impurities from the surface of the dielectric to the interior. This process can be aided by high humidity or an acidic environment. These impurities can eventually form a conductive path through the dielectric, resulting in thermal runaway or electrolytic breakdown.

Finally, even if all of the above mechanisms can be avoided, if the electric field is increased beyond a certain point, some electrons may be pulled out of the valence band (or from impurity sites) and accelerated by the field so that they gain sufficient energy to knock other electrons out of the valence band to form an avalanche breakdown. Polystyrene has one of the highest dielectric strengths with 140 MV/m. Ceramics such as alumina, BN, MgO, and ordinary window glass breakdown at $\sim$10–20 MV/m.

The very thin dielectric layer that separates the gate from the channel in a FET is very susceptible to breakdown from static charge before the chip is installed in the device. For this reason such chips are shipped in conductive static-proof packages and extreme care in handling such components must be exercised.

Unlike the Zener effect in which the breakdown voltage is engineered into the design of the dielectric, many of the above mechanisms are unpredictable and time-dependent, so the chances of breakdown increases with age.

23.2 Polarization in Dielectrics

As stated previously, the electrical and optical properties are primarily determined by the dielectric material's ability to form electric dipoles in the presence of an electric field. An electric dipole can be thought of as a positive and a negative charge q separated by some distance d and the dipole moment is defined as $\mathbf{p} = q\mathbf{d}$.

Macroscopically, the resulting electrical displacement, $\mathbf{D} = \varepsilon\mathbf{E} = \varepsilon_0\mathbf{E} + \mathbf{P}$, where ε is the permittivity, $\mathbf{P}$ is the polarization, which is defined as the dipole moment per unit volume, and ε_0 is the permittivity of a vacuum $= 8.85 \times 10^{-12}$ farads/m (Coulomb/V-m). For an anisotropic crystal, ε is a tensor of rank two, but here we consider only isotropic materials, so ε will be a scalar.

It is convenient to define a relative dielectric constant $\varepsilon_r = \varepsilon/\varepsilon_0$. Then $\varepsilon_r\mathbf{E} = \mathbf{E} + \mathbf{P}/\varepsilon_0$, $\mathbf{P} = \varepsilon_0(\varepsilon_r - 1)\mathbf{E} = \varepsilon_0\chi\mathbf{E}$, where the electric susceptibility $\chi = \varepsilon_r - 1$. The susceptibility relates the amount of polarization to the applied field, or as the name suggests, tells how susceptible a material is to being polarized by an electric field.

23.2.1 Capacitors

To illustrate how dielectrics interact electrically, we first consider the capacitor, a device for storing electric charge. Basically, a capacitor is simply a sandwich consisting of a dielectric surrounded by two conductive surfaces. When connected to a source, electrons will accumulate on one plate which will induce an equal but opposite charge on the opposite plate as shown in Figure 23.1. Once the potential difference between the plates become equal to the source, current ceases to flow in the circuit and the capacitor is fully charged.

The capacitance C is defined as the charge stored divided by the voltage applied, or $C = Q/V$. The unit of capacitance is the farad (from Michael Faraday) which, by definition, is 1 Coulomb/V. The stored charge is equal to the product of the area A, the electric field, and the dielectric constant of the material between the plates, or $Q = \varepsilon EA$. From this relationship, we see that the permittivity ε must have the units of Coulombs/Volt-m or Coulombs^2/Nt-m^2. Since the electric displacement $D = \varepsilon E$, we see that D is the charge displaced per area, or the surface charge on the capacitor. The electric field is the potential V divided by the distance d between the plates or $E = V/d$. Therefore, $C = \varepsilon A/d$.

Now if there were nothing in the space between the two capacitor plates, the permittivity ε would just be the permittivity of free space ε_0. But if we place a dielectric material

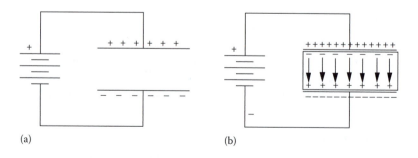

(a)   (b)

FIGURE 23.1

A fully charged capacitor with no dielectric between the plates (a). When a dielectric is inserted between the plates (b), the induced dipoles in the dielectric line up with their positive charge pointing toward the negative plate. The presence of these positive charges attracts more electrons from the battery increasing the negative charge on the plate. Similarly, more positive charge collects on the opposite plate, which increases the capacitance.

between the plates, the electric field, which runs from plus to minus, aligns the electric dipoles in the dielectric which produces a negative surface charge on the side of the dielectric facing the positive capacitor plate and a positive surface charge on the side facing the negative capacitor plate (think in terms of the charges on the nose and tail of the dipoles). These surface charges induce more positive charges to collect on the positive plate and more negative charges to collect on the negative plate, which increases the charge stored by ΔQ. The total charge can be written as $Q = Q_0 + \Delta Q = DA = \varepsilon_0 E + P$, where P is the polarization of the dielectric. Thus we see that the additional charge on the capacitor plates was brought about by the polarization of the dielectric and the capacitance is $C = \varepsilon_0 \varepsilon_r A / d$ is directly related to the relative dielectric constant, ε_r.

In an alternating current circuit, the current usually reverses before the capacitor is fully charged, thus the capacitor is continually being charged and discharged with reversing polarity. Thereby an alternating current can flow through a circuit containing a series capacitor whereas direct current cannot. As a result, capacitors are frequently used in electronic circuits to block unwanted DC in an AC circuit or to shunt unwanted AC ripple to ground in a DC circuit.

If a capacitor is discharged into a resistive load, the time to discharge $1/e$ of the charge is the product of the resistance and capacitance RC. This RC time constant can be set by choice of resistance and capacitance and is used in many timing circuits.

Another use of capacitors is to store energy that can be recovered quickly, much faster than from a battery. Dumping a large capacitor bank into a low-resistance load can result in a very large current flow and this technique is used for exploding wire experiments and other types of plasma research. The energy stored in a capacitor is $(1/2)CV^2 = (1/2)\varepsilon A V^2 / d$. One can see the need of materials with high dielectric constants and high dielectric strengths for high energy capacitor banks.

Recently a new type of capacitor called supercapacitor has been developed by Maxwell Technologies (San Diego, California). These supercapacitors store their charge in an electrolytic double layer and have capacitance measured in farads rather than microfarads, which is the typical capacitance of conventional capacitors. Unfortunately, they can only operate at low voltages ($\sim$2.5 V); therefore the energy storage is limited to 1–10 W h/kg, about 1/10 that of a nickel-metal hydride battery. Supercapacitors find uses in memory backups for short power interruptions, and because they can be charged and discharged much faster than batteries, they can be used for energy recovery system in electric and hybrid vehicles.

23.3 Dielectric Function

We are used to calling the relative permittivity, ε_r, the dielectric constant. Actually it is a function of frequency and should be referred to as the dielectric function. In order to be able to describe the electrical and optical properties of dielectrics, we must find the relative dielectric constant as a function of frequency or the dielectric function, $\varepsilon_r(\omega)$.

23.3.1 Electronic Polarization

All materials have some degree of polarization from the distortion of their electronic orbits by the presence of an electric field. This distortion causes a displacement between the center of negative charge and the nucleus as illustrated in Figure 23.2. The relative dielectric constant from electronic polarization is small, but its effect plays an important role in the optical properties of materials.

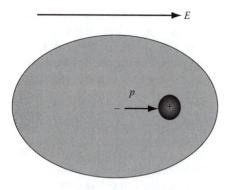

E

p

FIGURE 23.2
Electronic polarization. The presence of an electric field **E** distorts the electron's orbit and displaces its center from the positively charged nucleus creating a dipole moment **p**.

The magnitude of the displacement caused by an oscillating electric field can be obtained by treating the system as a damped harmonic oscillator in which the restoring force is characterized by ω_0, the resonant frequency of the system.

$$\ddot{x} + \gamma\dot{x} + \omega_0^2 x = -\frac{eE(\omega)}{m}e^{i\omega t}. \tag{23.2}$$

The well-known solution is written as

$$x(t) = \tilde{x}(\omega)e^{i\omega t}, \tag{23.3}$$

where $\tilde{x}(\omega)$ is the complex amplitude given by

$$\tilde{x}(\omega) = -\frac{e/m}{\omega_0^2 - \omega^2 - i\gamma\omega}E(\omega), \tag{23.4}$$

which can be verified by direct substitution.

The resulting dipole moment induced by the oscillating E-field is just $p = -e\tilde{x}(\omega)$ and the polarizability, α, is defined as $\chi\varepsilon_0$

$$\alpha(\omega) \equiv \frac{p}{E} = \frac{-e\tilde{x}(\omega)}{E(\omega)} = \frac{e^2/m}{\omega_0^2 - \omega^2 - i\gamma\omega}. \tag{23.5}$$

The resonant frequency ω_0 is material dependent and is typically 10^{15}–10^{16} rad/s (ultraviolet). The damping constant is generally small compared to ω_0 and is only important near resonance ($\omega \sim \omega_0$).

For $\omega < \omega_0$ (visible range),

$$\alpha(\omega) \approx \frac{e^2}{m(\omega_0^2 - \omega^2)} = \frac{(1.6 \times 10^{-19})^2}{9 \times 10^{-31}(10^{16})^2(1 - \omega^2/\omega_0^2)} = \frac{2.8 \times 10^{-40}}{(1 - \omega^2/\omega_0^2)}.$$

This is the contribution from a single electron. For a many electron atom, we need the effective α_i for each ion species so that we may sum over the various ionic species to obtain the dielectric function

$$\varepsilon_r(\omega) = 1 + \frac{P(\omega)}{\varepsilon_0 E} = 1 + \frac{1}{\varepsilon_0}\sum_i n_i\alpha_i(\omega). \tag{23.6}$$

It is customary to denote the static electronic dielectric function ($\omega \ll \omega_0$) as $\varepsilon(\infty)$, which seems strange. The only justification seems to be that $\varepsilon(0)$ is reserved for the total static dielectric constant (ionic + electronic) and the infinity simply means at frequencies far above those where the ionic contributions are no longer significant, but are still much smaller than ω_0.

23.3.2 Ionic Polarization

Ionically bonded compounds have oppositely charged ion cores that are displaced by the presence of an electric field as shown in Figure 23.3. This effect, called ionic polarization, plays an important role in the infrared absorption of materials. Even covalently bonded elements such as Si and Ge have large dielectric susceptibilities because the directional nature of the sp^3 bond places the center of the electronic charge between the ion cores that make up the lattice.

We now will compute the polarization and dielectric function of an ionic compound. As discussed in Chapter 16, there are three possible modes of vibration that can be stimulated by an electromagnetic wave passing through a chain of ions with alternating charges; two transverse modes (alternate ions oscillating perpendicularly to the chain), and one longitudinal mode (ions in the chain moving back and forth along the direction of the chain). Which mode get excited depends on the orientation of the wave **E**-vector relative to the chain. A photon passing from left to right in Figure 23.3 would present an oscillating electric field along the **E**-vector and would therefore excite a transverse oscillation in the plane of the paper.

The equations of motion for a diatomic chain of atoms oscillating at their natural frequency were derived in Chapter 16. We take Equations 16.11 and 16.12 in Chapter 16 and set them equal to a driving term, $-eE_0e^{i\omega t}$ for the anion and $eE_0e^{i\omega t}$ for the cation.

$$m\ddot{u}_{2n} - \beta(u_{2n+1} + u_{2n-1} - 2u_{2n}) = -eE_0e^{i\omega t} \tag{23.7}$$

$$M\ddot{u}_{2n+1} - \beta(u_{2n+2} + u_{2n} - 2u_{2n+1}) = eE_0e^{i\omega t}. \tag{23.8}$$

We assume the solutions can be written as in Chapter 16 (Equation 16.13)

$$u_{2n} = \xi e^{i(\omega t + 2nka)} \quad \text{and} \quad u_{2n+1} = \eta e^{i(\omega t + (2n+1)ka)} \tag{23.9}$$

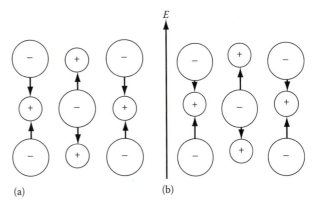

FIGURE 23.3

Ionic polarization. With no applied field (a), the dipoles between the anions and the cations all cancel each other. In the presence of an electric field (b), the cations are displaced in the direction of the field and the anions are displaced in the opposite direction, giving a net dipole moment in the direction of the field.

and put them back into Equations 23.7 and 23.8,

$$-\omega^2 m\xi - \beta\eta\left(e^{ika} + e^{-ika}\right) - 2\beta\xi = -eE_0 e^{i\omega t} \tag{23.10}$$

$$-\omega^2 M\eta - \beta\xi\left(e^{ika} + e^{-ika}\right) - 2\beta\eta = eE_0 e^{i\omega t}. \tag{23.11}$$

In the vicinity of $k=0$, $(e^{ika} + e^{-ika}) = 2$. We now have a set of simultaneous equations that can be solved for the complex amplitudes ξ and η,

$$\begin{aligned} \omega^2 m\xi + 2\beta(\eta - \xi) &= eE_0 e^{i\omega t} \\ \omega^2 M\eta - 2\beta(\eta - \xi) &= -eE_0 e^{i\omega t}. \end{aligned} \tag{23.12}$$

However, the quantity of interest is $(\xi - \eta)$ since this represents the oscillating change in the position of the two charges, hence, the induced oscillating dipole moment is $e(\xi - \eta)$. Multiply the top equation by M, the bottom by m, subtract and identify the reduced mass $\mu = mM/(m + M)$. After rearranging terms, we get the particular solution,

$$\left(\frac{2\beta}{\mu} - \omega^2\right)(\xi - \eta) = \frac{eE_0}{\mu}. \tag{23.13}$$

The complimentary solution is obtained by setting the driving term eE_0 to 0 in Equation 23.12. The terms may be rearranged to give the same indicial equation as Equation 16.16 (Chapter 16) for which the solution at $k=0$ was found to be

$$\omega_T^2 = \frac{2\beta(M + m)}{Mm} = \frac{2\beta}{\mu}. \tag{23.14}$$

We denote this natural frequency as ω_T, symbolizing this as the frequency of the transverse phonon spectrum in the optical branch. Putting this back into the particular solution (Equation 23.13),

$$\left(\omega_T^2 - \omega^2\right)(\xi - \eta) = \frac{eE_0}{\mu}. \tag{23.15}$$

Now the induced dipole is just the relative displacement between the ions $(\xi - \eta)$ times the unit of charge and the ionic contribution to the polarization obtained from Equation 23.15 (with damping terms omitted) is just

$$P_{\text{ionic}} = N_{\text{ions}}(\xi - \eta)e = \frac{N_{\text{ions}}e^2 E_0}{\mu\left(\omega_T^2 - \omega^2\right)}. \tag{23.16}$$

It is customary to add the electronic contribution to the ionic contribution and write

$$\varepsilon_r(\omega) = \varepsilon(\infty) + \frac{N_{\text{ions}}e^2}{\varepsilon_0\mu\left(\omega_T^2 - \omega^2\right)}. \tag{23.17}$$

For $\omega \ll \omega_T$, the static dielectric constant $\varepsilon(0)$ is real and positive and is given by

$$\varepsilon(0) = \frac{N_{\text{ions}}e^2}{\varepsilon_0\mu\omega_T^2} + \varepsilon(\infty). \tag{23.18}$$

23.3.3 Lyddane, Sachs, Teller (LST) Relationship

From Equations 23.17 and 23.18 it can be seen that the dielectric function for an ionic solid can be written as

$$\varepsilon(\omega) = \varepsilon(\infty) + \frac{(\varepsilon(0) - \varepsilon(\infty))\omega_T^2}{\omega_T^2 - \omega^2}. \tag{23.19}$$

We see in Figure 23.4 that $\varepsilon(\omega)$ blows up at ω_T, then goes negative and eventually comes back up to $\varepsilon(\infty)$. We define ω_L as that value of ω when it crosses back into positive territory or the value for ω that makes $\varepsilon(\omega_L) = 0$ so that

$$\frac{(\varepsilon(0) - \varepsilon(\infty))\omega_T^2}{\omega_T^2 - \omega_L^2} = -\varepsilon(\infty) \tag{23.20}$$

or

$$\omega_L^2 = \omega_T^2 \left[1 - \frac{(\varepsilon(0) - \varepsilon(\infty))}{\varepsilon(\infty)} \right] = \omega_T^2 \frac{\varepsilon(0)}{\varepsilon(\infty)}. \tag{23.21}$$

This is the LST relationship, which relates the ratio of the squares of the longitudinal to the transverse resonance frequencies to the ratio of the static to electronic dielectric function.

The behavior of the total dielectric function (ionic + electronic) in the vicinity of ω_T is shown in Figure 23.4. The ω_T was taken to be 10^{13}/s and the static ionic contribution, $\varepsilon(0) - \varepsilon(\infty)$ was set to 5. The γ was set to 0.05 as before.

23.3.4 Polarization in Polar Liquids

Polar materials, in which the molecules lack centosymmetric symmetry, can have permanent dipole moments and consequently would have quite large static dielectric constants. Water is an excellent example of a molecule with a permanent dipole moment. The bond angle of 107° between the two H ions places these virtually naked protons on one side of the doubly charged O ion, giving water a static relative dielectric constant of ~80 and is responsible for the unusual properties of water and its frozen counterpart. The forces on a dipole are illustrated in Figure 23.5.

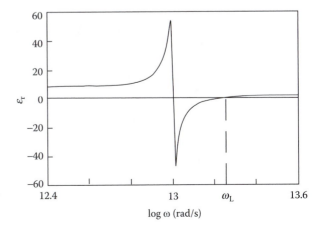

FIGURE 23.4
Ionic dielectric function in the vicinity of ω_T. The ω_L is defined as the value for ω when the dielectric function becomes positive again.

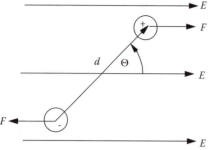

FIGURE 23.5
Forces on an electric dipole in an electric field **E**. The dipole moment **p** is the product of the charge and the separation or qd. The **E**-field produces a torque given by $\mathbf{p} \times \mathbf{E} = pE \sin\theta$ that tends to align the dipole with the field. The work required to rotate the dipole through angle θ is the integral of the torque from 0 to θ, or $-pE\cos\theta$ (assuming $U = 0$ at $\theta = \pi/2$).

The presence of an electric field tends to align the individual dipole moments in the material, but thermal effects work to destroy this alignment as will be shown below. Using Maxwell–Boltzmann statistics, we can calculate the average dipole moment by setting the energy required to move from the equilibrium to a higher energy equal to $-pE\cos\theta$.

At thermal equilibrium, the average dipole moment will be given by

$$\langle p \rangle = \frac{\int\limits_0^\pi (p\cos\theta)e^{pE\cos\theta/kT}2\pi\sin\theta d\theta}{\int\limits_0^\pi e^{pE\cos\theta/kT}2\pi\sin\theta d\theta} = pL(x), \tag{23.22}$$

where $x = pE/kT$ and the Langevin function $L(x) = \coth x - 1/x$. At ambient temperatures, $kT \gg pE$. Since $\coth x = 1/x + x/3 - x^3/45 + \cdots$, for small x, $L(x) \to x/3$. For very low temperatures, $x \gg 1$, and $L(x) \to 1$. At ambient temperatures ($x \ll 1$)

$$\langle p \rangle = px/3 = p^2E/3kT \tag{23.23}$$

and since the polarization P is $N\langle p \rangle$,

$$P = \frac{Np^2E}{3kT} = \frac{C}{T}E. \tag{23.24}$$

This is similar to the Curie law of paramagnetism, which relates the magnetic susceptibility to the Curie constant for a given material times the applied magnetic field divided by the absolute temperature (see Chapter 25). This type of behavior in dielectrics is known as paraelectric behavior. In both paramagnetic and paraelectric materials, the applied field tends to order the alignment of the dipole moments of individual atoms or molecules while thermal energy tends to randomize them. Both cases follow the Curie law in which the magnitude of the induced dipole moment is directly proportional to the applied field and inversely proportional to the temperature. The constant of proportionality, or Curie constant, is a material property.

23.3.5 Polarization in the Dipolar Solid

The previous treatment assumed that the dipoles were free to rotate, which is true only in liquids and gases. In most dipolar solids, some form of potential barrier must be overcome in order to rotate a dipole. Assume a solid in which dipoles can only point to the right or to the left and let there be a potential barrier ϕ that must be overcome in flipping a dipole back and forth. The rate at which dipoles on the left flip to the right is given by

$$\frac{dN_R}{dt} = \nu N_L e^{-(\phi - pE)/kT}, \tag{23.25}$$

where ν is a jump frequency. Similarly, the rate at which dipoles flip from right to left is

$$\frac{dN_L}{dt} = \nu N_R e^{-(\phi+pE)/kT}. \tag{23.26}$$

At equilibrium, the two rates must be equal and

$$\frac{N_L}{N_R} = e^{-2pE/kT}. \tag{23.27}$$

The net polarization is

$$P = (N_R - N_L)p = Np\frac{(N_R - N_L)}{(N_R + N_L)} = Np\frac{1 - e^{-2pE/kT}}{1 + e^{-2pE/kT}} = Np\tanh\left(\frac{pE}{kT}\right) \approx \frac{Np^2E}{kT}. \tag{23.28}$$

In this case, the average polarizability is three times greater than that for a gas or a liquid because the dipoles can only be either parallel or antiparallel. It is interesting to note that the barrier height cancels out and does not appear in the final result. This result assumes that equilibrium has been reached, but does not say how long it takes to come to equilibrium.

With no field present, the rate of change of the polarization is

$$\dot{P} = p\left(\dot{N}_R - \dot{N}_L\right) = p\nu e^{-\phi/kT}(N_L - N_R) = -P(t)\nu e^{-\phi/kT}. \tag{23.29}$$

The solution is

$$P = P_0 e^{-t/\tau}; \quad \text{where } 1/\tau = \nu e^{-\phi/kT}. \tag{23.30}$$

23.3.6 Dipolar Materials in an Alternating Field

For an alternating field, we add a driving term to Equation 23.29

$$\dot{P}(t) + \frac{P(t)}{\tau} = \frac{\varepsilon_0\chi_0}{\tau}E_0 e^{i\omega t}. \tag{23.31}$$

Now replace $P(t)$ with $\varepsilon_0\tilde{\chi}E_0 e^{i\omega t}$, where $\tilde{\chi}$ is the complex susceptibility.

$$i\omega\tilde{\chi} + \tilde{\chi}/\tau = \chi_0/\tau \tag{23.32}$$

from which

$$\tilde{\chi} = \frac{\chi_0}{1 + i\omega\tau} = \frac{\chi_0}{1 + \omega^2\tau^2}(1 - i\omega\tau). \tag{23.33}$$

Separating real and imaginary parts, we define

$$\chi'(\omega) = \text{Re}(\chi(\omega)) = \frac{\chi_0}{1 + \omega^2\tau^2}; \quad \chi''(\omega) = \text{Im}(\chi(\omega)) = \frac{\chi_0\omega\tau}{1 + \omega^2\tau^2}, \tag{23.34}$$

where $\chi_0 = \chi(0)$ and is given by Equation 23.28,

$$\chi(0) = \frac{Np^2}{\varepsilon_0 kT}. \tag{23.35}$$

Notice that the susceptibility contributions from the permanent dipoles vanish for $\omega \gg 1/\tau$.

A similar result was obtained by Debye for liquids and gases in which there is no potential barrier to inhibit the dipole motion. In this case the relaxation time results from the viscosity of the medium and is given by $\tau = 4\pi\eta R^3/kT$, where R is the radius of the molecule and the susceptibility is obtained from Equation 23.24 as

$$\chi(0) = \frac{Np^2}{3\varepsilon_0 kT}. \tag{23.36}$$

Typical values for τ range from 10^{-9} to 10^{-11} s.

23.3.7 Dielectric Losses

The $1/\tau$ term acts as a frictional or drag term making it difficult for the dipole to change states. As the dipole is oscillated in the presence of an alternating field, energy is being dissipated in the form of heat into the material. The energy lost per unit volume per cycle is given by

$$Q = \oint \dot{P}(t)E(t)dt. \tag{23.37}$$

Let $E(t) = E_0\cos(\omega t)$ and $P(t) = \varepsilon_0\tilde{\chi}E_0\cos(\omega t)$. The complex susceptibility can be written as $\tilde{\chi} = \chi(0)e^{i\phi}$, where $\chi' = \chi(0)\cos(\phi)$ and $\chi'' = \chi(0)\sin(\phi)$. The ratio χ''/χ' is called the loss tangent. $P(t)$ becomes $\varepsilon_0\chi(0)E_0\cos(\omega t - \phi)$ as the polarization lags behind the field by the phase angle ϕ. The product $P^Y(t)E(t)$ becomes

$$\dot{P}(t)E(t) = \omega\varepsilon_0\chi(0)E_0^2\left[-\sin(\omega t)\cos(\omega t)\cos(\phi) + \cos^2(\omega t)\sin(\phi)\right]. \tag{23.38}$$

Taking the time average

$$Q = \frac{1}{2}\varepsilon_0\chi(0)E_0^2\sin(\phi) = \frac{1}{2}\varepsilon_0 E_0^2\chi''(\omega). \tag{23.39}$$

Replacing χ'' with $\chi_0\omega\tau/1+\omega^2\tau^2$, the loss term is finally

$$Q = \frac{1}{2}\varepsilon_0 E_0^2\chi(0)\frac{\omega\tau}{1+\omega^2\tau^2}. \tag{23.40}$$

These losses provide the heating of foods in a microwave oven.

23.3.8 Clausius–Mossotti Equation

Thus far we have assumed that the applied field or external field was the local field inside the dielectric. This is not quite true because the aligned dipoles produce a local field that adds to the applied field. We now set out to determine how this local field relates to the applied field.

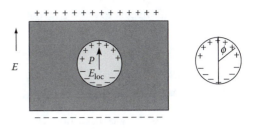

FIGURE 23.6
Geometry for obtaining the local field using the Clausius–Mozzotti model.

Consider the local field inside a dielectric medium in an applied field E_0 illustrated in Figure 23.6.

If we cut out a small spherical cavity around the point in question, the field in the center of this sphere will be given by the sum of the applied field E, the field due to the charges on the inside surface of the sphere, the field due to the dipoles outside the sphere and the field due to the dipoles inside the sphere. For materials with cubic symmetry, the latter contribution vanishes by symmetry, as does the field from the dipoles outside of the cavity. The charge density on the surface of the cavity is given by $P \cos \phi$. The field due to the surface charges inside the cavity is then given by

$$E_{\text{in}} = 2\pi \int_0^{\pi} R^2 \sin \phi d\phi \cos \phi \frac{P \cos \phi}{4\pi \varepsilon_0 R^2} = \frac{P}{3\varepsilon_0}. \tag{23.41}$$

Therefore, the local field is given by

$$E_{\text{loc}} = E + \frac{P}{3\varepsilon_0} = E + \frac{N\alpha E_{\text{loc}}}{3\varepsilon_0}. \tag{23.42}$$

Solving for the local field

$$E_{\text{loc}} = \frac{E}{1 - N\alpha/3\varepsilon_0}. \tag{23.43}$$

Putting this back into Equation 23.42 and solving for P,

$$P = \frac{N\alpha E}{(1 - N\alpha/3\varepsilon_0)}. \tag{23.44}$$

Since the electric susceptibility χ is given by

$$\chi = \frac{P}{\varepsilon_0 E} = \frac{N\alpha}{\varepsilon_0 (1 - N\alpha/3\varepsilon_0)}. \tag{23.45}$$

The relative dielectric constant is given by $1 + \chi$ or

$$\varepsilon_{\text{r}} = 1 + \chi = 1 + \frac{N\alpha}{\varepsilon_0 (1 - N\alpha/3\varepsilon_0)} = \frac{1 + 2N\alpha/3\varepsilon_0}{1 - N\alpha/3\varepsilon_0}, \tag{23.46}$$

which can also be written as

$$\frac{\varepsilon_R - 1}{\varepsilon_R + 2} = \frac{N\alpha}{3\varepsilon_0}. \tag{23.47}$$

This is known as the Clausius–Mossotti equation.

23.3.9 Corrections for Local Fields

For the case of electronic polarization, in deriving Equation 23.6, we wrote $P(\omega) = (E/\varepsilon_0) \sum n_i \alpha_i(\omega)$ when we should have used E_{Loc} in place of the applied field. Therefore Equation 23.6 should be written as

$$\varepsilon_r(\omega) = 1 + \frac{P(\omega)}{\varepsilon_0 E} = 1 + \frac{E_{Loc}}{\varepsilon_0 E} \sum_i n_i \alpha_i(\omega). \tag{23.48}$$

Using Equation 23.43 for E_{Loc}/E, Equation 23.48 becomes

$$\varepsilon_r(\omega) = \frac{1 + (2/3\varepsilon_0) \sum\limits_i n_i \alpha_i(\omega)}{1 - (1/3\varepsilon_0) \sum\limits_i n_i \alpha_i(\omega)} = \frac{1 + (2/3\varepsilon_0)N\alpha(\omega)}{1 - (1/3\varepsilon_0)N\alpha(\omega)} \quad \text{for a single component.} \tag{23.49}$$

Values for α are measured for various ionic species and are typically $O(10^{-40})$ at optical frequencies for the various ionic species (see Chapter 24), hence

$$\frac{n\alpha}{\varepsilon_0} = \frac{10^{29} 10^{-40}}{9.9 \times 10^{-12}} = O(1).$$

It is possible by treating the atom as a sphere of uniform charge and using Gauss' law to find the displacement between the nucleus and the center of the charge resulting from an applied electric field. Using this model the static electronic polarizability $\alpha(0)$ can be estimated to be $4\pi\varepsilon_0 R^3$, where R is the atomic radius.

So we see that the effect of electronic polarization in the visible region is to increase the dielectric constant by a quantity that is on the order of 1. For the more general case, the solution for $\alpha(\omega)$ given by Equation 23.5 can be separated into real and imaginary parts by multiplying top and bottom by the complex conjugate,

$$\alpha(\omega) = \frac{e^2/m}{\left(\omega_0^2 - \omega^2\right)^2 + \gamma^2 \omega^2} \left[(\omega_0^2 - \omega^2) + i\gamma\omega\right]. \tag{23.50}$$

As $\omega \to \omega_0$, the Re(α) gets larger until $\omega = \omega_0$, at which point it reaches its peak value, limited only by the damping term. As $\omega > \omega_0$, the sign of the Re(α) becomes negative and then eventually rises and approaches 1, the free space value. The imaginary part is only significant in the region where the real part goes through zero.

To illustrate the effect of the local field correction, the dielectric function is plotted using Equation 23.6 (uncorrected) and Equation 23.49 (corrected) in Figure 23.7. The quantity $N\alpha$ $(0)/\varepsilon_0$ was taken to be 1.01 and ω_0 was set to 10^{16}/s. The γ in Equation 23.50 was set to 0.05 s to provide some damping to keep the dielectric function from blowing up.

There is a small difference in the ε_r for $\omega \ll \omega_0$, 2.523 for the corrected versus 2.01 for the uncorrected. This difference will be greater for larger values of $N\alpha(0)/\varepsilon_0$. Also note the

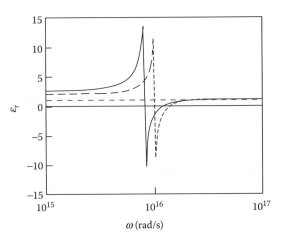

FIGURE 23.7
The electronic dielectric function in the vicinity of ω_0. The dashed line is the uncorrected dielectric function and the solid line is the dielectric function corrected for the local field. Both functions tail off to 1 (the dotted line) for $\omega > \omega_0$.

corrected function blows up before the resonant frequency. This can be understood from the following analysis. The real part of the electronic function from Equation 23.50, ignoring the damping term, can be written as

$$\alpha(\omega) = \frac{\alpha(0)\omega_0^2}{\left(\omega_0^2 - \omega^2\right)}. \tag{23.51}$$

From Equation 23.49. the dielectric function blows up when $N\alpha(\omega)/3\varepsilon_0 = 1$. (This is sometimes referred to as the polarization catastrophe.) The denominator in Equation 23.49 can be written using Equation 23.51 as

$$1 - \frac{N}{3\varepsilon_0} \frac{\alpha(0)}{\left(1 - \omega_1^2/\omega_0^2\right)} = 0 \tag{23.52}$$

and solved for ω_1, the actual frequency where the dielectric function becomes singular, to obtain

$$\omega_1 = \omega_0 \sqrt{1 - \frac{N\alpha(0)}{3\varepsilon_0}}. \tag{23.53}$$

Thus, we see that the effect of the local field is to reduce the natural frequency of the oscillations of the nucleus and the electron shell.

There appears to be some controversy in the literature over how to (or if to) correct the ionic dielectric function for the local field. This issue will be discussed in the Appendix.

23.3.10 Total Dielectric Function

If we have some fictitious material that possesses a permanent dipole moment along with an ionic and electronic dielectric function, the total real dielectric function could be written as the real part of the electronic contribution (Equation 23.49), plus the real part of the ionic susceptibility (Equation 23.19 with damping terms added), plus the real part of the dipolar susceptibility (Equation 23.34),

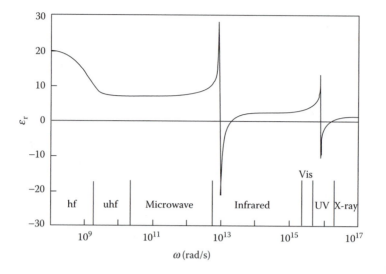

FIGURE 23.8

Total dielectric function for a hypothetical material with a static dielectric constant of 20, $\tau = 10^{-9}$ s, $\varepsilon(0) = 5$ and $\varepsilon(\infty) = 2.523$. The damping constant γ was set to 0.1 s. Note the resonances in the far infrared and in the near ultraviolet.

$$\varepsilon_r(\omega) = \frac{\chi_{\text{perm}}(0)}{1 + \omega^2 \tau^2} + \frac{(\varepsilon(0) - \varepsilon(\infty))(1 - \omega^2/\omega_T^2)}{(1 - \omega^2/\omega_T^2)^2 + \gamma^2/\omega^2} + \frac{1 + (2/3\varepsilon_0)N\alpha(\omega)}{1 - (1/3\varepsilon_0)N\alpha(\omega)}, \qquad (23.54)$$

where the $\alpha(\omega)$ is given by Equation 23.50. The total dielectric function is displayed in Figure 23.8.

23.4 Ferroelectrics

Some of dielectric materials have a spontaneous dipole moment that does not require the presence of an applied electric field, just as ferromagnetic materials have a spontaneous magnetic field in the absence of an applied magnetic field. Such materials are called ferroelectrics because they exhibit several similarities to ferromagnetic materials in that they form domains of like polarization and they exhibit hysteresis in response to alternating electric fields as illustrated in Figure 23.9a and b. Also as in ferromagnets, the spontaneous polarization in ferroelectrics decreases with temperature and vanishes at some point called the dielectric Curie temperature, T_C (Figure 23.11). Above this temperature, they become paraelectric, meaning their dipole moment is proportional to the applied electric field and now the dielectric susceptibility follows the Curie–Weiss law, $\chi = C/(T - T_C)$ for $T > T_C$, where C and T_C are material-related constants (Figure 23.12).

Notice that the area under the hysteresis curve is represented by the product $D-E$ or $\varepsilon_0\varepsilon_r E^2$ which has units of $(C^2/N \cdot m^2)(N^2/C^2) =$ Joules. So, as in a magnetic hysteresis curve, the area enclosed by the curve represents energy lost per cycle (see Section 23.3.2).

In order for a material to have a permanent dipole moment, there must be a single rotation axis that is along the dipole, there can be no mirror planes perpendicular to the rotation axis, and there must be no center of inversion. Of the 32 point groups, only 10 meet this criteria, $c_{1,2,3,4,6}$, $c_{2v,3v,4v,6v}$ and c_{1h} or 1, 2, 3, 4, 6, 2mm, 3m, 4mm, 6mm, and m in the international symbols. Crystals possessing a permanent dipole are called pyroelectrics

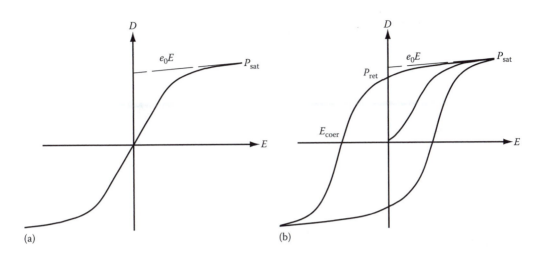

FIGURE 23.9
(a) Hysteresis curve for a dipolar material for $\omega\tau \ll 1$. The depolarization curve simply retraces the polarization curve and the polarization vanishes at $E = 0$. The $\varepsilon_0 E$ is the component of D after all the dipoles become aligned (P_{sat}). If $\omega\tau \geqslant 1$, a hysteresis will be observed. (b) Hysteresis curve for a ferroelectric material. P_{ret} is the polarization retained when the field is removed. E_{coer} is the coercive field required to destroy the polarization. The area enclosed is lost to heat during each cycle.

because their dipole moment will change with temperature (due to thermal expansion) which induces an additional charge on their surfaces. Some pyroelectrics have a permanent dipole moment that can be reversed by applying a strong electric field allowing them to exhibit a hysteresis effect. This class of materials is the ferroelectrics. Pyroelectrics that are not ferroelectrics may have a coercive field greater than their breakdown field or their Curie temperature may be above their melting temperature. Of course, all ferroelectrics are also pyroelectrics, but not the other way around.

In addition to the 10 point groups mentioned above there are 10 more point groups that lack a center of inversion but are not polar. These are c_{3h}, $d_{2,3,4,6,2d,3h}$, s_4, t and t_d. or $\bar{6}$, 222, 32, 422, 622, $\bar{4}2m$, $\bar{6}2m$, $\bar{4}$, 23, and $\bar{4}3m$. Crystals in this class, when distorted, may produce a dipole and are called piezoelectrics. Therefore, all ferroelectrics and pyroelectrics are also piezoelectrics, but not all piezoelectrics are pyroelectrics or ferroelectrics.

23.4.1 Types of Ferroelectric Materials

The first ferroelectric material to be discovered was some sodium–potassium tartrate tetra-hydrate crystals prepared in 1672 by a pharmacist, Elie Seignette, who lived in La Rochelle, France—hence the name Rochelle salts. However, their piezoelectric properties were not discovered until 1880 and their ferroelectric properties not until the 1920s. This material is only a ferroelectric between the temperatures of $-18°C$ and $23°C$ where it transforms from orthorhombic to monoclinic and back to orthorhombic. The Curie temperature as well as its polarization is increased when the hydrogen is replaced with deuterium.

Later ferroelectricity was discovered in KH_2PO_4 (KDP) below its Curie temperature of $123\,K$ where it transforms from its piezoelectric tetragonal d_{2d} ($\bar{4}2m$) form to the ferroelectric c_{2v} (2 mm) symmetry. RbH_2PO_4 and KH_2AsO_4 (KDA) also exhibit ferroelectricity at low temperatures. Another member of this class is triglycene sulfate (TGS). It transforms from its monoclinic form c_2 (2) to its centrosymetric form c_{2h} (2/m) at its Curie temperature of $49.7°C$. The Curie temperatures of these systems are also increased in their dueterated

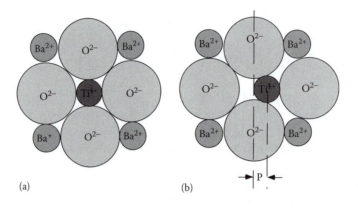

FIGURE 23.10

$BaTiO_3$ molecule above the 120°C transition temperature (a). The thermal expansion of the lattice provides sufficient space for the Ti^{4+} ion the remain in the center of the cubic unit cell with the perovskite structure ($Pm\bar{3}m$). Below 120°C (b) the thermal expansion can no longer maintain the cubic symmetry and a solid state transition to a tetragonal lattice (P4 mm) occurs. The four O^{2-} ions squeeze the Ti^{4+} to one side or the other creating a permanent electric dipole.

forms. These water soluble molecules have hydrogen-bonded PO_3 at the corners of the unit cell. The fact that replacement of the H by D strongly suggests that these bonds play an important role in the ferroelectric behavior of these compounds, perhaps in the position of the H atom in the H-bond. The fact that the ferroelectric transition appears to be second-order also suggests an order–disorder transition in the position of these H (or D) atoms.

Another class of ferroelectrics is the perovskites such as the titanates ($BaTiO_3$, $PbTi0_3$, $SrTiO_3$, $CaTiO_3$), the niobates ($KNbO_3$, $NaNbO_3$), the ilmentites, ($LiNbO_3$, $LiTaO_3$)), the ternary lead zirconate titanate (PZT), and the quaternary lead lanthanum zirconate titanate (PLZT). These materials undergo a displacive-type phase transformation at their Curie temperature.

For example, the oxygen perovskites such as $BaTiO_3$ transforms from cubic perovskite to tetragonal at 120°C. In the perovskite unit cell, the Ti^{4+} sits in the middle of the cube, the Ba^{2+} on the four corners, and the O^{2-} on the six sides as illustrated in Figure 23.10. The point group is O_h or mm and because of the cubic symmetry, there is no electric dipole. Below 120°C, however, the structure collapses into tetragonal as the bond lengths shorten at the lower temperatures. There is not enough room for the Ti^{4+} as it is being squeezed by the O^{2-} on the sides, so it is pushed in one direction while the four O^{2-} ions are pushed in the other direction. The resulting crystal must then be described by C_{4v} in the Schoenflies notation or by 4 mm in the international notation indicating a tetragonal lattice with fourfold symmetry about the c-axis and two sets of vertical mirror planes. It should be mentioned that $BaTiO_3$ also undergoes two other ferroelectric transitions, one at −5°C where it becomes orthorhombic, and another at −90°C where it becomes tetragonal.

The order–disorder transitions typical of the water-soluble hydrogen bonded ferroelectrics are second-order transformations as illustrated in Figure 23.11, while the displacive transformations are first-order with some second-order tendencies (a softening of the spontaneous polarization and a rise in heat capacity).

23.4.2 $BaTiO_3$ System

We will examine some properties of $BaTiO_3$ to get a feeling for the numbers involved. We will assume the following measured properties (there are some variations in the literature).

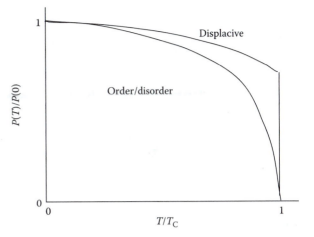

FIGURE 23.11
Typical behavior of the retained polarization as a function of temperature. Order–disorder transitions vanish continuously at the Curie temperature and are second-order transitions. Displacive transitions vanish discontinuously at the Curie temperature and are first-order transitions.

$T_C = 393\,\text{K}$, $P_{sat} = 0.26\,\text{C/m}^2$ (from Ashcroft and Merman), $a = 0.401\,\text{nm}$, $\varepsilon_r(T_C) = 35000$ (from Fischer-Cripps).

The number of dipoles $N = a^{-3} = 1.551 \times 10^{28}/\text{m}^3$. The dipole moment $p = P_{sat}/N = 1.67 \times 10^{-29}\,\text{C} \cdot \text{m}$ (accepted value 7.68×10^{-30} from Fischer-Cripps). Since the charge on the Ti ion is +4, the displacement $d = p/4e = 0.026\,\text{nm}$ (accepted value 0.012 nm from Fischer-Cripps). These numbers are not entirely consistent and seem to suggest a smaller value for N, but the value $1.551 \times 10^{28}/\text{m}^3$ is consistent with the value $1.555 \times 10^{28}/\text{m}^3$ computed from a molecular weight of 233.192 and a density of 6.02 g/cm³.

From the Arrhenius slope of the paraelectric curve taken from Kittel, the Curie constant C is found to be $1.24 \times 10^5\,\text{K}$ and the dielectric constant can be expressed as $\varepsilon_r = C/(T - T_0)$, where T_0 is the Curie point, not the Curie temperature. The Curie point = 389 K is chosen so that the ε_r matches the measured value at T_C shown in Figure 23.12. This is done to avoid the singularity at $T = T_C$.

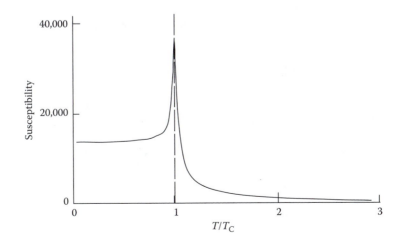

FIGURE 23.12
Typical dielectric susceptibility of a displacive ferroelectric transition versus the ratio of temperature to the Curie temperature.

23.4.3 Theories of Ferroelectrics

As in ferromagnetic materials, the presence of a spontaneous field in ferroelectrics has generated intensive theoretical interest. The behavior of ferroelectrics is fairly straightforward qualitatively, at least for the oxygen perovskites. In the ferroelectric state, the dipole is locked into one of six possible orientations. If these are randomly oriented, there will be no polarization. The molecules can be poled by heating to near the Curie temperature and applying a strong electric field to align the dipoles and then quenching them in place. A strong electric field is required to reverse the polarization. At the Curie temperature, the lattice expands so that the cubic configuration becomes the stable phase and the material becomes paraelectric since there is no longer a permanent dipole.

The second-order ferroelectric transitions can be modeled using the Curie–Weiss theory of ferromagnetism. It is assumed that there is a local field that is proportional to the polarization or $E_{Loc} = E + \gamma P / \varepsilon_0$, where γ is the constant of proportionality. For $T > T_C$, Equation 23.28 gives $P = N p^2 E_{Loc} / kT$ and the susceptibility is given by

$$\chi = \frac{P}{\varepsilon_0 E} = \frac{N p^2 / \varepsilon_0 kT}{1 - \gamma N p^2 / \varepsilon_0 kT} = \frac{T_C / \gamma T}{1 - T_C / T} = \frac{T_C / \gamma}{T - T_C}, \qquad (23.55)$$

where $T_C = \gamma N p^2 / \varepsilon_0 k$. Thus we have the susceptibility in the form of the Curie–Weiss law where the Curie constant is $C = N p^2 / \varepsilon_0 k$, which is the same as the previous result. Letting the applied field go to zero, we go back to Equation 23.28 and write

$$P = N p \tanh(p E_{Loc} / \varepsilon_0 kT) = N p \tanh(p \gamma P / \varepsilon_0 kT), \qquad (23.56)$$

which can be solved for P. Since tanh(x) can be no greater than 1, as T goes to T_C, γ $P p / \varepsilon_0 kT_C = P / N p$ if there is to be a solution. From this

$$\gamma = \frac{\varepsilon_0 kT_C}{N p^2} \qquad (23.57)$$

and

$$\frac{P}{N p} = \frac{P}{P_{Sat}} = \tanh\left(\frac{P}{P_{Sat}} \frac{T_C}{T}\right). \qquad (23.58)$$

The ratio of T/T_C to P/P_C may be written as

$$T/T_C = \frac{P/P_C}{a \tanh(P/P_C)}. \qquad (23.59)$$

Figure 23.13 is the resulting plot.

23.4.4 Antiferroelectrics

Continuing the analogy between ferroelectricity and ferromagnetism, there also exists materials such as ammonium dihydrogen phosphate NH_4PO_4 (ADP), $PdZrP0_3$, and $NaNbO_3$ that behave as antiferroelectrics. Like the ferroelectrics, they form spontaneous dipoles below their transition temperature, but the dipoles are antiparallel, thus they cancel out their net dipole moment. Their structure resists being polarized by an applied electric

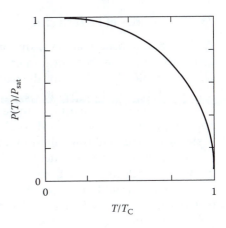

FIGURE 23.13
Polarization versus temperature predicted using the Curie–Weiss model for spontaneous polarization.

field until a transition temperature is reached at which point their dipoles become mobile and the material becomes paraelectric.

23.4.5 Piezoelectrics

Noncentrosymmetric materials, when distorted, will form dipole moments as the center of one charge is moved away from the center of the opposite charge. These dipoles will produce a surface charge and the voltage between the two surfaces will be given by this charge divided by the capacitance. Conversely, applying a voltage across the material will cause the material to expand or contract, a property called electrostriction. Such materials are called piezoelectrics ("piezo" meaning push) and convert mechanical energy directly to electrical energy and vice versa.

As discussed previously, many materials are piezoelectric without being ferroelectric, quartz being a prime example (see Figure 23.14). But since ferroelectrics are also piezo-electric and since the piezoelectric property is stronger in some of these ferroelectrics such as barium titanate, PZT, and PLZT, these materials are primarily used as piezoelectrics. Composites of PZT imbedded in polymers are used to make flexible transducers that can be curved to suit the application.

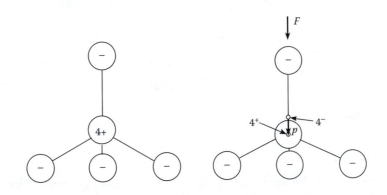

FIGURE 23.14
The SiO_4^{-4} building block of quartz. When unstressed, the center of charge resides in center of the Si^{4+} ion. When compressed, the bonds to the bottom three oxygen ions bend, shifting the center of negative charge above the center of the Si^{4+} ion, thus creating a dipole **p**.

23.5 Applications

23.5.1 Applications of Ferroelectrics

Because of the large dielectric constant of barium titanate at ambient temperature, it finds application in ceramic disc capacitors. The large dielectric constant allows higher capacitance with less volume.

Because ferroelectrics can retain their polarization indefinitely until their polarity is switched by the application of an electric field, they can serve as nonvolatile memory elements or ferroelectric RAMs (FRAMs). The storage capacitors in dynamic RAMs (DRAMs) can be replaced by ferroelectrics and eliminate the need to continually refresh the memory.

Thin ferroelectric films are finding applications as optical waveguides to carry light along a substrate. PLZT is particularly interesting because it is transparent and has a high electro-optical coefficient which makes it a candidate for applications for optical switching, optical memories, and display devices.

23.5.2 Applications of Pyroelectrics

Pyroelectric devices can detect very small changes in temperature by the change in polarization. This makes them useful for detecting radiation, especially far infrared radiation where the photons have insufficient energy to be detected by photoconductive or photovoltaic sensors. TGS has been used extensively for such detectors although now thin film PZT and other ferroelectric materials are beginning to replace single crystal TGS for such applications.

Pyroelectric crystals are capable of generating enormous electric fields (GV/m) by heating or cooling. Recently several groups have rapidly heated $LiTaO_3$ crystals from $-30°C$ to $45°C$ and used the high voltage to accelerate D atoms into erbium deuteride and produce D–D fusion reactions accompanied by the production of He^3 and 2.45 MeV neutrons (see Naranjo et al., 2005). The number of reactions is far too small for potential power production, but could provide a simple laboratory source of neutrons. The same technique also provides a simple source for high energy x-rays.

23.5.3 Applications of Piezoelectrics

As electromechanical transducers, piezoelectrics find many uses in ultrasonics such as sonar transmitters and receivers, ultrasound imaging, and nondestructive testing. They are also used in microphones and speakers. Their extreme sensitivity to small changes in pressure makes them useful in strain gages, pressure gages, and accelerometers. Coating metal strips with a piezoelectric makes a very sensitive temperature gage used in thermostats.

Organic piezo-polymers such as polyvinylidene fluoride along with trifluoroethylene and their copolymers can be made ferroelectric by poling the crystalline regions of these polymers. Because the density of these materials is close to that of human tissue, they have found use as transducers for ultrasound imaging with no acoustic mismatch.

When compressed, piezo crystals can generate a healthy charge separation, which produces a very high voltage across the faces. The spark generated by the discharge of this voltage is used to ignite gas appliances such as lighters and stoves.

Although the expansion of piezoelectric crystals is small, the force they can exert is large and, when stacked together, they can serve as linear actuators. Also they can provide very small movements with extreme precision which makes them ideal for positioning atomic force microscopes.

Quartz, since it is a piezoelectric and not a ferroelectric, has no hysteresis loss when it oscillates, thus quartz crystal oscillators are widely used as frequency control devices in radios, computers, and watches. Since the frequency is a function of the mass of the crystal, they can serve as deposition monitors (quartz crystal microbalances) with sensitivities of less than 1 ng. By functionalizing the surface to absorb specific gases, they can also act as chemical sensors. The temperature sensitivity of a quartz crystal oscillator can be minimized by choosing the cut of the crystal relative to the optical axis, which is necessary for its use as a frequency standard. On the other hand, a cut can be chosen to maximize the frequency dependence on temperature and quartz crystal thermometers with millikelvin resolution are available.

23.5.4 Electrets

Electrets are materials that can retain a high surface charge for a long period of time. Electrets can be made from ferroelectric materials that have been poled (heated above their Curie temperatures in the presence of a strong electric field to align all of the dipoles and then quenched). Or, charge may be sprayed on a dielectric by a corona discharge or by electron bombardment. Electrets are made in this fashion from polymers such as Mylar, Teflon, or carnauba wax. Obviously, the material must be a good insulator (σ for Teflon is $10^{-22}/\Omega$ cm compared to $10^{-14}/\Omega$ cm for quartz). With their frozen-in electric field, electrets find applications in speakers and microphones, in hearing aids, in particle removal from air, and in electrostatic printing processes such as xerography.

23.6 Summary

As nonconductors of electricity, dielectrics interact with electric fields through their ability to form dipoles, which is called polarization. There are three sources of polarization: electric, ionic, and dipolar.

Electronic polarization results from the orbits of the electrons being displaced from the nuclear charge by the electric field. The dielectric constant from the electronic polarization is denoted as $\varepsilon(\infty)$. The resonance frequency occurs in the ultraviolet.

Ionic polarization occurs in ionically and covalently bonded materials on which there is some charge separation. The ionic dielectric constant plus the electronic dielectric constant is denoted by $\varepsilon(0)$. The resonance frequency for ionic polarization is in the infrared.

Polar liquids and gases possess permanent dipoles which can be aligned by an applied electric field. The aligning energy provided by the field is in competition with thermal motion which tends to randomize the orientation of the dipoles. Such materials are called paraelectrics. When the field is removed, the dipoles relax with a time constant τ, which is on the order of 10^{-11} s for liquids such as water where the dipoles are free to rotate. These dipoles can couple with microwave radiation to cause heating of the material, which is the principle on which microwave ovens operate.

In dipolar solids, frozen polar liquids or solids with built-in dipole moments, the dipoles can also be aligned by an external field and are not free to rotate. In this case, the relaxation time is proportional to $\exp(-\phi/kT)$, where ϕ is the potential barrier that must be overcome for the dipole to rotate. If the potential barrier is on the order of the chemical bond, the relaxation time becomes virtually infinite. Such solids are said to have spontaneous or permanent dipole moments and are called pyroelectrics. If the dipole moment can be reversed by an electric field, the material is also a ferroelectric.

Ferroelectric materials behave similarly to ferromagnetic materials. They form domains in which a number of adjacent cells have the same dipole orientation, they exhibit a hysteresis when subjected to an alternating electric field, they have a critical temperature called the Curie temperature, above which their spontaneous dipole moment is lost, and they obey the Curie–Weiss law, $P = C/(T - T_C)$ in which they become paraelectric above the Curie temperature, T_C.

Ferroelectrics belong to two general classes, those that undergo order–disorder transition such as Rochelle salts, and those that undergo displacive transitions such as the cubic perovskites. Ferroelectric behavior is restricted to 10 of the 32 point groups that are dipolar. The Curie temperature is also the transition temperature from a point group that permits ferroelectric behavior to one that does not.

Since the permanent dipole moment in pyroelectrics changes with temperature, they are used as infrared and far infrared detectors. Ferroelectrics find use in capacitors and in computer memories.

Ferroelectrics are also members of the 20 point groups that have no inversion center. Noncentrosymmetric crystals may not have a dipole moment in their relaxed state, but when distorted, the center of negative charge is separated from the center of positive charge and a dipole moment results. Such crystals are piezoelectrics. All ferroelectrics and pyroelectrics are also piezoelectrics, but not all piezoelectrics are ferroelectrics. Piezoelectrics are used in strain and pressure gages, accelerometers, microphones and speakers, and linear actuators.

Appendix: Internal Field Correction for Ionic Dielectric Function

Some solid-state physics texts (e.g., Kittel) include the correction for internal fields in calculating the electronic dielectric function but ignore it when calculating the ionic dielectric function and obtain the same result as Equation 23.17. Instead of computing the internal field corrections for ionic and electronic contributions to the dielectric function separately and then adding them, Ashcroft and Mermin combine the polarizabilities of the ions and the electrons and then apply the Clausius–Mossotti equation to the sum:

$$\frac{\varepsilon(\omega) - 1}{\varepsilon(\omega) + 2} = \frac{N_{ion} e^2}{3\varepsilon_0 \mu (\varpi^2 - \omega^2)} + \frac{\sum n_i \alpha_i}{3\varepsilon_0}, \tag{A.23.1}$$

where ϖ is the natural lattice frequency. For $\omega \gg \varpi$ (but $\ll \omega_0$) the ionic term drops out and the electronic contribution is seen to be

$$\frac{\sum n_i \alpha_i}{3\varepsilon_0} = \frac{\varepsilon(\infty) - 1}{\varepsilon(\infty) + 2}, \tag{A.23.2}$$

which is the result we obtained in Equation 23.49.

Now for $\omega \ll \varpi$,

$$\frac{\varepsilon(0) - 1}{\varepsilon(0) + 2} = \frac{N_{ion} e^2}{3\varepsilon_0 \mu \varpi^2} + \frac{\varepsilon(\infty) - 1}{\varepsilon(\infty) + 2}. \tag{A.23.3}$$

Put this back into Equation A.23.1 and write

$$\frac{\varepsilon(\omega) - 1}{\varepsilon(\omega) + 2} = \frac{1}{1 - \omega^2/\varpi^2} \left[\frac{\varepsilon(0) - 1}{\varepsilon(0) + 2} - \frac{\varepsilon(\infty) - 1}{\varepsilon(\infty) + 2} \right] + \frac{\varepsilon(\infty) - 1}{\varepsilon(\infty) + 2}. \tag{A.23.4}$$

It would be more convenient to be able to write Equation A.23.4 in the form

$$\varepsilon(\omega) = \varepsilon(\infty) + \frac{\varepsilon(0) - \varepsilon(\infty)}{1 - \omega^2/\omega_T^2}, \tag{A.23.5}$$

where ω_T is the observed frequency where the ionic contribution causes $\varepsilon(\omega)$ to become singular. Equation A.23.5 is consistent with 23.A4 if

$$\omega_T^2 = \varpi^2 \frac{\varepsilon(\infty) + 2}{\varepsilon(0) + 2} \text{ (proof left to the reader).} \tag{A.23.6}$$

Since $\varepsilon(\infty) < \varepsilon(0)$, the effect of the internal field is to lower the resonance frequency.

The final result, Equation A.23.5 is the same as the real part of the electronic and ionic dielectric function found in Equation 23.19. The question is whether it is more accurate to calculate the ionic contribution with or without the correction for the local field (see Problems 6–8).

Bibliography

Ashcroft, N.W. and Mermin, N.D., *Solid State Physics*, Brooks Cole, Philadelphia, 1976.

Barsoum, M.W., *Fundamentals of Ceramics*, Institute of Physic Publishing, Bristol and Philadelphia, 2003.

Dekker, A.J., *Solid State Physics*, Prentice-Hall, Englewood Cliffs, NJ, 1962.

Fischer-Cripps, A.C., *The Materials Physics Companion*, Taylor & Francis, New York, 2008.

Frohlich, H., Dielectric properties of dipolar solids, *Proc. Royal Soc. Lond. A.*, 185/1003, 399–414, 1946.

Hummel, R.E., *Electronic Properties of Materials*, 3rd edn., Springer, New York, 2000.

Kittel, C., *Introduction to Solid State Physics*, 7th edn., John Wiley & Sons, New York, 1966.

Naranjo, O.B., Gimzewski, J.K., and Putterman, S., Observation of nuclear fusion driven by a pyroelectric crystal. *Nature*, April 28, 2005.

Omar, M.A., *Elementary Solid State Physics*, Addison-Wesley, Reading, MA, 1975.

Yacoby, Y. and Girshberg, Y., Theory of ferroelectric phase transitions in pure and mixed perovskites, *Mat. Res. Soc. Symp. Proc.*, Vol. 718, Materials Research Society, Warrendale, Pennsylvania, 2002.

Problems

1. How would you define the dielectric constant of a ferroelectric material? How would you go about measuring it?

2. The relaxation time for liquid water is 10^{-11} s. What microwave frequency would cause maximum heating? Actual microwave ovens operate at 2.45 GHz, somewhat less than the frequency from maximum heating. Why do you suppose this is?

3. Can ferroelectric behavior be described in terms of the polarization catastrophe in the Clausius–Mossotti equation? Use Equation 23.42 to relate the Clausius–Mossotti local field to the polarization in the medium. Then use Equation 23.28 to relate the temperature to the polarization assuming E as the local field. Now use a procedure similar to that used to develop the Curie–Weiss law (Equation 23.55) to obtain the Curie temperature and the Curie constant for $BaTiO_3$. Does the result agree with observed behavior?

TABLE A.23.1

Static and Electronic Dielectric Constants
for Selected Ionic Compounds

Material	$\varepsilon(0)$ obs	$\varepsilon(\infty)$ obs
LiF	8.9	1.9
LiCl	12	2.7
NaCl	5.9	2.25
NaBr	6.4	2.6
KCl	4.85	2.1
KI	5.1	2.7

Source: Data from Kittel, C., *Introduction to Solid State Physics*, 7th edn., John Wiley & Sons, New York, 1966.

4. The "g" piezoelectric coefficient is the electric field produced by an applied stress in units of volts meter per Newton. Derive an expression relating the electric displacement, D, to the strain, e_x, produced in the crystal by the applied stress.

5. Show that substituting $\omega_T^2 = \varpi^2(\varepsilon(\infty) + 2)/(\varepsilon(0) + 2)$ into Equation A.23.4 results in Equation A.23.5.

6. Use the calculated values for the lattice frequency ω_T in Table 16.1 to compute the ionic contribution to the dielectric function without the Clausius–Mossotti correction for internal fields (Equation 23.18) and compare with observed measurements in Table A.23.1. Now do the same using the correction developed by Ashcroft and Mermin. Does the correction improve the result?

7. Repeat Problem 6 using the observed values for ω_T.

8. Now use the $\varepsilon(0)$ and $\varepsilon(\infty)$ values in Table A.23.1 with Equation A.23.6 to find ϖ. Then use Equation A.23.6 to to reduce ϖ to ω_T and compare with the observed values. What can you conclude about the need to correct the ionic contribution to the dielectric function for internal fields with the Clausius–Mossotti equation?

24

Optical Properties of Materials

Having developed the dielectric function for dielectric materials in Chapter 23, we are now in position to explore the optical properties of nonconductive materials. We will then modify the dielectric function to account for free electrons and will explore the optical properties of metals and other conductive media.

24.1 Review of Electricity and Magnetism

In order to understand the optical properties of materials, it may be useful to review some basic electromagnetic theory that governs the interaction of electromagnetic waves with matter. We will start with Maxwell's equations.

24.1.1 Maxwell's Equations

Around 1865, James Clerk Maxwell reformulated what was known about electricity and magnetism into four equations, known as Maxwell's equations. These can be written in differential form using vector notation as

$$\nabla \cdot \mathbf{D} = \rho \quad \nabla \times \mathbf{E} = -\dot{\mathbf{B}}$$
$$\nabla \cdot \mathbf{B} = 0 \quad \nabla \times \mathbf{H} = \mathbf{J} + \dot{\mathbf{D}}, \tag{24.1}$$

where $\mathbf{D} = \varepsilon \mathbf{E}$, $\mathbf{B} = \mu \mathbf{H}$, $\mathbf{J} = \sigma \mathbf{E}$. This set of equations describes all known electromagnetic phenomena (which is quite remarkable considering that the electron had yet to be discovered). The $\nabla \cdot \mathbf{D} = \rho$ equation is a differential form of Gauss's law, from which Coulomb's law can be derived as a special case. The $\nabla \times \mathbf{E} = -\dot{\mathbf{B}}$ equation is a generalized form of Faraday's law and $\nabla \times \mathbf{H} = \mathbf{J} + \dot{\mathbf{D}}$ is a generalized form of Ampere's law. The $\nabla \cdot \mathbf{B} = 0$ describes the fact that magnetic field lines are closed loops, i.e., they do not start and end on charges as do electric field lines because isolated magnetic charges (magnetic monopoles) do not seem to exist. If one is discovered, this equation would have to be modified to add a magnetic charge density like the electric charge density in the $\nabla \cdot \mathbf{D} = \rho$ equation and a magnetic current term would have to be added to the $\nabla \times \mathbf{E}$ equation to match the $\mathbf{J}$ in the $\nabla \times \mathbf{H}$ equation. It is this added symmetry to the equations that has prompted many researchers to look for magnetic monopoles, but nobody has found one yet.

In vacuum $\rho=0$, $\sigma=0$, $\varepsilon=\varepsilon_0$, and $\mu=\mu_0$. The four equations simplify to

$$\nabla \cdot \mathbf{D} = 0 \quad \nabla \times \mathbf{E} = -\dot{\mathbf{B}}$$
$$\nabla \cdot \mathbf{B} = 0 \quad \nabla \times \mathbf{H} = \dot{\mathbf{D}}. \tag{24.2}$$

24.1.2 Wave Equation

Take the curl of the $\nabla \times \mathbf{E}$ equation,

$$\nabla \times (\nabla \times \mathbf{E}) = -\nabla \times \dot{\mathbf{B}} \tag{24.3}$$

and use the vector identity for curl-curl

$$\nabla(\nabla \cdot \mathbf{E}) - \nabla^2 \mathbf{E} = -\nabla \times \dot{\mathbf{B}} = -\mu_0 \left(\nabla \times \dot{\mathbf{H}}\right). \tag{24.4}$$

But $\nabla \cdot \mathbf{E} = 0$ and $\nabla \times \dot{\mathbf{H}} = \dot{\mathbf{D}} = \varepsilon_0 \dot{\mathbf{E}}$, therefore,

$$\ddot{\mathbf{E}} = \frac{1}{\mu_0 \varepsilon_0} \nabla^2 \mathbf{E}, \tag{24.5}$$

which we recognize as a three-dimension vector wave equation for the E-vector. A similar equation can be derived for the **H**-vector. For an electromagnetic wave traveling in the x-direction and oscillating in the y-direction, the wave equation is

$$\frac{\partial^2 E_y(x,t)}{\partial t^2} = \frac{1}{\mu_0 \varepsilon_0} \frac{\partial^2 E_y(x \cdot t)}{\partial x^2}, \tag{24.6}$$

which has a solution in the form

$$E_y(x,t) = E_0 e^{i(kx+\omega t)}. \tag{24.7}$$

Carrying out the differentiation, we see that

$$\frac{\partial^2 E_y(x,t)}{\partial t^2} = -E_0 \omega^2 e^{i(kx+\omega t)} \quad \text{and} \quad \frac{\partial^2 E_y(x \cdot t)}{\partial x^2} = -E_0 k^2 e^{i(kx+\omega t)}. \tag{24.8}$$

Therefore, putting this back into the wave equation requires that

$$\omega = \frac{k}{\sqrt{\mu_0 \varepsilon_0}}. \tag{24.9}$$

Since $\omega = 2\pi\nu$, $k = 2\pi/\lambda$, and since the wavelength of a wave is given by its speed divided by its frequency, the speed of an electromagnetic wave in vacuum must be

$$c = \frac{1}{\sqrt{\mu_0 \varepsilon_0}}. \tag{24.10}$$

This is a truly remarkable result. We not only see that the speed of light c is a universal constant that is interrelated to the universal constants μ_0 and ε_0 but that this speed is independent of any reference system. You may recall that this was the key assertion in Einstein's theory of special relativity. Here this was predicted from Maxwell's equations some 30 years before.

24.1.3 Index of Refraction

If the electromagnetic wave is traveling through a medium with no free charge whose permittivity is ε and permeability is μ, its speed will be given by

$$v = \frac{1}{\sqrt{\mu\varepsilon}}. \tag{24.11}$$

If this medium is nonmagnetic ($\mu = \mu_0$), the relative speed

$$n = \frac{c}{v} = \sqrt{\frac{\varepsilon}{\varepsilon_0}} = \sqrt{\varepsilon_r}. \tag{24.12}$$

This relative speed is called the index of refraction, which is given by the square root of the dielectric constant.

24.1.4 Reflection and Transmission of Electromagnetic Waves

We can show that the *E*- and *H*-vectors are perpendicular to one another and obtain a relation between their relative magnitudes from the $\nabla \times \mathbf{E} = -\dot{\mathbf{B}}$ equation. Assume the direction of propagation is in the *x*-direction and that the *E*-vector oscillates in the *x*–*y* plane. Expanding the curl $\nabla \times \mathbf{E}$ and retaining only the *x*-variation of the *y*-component of $\mathbf{E}$, we get

$$\frac{\partial E_y}{\partial x} = -\frac{\partial B_z}{\partial t} = -\mu \frac{\partial H_z}{\partial t} \tag{24.13}$$

or the *z*-component of $\mathbf{H}$. Now expand the $\nabla \times \mathbf{H} = \mathbf{J} + \dot{\mathbf{D}}$ equation with $\mathbf{J} = 0$, keeping only the *x*-variation of the *z*-component,

$$-\frac{\partial H_z}{\partial x} = \varepsilon \frac{\partial E_y}{\partial t}. \tag{24.14}$$

Now take the indicated derivatives of the solutions to the wave equations, $E_y = E_0 e^{i(kx + \omega t)}$ and $H_z = H_0 e^{i(kx + \omega t)}$ and put them into the Equations 24.13 and 24.14 above,

$$ikE_0 = -i\omega\mu H_0$$
$$-ikH_0 = i\omega\varepsilon E_0. \tag{24.15}$$

Multiply $ikE_0 = -i\omega\mu H_0$ by εE_0, the $-ikH_0 = i\omega\varepsilon E_0$ by μH_0 and adding both, we get $\varepsilon E_0^2 = \mu H_0^2$. Thus we see that the amplitudes of *E* and *H* are related by

$$\frac{|\mathbf{E}|}{|\mathbf{H}|} = \sqrt{\frac{\mu}{\varepsilon}}. \tag{24.16}$$

The energy flux propagated by the electromagnetic wave is give by $\mathbf{P}$, the Poynting vector defined as $\mathbf{P} = \mathbf{E} \times \mathbf{H}$, or

$$|P| = E^2 \sqrt{\frac{\varepsilon}{\mu}} = H^2 \sqrt{\frac{\mu}{\varepsilon}}. \tag{24.17}$$

Now let us consider what happens when an electromagnetic wave traveling through vacuum encounters a material with different μ and ε. For simplicity, we consider only the case of normal incidence. The E and H vectors must be conserved so that, if part of the incident wave is reflected, the remainder must be transmitted. This can be expressed as

$$E_i - E_R = E_T \quad \text{and} \quad H_i - H_R = H_T. \tag{24.18}$$

Using the above relations between E and H, we can write

$$\frac{H_i}{E_i} = \sqrt{\frac{\varepsilon_0}{\mu_0}}, \quad \frac{H_R}{E_R} = -\sqrt{\frac{\varepsilon_0}{\mu_0}}, \quad \frac{H_T}{E_T} = \sqrt{\frac{\varepsilon}{\mu}}, \tag{24.19}$$

which can be used to replace the components of H in the above conservation relationship to give

$$E_i \sqrt{\frac{\varepsilon_0}{\mu_0}} + E_R \sqrt{\frac{\varepsilon_0}{\mu_0}} = E_T \sqrt{\frac{\varepsilon}{\mu}} = (E_i - E_R) \sqrt{\frac{\varepsilon}{\mu}}, \tag{24.20}$$

from which

$$\frac{E_R}{E_i} = \frac{\sqrt{\varepsilon/\mu} - \sqrt{\varepsilon_0/\mu_0}}{\sqrt{\varepsilon/\mu} + \sqrt{\varepsilon_0/\mu_0}} \longrightarrow \mu \longrightarrow \mu_0 \frac{\sqrt{\varepsilon} - \sqrt{\varepsilon_0}}{\sqrt{\varepsilon} + \sqrt{\varepsilon_0}} = \frac{n-1}{n+1}. \tag{24.21}$$

Since the reflected intensity is $|E_R|^2$, the reflection coefficient can be written as

$$R = \left|\frac{E_R}{E_i}\right|^2 = \left|\frac{n-1}{n+1}\right|^2 = \frac{(n-1)^2}{(n+1)^2}. \tag{24.22}$$

This can be generalized to account for an electromagnetic wave in medium with n_1 incident on medium n_2 by writing

$$R = \left|\frac{n_1 - n_2}{n_1 + n_2}\right|^2. \tag{24.23}$$

The transmission coefficient is

$$T = 1 - R = \frac{4 n_1 n_2}{(n_1 + n_2)^2}. \tag{24.24}$$

24.1.5 Complex Index of Refraction

However, we shall soon see that the index of refraction is a complex quantity for most materials, which we will write as $\tilde{n} = N + iK$ where N is the real part and K is the imaginary

part. These quantities, which are sometimes called the optical constants of a material (although they do depend on ω), characterize the optical properties of any material. (Usually, in the literature, you will see these as lowercase n and k, but we will use upper case here to keep from confusing these quantities with the other quantities we use n and k to denote.) Therefore, we must replace the values for n in the above equations for R and T with the complex $\tilde{n}$.

$$R = \left|\frac{\tilde{n}_1 - \tilde{n}_2}{\tilde{n}_1 + \tilde{n}_2}\right|^2 = \left(\frac{\tilde{n}_1 - \tilde{n}_2}{\tilde{n}_1 + \tilde{n}_2}\right)\left(\frac{\tilde{n}_1^* - \tilde{n}_2^*}{\tilde{n}_1^* + \tilde{n}_2^*}\right) = \frac{N_1^2 + K_1^2 + N_2^2 + K_2^2 - 2N_1N_2 - 2K_1K_2}{N_1^2 + K_1^2 + N_2^2 + K_2^2 + 2N_1N_2 + 2K_1K_2}, \quad (24.25)$$

where the * denotes the complex conjugate. If either the incident or the emerging wave is in air, for which $n = 1.0007$, this equation reduces to

$$R = \frac{(N - 1)^2 + K^2}{(N + 1)^2 + K^2}. \quad (24.26)$$

This is the form we use in our following discussions.

24.1.6 Absorption of Electromagnetic Radiation

Now let us examine the consequences of the imaginary component of the complex index of refraction. Recall that the solution to the wave equation required that $k = \omega\sqrt{\mu_0\varepsilon}$. Multiplying the top and bottom $k = \omega\sqrt{\mu_0\varepsilon}$ by c, and using the fact that $c = 1/\sqrt{\mu_0\varepsilon_0}$, we get

$$\tilde{k} = \frac{\omega\sqrt{\mu_0\varepsilon}}{c\sqrt{\mu_0\varepsilon_0}} = \frac{\omega\tilde{n}}{c} = \frac{\omega(N + iK)}{c}. \quad (24.27)$$

So we see that k is also complex. Now the solution to the wave equation becomes

$$E = E_0 e^{i(\tilde{k}x + \omega t)} = E_0 e^{i(\omega Nx/c + \omega t)} e^{-\omega Kx/c}. \quad (24.28)$$

This is the equation for a damped wave or an exponentially decaying wave. Thus the presence of an imaginary component in the complex index of refraction is to cause the propagating wave to lose energy. Thus the K is sometimes called the extinction coefficient. Since the intensity of the radiation $I \sim EE^*$, we can write

$$I(x) = I_0 e^{-2\omega Kx/c} = I_0 e^{-\alpha x}, \quad (24.29)$$

which is known as Beer's law and where we have identified the absorption coefficient α as

$$\alpha = \frac{2\omega K}{c}. \quad (24.30)$$

24.2 Optical Properties of Dielectric Materials

For the most part, dielectrics are transparent throughout the visible spectrum, assuming their bandgap is greater than $\sim$3.8 eV (the energy of a near-ultraviolet [UV] photon).

Having said that, it should be understood that dielectrics are transparent only as single crystals or in the amorphous state (glasses). In a polycrystalline material, the light is lost due to multiple scattering at the grain boundaries unless the wavelength is much greater than the average crystal size (e.g., polycrystalline dielectric materials with crystals on the order of micrometers are transparent to microwave (λ ranges from millimeter to centimeter) radiation, but not to visible radiation).

In the previous section we showed (Equation 24.12) that the index of refraction was simply the square root of the dielectric constant and, since the dielectric constant is a function of frequency and can become negative, the index of refraction is a function of frequency and may be complex. Summarizing,

$$\tilde{n}(\omega) = N(\omega) + iK(\omega) = \sqrt{\varepsilon_r(\omega)}. \tag{24.31}$$

Having spent a fair amount of effort in Chapter 23 determining the dielectric function of dielectric materials, we are now in a position to obtain their N and K values at various wavelengths, which describe their optical properties.

24.2.1 Visible and Near-Ultraviolet

Having found the electronic dielectric function, Equation 23.49 to be given by

$$\varepsilon_r(\omega) = \frac{1 + (2/3\varepsilon_0) \sum_i n_i \alpha_i(\omega)}{1 - (1/3\varepsilon_0) \sum_i n_i \alpha_i(\omega)} \tag{24.32}$$

and the polarizability to be given by Equation 23.50 (we can ignore the imaginary part for the time being)

$$\alpha(\omega) = \frac{e^2/m}{\omega_0^2 - \omega^2}. \tag{24.33}$$

The α values are given for $\omega \ll \omega_0$, so ω_0 can be found from Equation 24.33.

We can now determine the optical properties of a material in the visible range. But first we have to know the polarizabilities of the components. Fortunately, these have been determined and the values for some common optical materials are given in Table 24.1. By assuming the electron density surrounding the nucleus is uniform, one can show using Gauss law that the polarization should be given by $4\pi\varepsilon_0 r^3$, where r is the ion radius. This value is also included in Table 24.1 for comparison with the measured polarizability.

One can see that using $4\pi\varepsilon_0 r^3$ estimates the polarizability to within an order of magnitude, but is not accurate enough to make useful predictions of the optical properties of materials.

As an example, we will plot the N and K values for the SiO_2 molecule, the building block of quartz and the silicate glasses. Using the values in Table 24.1, from Equation 24.33 we find the resonance frequencies for the Si^{4+} ion to be 1.251×10^{17} rad/s and for the O^{2-} to be 8.075×10^{15} rad/s. There are 2.205×10^{28} ion pairs/m^3, one third of which are Si and two third are O ions. Forming the sums and inserting them in Equation 24.33, the $N(\omega) = Re\left[\sqrt{\varepsilon_r(\omega)}\right]$ and $K(\omega) = Im\left[\sqrt{\varepsilon_r(\omega)}\right]$. The $\varepsilon(0) = 1.946$ and $N(0) = 1.395$ and N at 3.543×10^{15} rad/s (532 nm) is 1.505. The measured index of refraction of crystalline quartz at 532 nm is 1.55.

TABLE 24.1

Polarizability of Selected Ions

Ion	α (Measured)	$4\pi\varepsilon_0 r^3$
C^{4+}	0.001	0.005
B^{3+}	0.003	0.014
Si^{4+}	0.018	0.071
Li^+	0.033	0.350
Ti^{4+}	0.206	0.350
Na^+	0.222	1.18
Zr^{4+}	0.411	0.548
Ca^{2+}	0.523	1.11
Y^{3+}	0.612	0.894
K^{2+}	1.00	2.92
F^-	1.33	2.62
Cl^-	4.07	6.59
O^{2-}	4.32	3.05

Source: Polarizability values were taken from Kittel, C., *Introduction to Solid State Physics*, 7th edn., John Wiley & Sons, New York, 1966.

Note: Polarization values are in 10^{-40} MKS units.

The N and K values are plotted over the visible and near UV spectrum in Figure 24.1. The plot is separated to show the detail near resonance at 7.043×10^{15} rad/s and where the N becomes greater than zero at 9.8×10^{15} rad/s. Between these frequencies, the N is zero and from Equation 24.26 the reflectivity will have the value of 1. Therefore, the high frequency limit of quartz is 7.043×10^{15} rad/s, which is equivalent to 267 nm (UV). For very thin films (<267 nm), some radiation can penetrate, but it will be quickly absorbed because of the very large K in this region. For $\omega \gg 9.8 \times 10^{15}$ rad/s, the electronic vibrations are no longer able to keep up with the oscillating E-field of the photons, thus there will little

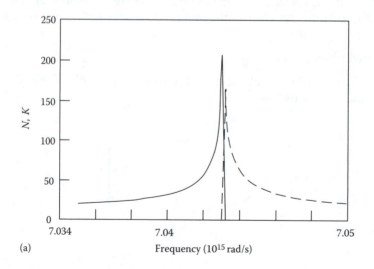

(a)

FIGURE 24.1

(a) N (solid line) and K (dashed line) values for SiO_2 (quartz) in the vicinity of the resonant frequency. A small damping factor was used in these calculations to keep the N finite.

(continued)

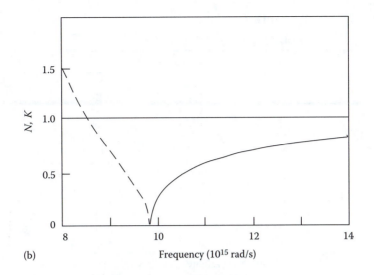

(b)   Frequency (10^{15} rad/s)

FIGURE 24.1 (continued)

(b) N (solid line) and K (dashed line) values for SiO_2 (quartz) in the vicinity of the frequency, where ε becomes positive again (see Figure 23.7). Here the N becomes finite as K goes to 0.

interaction between the photons and the dielectric and the N will return to its vacuum value of 1. Note that in this region $N < 1$, which implies the phase velocity $= c/N$ will be greater than the velocity of light.

Notice that the N increases with frequency until the resonant frequency is reached. This increasing index with frequency is called dispersion and is responsible for blue light being refracted more than red light as it passes through a prism. The N decreases with frequency in the region between the resonant frequency and the point where the dielectric function becomes positive again. This decrease of N with increasing frequency is called anomalous dispersion.

The dispersion in the visible is seen in Figure 24.2, where N is plotted against λ. Notice the $N = 1.505$ at $\lambda = 532$ nm, very close to the observed value of 1.55 as mentioned previously.

24.2.2 Infrared Absorption

We saw from Equation 23.19 that the dielectric function for ionic or mixed ionic-covalent compounds could be written as

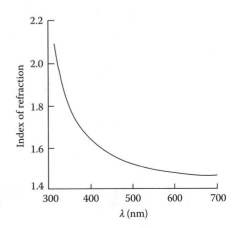

FIGURE 24.2

Calculated index of refraction for quartz in the visible and near-UV.

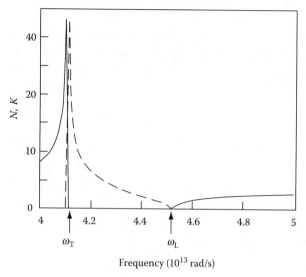

FIGURE 24.3

N (solid) and K (dashed) values for InAs. The transverse and longitudinal frequencies are indicated.

$$\varepsilon(\omega) = \varepsilon(\infty) + \frac{[\varepsilon(0) - \varepsilon(\infty)]\omega_T^2}{\omega_T^2 - \omega^2}. \tag{24.34}$$

Taking the real and imaginary parts of the square root of Equation 24.34, a typical plot of the optical constants for an ionic or mixed covalent/ionic compound is shown in Figure 24.3. This particular plot was calculated for InAs using $\varepsilon(0) = 14.9$, $\varepsilon(\infty) = 12.3$, and $\omega_T = 4.1 \times 10^{13}$ rad/s, which corresponds to a wave number of 218 cm^{-1} (data from Kittel, 1996). Notice that the $N > 0$ at 4.5×10^{13} rad/s in accordance with the Lyddane, Sachs, Teller (LST) relationship.

Figure 24.4 is a plot of the reflectivity of InAs calculated from Equation 24.26 compared with the observed value. The observed reflectivity was computed from the N and K values from Palik, 1997.

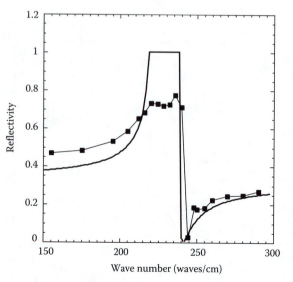

FIGURE 24.4

Calculated (solid) versus measured (squares) reflectivity for InAs in the IR. (Measured values calculated from Palik, E.D., Ed., *Handbook of Optical Constants of Solids*, Vols. 1–3, Academic Press, 1997.)

24.2.3 Phonon–Photon Coupling

Previously we had only considered the optical modes at $k=0$ because the large difference between the momentum of a photon and a phonon would not permit energy and momentum to be conserved otherwise. Obviously, k cannot be exactly zero, so let us explore the region of small k by developing the dispersion relation ω versus k where the photons can couple with phonons. Rewriting Equation 23.34,

$$\varepsilon(\omega) = \varepsilon(\infty) + \frac{\varepsilon(0)\omega_T^2}{\omega_T^2 - \omega^2} - \frac{\varepsilon(\infty)\omega_T^2}{\omega_T^2 - \omega^2} = \varepsilon(\infty)\left[1 - \frac{\omega_T^2}{\omega_T^2 - \omega^2}\right] + \frac{\varepsilon(0)\omega_T^2}{\omega_T^2 - \omega^2}. \tag{24.35}$$

Using the LST relationship,

$$\varepsilon(\omega) = \varepsilon(\infty)\left[\frac{\omega^2}{\omega_T^2 - \omega^2}\right] + \frac{\varepsilon(\infty)\omega_L^2}{\omega_T^2 - \omega^2} = \varepsilon(\infty)\left[\frac{\omega_L^2 - \omega^2}{\omega_T^2 - \omega^2}\right]. \tag{24.36}$$

The index of refraction $N = c/v = ck\omega = \varepsilon(\omega)^{1/2}$. We can write the dispersion relationship as

$$\frac{c^2 k^2}{\omega^2} = \varepsilon(\infty)\left[\frac{\omega_L^2 - \omega^2}{\omega_T^2 - \omega^2}\right]. \tag{24.37}$$

Solving this equation for ω and plotting ω versus k provides a dispersion curve for NaCl as shown in Figure 24.5. Notice that the curve has two branches; the upper branch is the phonon dispersion relation for the longitudinal mode and approaches ω_L at $k=0$ while the lower branch is the dispersion curve for the transverse mode, which approaches ω_T as k increases. No frequencies can propagate between ω_T and ω_L, which causes an energy gap in this region.

The group velocity of light $(\partial\omega/\partial k)$ is zero at ω_T and ω_L. The index of refraction N can be found from Figure 24.5 by dividing the velocity of light by the phase velocity (ω/k) and plotting this against ω. Since the ω in the lower branch can never exceed ω_T as k increases, the phase velocity gets smaller and the index of refraction increases dramatically as ω

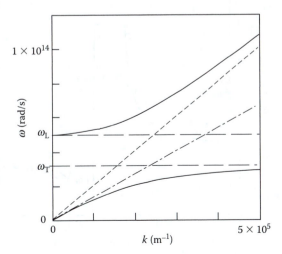

FIGURE 24.5
Dispersion relationship (ω vs. k) for NaCl in the vicinity of small k. The dashed diagonal line represents the speed of light in the medium with $\varepsilon(\infty)$ and the dash-dot diagonal line represents the speed of light in a medium with $\varepsilon(0)$. The horizontal lines at ω_T and ω_L are the energy gap. The lattice cannot propagate frequencies between these frequencies.

approaches ω_T, as in Figure 24.3. In the upper branch, the phase velocity is infinite at $k = 0$ and quickly approaches the velocity of light in a medium with dielectric constant $\varepsilon(\infty)$.

The lower curve in Figure 24.5 represents the coupling of photons with the transverse lattice vibrations. In this case, the photon and phonon travel together with the same frequency and wavelength. The quanta of this coupled wave are called polaritons. The absorption edge is ω_L at 5×10^{13} rad/s ($\lambda = 60.8$ μm). The ω_T is 3.1×10^{13} rad/s ($\lambda = 37.7$ μm in free space). A photon just below this frequency represented by the end of the lower branch in Figure 24.5 would have a k of 5×10^5/m or $\lambda = 12.5$ μm in the medium giving it a velocity of 6.7×10^7 m/s corresponding to $N = 4.8$. One can see that the velocity of the photon in the medium rapidly falls with increasing k as the $N = k/\omega$ increases.

Because there is an energy gap between ω_T and ω_T, no photon can enter and propagate in the medium, the material becomes reflective as shown in Figure 24.4. This region is sometimes called the *reststrahlen* (residual ray) band and can be used to obtain a narrow band of infrared (IR) radiation by reflection from a blackbody source.

As mentioned in Chapter 16, a photon at normal incidence to the crystal can only excite a transverse wave as illustrated in Figure 16.5. To excite a longitudinal wave along a line of diatomic ions, there must be a component of the E-vector along the line as illustrated in Figure 24.6. Vertically polarized photons impinging obliquely on a thin film of a material such as LiF deposited on a reflective metal substrate such as Au or Ag will be strongly absorbed at the longitudinal frequency while virtually no absorption is observed at the transverse frequency or for horizontally polarized light at the longitudinal frequency (see Chapter 10 in Kittel, 1996). The IR reflective film doubles the path and enhances the absorption.

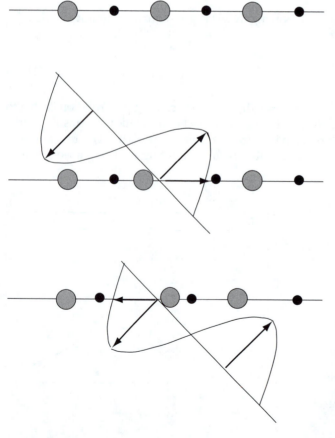

FIGURE 24.6
Vertically polarized light exciting a longitudinal vibration in a line of diatomic ions. Undisturbed lattice (top) is excited by a vertically polarized photon causing the cations (black) and the anions (gray) to oscillate back-and-forth relative to each other.

24.2.4 Dipolar Absorption

The dielectric function for a medium containing dipolar molecules that are free to rotate can be written as $\varepsilon_{dip}(\omega) = \varepsilon(0) + \chi_{dip}(\omega)$, where $\varepsilon(0)$ is the dielectric constant without the dipolar molecules. The dielectric susceptibility for such dipolar molecules was found in Equation 23.33 to be

$$\tilde{\chi}_{dip}(\omega) = \frac{\chi_{dip}(0)}{1 + \omega^2 \tau^2}(1 - i\omega\tau), \tag{24.38}$$

where $\chi_{dip}(0)$ is given by Equation 23.36 as $\chi_{dip}(0) = N_{dip}(p_{dip}^2/3\varepsilon_0 kT)$. Since $\tilde{\chi}_{dip}(\omega)$ is complex, the dielectric function will also be complex and can be written as the sum of a real and imaginary part, $\tilde{\varepsilon}(\omega) = \varepsilon'(\omega) + i\varepsilon''(\omega)$, where

$$\varepsilon'_{dip}(\omega) = \varepsilon(0) + \frac{\chi_{dip}(0)}{1 + \omega^2 \tau^2}; \quad \varepsilon''_{dip}(\omega) = \frac{\chi_{dip}(0)}{1 + \omega^2 \tau^2}(\omega\tau). \tag{24.39}$$

These two equations are sometimes referred to as the Debye equations. Note that ε''_{dip} is maximized when $\omega = 1/\tau$ and ε'_{dip} is reduced by ½, thus $\omega = 1/\tau$ could be considered the damping frequency and plays a similar role as the plasma frequency in a conductive medium.

When the dielectric function itself is complex, we must take the square root of a complex number or $\tilde{n} = \sqrt{\tilde{\varepsilon}}$ where the tildes represent complex quantities. We can write

$$\tilde{n}^2 = (N + iK)^2 = N^2 + 2iNK - K^2 = \varepsilon' + i\varepsilon'' \tag{24.40}$$

and identify $\varepsilon' = N^2 - K^2$ and $\varepsilon'' = 2NK$. Solving these last two equations simultaneously for N and K, we get

$$N^2 = \frac{\varepsilon'}{2} + \frac{1}{2}\sqrt{\varepsilon'^2 + \varepsilon''^2} \quad \text{and} \quad K^2 = -\frac{\varepsilon'}{2} + \frac{1}{2}\sqrt{\varepsilon'^2 + \varepsilon''^2}. \tag{24.41}$$

At very low frequencies ($\omega\tau \ll 1$), $N^2 \rightarrow \varepsilon' = \varepsilon(0) + \chi_{dip}(0)$ because dipole rotation is able to follow the changing **E**-vector and the polarization approaches the static value resulting in a high index of refraction and high reflectivity. However, as ω approaches $1/\tau$, the damping causes the extinction coefficient K to rise as the N falls, resulting in strong absorption of radiation as shown in Figure 24.7.

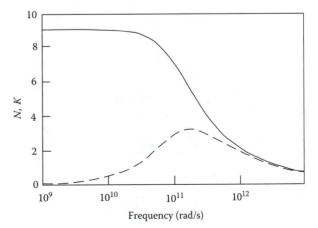

FIGURE 24.7
N and K values for a dipolar medium with a static dielectric constant of 80 and a relaxation time $\tau = 10^{-11}$ s, values typical of water.

24.2.5 Other Absorption Phenomena

Generally the optical transmission of dielectric media is limited at high frequencies either by the bandgap energy or by the electronic polarization near ω_0 and at the lower frequencies by the *reststrahlen* region or by molecular absorption. However, there are other absorptive processes that can occur in the intervening region.

24.2.5.1 Inherent Absorption

So far we have treated the optical properties of dielectrics as if they were perfect insulators. However, as seen in Chapter 23, there are several mechanisms by which some small electrical conduction can take place. Consequently, as will be shown in the next section, the small but finite conduction will result in some losses in optical transmission.

24.2.5.2 Optical Scattering

Scattering from grain boundaries makes polycrystalline dielectrics translucent but not transparent. Optical transparency requires either single crystals or amorphous material (glasses) as discussed previously. Nanostructured materials can have grain boundaries much smaller than the wavelength of light and can be considered amorphous as far as optical transparency is concerned.

Inclusions or other defects in single crystals or glasses that are on the order of optical wavelengths are effective sources of scattering losses and can be treated using the Mie theory of scattering. A suspension of small, uniform metallic particles can scatter selective wavelengths and produce striking colors in glasses. For example, the ancient Romans were able to suspend nanoclusters of gold in a glass by controlled heating, which imparted a beautiful ruby color to it.

Even in the absence of such defects, Rayleigh scattering from the atoms or molecules in the glass will cause some scattering losses. Such scattering represents the theoretical minimum loss due to scattering. Optical fibers fabricated from high-purity silica have achieved losses close to this theoretical limit. Since Rayleigh scattering is inversely proportional to the fourth power of wavelength, there is great interest is developing optical communication links that operate further into the IR to reduce these losses.

24.2.5.3 Color Centers

Color centers result from electrons trapped in anion vacancies. Those involving single anion vacancies are called F-centers (from Fabre meaning color in German). Two adjacent anion vacancies produce what is known as an M-center, and three such adjacent vacancies is an R-center. The trapped electron has different quantum states that can be excited by photons in the visible spectrum resulting in an intense coloration in an otherwise transparent colorless crystal. The absorption frequency, hence the color, is a characteristic of the material.

To maintain charge neutrality, anion vacancies are normally compensated by cation interstitials (Frenkel defects), cation vacancies (Schottky defects), or by the vacancy caused by the substitution of a nonstoichiometric ion (cation with fewer electrons or anion with more electrons). Electron traps can be created by bombarding an alkali halide with radiation or neutrons to knock anions out of the lattice or by heating the material in the presence of the vapor of an alkali metal (for example, heating NaCl in a vapor of metallic Na).

Electron traps may also be created in other systems by disrupting the lattice by energetic radiation bombardment. Diamonds and other gemstones are artificially colored by exposing them to various forms of radiation followed by a partial anneal.

24.2.5.4 Impurity Absorption

The presence of impurities, especially transition metals or rare earths, can provide very strong absorption of selective wavelengths. For example, the presence of a few parts per million of Cr in otherwise colorless Al_2O_3 produces the red color in rubies while the presence of Ni produces the blue in sapphire. The unfilled *d*-electron shells in the transition metals (and the *f*-electrons in the rare earths) are degenerate in isolated atoms, but they can interact with the ligand fields of the surrounding atoms in a crystal causing a splitting of the energy levels. If the ΔE between the split levels happens to lie in the visible part of the spectrum, it can cause strong absorption of certain wavelengths.

24.2.5.5 Excitons

If a photon with energy equal or greater than the bandgap energy is incident on a semiconductor (or insulator), it will be absorbed by giving its energy to produce a hole in the valence band and a free electron in the conduction band. Although we tend to think of semiconductors being transparent to frequencies below the bandgap energy, it is possible for a photon with less than the bandgap energy to lift an electron from the valence band, but instead of going all the way to the conduction band, the electron is attracted to the hole it created to form a bound pair. Such a bound electron–hole pair is called an exciton and can be described by hydrogen-like wave functions. Since excitons can have several energy states, they are observed as small absorption lines at wavelengths slightly longer than the cut-on wavelength, the wavelength at which the semiconductor begins transmitting light. Excitons can move through the crystal carrying energy (but not charge) and eventually decay, giving up their energy as a photon.

24.2.6 Heavy Metal Glasses

There is considerable technological interest in IR transmissive materials with applications ranging from IR detectors and imaging systems, to fiber optics for telecommunications, chemical sensors, and for piping radiation from large CO_2 or Er:YAG lasers to the point where it is needed for surgery or industrial cutting operations.

Silica optical fibers, operating in the 1.2–1.6 μm range, are the backbone of the optical fiber telecommunications industry and were believed to have reached their theoretical limit in which optical losses are primarily due to Rayleigh scattering from the silica molecules. Since Rayleigh scattering is proportional to the inverse fourth power of the wavelength, and since the Si-O-Si molecular absorption limits silica to <2 μm, there was a great effort to develop optical fibers that could operate in the near-IR at wavelengths >2 μm. We saw in Equation 23.14 that the resonant frequency is proportional to the square root of the spring constant divided by the reduced mass of the ions. (Note: It can be shown that the frequency of the IR absorption edge $\omega_L^2 \sim \omega_{Debye}^2 (M + m)^2 / Mn$.) This fact created interest in increasing the mass of the ions in the system in order to move the resonant frequency further into the IR, which led to the development of nonoxide heavy metal glasses.

Much effort has gone into the development of ZBLAN (ZrF_4-BaF_2-LaF_3-AlF_3-NaF) and its derivatives to push the operational frequency out to 4 μm. Other efforts were made to develop chalcogenide glasses such as As_2S_3 and AsGeTeSe, which transmit out to 11 μm. Unfortunately, these systems are difficult to work with and are not as durable or robust as

silica. Also their losses are considerably higher than those in silica. Despite some excellent fundamental research into these heavy metal glasses, they are a long way from replacing silica optical fibers for telecommunications applications.

The chalcogenide fibers show promise for temperature sensors that carry blackbody radiation to a detector, for IR imaging using coherent fiber bundles, and for chemical sensors using evanescent wave spectroscopy (a sensing technique based on the inter-action of the internally reflected light in an optical fiber with its chemical environment). Hollow glass waveguides made from AgI-coated silica glass are showing the greatest promise for laser power delivery (Harrington, J.A.).

24.2.7 Birefringence

Crystals with cubic symmetry have three equivalent axes that interact with light in the same manner. This means that light passing through the crystal is refracted in the same manner without change in polarization regardless of its orientation. Such crystals are considered to be optically isotropic. The atoms or molecules in crystals with lower sym-metry will not have the same nearest neighbor interactions in each direction and are optically anisotropic. The c-axis in hexagonal, trigonal (rhombohedral), and tetragonal crystals has a unique symmetry and is called the optical axis. Light propagating along this axis behaves the same as in an isotropic crystal and is called the ordinary ray. However, light that enters the crystal at an angle to the optical axis is split into two rays, the ordinary ray with index of refraction n_O and the extraordinary ray with index of refraction n_E. The measure of birefringence is the absolute value of the difference between the two indices of refraction.

When the incident ray is split into the ordinary and extraordinary rays, the extraordinary ray vibrates in the plane that includes the optical axis while the ordinary ray vibrates in a perpendicular plane. Since the two rays traverse the crystal at different speeds and with different angles, waves with a specific polarization can be selected in a Nicol prism by removing one of the waves by internal reflection.

24.2.8 Nonlinear Optical Effects

The polarization is not truly a linear function of the electric field as we have assumed thus far. A more general relation can be written as

$$P = \varepsilon_0 \left(\chi^{(1)} E + \chi^{(2)} E^2 + \chi^{(3)} E^3 \cdots \right), \tag{24.42}$$

where the $\chi^{(2)}$ and $\chi^{(3)}$ are the nonlinear susceptibilities. These higher order terms in the polarization can produce some very interesting and useful effects such as frequency doubling ($\chi^{(2)}$) and optical switching with the possibility of optical computing ($\chi^{(3)}$). Chi-2 ($\chi^{(2)}$) materials are noncentrosymmetric, meaning they have no inversion symmetry. Single crystals of potassium dihydrogen phosphate (KDP), triglycine sulfate (TGS), potas-sium niobate ($KNbO_3$), and lithium niobate ($LiNbO_3$) are widely used for converting the IR output from YAG or Ti:sapphire lasers to visible light. Organic materials have also been found to possess high $\chi^{(2)}$ and $\chi^{(3)}$ values.

24.2.9 Raman Scattering

Most scattering events between photons and atoms or molecules are elastic (Rayleigh scattering). However, inelastic or Raman scattering can result when photons either

give energy to or receive energy from the system. In the case of molecules, some of the photon energy can go into the excitation of vibrational modes, resulting in a decrease in frequency of the photon. This is known as Stokes scattering. Conversely, an excited molecule can give some of its energy to the photon, increasing its frequency, which is called anti-Stokes scattering. The difference between the original photon frequency and the Stokes or anti-Stokes frequency is the vibrational frequency of the molecule. These effects were first reported by C.V. Raman in 1928 for which he received a Nobel Prize in 1930. Being able to observe molecular vibrational spectra by means of visible light is a valuable tool for analyzing the composition of materials. Also, since the anti-Stokes line depends on the number of molecules in the excited state, which in turn depends on the temperature, the ratio of anti-Stokes to Stokes intensities can be used as a noncontact temperature measurement, which is especially useful in combustion diagnostics. Raman scattering is also used in remote sensing. A laser beam is used to illuminate an unknown molecular solid, liquid, or vapor, which can be identified remotely from the Raman scattered molecular spectra.

In solids, Raman scattering results from inelastic interactions between photons and phonons in the optical mode. (Similar inelastic scattering from phonons in the acoustic mode is called Brillouin scattering.) Solids have characteristic phonon modes, which can be identified by Raman scattering.

24.3 Optical Properties of Conductive Media

In the previous sections, we derived the equations for reflection and transmission coefficients using the assumption that no free charge was present. Now we would like to use these results to examine what happens when an electromagnetic wave encounters a conductive media, say a sheet of metal, a semiconductor, or an ionized gas or plasma such as that surrounding a spacecraft during reentry. We can use the previous arguments to assign appropriate N and K values to conductive media if we modify the dielectric function to include the conductivity. We do this by first considering the interaction of the E-vector with the free electrons present.

24.3.1 Conductivity at High Frequencies

The equation of motion for the electrons in an oscillating E-field can be written as

$$\dot{v} + \frac{v}{\tau} = -\frac{e}{m} E_0 e^{i\omega t}, \tag{24.43}$$

the solution is given by

$$v = -\frac{eE_0\tau}{m(1 + i\omega\tau)} e^{i\omega t} \tag{24.44}$$

and the current density is

$$J = -nev = -\frac{ne^2 E_0 \tau}{m(1 + i\omega\tau)} e^{i\omega t}, \tag{24.45}$$

from which we can identify the complex conductivity as

$$\tilde{\sigma} = \frac{ne^2\tau}{m(1 + i\omega\tau)} = \frac{ne^2\tau}{m(1 + \omega^2\tau^2)}(1 - i\omega\tau) = \frac{\sigma_0(1 - i\omega\tau)}{(1 + \omega^2\tau^2)}. \tag{24.46}$$

Note that in the limit $\omega \to 0$, the real part is just the DC conductivity and the imaginary part vanishes.

24.3.2 Modification of Dielectric Function to Account for Conductivity

Next, we go back to the Maxwell equation, $\nabla \times \mathbf{H} = \mathbf{J} + \dot{\mathbf{D}}$, which can be written as

$$\nabla \times \mathbf{H} = \mathbf{J} + \dot{\mathbf{D}} = \tilde{\sigma}\mathbf{E} + \varepsilon\dot{\mathbf{E}} \approx \tilde{\sigma}\mathbf{E} + \varepsilon_0\dot{\mathbf{E}}. \tag{24.47}$$

Here, we have ignored the polarization induced by the interaction of the wave with the bound charge by setting ε to ε_0.

Since both $\mathbf{E}$ and $\mathbf{H} \sim e^{i\omega t}$, the right hand side of this equation can be written as

$$(i\varepsilon_0\omega + \tilde{\sigma})E_0 = \left[i\varepsilon_0\omega + \sigma_0\frac{(1 - i\omega\tau)}{(1 + \omega^2\tau^2)}\right]E_0 = \left\{\frac{\sigma_0}{(1 + \omega^2\tau^2)} + i\omega\left[\varepsilon_0 - \frac{\sigma_0\tau}{(1 + \omega^2\tau^2)}\right]\right\}E_0. \tag{24.48}$$

Now we want to represent $(i\varepsilon_0\omega + \tilde{\sigma})$ in the above expression with $i\omega\varepsilon_0\tilde{\varepsilon}$, where $\tilde{\varepsilon}$ is a complex dielectric constant that incorporates the effects of the free electrons in the medium. We set $\tilde{\varepsilon} = \varepsilon' + i\varepsilon''$ where the prime and double prime are the real and imaginary parts, respectively. Equating these to the real and imaginary parts of $(i\varepsilon_0\omega + \tilde{\sigma})$ from above, we obtain

$$i\omega\varepsilon_0\tilde{\varepsilon} = i\omega\varepsilon_0\varepsilon' - \omega\varepsilon_0\varepsilon'' = \frac{\sigma_0}{(1 + \omega^2\tau^2)} + i\omega\left[\varepsilon_0 - \frac{\sigma_0\tau}{(1 + \omega^2\tau^2)}\right], \tag{24.49}$$

where we identify

$$\varepsilon' = 1 - \frac{\sigma_0\tau}{\varepsilon_0(1 + \omega^2\tau^2)} \quad \text{and} \quad \varepsilon'' = -\frac{\sigma_0}{\omega\varepsilon_0(1 + \omega^2\tau^2)}. \tag{24.50}$$

24.3.3 Plasma Frequency

Going back to the Equation 24.44 for velocity of the electrons, note that as $\omega\tau \gg 1$, the electron velocity starts to fall with increasing frequency because the electrons are not sufficiently mobile to keep up with the changing field. Note also that if we write $\sigma_0/\varepsilon_0 = ne^2\tau/m\varepsilon_0$, the group $ne^2/m\varepsilon_0$ has the dimensions of s^{-2}, hence we define this as the square of the plasma frequency, or

$$\omega_P^2 = \frac{ne^2}{m\varepsilon_0}. \tag{24.51}$$

The real and imaginary parts of the complex dielectric constant can be written in terms of the plasma frequency:

TABLE 24.2

Properties of Various Conductive Media

Material	n (Electrons/m^3)	σ (S/m)	ω_p (rad/s)	τ(s)	$\omega_p\tau$	λ_p(μm)
Ag	5.85×10^{28}	6.21×10^7	1.36×10^{16}	3.77×10^{-14}	514.21	0.138
Cu	8.95×10^{28}	5.88×10^7	1.69×10^{16}	2.33×10^{-14}	393.63	0.112
Al	18.1×10^{28}	3.64×10^7	2.40×10^{16}	7.15×10^{-15}	171.54	0.079
Zn	13.1×10^{28}	1.69×10^7	2.04×10^{16}	4.58×10^{-15}	93.51	0.092
Inconel	8.95×10^{28}	7.75×10^5	1.69×10^{16}	3.07×10^{-16}	5.19	0.112
Si*	1.00×10^{25}	1.60×10^2	5.64×10^{14}	5.69×10^{-13}	101.46	10.56
Si**	1.00×10^{19}	1.60×10^{-1}	1.78×10^{11}	5.69×10^{-13}	0.10	10565
GaAs*	1.00×10^{25}	1.28×10^3	5.64×10^{14}	4.55×10^{-12}	811.67	10.56
GaAs**	1.00×10^{19}	1.28×10^0	1.78×10^{11}	4.55×10^{-12}	0.81	10565

* Lightly doped.
** Heavily doped.

$$\varepsilon' = 1 - \frac{ne^2\tau^2}{m\varepsilon_0(1 + \omega^2\tau^2)} = 1 - \frac{\omega_p^2\tau^2}{(1 + \omega^2\tau^2)} \quad \text{and} \quad \varepsilon'' = -\frac{\omega_p^2\tau^2}{\omega\tau(1 + \omega^2\tau^2)}. \tag{24.52}$$

Before we proceed, it would be useful to examine the magnitudes of the parameters ω_p and τ that determine the optical constants. These are displayed in Table 24.2 for several metals and semiconductors.

From Table 24.2, we see the $\omega_p\tau$ product is $\gg 1$ for the metals. Taking $\omega_p\tau = 100$, we can plot in Figure 24.8 the ε' and ε'' as a function of λ/λ_p, where λ_p is the wavelength corresponding to the plasma frequency.

At short wavelengths, the real part of the dielectric approaches 1 and the imaginary part approaches 0. Near the plasma frequency, the real part crosses zero and becomes increasingly negative. The imaginary part starts to become increasingly negative.

24.3.4 Optical Constants

We now take the real and imaginary part to the dielectric function plotted in Figure 24.8 and, using Equation 24.41, plot out the N and K values in Figure 24.9a and b and the reflectance in Figure 24.10 as a function of the wavelength ratio.

One can see that the definition of the ratio $ne^2/m\varepsilon_0$ as the plasma frequency becomes the dividing line where the metal goes from being transparent to an almost perfect reflector of

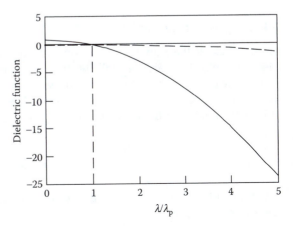

FIGURE 24.8
Real (solid) and imaginary (dashed) parts of the dielectric function for a conductive media with free electrons.

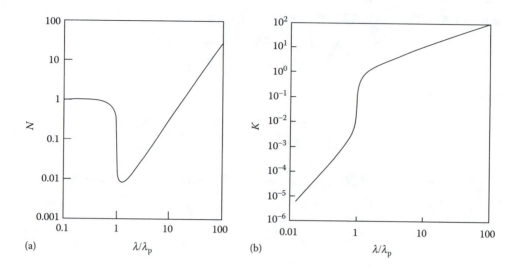

FIGURE 24.9
(a) Theoretical index of refraction N for a free electron metal as a function of $\lambda/\lambda_\mathrm{p}$ and (b) the theoretical extinction coefficient K for a free electron metal as a function of $\lambda/\lambda_\mathrm{p}$.

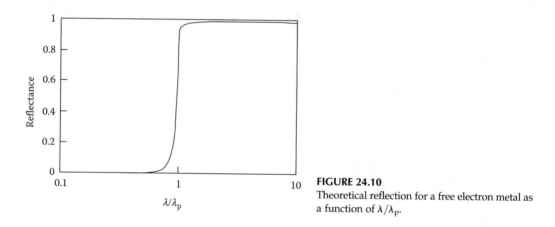

FIGURE 24.10
Theoretical reflection for a free electron metal as a function of $\lambda/\lambda_\mathrm{p}$.

electromagnetic radiation. From Table 24.2, we find for metals this line is O(100 nm) in the far-UV. One can gain more insight into the optics of conducting media by considering a couple of limiting cases.

24.3.5 Low Frequency Case

For $\omega \ll \omega_\mathrm{p}$ and $\omega\tau \ll 1$, the real and imaginary parts of the dielectric function may be written as

$$\varepsilon' \approx -\omega_\mathrm{p}^2 \tau^2, \quad \varepsilon'' \approx -\frac{\omega_\mathrm{p}^2 \tau}{\omega}, \quad \text{and} \quad \frac{\varepsilon'}{\varepsilon''} \approx \omega\tau \ll 1. \tag{24.53}$$

Since $\varepsilon'' \gg \varepsilon'$, Equation 24.41 may be approximated as

$$N^2 \approx K^2 \approx \frac{\varepsilon''}{2} = \frac{\omega_\mathrm{p}^2 \tau}{2\omega}. \tag{24.54}$$

The reflectance (Equation 24.27) may be written as

$$R = \frac{N^2 + 2N + 1 + K - 4N}{N^2 + 2N + 1 + K} = 1 - \frac{4N}{N^2 + 2N + 1 + K}. \tag{24.55}$$

But since $N = K$,

$$R = 1 - \frac{4N}{2N^2 + 2N + 1} \approx 1 - \frac{2}{N} = 1 - 2\sqrt{\frac{2\omega}{\omega_p^2 \tau}}. \tag{24.56}$$

Using Equations 24.46 and 24.51 to express ω_p and τ in more familiar terms, we can write the reflectance as

$$R = 1 - 2\sqrt{\frac{2\omega\varepsilon_0}{\sigma}}. \tag{24.57}$$

This expression for the reflectivity of metals at low frequencies is known as the Hagen–Rubens relation.

24.3.6 High Frequency Case

For the case where $\omega > \omega_p$ and $\omega_p^2 \tau^2 \gg 1$ for good conductors such as Cu or Ag, Equation 24.52 can be written as $\varepsilon' \approx 1 - (\omega_p^2/\omega^2)$ and $|\varepsilon''| \approx \omega_p^2 \tau^2 / \omega^3 \tau^3 \ll 1$.
   Using Equation 24.41,

$$N^2 \approx \varepsilon' \approx 1 - \frac{\omega_p^2}{\omega^2}; \quad K^2 \approx \varepsilon'\left[-\frac{1}{2} + \frac{1}{2}\left(1 + \frac{\varepsilon''^2}{\varepsilon'^2}\right)^{1/2} \right] \approx \frac{\varepsilon''^2}{4\varepsilon'}. \tag{24.58}$$

Thus $N \approx \sqrt{\varepsilon'} \approx 1 - [(1/2)(\omega_p^2/\omega^2)]$ and $K \approx (|\varepsilon''|/2\sqrt{\varepsilon'})$. Since ε' approaches 1 and ε'' becomes small, N approaches 1 and K will be $\ll 1$. The reflectivity falls as the fourth power of ω/ω_p as may be seen from Equation 24.26

$$R = \frac{(1 - N)^2 + K^2}{(1 + N)^2 + K^2} \approx \left(\frac{\omega_p^2}{2\omega^2}\right)^2 \tag{23.59}$$

and metals become transparent to radiation whose frequency is greater than their plasma frequency ω_p. From Table 24.2, we see that plasma frequencies are $O(10^{16} \text{ rad/s})$, which corresponds to a wavelength of 188 nm in the UV region. Thus metals start to become transparent in the UV and are quite transparent in the x-ray portion of the spectrum. (The partial opacity of metals to x-rays is due to scattering of the x-rays from the electrons bound to the ion cores, not from the interactions with the free electrons.)

24.3.7 Reflectance Spectra of Real Metals

The optical constants derived for a free electron conductive media do not tell the whole story, however, as will become evident when we examine the reflection spectra of real metals.
   First consider Al, whose band structure resembles that of free electrons. The theoretical values for N and K are compared to measured values in Figure 24.11a and b. The plasma

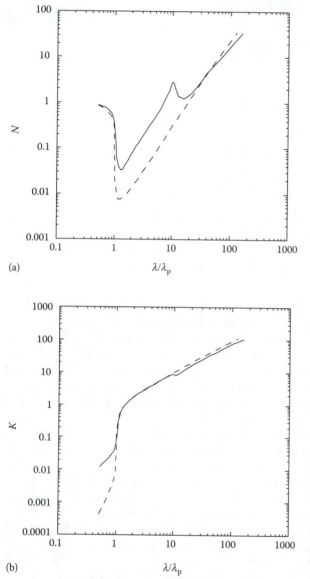

(a) λ/λ_p

(b) λ/λ_p

FIGURE 24.11
Observed values of (a) N and (b) K for vapor deposited Al under UHV conditions (solid lines) compared with theoretical free electron calculations (dashed lines). (N and K values taken from Smith, D.Y., Shiles, E. and Inokuti, M. in Palik, E.D., Ed., *Handbook of Optical Constants of Solids*, Vols. 1–3, Academic Press, 1997.)

frequency for Al is 2.40×10^{16} s^{-1}, which corresponds to 15.7 eV or $\lambda_p = 79$ nm. There is a strong interband transition at 1.5 eV, which corresponds to $\lambda/\lambda_p = 10.12$. The effect of this transition shows up in both the N and K plot, but otherwise the measured values closely follow the theoretical simple Drude free electron theory.

Similarly, the reflectance closely follows the Drude free electron model except for the region in the vicinity of the interband transition as seen in Figure 24.12.

Despite the small loss in reflectance in the visible region, pure Al is the best reflector for the far UV because its high electron density (three valence electrons per atom) extends its plasma frequency to shorter wavelengths and because there are no additional interband transitions in this region. Ag maintains its high reflectance throughout the visible spectrum, but undergoes an interband transition at ~4.0 eV (310 nm), which kills its UV reflectance.

However, we virtually never see a pure Al surface. Al has a high affinity for oxygen and an oxide layer will quickly form when exposed to any O_2 that may be in the atmosphere.

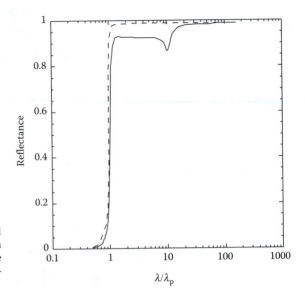

FIGURE 24.12
Measured reflectance of vapor deposited Al under UHV (solid) compared with free electron theory (dashed line). (The observed reflectance was computed from the measured N and K values shown in Figure 24.11a and b.)

The data shown in Figures 24.11 and 24.12 were taken under ultrahigh vacuum (UHV) conditions (10^{-9} to 10^{-10} Torr) and are presumed to be free of any contamination. The oxide film becomes absorbent in the vacuum UV region ($\sim$120 nm). Consequently, mirrors and gratings for use in this spectral region must be coated with transparent protective coatings such as LiF before they are removed from the UHV environment where they were prepared. It is also worth mentioning that evaporated Al films have higher reflectance than polished bulk Al because of the increased surface scattering created by the polishing process.

Next we consider the observed optical spectrum of Cu, a metal known for its reddish color. We see from Figure 24.13a and b that the N and K values approach the free electron theory in the long wavelength limit but differ significantly in the vicinity of λ_p. These departures are manifested in the reflectance spectrum shown in Figure 24.14a and b.

From Table 24.2, we see that $\lambda_p = 112$ nm, and from Figure 24.14a, we see the reflectance of Cu starts to fall off at around $8\lambda_p$, which corresponds to the beginning of the visible spectrum. Since the reflectance falls off more rapidly at the shorter wavelengths, the reflectance of blue light is much less than that of red, hence the reddish color of Copper. Adding Zn to the Cu shifts these transitions to longer wavelengths giving the alloy a redder color. An experienced metallurgist can judge the amount of Zn in a brass alloy by its color. Notice the bumps in the reflectance curve as the wavelength decreases. These are due to additional interband transitions and can be better resolved by plotting the log of reflectance against the log of the photon energy as shown in Figure 24.14b.

24.3.8 Interband Transitions in Metals

We saw in Chapter 20 that direct or vertical interband transitions from a full band to an empty band were possible even in indirect bandgap semiconductors provide the photon energy $\hbar\omega = E_{\text{empty}}(k) - E_{\text{full}}(k)$. Similar interband transitions occur in metals where electrons can be excited from full bands below the Fermi level to empty bands just above the Fermi level or from full bands just below the Fermi level to empty bands above the Fermi level as illustrated in Figure 24.15. The flat d-bands in Cu lie a couple of volts below the Fermi level and transitions from these bands to the empty parabolic s-bands just above the Fermi level are responsible for the loss of copper's reflectance in the visible spectrum.

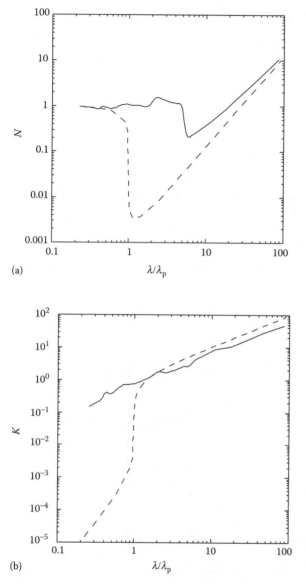

FIGURE 24.13
Observed values of (a) N and (b) K for vapor deposited Cu (solid lines) compared with theoretical free electron values (dashed lines). (N and K values taken from Lynch, D.W. and Hunter, W.R. in Palik, E.D., Ed., *Handbook of Optical Constants of Solids*, Vols. 1–3, Academic Press, 1997.)

Gold's yellow color is due to a similar mechanism. Silver has a band structure similar to Cu except the d-bands are $\sim$4 eV below the Fermi level; hence the loss of reflectance occurs in the UV region.

24.3.9 Semiconductors

Intrinsic semiconductors or insulators whose bandgap is above the visible will have low enough electron densities to place their plasma frequencies well below the visible region of the electromagnetic spectrum. As shown in Chapter 20, such materials can be made electrically conductive by adding impurities to create donor or acceptor states. Even at doping levels as high as 10^{25} m^{-3}, the plasma frequency is well below the lowest part of the visible spectrum. The ability to control the electron density by impurity doping makes it possible to create reasonably good electrical conductors that are transparent to visible light.

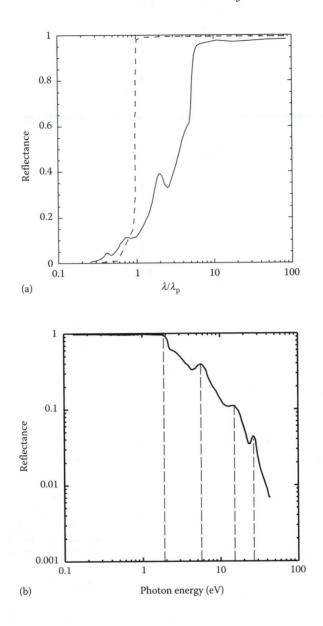

(a)

(b)

FIGURE 24.14

(a) Measured reflectance of vapor deposited Cu (solid) compared with free electron theory (dashed line). The reflectance was computed from the measured N and K values shown in Figure 24.13a and b. (b) Log plot of the measured reflectance of Cu versus photon energy. Peaks are associated with interband transitions.

Thin transparent conductive films such as indium tin oxide are crucial to the fabrication of electroluminescent panels and to liquid crystal displays.

24.3.10 Plasmas

The sharp transition from transparency to reflection at the plasma frequency predicted by the Drude free electron model can be better understood by considering a free electron gas. The undamped motion of an electron responding to a time varying electric field is $m\ddot{x} = -eE_0 e^{i\omega t}$. The solution is $x(\omega) = eE_0/m\omega^2$, the dipole moment of a single

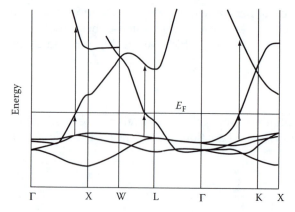

FIGURE 24.15
Selected bands in Cu showing possible inter-band transitions in the visible and near-UV. (Adapted from Figure 19.9)

electron is $p(\omega) = -ex(\omega) = -e^2 E_0/m\omega^2$ and the dipole moment per unit volume is $P(\omega) = -ne^2 E_0/m\omega^2$. The susceptibility becomes $\chi(\omega) = -ne^2/\varepsilon_0 m\omega^2$ and the dielectric function may be written as

$$\varepsilon_r(\omega) = 1 + \chi(\omega) = 1 - \frac{ne^2}{\varepsilon_0 m\omega^2} = 1 - \frac{\omega_p^2}{\omega^2}, \quad (24.60)$$

where ω_p is the frequency at which $\varepsilon_r(\omega) = 0$. Notice that this is also the plasma frequency we defined by Equation 24.51 and that Equation 24.60 is the high frequency limit of Equation 24.52. For $\omega < \omega_p$, the electrons behave individually and can oscillate with the electric field and reflect the radiation as assumed by the Drude theory. However, when the electron gas becomes dense enough, the electrons in the gas start to behave collectively and begin to oscillate together in a longitudinal mode as shown in Figure 24.16. These collective oscillations are quantized with energies given by $\hbar\omega_p$ and are referred to as plasmons. Since there is no damping, $N(\omega) = \sqrt{\varepsilon_r(\omega)}$ and $K = 0$. When the applied frequency ω is less than the plasma frequency, $\varepsilon_r < 0$ and $N = 0$. For $\omega > \omega_p$, N approaches 1 as ω increases.

Plasmons can be excited by photons entering the metal obliquely (Figure 24.6) so that there is a component of the E-vector in the longitudinal direction or by an electron passing through the metal. Plasmon energies are measured by measuring the energy lost by a beam of electrons passing through a thin metal foil, a technique known as electron energy loss spectroscopy (EELS).

The plasma frequency derived in Equation 24.60 also applies to electrons in the iono-sphere. Radio waves below the plasma frequency (amplitude modulated [AM] and short-wave broadcast) are reflected back to earth by the ionosphere and can carry over long distances. Frequency modulated (FM) and TV frequencies ($\gtrsim 50$ MHz) penetrate the iono-sphere and reception is limited to line-of-sight distances. Spacecraft and direct broadcast

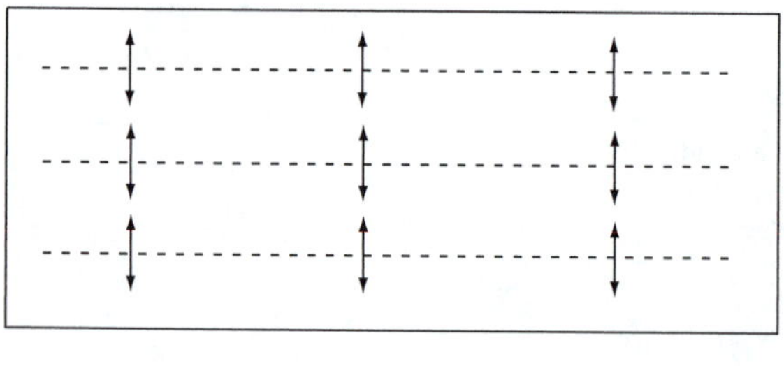

FIGURE 24.16
Plasmons consist of layers of electrons in a metal that oscillate collectively at the plasma frequency. The electrons interact with the ion cores causing them to oscillate at their longitudinal frequency ω_L.

satellites use UHF and microwave channels to penetrate the ionosphere. During spacecraft reentry, the electron density gets so large because of the intense heat that even these highest frequencies cannot penetrate the plasma, which causes the communication blackout.

24.3.11 Dispersion Relation

A dispersion relation relating ω to k for photons propagating in metals can be obtained by starting with the index of refraction N for $\omega > \omega_p$ (Equation 24.58). Since N is by definition $N = c/\nu$ the phase velocity $\nu = \omega/k, = k\nu$,

$$N^2 = 1 - \frac{\omega_p^2}{\omega^2} = \frac{c^2}{\nu^2} = \frac{c^2 k^2}{\omega^2}, \tag{24.61}$$

from which we obtain

$$c^2 k^2 = \omega^2 - \omega_p^2. \tag{24.62}$$

This dispersion relation shown in Figure 24.17 is similar to the upper or longitudinal branch of Figure 24.5. No wave can propagate if $\omega < \omega_p$ and above ω_P, only the longitudinal mode can be excited. Recall that we defined ω_L as the frequency at which the dielectric constant became positive again. So we see that for conductive media, the plasma frequency corresponds to the frequency of the longitudinal optical mode.

Even though the phase velocity $\omega/k > c$, it can be seen that the group velocity $\partial\omega/\partial k$ approaches c as $\omega \gg \omega_p$ but is always less than c as can be verified by differentiating Equation 24.62 with respect to k.

$$2kc^2 = 2\omega\nu_g, \tag{24.63}$$

from which we see that

$$\nu_g = c^2 k/\omega = c^2/\nu. \tag{24.64}$$

Since $\nu > c$, then $\nu_g < c$.

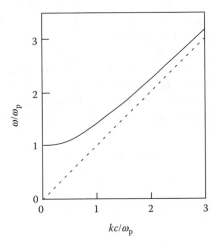

FIGURE 24.17
Dispersion relation for radiation passing through a conductive medium with frequency $> \omega_p$. The dashed line represents the velocity of light. No propagation is possible for $\omega < \omega_p$.

24.4 Summary

The optical properties of a material are characterized by the real and imaginary parts of the index of refraction $(N + iK)$ as a function of wavelength. These optical parameters are obtained from the square root of the dielectric function, which may also be complex.

Most dielectrics are transparent from the near IR to the near-UV. The short wavelength limit is usually determined by electronic polarization unless the band edge is encountered first. The band edge is related to the photon energy by λ (μm) $= 1.24/E$(eV). Since the polarizability $\alpha(0)$ goes roughly as r^3 and the resonant frequency ω_0 goes as $\alpha(0)^{-1/2}$, the lightest molecules such as LiF will have the shortest wavelength cutoff. The increase in the N with frequency as the resonant frequency is approached is called dispersion and is responsible for blue light being refracted more than red light in a prism.

The long wave limit for dielectrics is usually set by photons coupling with the optical modes of the molecular vibrations as discussed in Chapter 16. The IR absorption edge goes as the square root of the spring constant divided by the reduced mass of the system, so to extend the IR cutoff, one looks for heavy molecules with a weak spring constant, or a system with a low sound velocity since the velocity of sound goes as the square root of the spring constant divided by the density.

Photons can couple with transverse optical phonons near $k = 0$ in dielectric materials. This quantized photon–phonon has energy $\hbar\omega$ and is call a polariton. As the energy of the photon approaches the transverse resonant frequency, the N gets large and the velocity of propagation approaches zero. The N drops suddenly to 0 at ω_T and remains 0 until ω_L and no propagation can occur between these two frequencies. The crystal becomes almost a perfect reflector between these two frequencies and this band is called the *reststrahlen* region. Above ω_L only longitudinal optical modes can couple with photons. Longitudinal modes can only be excited by photons traveling obliquely to the lattice planes, but this does not mean that photons at normal incidence cannot propagate; they simply do not interact with the phonons.

Media with dipolar molecules will have a large N at frequencies below their damping frequency given by $1/\tau$, where τ is their relaxation time. Around this damping frequency, the N drops and K increases as energy is being absorbed by the oscillating dipoles. This absorption is the mechanism responsible for cooking foods in a microwave oven. It is also responsible for the reflection of microwave radar from clouds. Above the damping frequency, the dipoles cannot oscillate as fast as the oscillating E-field of the phonon and there is no interaction.

The optical behavior of conductive media is characterized by the plasma frequency, $\omega_p^2 = ne^2/m\varepsilon_0$ which is a function only of the electron concentration. Plasma frequencies for metals are $O(10^{16}/\text{s})$ which corresponds to the vacuum UV. Below the plasma frequency, the N and K values are large making the material highly reflective (except for inner band transitions which are responsible for the color of Cu and Au). Above the plasma frequency, metals start to becomes transmissive, but they will still absorb strongly until the $\omega \gg \omega_P$.

The plasma frequency of semiconductors is determined by the doping level. By selecting an appropriate doping level, it is possible to make a conducting film that is still transparent to visible radiation. Such films are widely used in optoelectronics devices.

The free electrons in the ionosphere will reflect short wave radio waves at frequencies below their plasma frequency ($\sim$50 MHz) but will transmit FM and TV frequencies. Spacecraft reentry creates a plasma whose frequency is much higher than that of the atmosphere, which accounts for the communication blackout during this phase of a space mission.

Bibliography

Ashcroft, N.W. and Mermin, N.D., *Solid State Physics*, Brooks Cole, Philadelphia, 1976.

Barsoum, M.W., *Fundamentals of Ceramics*, Institute of Physic Publishing, Bristol/Philadelphia, PA, 2003.

Gersten, J.I. and Smith, F.W., *The Physics and Chemistry of Materials*, John Wiley & Sons, New York, 2001.

Harrington, J.A., *Infrared Fiber Optics*, OSA Handbook, Vol. III, McGraw Hill, New York.

Henning, T.H., et al., *Reference & Data Base of Optical Constants*, Astron. Astrophys. Suppl., 136, 1999. Available at www.astro.spbu.ru/JPDOC/f-dbase.html.

Hummel, R.E., *Electronic Properties of Materials*, 3rd edn., Springer-Verlag, New York, 2000.

Ibach, H. and Lüth, H., *Solid State Physics*, 3rd edn., Springer-Verlag, New York, 1990.

Kingery, W.D., Bowen, H.K. and Uhlmann, D.R., *Introduction to Ceramics*, 2nd edn., John Wiley & Sons, New York, 1976.

Kittel, C., *Introduction to Solid State Physics*, 7th edn., John Wiley & Sons, New York, 1966.

Palik, E.D., Ed., *Handbook of Optical Constants of Solids*, Vols. 1–3, Academic Press, San Diego, California, 1997.

Wooten, F., *Optical Properties of Solids*, Academic Press, New York, 1972.

Problems

1. NaCl is often used as an IR window. What would you estimate its short wavelength cutoff to be? $\alpha = 3.0 \times 10^{-40}$ for Cl^- and 0.18×10^{-40} for Na^+ (MKS units).

2. Show that $\omega_L^2 \sim \omega_{Debye}^2 (M + m)^2 / Mn$.

3. You are called upon to design a solar blind filter for a far-UV space experiment. The sun emits a strong L-alpha line at 121.6 nm, which you must block, but you would like to transmit all wavelengths shorter than this as much as possible. You propose to use a thin metallic foil as the filter. Consider an appropriate choice for the material. Plot the transmission curve from 100 to 10 nm. (Better use log plots.)

4. Spectrophotometry is often used in qualitative and quantitative analysis to determine the species and concentration of an unknown solute. A cuvette of known width x is filled with the solution and white light is passed through the solution. The transmitted light is analyzed with a monochromator to obtain the absorption spectra, which can be used to identify the solute. The concentration of the solute can be determined by measuring the absorbance $A = \log_{10}(I_{ref}/I_{sample})$ at a wavelength where the solute absorbs strongly. For dilute solutions, the absorptivity α is directly related to the concentration so the measured A is directly related to the log of the concentration.

   (a) Let the light with intensity I_0 be partially reflected at the first air–cuvette interface with reflectivity R and partially reflected with the same R as it emerges from the cuvette. The index of refraction of the solution in the cuvette is matched to the cuvette so there is no reflection at the solution–cuvette interfaces. Show that the intensity of the emerging ray is given by $I_0(1 - R)^2 \exp(-\alpha x)$.

   (b) Now trace the ray reflected from the back of the sample back through the solution Nuntil it is reflected from the front surface, then back through the solution until it emerges from the back. Show that the intensity of this ray is $I_0 R^2 (1 - R)^2 \exp(-3\alpha x)$.

      Continue to trace the internally reflected ray back-and-forth through the sample and sum the total intensity of the transmitted ray. (After a few terms you should be

able to express the sum in terms of a geometric series. $\sum_{n=0}^{\infty} s^n = 1/(1-s)$ provided $s < 1$.)

(c) Show that the transmitted intensity is $I_0(1-R)^2 e^{-\alpha x}/1 - R^2 e^{-2\alpha x}$.

(d) This correction for multiple paths is generally ignored in spectrophotometric measurements. Comment on when this is and is not justified. Does the fact that you are comparing the transmitted light through the unknown sample to that of a reference sample cancel out this source of error? Support your answer with relative error calculations for various values of α.

(e) Using the procedure in (c), sum the total reflected intensity, the total absorbed intensity, and show that the sum of transmitted, absorbed, and reflected intensities is equal to the incident intensity.

5. You are given the task of designing a smooth absorptive coating to go on a reflective metal surface. You are restricted to a thickness of one wavelength of the radiation you want to absorb.

(a) What strategy would you adopt in looking for materials in terms of their N and K (real and imaginary index of refraction) values to minimize the reflectance? You have come up with a binder that has an index of refraction of $N = 1.2$ and you have found that you can control the K-value by doping it with carbon nanotubes.

(b) What value of K would you choose to minimize the total reflectivity? What would be the value of this reflectivity?

(Trace a ray at normal incidence on the front surface. Let the transmitted fraction of light enter the coating and reflect from the rear surface assuming $R = 1$. Then let it re-emerge through the front surface. The total reflectance is the portion of the incident wave reflected from the front surface plus the ray that was reflected from the rear surface. It is not necessary to consider multiple reflections within the coating.)

6. You are asked to design a conductive coating that is transparent in the visible region. You have chosen a semiconductor with a bandgap well above optical frequencies.

(a) What dopant level would you choose to maximize the conductivity and still be transparent in the visible (>450 nm)?

(b) If the mobility $\mu = 0.1$ m^2/V s, what would the conductivity be? How does this compare with Cu?

7. AM radio waves (540–1600 KHz) as well as shortwave broadcasts (up to 20 MHz) can be heard over great distances because their waves are reflected back to Earth by the ionosphere. However, TV (Channel 2 is 55 MHz) and FM broadcasts (88–108 MHz) are limited to line of sight between the transmitter and receiver. With this information, what limits can you place on the electron density of the ionosphere?

8. Several space shuttle mapping missions using L-band (20 cm) side-looking synthetic aperture radar were able to see rock formations and water channels buried 3–4 m under the sands of the Arabian Peninsula. Explain how this is possible and estimate the conductivity of the sands that would allow such observations.

9. Relaxation time for H_2O molecules is 10^{-11} s. What is the frequency used in a microwave oven to maximize the heating of foods containing water? What frequency would you set your weather radar to in order to image clouds? Why does snow produce less radar return than rain?

25

Magnetism and Magnetic Materials

Magnetism may be thought of as electricity in motion. It is actually a relativistic effect of moving charges according to Einstein's special theory of relativity. One observes a magnetic field in a reference frame that electrons are flowing through, but in a reference frame that moves with the electrons, one observes only an electric field. Natural ferromagnets in the form of loadstones were known to the ancient Chinese who used them for navigation. Hans Christian Ørsted was the first person to connect magnetism with electricity when he noticed that a flowing current influenced a compass needle.

Magnetic materials continue to play an ever increasing role in our modern technical society. For example, the recent discovery of low-cost iron-neodymium-boron magnets has made it possible to build highly efficient permanent magnet motors that are used in hybrid vehicles and will be used in future electric vehicles. Another example is the continued development of magnetic storage media, which have extended hard drive storage capacities far beyond what anyone would have expected a decade ago and have made high performance computers affordable to almost everyone. In order to understand how these materials function, we need to start with some basic principles.

25.1 Basic Relationships

Just as the capacitance was increased when a dielectric with polarization $\mathbf{P}$ was placed between the plates of a capacitor, the magnetic field in, or in the vicinity of, a material is altered by the magnetization $\mathbf{M}$ of the material. By analogy with the electric field case in which we wrote $\mathbf{D} = \varepsilon \mathbf{E}$, we can write for the case of the magnetism $\mathbf{B} = \mu \mathbf{H}$, where $\mathbf{H}$ is the applied magnetic field (A/m), $\mathbf{B}$ is the resulting flux density (V s/m^2 = Wb/m^2 = T), and μ is the permeability of the medium. The flux density can also be written as $\mathbf{B} = \mu_0 (\mathbf{H} + \mathbf{M})$, where μ_0 is the permeability of free space ($\mu_0 = 4\pi \times 10^{-7}$ H/m = V s/A m) and $\mathbf{M}$ is the number of magnetic dipoles/volume or the magnetization (A/m). The magnetic susceptibility χ is defined such that

$\mathbf{M} = \chi \mathbf{H}$ so $\mathbf{B} = \mu_0 (\mathbf{H} + \mathbf{M}) = \mu_0 (\mathbf{H} + \chi \mathbf{H}) = \mu_0 \mathbf{H}(1 + \chi) = \mu_0 \mu_R \mathbf{H}$, where the relative permeability $\mu_R = 1 + \chi$.

25.1.1 Types of Magnetism

We identify three basic types of magnetic interaction. If the flux density is reduced by the presence of a material in proportion to the applied field ($\chi < 1$), the material is diamagnetic. If the flux density is increased by the presence of the material in proportion to the applied field ($\chi > 1$), the material is paramagnetic. If the magnetic field is present even in the

absence of an applied field (spontaneous magnetism), the material is ferromagnetic. Ferrimagnetism and antiferromagnetism are special cases of ferromagnetism.

25.2 Origin of Magnetism

What gives rise to magnetism? We know that a current flowing through a wire creates a magnetic field around the wire. The magnitude of the field can easily be found by using the circuit form of Ampere's law:

$$\oint \mathbf{B} \cdot d\ell = \mu I. \tag{25.1}$$

Taking the path of integration at a constant radius R around the wire, $B = \mu_0 I / 2\pi R$ or $H = I / 2\pi R$ (A/m) in free space. Similarly, the B inside of a long solenoid may be found by taking the path of integration as a rectangle of length L parallel to the axis of the solenoid that encloses one side of the windings. Since the field inside of a long solenoid is contained inside the windings, only the side of the rectangle inside the solenoid will be along field lines. Thus from Ampere's law we find $B = \mu I(N/L)$, where I is the current flowing through the wires and (N/L) is the number of turns per length.

A current flowing around a single loop with radius R constitutes a magnetic dipole with a magnetic dipole moment with magnitude μ_M which equals the current times the area enclosed by the current loop, or in this case, $\mu_M = \pi I R^2$. The circuit form of Ampere's law is useful only in cases of high symmetry, which does not apply in this example. Thus to find the resultant magnetic field, we must go to the more general law developed by Biot and Savart, i.e.,

$$d\mathbf{B}_\perp = \frac{\mu_0 I}{4\pi} \left(\frac{d\ell \times R}{r^3} \right), \tag{25.2}$$

where r is the distance from the element of length $d\ell$ to the observation point. Integrating this equation around a circular loop with radius R, the magnetic field along the axis can be found to be

$$B(x) = \frac{\mu_0 I R^2}{2(R^2 + x^2)^{3/2}}. \tag{25.3}$$

For $x \gg R$, this becomes

$$B(x) = \frac{\mu_0 I R^2}{2x^3} = \frac{\mu_0 \mu_M}{2\pi x^3}, \tag{25.4}$$

while at the center of the loop,

$$B(0) = \frac{\mu_0 I}{2R} = \frac{\mu_0 \mu_M}{2\pi R^3}. \tag{25.5}$$

25.2.1 Bohr Magneton

What goes on at the microscopic level inside a material that determines its magnetic properties? Macroscopically, we identified magnetism with current flow or moving charge.

Electrons moving in their orbits in the atoms making up the material constitute current loops, which should have magnetic moments (if they do not all cancel out). Classically, an electron moving around the nuclei with velocity v represents a current $qv/2\pi R$, where q is the electronic charge and $2\pi R$ is the circumference of the orbit. The magnetic moment of an electron in orbit is then the product of this current and the area enclosed, or

$$\mu_M = -\frac{qv}{2\pi R}\pi R^2 = -\frac{q\omega R^2}{2} = -\frac{q}{2m}|L| = -\frac{q\hbar}{2m}m_e, \qquad (25.6)$$

where
   L is the orbital angular momentum
   m_e is the projection quantum number (the negative sign implies the μ_M is opposite to the angular momentum)

(Recall that $|L| = \hbar\sqrt{\ell(\ell+\ell)}$ and $\ell = 0, 1, 2\cdots, n-1$, and $m_e = -\ell, -\ell+1\cdots+\ell$. L never lines up with the magnetic field direction, but projects a component, $\hbar m_\ell$ on the B-direction.) The quantity

$$\frac{q\hbar}{2m} = \mu_B = 9.27 \times 10^{-24} \text{ A m}^2 \qquad (25.7)$$

is called the Bohr magneton. The orbital contribution to the magnetic moment is in integral units of Bohr magnetons. The electron spin also contributes to magnetic moment in units of $2\mu_B m_s$ where $m_s = \pm 1/2$.

25.2.2 Net Magnetic Moment

The net magnetic moment of an atom (or ion) consists of contributions from both the electron spin and orbital angular momentum. It may be found by adding the total spin angular momentum S to the total orbital angular momentum L according to

$$\boldsymbol{\mu}_M = -\frac{q}{2m}(2S + L). \qquad (25.8)$$

A factor of 2 in front of the spin angular momentum is the experimentally determined gyromagnetic ratio of the electron (actually this is 2.0023). It is more convenient to express this in terms of the total angular momentum $J = L + S$ by introducing the Lande g-factor,

$$\boldsymbol{\mu}_M = -\frac{q}{2m}gJ. \qquad (25.9)$$

The appropriate expression for g can be found by taking dot product with J in Equations 25.8 and 25.9, which gives

$$gJ^2 = J \cdot (2S + L) = (L + S) \cdot (2S + L) = L^2 + 2S^2 + 3L \cdot S. \qquad (25.10)$$

From the law of cosines, $J^2 = L^2 + S^2 + 2L \cdot S$ which when put into the above equation yields

$$g = \frac{L^2 + 2S^2 + 3/2(J^2 - L^2 - S^2)}{J^2} = 1 + \frac{J^2 + S^2 - L^2}{2J^2}, \qquad (25.11)$$

where

$$J^2 = J'(J' + 1)\hbar$$
$$S^2 = S'(S' + 1)\hbar$$
$$L^2 = L'(L' + 1)\hbar$$

J', S', and L' are the sums of the individual quantum number of the electrons making up the atoms

These values are determined by Hund's rules, which were empirically determined by examining spectra.

25.2.3 Hund's Rules

Hund's rules determine how the electrons go into the various shells in the ground state.

Rule 1. Maximize S' by placing as many electrons in the spin-up position as allowed by the Pauli exclusion principle.

Rule 2. Maximize L' consistent with the Pauli principle and Rule 1.

Rule 3. If shell is less than half full, choose $J' = L' - S'$. If more than half full, $J' = L' + S'$.

For Fe^{2+}, the only electrons not in closed shells are the $3d^6$. Since the d-shell can hold 10 electrons, the Pauli principle allows only 5 to be put in with spin up. Therefore, for $3d^6$, the configuration must be ↑↑↑↑↓↑. Each of these contributes spin $1/2$, so $S' = 2$. To maximize L', we put the ↑↓ pair in $m_e = 2$ and each of the four remaining electrons in $m_e = 1$, $m_e = 0$ $m_e = -1$, $m_e = -2$ respectively. This arrangement effectively cancels out all but one electron in $m_e = 2$, making $L' = 2$. Since the shell is more than half full, $J' = L' + S' = 4$.

25.3 Diamagnetism

Earnshaw's theorem states that it is not possible to arrange a static combination of electrostatic charges or magnets such that a charge or a magnet can be levitated in a stable configuration (Earnshaw, 1842). However, diamagnetic materials are repelled by magnetic fields and Earnshaw's theorem does not apply. Superconductors are perfect diamagnets ($\chi = -1$) and can be levitated above permanent magnets. The diamagnetic moment of most materials is too small to be of practical use and is often exceeded by the paramagnetic susceptibility. But there are a few normal materials that have unusually high diamagnetic susceptibilities that can be levitated by static magnetic fields.

25.3.1 Langevin Diamagnetism (Classical Approach)

Atoms with full electron shells will have no intrinsic magnetic moment. (All orbital and spin magnetic moments have opposite pairs and hence cancel each other.) They still have a slight magnetic interaction due to the interaction of their orbits with the applied magnetic field (analogous to the electronic contribution to polarization in dielectrics). Probably the easiest way to understand diamagnetism is to consider the effect of an applied magnetic field on the orbit of an electron. Consider an electron orbiting a nucleus with angular velocity ω_0 at radius r_0. From Equation 25.6, the magnetic moment will be $\mu_M = - qr^2\omega_0/2$ where ω_0 is the unperturbed orbital frequency in the right-hand sense. The Coulomb force on the electron is $-mr\omega_0^2$. In the presence of a magnetic field, the electron will feel a Lorentz force $-q(v \times B) = qr\omega B$ if the B is aligned with μ_M. The effect of the magnetic field is to oppose the Coulomb attraction, thus slowing the ω of the electron which reduces its

magnetic moment (this is a consequence of Lenz's law which states that the current induced by a magnetic field creates an opposing field). Quantitatively, the new angular frequency is

$$\omega^2 = \omega_0^2 - \frac{qB}{m}\omega. \tag{25.12}$$

Solving for ω

$$\omega = \omega_0 - \frac{qB}{2m} + O(B^2). \tag{25.13}$$

The new magnetic moment is

$$\mu'_M = \frac{qr^2}{2}\left(\omega_0 - \frac{qB}{2m}\right) = \mu_M - \frac{q^2 r^2}{4m}B. \tag{25.14}$$

Had the **B**-field been opposite the μ_M, the effect would be to speed up the electron and make the magnetic moment more negative and Equation 25.14 would be

$$\mu'_M = -\frac{qr^2}{2}\left(\omega_0 + \frac{qB}{2m}\right) = -\mu_M - \frac{q^2 r^2}{4m}B. \tag{25.15}$$

If we sum over Z electrons per atom and N atoms/m^3, the μ_M cancel out and the magnetic susceptibility becomes

$$\chi = \frac{M}{H} = \frac{\mu_0 ZN\Delta\mu_m}{B} = -\frac{\mu_0 ZNq^2 r^2}{4m}. \tag{25.16}$$

Since the susceptibility <0, the response is diamagnetic. We estimate the susceptibility by taking $Z=10$, $N=10^{28}$ m^{-3}, $r=10^{-10}$ m, $m=10^{-30}$ kg, and $\mu_0=10^{-6}$, $q \sim 10^{-19}$, $\chi \sim 10^{-5}$.

25.3.2 Larmor Diamagnetism (Quantum Approach)

Quantum mechanics does not allow the orbital angular momentum to line up with the magnetic field as assumed in the previous example. Instead **L** projects a component, $\hbar m_\ell$ on the B-direction as was shown in Chapter 2. A magnetic field exerts a torque on a magnetic moment according to $\dot{\mathbf{L}} = \boldsymbol{\mu}_M \times \mathbf{B}$. This torque is perpendicular both to the field **B** and the **L**, which is along the $\boldsymbol{\mu}_b$; therefore, the effect is to cause a precession of **L** about **B**. This precession is known as Larmor precession and takes place with the Larmor frequency

$$\omega_L = \frac{qB}{2m}. \tag{25.17}$$

This precession represents a shell of charge rotating about the B-direction. The resulting magnetic moment is

$$\mu_M = -\frac{q\omega_L r^2}{2} = -\frac{q^2 B r^2}{4m}, \tag{25.18}$$

which is the same as the $\Delta\mu_M$ found in Equation 25.16.

The effect is small, but all atoms have a small amount of diamagnetism. This includes atoms in gases and liquids, atoms in van der Waals' bonded crystals, ions in ionic salts, ion cores in metals, inner core electrons in covalently bonded crystals, and in molecules of living systems.

There are, however, materials in which the diamagnetic susceptibility is considerably more than $\sim 10^{-5}$. Bismuth and pyrolytic graphite have high enough diamagnetic susceptibilities that they can be floated above an array of small permanent magnets. There are some interband effects in Bi and graphite that account for their unusually large diamagnetic susceptibilities. See references for discussions of the unusually large dielectric coefficients of water and graphite.

Water also has a diamagnetic susceptibility about 20 times smaller that Bi and graphite, but high enough so that small living creatures can be levitated in the core of a large superconducting magnet. A photograph of a frog being levitated in a 16 T magnet is available at www.hfml.science.ru.nl/froglev.html. (The frog was not harmed.)

25.4 Paramagnetism

25.4.1 Langevin Paramagnetism (Classical Approach)

If we have unpaired electrons, it is possible to have a net magnetic moment. The energy of a magnetic dipole in a **B**-field is given by $U = -\mathbf{B} \cdot \boldsymbol{\mu}_M$. The **B**-field will try to line up the magnetic dipoles, but thermal motion will try to disrupt this ordering. At thermal equilibrium, the average dipole moment will be given by

$$\langle \mu_M \rangle = \frac{\int_0^{\pi} (\mu_M \cos \theta) e^{\mu_M B \cos \theta / kT} \, 2\pi \sin \theta d\theta}{\int_0^{\pi} e^{\mu_M B \cos \theta / kT} \, 2\pi \sin \theta d\theta} = \mu_M L(\alpha), \tag{25.19}$$

where $\alpha = \mu_M B / kT$ and the Langevin function $L(\alpha) = \coth \alpha - 1/\alpha$. At ambient temperatures, $kT \gg \mu_M B$. Since $\coth \alpha = 1/\alpha + \alpha/3 - \alpha^3/45 + \cdots$, for small α, $L(\alpha) \to \alpha/3$. For very low temperatures, $\alpha \gg 1$, and $L(\alpha) \to 1$.

At ambient temperatures ($\alpha \ll 1$), $\langle \mu_M \rangle = \mu_M \alpha / 3 = \mu_M^2 B / 3kT$ and, since the magnetism M is $N \langle \mu_M \rangle$,

$$M = \frac{N \mu_M^2 B}{3kT} = \frac{C}{T} H. \tag{25.20}$$

This is known as the Curie law of paramagnetism and $C = N \mu_0 \mu_M^2 / 3k$ is the Curie constant for a given material. The magnetic susceptibility is

$$\chi = \frac{M}{H} = \frac{N \mu_M^2 \mu_0}{3kT}, \tag{25.21}$$

which is positive.

What materials would be expected to exhibit paramagnetism? Even-numbered elements will always have electrons in pairs; hence, their magnetic moments will cancel each other. Odd-numbered elements, could have unpaired electrons, but generally this odd electron has formed either an ionic or covalent bond to form a full shell in which the electron pairs

cancel themselves. Thus paramagnetic materials are generally salts of the transition elements with unfilled *d*-shells, or of the rare earth elements with unfilled *f*-shells.

However, there is one important exception—the O_2 molecule. Recall from Figure 3.7, the electron bonding scheme for N_2. For O_2, we need to place two more electrons and by Hund's rules they will have to go into the two p_y antibonding sites, giving us two unpaired electron spins. Liquid O_2 will cling to the poles of a strong magnet. Moreover, there are commercially available process stream oxygen sensors based on the paramagnetic susceptibility of the O_2 molecule.

The large change in susceptibility with temperature of certain rare earth salts at cryogenic temperatures has been used to make highly sensitivity thermometers, especially with the ability to detect minute changes in the magnetic field using superconducting quantum interference detectors. See Chapter 26.

25.4.2 Paramagnetism (Quantum Approach)

In the classical picture, we assumed that the magnetic moments of the electrons were randomly distributed in all possible orientations relative to the magnetic field. Of course, only discrete orientations are allowed quantum mechanically. Therefore, we must sum over these orientations rather than integrate over a continuum. The z-component of the magnetic dipole moment $\mu_z = -g\mu_B J_z/\hbar = -g\mu_B M_j$ and the energy is $U = -\mu_z B = -g\mu_B M_j$, where M_J is the total angular momentum projection quantum number, $M_J = -J', -J' + 1 \cdots J'$. The average value for μ_z in thermal equilibrium is given by

$$\langle \mu_z \rangle = -\frac{\sum_{-J'}^{J'} M_J g\mu_B e^{M_J g\mu_B B/kT}}{\sum_{-J'}^{J'} e^{M_J g\mu_B B/kT}} = g\mu_B J' B_{J'}(y), \tag{25.22}$$

where $y = J' g\mu_B B/kT$ and the Brillouin function $B_{J'}(y)$ is given by

$$B_{J'}(y) = \frac{2J' + 1}{2J'} \coth\left[\frac{(2J' + 1)y}{2J'}\right] - \frac{1}{2J'} \coth\left[\frac{y}{2J'}\right]. \tag{25.23}$$

For small values of y, the magnetization reduces to

$$M = \frac{Ng^2\mu_B^2 J'(J' + 1)}{3kT} B, \tag{25.24}$$

which is the classical result if μ_M^2 in Equation 25.21 is replaced by $g^2\mu_B^2 J'(J' + 1)$.

25.4.3 Effective Magnetic Moment

We can define an effective dipole moment as $\mu_M = \mu_B p_{eff}$ where p_{eff} is experimentally determined. For the rare earth ions, $p_{eff} = g\sqrt{J'(J' + 1)}$, where J' is calculated from the Hund's rules as outlined above. However, for the transition elements, $p_{eff} = g\sqrt{S'(S' + 1)}$ where the orbital component apparently makes no contribution. Here we say the orbital angular momentum is quenched. For the transition elements, the *d*-electrons lie close to the outer regions of the atom and are acted upon by neighboring crystal fields that perturb the orbits so as to eliminate its contribution to the magnetic moment. The *f*-electrons in the

rare earth elements lie further inside the atom; hence, they are less perturbed and contribute according to the above theory.

25.4.4 Paramagnetism of Conduction Electrons (Pauli Paramagnetism)

If the electrons in a metal obeyed Boltzmann statistics, one would expect their spins to line up in the presence of a magnetic field and exhibit Langevin paramagnetism. Setting $\mu_M = 2\mu_B m_s$ where $m_s = (1/2)$ for electrons and using Equation 25.21,

$$\chi = \frac{N\mu_0\mu_B^2}{3kT}. \tag{25.25}$$

We estimate the order of magnitude of this susceptibility: $N \sim 10^{29}$, $\mu_0 \sim 10^{-6}$, $\mu_B \sim 10^{-23}$, $3kT \sim 10^{-20}$ (all MKS units); $\chi \sim 10^{-3}$. Some metals do show paramagnetism, but the observed χ is $O(10^{-5})$. This discrepancy was one of the major problems with the old Drude theory of a free electron gas in metals. Of course we now know that only the electrons near the top of the Fermi level are able to respond to a magnetic field and we have to write the susceptibility as

$$\chi = \mu_0\mu_B^2 N(E_F), \tag{25.26}$$

where $N(E_F)$ is the density of states at the Fermi level. Since the Fermi energy $\sim N^{2/3}$, the density of state can be written as $N(E_F) = dN/dE_F = 3N/2E_F$ and Equation 25.26 becomes

$$\chi = \frac{3N\mu_0\mu_B^2}{2E_F} = \frac{3N\mu_0\mu_B^2}{2kT_F}. \tag{25.27}$$

This equation looks similar to the classical result except the inverse temperature has been replaced with the Fermi temperature. Since $T_F/300 \sim 10^2$, $\chi \sim 10^{-5}$, which is typical of the observed values. Since the diamagnetic and paramagnetic susceptibilities are of the same order of magnitude, metals can be either paramagnetic or diamagnetic, depending on which susceptibility is the larger. Most metals are paramagnetic; Cu, Ag, and Au are diamagnetic.

25.5 Ferromagnetism

By far the most useful and fascinating aspect of magnetism is the mysterious ability of certain materials to attract certain other materials with great force. Elements such as Co, Fe, Ni, Gd, and Dy and compounds such as MnBi, MnSb, and MnAs exhibit spontaneous magnetization, a phenomena known as ferromagnetism. They maintain this property until they reach a certain temperature known as their Curie temperature (T_C), at which point their magnetism is destroyed. Cooling below the Curie temperature does not automatically restore their magnetism, but they can be remagnetized by placing them in a strong magnetic field. How do we explain such behavior?

25.5.1 Curie–Weiss Theory (Classical Approach)

We suspect that spontaneous magnetism must have something to do with the d-electrons in the transition metals and the f-electrons in the rare earth metals since they have unpaired spins. We can postulate that there must be some self-generated ordering force that keeps

these spins aligned in adjacent atoms so that they remain aligned until they are disrupted by thermal disordering at their Curie temperature, which in the case of Fe, is over 1000 K. What could be the source of this ordering energy?

Weiss assumed that there was an internal H-field that was somehow proportional to the magnetization M. Recall that in dielectrics, the local electric field was enhanced by the electric dipoles in the surrounding media. Perhaps once the electron spins were lined up, their magnetic field would be strong enough to keep them aligned. Weiss set this local field $B = \mu_0(H + \lambda M)$, where λ is called the Weiss constant. Now consider the quantum equation for paramagnetism (Equation 25.22). For electrons, $M_J = 1/2, -1/2$ and the average value for μ_z for $H = 0$ is given by

$$\langle \mu_z \rangle = \frac{\mu_B \left(e^{\mu_B B/kT} - e^{-\mu_B B/kT} \right)}{e^{\mu_B B/kT} + e^{-\mu_B B/kT}} - = \mu_B \tanh(\mu_B B/kT) = \mu_B \tanh(\mu_B \mu_0 \lambda M/kT) \qquad (25.28)$$

and the magnetization M can be written as

$$M = N_{\text{eff}} \langle \mu_z \rangle = N \mu_B \tanh(\mu_B \mu_0 \lambda M/kT), \qquad (25.29)$$

which can be solved for M. (Note the similarity to Equation 23.28 for dipolar solids.) As T goes to zero, the tanh goes to 1 and $M = M_{\text{sat}} = N_{\text{eff}} \mu_B$ where N_{eff} is the effective number of electrons with aligned spins. As T gets large so that $\mu_0 \mu_B \lambda M/kT \ll 1$, Equation 25.29 becomes

$$M = N_{\text{eff}} \mu_B^2 \mu_0 \lambda M/kT_C \qquad (25.30)$$

and a solution to Equation 25.29 is possible only if $T \leq T_C$, the Curie temperature. We then can use Equation 25.30 to find the Weiss constant λ:

$$\lambda = kT_C/N_{\text{eff}} \mu_B^2 \mu_0 = kT_C/\mu_B \mu_0 M_{\text{sat}}. \qquad (25.31)$$

Putting this result into Equation 25.29, we can solve for the magnetization as a function of temperature:

$$\frac{T}{T_C} = \frac{M/M_{\text{sat}}}{\text{atanh}(M/M_{\text{sat}})}. \qquad (25.32)$$

The universal reduced magnetization as a function of reduced temperature curve is calculated from Equation 25.32 and shown in Figure 25.1.

Solutions exist for $T > T_C$ only if the applied field $H > 0$. Equation 25.29 may be approximated by

$$M = N \mu_B \tanh \left[\frac{\mu_B \mu_0 (\lambda M + H)}{kT} \right] \approx \frac{N \mu_B^2 \mu_0 H}{kT} + \frac{N \mu_B^2 \mu_0 \lambda M}{kT}, \qquad (25.33)$$

which can be solved for M to give the Curie–Weiss law,

$$M = \frac{N_{\text{eff}} \mu_0 \mu_B^2 H/kT}{1 - N_{\text{eff}} \mu_0 \mu_B^2 \lambda/kT} = \frac{C}{T - T_C} H \qquad (25.34)$$

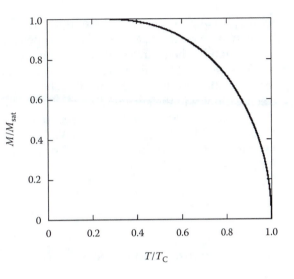

FIGURE 25.1
Universal magnetization versus reduced temperature curve. The model gives very good agreement with measured data.

or

$$\chi = \frac{M}{H} = \frac{C}{T - T_C}; \quad T > T_C, \tag{25.35}$$

where $C = N_{eff}\mu_0\mu_B^2/k$ is the Curie constant for the material. Note that for $T > T_C$, the spontaneous magnetism is destroyed and the magnetism is directly proportional to H, which shows paramagnetic behavior.

For Fe, the saturation magnetization near 0 K is observed to be 1.75×10^6 A/m and the $N_{eff} = 1.75 \times 10^6/\mu_B = 1.882 \times 10^{29}/m^3$. The number density of atoms is $N = 8.5 \times 10^{28}$ atoms/m^3. Therefore, the $p_{eff} = N_{eff}/N = 2.214$. The Curie constant is 1.482 K and since $T_C = 1043$ K, $\lambda = 703$. This would imply an internal M that is 700 times the saturation magnetization. Clearly, this ordering energy cannot be magnetic in nature. The Curie–Weiss law correctly describes paramagnetic behavior for $T > T_C$; it gives a fairly accurate description of the magnetization as a function of temperature, and it estimates the magnitude of the ordering energy required to the keep the electron spins aligned up to the T_C. It does not give a clue as to the source of the ordering energy or why some materials exhibit spontaneous magnetization and others with unpaired spins do not. For these explanations, we must turn to a quantum mechanical model.

It should also be mentioned that p_{eff} obtained from the saturation magnetism was only 2.22, whereas the value for Fe^{2+} with four unpaired spins estimated from $2\sqrt{S'(S'+1)} = 4.90$. There are several possible reasons for this lower value of p_{eff} in the pure metal: spin–orbit interactions, interference with the conduction electrons, and the Fermi level cutting off some of these higher energy states. Also some fraction of the d-electrons participate in forming covalent bonds. It is not a perfect theory.

25.5.2 Heisenberg Exchange Theory (Quantum Approach)

When two particles are far apart, their spins do not interact and it makes no difference if their spins are parallel or antiparallel. When they are brought close together and the wavefunctions overlap, there will be a strong repulsion if their spins are parallel due to Pauli exclusion. Hence the electron spins are antiparallel in the ground state of the H_2 molecule. There is however, an intermediate distance where the exchange interactions between the two particles favor the parallel spins. This exchange energy can be calculated

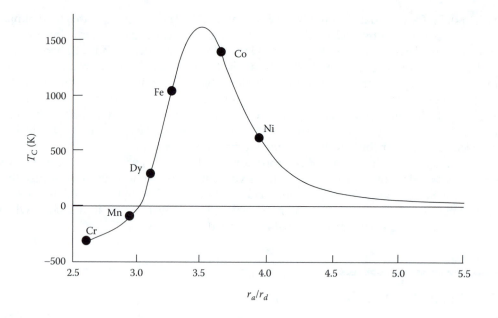

FIGURE 25.2
Bethe-Slater type curve in which the Curie temperature rather than the exchange energy is plotted against the ratio of atomic radius to the radius of the *d*-electron shell (*f*-electron shell in the case of Dy). Negative temperatures for Cr and Mn correspond to their Néel temperatures. (Data taken from Kittel, 7th edn., 1996).

using the Heisenberg exchange theory (the reader is referred to Ashcroft and Mermin [1976] for details). Bethe and Slater showed that when the ratio of the distance between nearest neighbors to the diameter of the orbits of the *d*-electrons > 3, the parallel spin state is favored. Figure 25.2 is a plot of the Curie temperature, which is a measure of the exchange energy, as a function of the radius of the atom r_a divided by the radius of the *d*-electron shell r_d. The Néel temperatures of Cr and Mn were plotted as negative T_C since these elements are antiferromagnetic. The r_a/r_f for Dy was included to see if the model could be extended to the rare earth series. Cu would fit on the curve past Ni, but it has a filled *d*-shell.

25.5.3 Antiferromagnetism

As mentioned in Section 25.1.1, certain materials exhibit what is called antiferromagnetism. In these cases, the exchange energy favors antiparallel spins so that there is no net magnetism. This negative ordering energy also resists paramagnetism until this energy is destroyed by heating above the Néel temperature (sort of a negative Curie temperature). The Curie–Weiss law can be written as

$$\chi = \frac{C}{T + T_N}; \quad T > T_N. \tag{25.36}$$

25.5.4 Ferrimagnetism

Ceramic systems involving ferromagnetic elements may exhibit ferrimagnetism. Ferrimagnetism occurs when the magnetic moments of the different constituents oppose one another and cause a partial cancellation of the total magnetization. A good example of this phenomenon is Fe_3O_4 (magnetite, the original lodestone used for navigation by ancient

mariners). Fe_3O_4 is actually a mixture of FeO and Fe_2O_3. The structure of this (and similar cubic ferrites denoted by $MO(Fe_2O_3)$ is inverse spinel in which the O atoms form a face-centered cubic lattice with the Fe^{3+} occupying both the octahedral and tetrahedral sites, and the Fe^{2+} occupying only tetrahedral sites. The O atom shares one electron with an Fe^{3+} atom and the other with an adjacent Fe^{3+} atom. Since the spin arrangement of Fe^{3+} is $\uparrow\uparrow\uparrow\uparrow\uparrow$, the added electron cannot go in $\uparrow$ without violating the Pauli principle, so it must go in as $\uparrow\uparrow\uparrow\uparrow\uparrow\downarrow$. But the other electron from the O atom is $\uparrow$, so it must go into the other Fe^{3+} atom as $\uparrow\downarrow\downarrow\downarrow\downarrow\downarrow$. Thus the magnetic moments of the Fe^{3+} ions all cancel out, leaving just the Fe^{2+} electrons to contribute to the permanent magnetic moment.

Even though they do not form as strong a magnet as their ferromagnetic cousins, the ferrites, as this class of ceramics is called, are very important technologically because they are also insulators. Because they do not conduct electricity, they may be used as cores for transformers or choke coils in high frequency radio or microwave applications where losses from induced eddy currents must be avoided.

25.6 Magnetic Domains

We all know that if we heat a permanent or ferromagnetic magnet to high temperatures and allow it to cool, the magnetism disappears. What happens to it?

The exchange forces are still active in keeping the unpaired spins aligned, so the magnetism is still there, only it has become rearranged. The material has become an array of small magnetic segments arranged in such a way that the magnetism stays within the material. A magnetic field contains energy given by $B^2/2\mu_0$ per unit volume. If a magnet could be broken up into a bunch of smaller individual magnets, they could be arranged in such a way that all the flux stayed inside the magnet and thus would reduce the energy of the system significantly. Anyone who has played with small bar magnets will remember that they have a tendency to arrange themselves as pairs with opposite polarity to lower their overall energy.

But how can a large magnet break up into many small magnets and still remain intact? It does this by forming magnetic domains; small isolated regions of the material that are separated from each other by a Bloch wall. The domain boundary (or Bloch wall) is a region over which the dipoles rotate by 180°, which decouples the exchange forces on either side of the wall and allows the system to reduce its overall energy as shown in Figure 25.3.

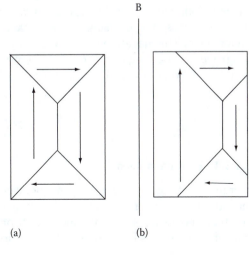

FIGURE 25.3
Magnetic domains in a demagnetized state (a). The domains are arranged so that the magnetic field is internal to the material in order to minimize the magnetic energy. When an external field is applied (b), the domains that are favorably oriented grow at the expense of those less favorably oriented by moving the Bloch walls.

(a)   (b)

The number of domains that can form is limited by the fact that it also costs energy to form the domain wall. This energy is a combination of the anisotropy energy and the exchange energy. In Fe, it is easier to magnetize along the [100] direction than along the [110] or [111] directions. Therefore, it would be energetically favorable to change the orientation to the magnetization in as short a distance as possible to minimize the number of orientations away from the easy direction. However, a large change in orientation costs exchange energy since the exchange force wants to keep the electron spins aligned. The thickness of the domain wall is an optimization of the two competing energies and is generally $O(0.1 \ \mu m)$. The number of domains is a balance between the bulk magnetic energy saved by subdividing the material into smaller domains and the energy cost of the additional wall required.

25.7 Magnetic Hysteresis

When a demagnetized ferromagnet is magnetized by placing it in a strong magnetic field, domains aligned along the field can grow by moving their Bloch Walls to increase their size at the expense of domains that are not favorably aligned as illustrated in Figure 25.3. As the applied magnetic field H is increased, the magnetization M increases until all the favorably oriented domains have grown to their maximum size and the material becomes saturated. The flux density $\mathbf{B}$ will continue to increase according to $\mathbf{B} = \mu_0 \ (\mathbf{H} + \mathbf{M}_{sat})$ even though there is no further increase in $\mathbf{M}$. As the H is decreased to zero, the remaining B_{rem} is called the remanence. The value of H required to demagnetize the material is the coercive strength H_C.

Because the product $\mathbf{H} \cdot \mathbf{B}$ is work, the product $B_{rem}H_{coer}$ is referred to as the magnetic energy of the material. The area of the box inside the demagnetization loop labeled $(BH)_{max}$ is also quoted as an energy product of the magnet. A good permanent magnet should have a high magnetic energy product or $B_{rem}H_{coer}$ product which requires a broad, almost square hysteresis loop as shown in the Figure 25.4a. Such materials are called hard magnetic materials. On the other hand, if the material is to be used as a transformer core, the area enclosed by the hysteresis loop represents energy that goes to heat for each cycle. Therefore, materials with a narrow hysteresis loop or small $B_{rem}H_{coer}$ product such as illustrated in Figure 25.4b are sought for such applications.

25.8 Magnetic Materials

25.8.1 Hard Magnetic Materials

The magnetic properties are determined by the ease at which the Bloch wall can move. Since a Bloch wall cannot move across grain boundaries, a material with a fine grain structure would be more difficult to magnetize and to demagnetize. This would be considered a hard magnetic material and would make a good permanent magnet. Some of the strongest magnets are alloys of cobalt and a rare earth (CoRE magnets) such as Co_5Sm, which has a $(BH)_{max} = 159 \ kJ/m^3$. A commonly used, low cost, magnetic alloy is Alnico V $(Fe_{51}AL_8Ni_{14}Co_{24}Cu_3)$ which has a $(BH)_{max} = 40 \ kJ/m^3$. A second phase precipitates out of this in the form of long needle-like crystals of ferromagnetic material. Solidifying under the presence of a strong magnetic field aligns these needles and locks in the magnetization. The combination of this needle-like geometry and the built-in lattice strain makes a material that is extremely hard, both magnetically and mechanically. The recent development of

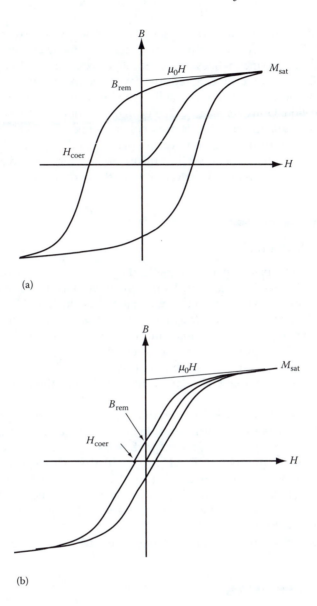

(a)

(b)

FIGURE 25.4
Examples of magnetic hysteresis loop for (a) a hard magnetic material and (b) a soft magnetic material. The flux density remaining when the applied field is removed is the remanence, B_{rem}. The reverse applied field required to destroy the magnetism is the coercive strength, H_{coer}. The area enclosed by the hysteresis loop represents energy lost to heat during the magnetization–demagnetization cycle.

iron-neodymium-boron ($Fe_{14}Nd_2B$) magnets with a $(BH)_{max} = 318$ kJ/m^3 for use in compact brushless DC motors has found its way into the auto industry. These motors are used for power windows and their smaller size made it possible to reduce the thickness of the doors to give more interior room with the same exterior dimension. Their high efficiency also makes them ideal motors for hybrid and electric drive vehicles.

Magnetic tapes require magnets with a high enough coercive strength to not be easily demagnetized, but low enough for the recording head to easily change the state when recording. Ferromagnetic CrO_2 has largely replaced the ferrites used in earlier tapes

because of its higher coercive strength. Saturated single domain grains are embedded in a nonmagnetic binder and aligned with their long axes along the direction of the tape motion. As they move under the recording head, their magnetization is altered according to the amplitude of the signal to be recorded.

25.8.2 Soft Magnetic Materials

Applications such as transformer cores that must be rapidly magnetized and demagnetized with minimum hysteresis loss require a soft magnetic material—a material with a large μ but small remanence. Extremely pure iron with large grain structure is a good candidate for such applications. The addition of 4% Si allows the grains to be textured in order to reduce the anisotropy and further reduce the coercive force. The addition of Ni also reduces the anisotropy and alloys of Fe and Ni, called permalloys, have very low hysteresis loss. Permalloy 78, $(Fe_{21.5}Ni_{78.5})$ has a permeability on the order of 100,000 and a hysteresis loss of ~2 W/kg. Better yet, the new metallic glasses such as Metglass 2605 eliminates the grain structure altogether, so there is nothing to impede the domain walls. They have a loss factor of only 0.3 W/kg. The efficiency of transformers can be improved by as much as 5% by using the metallic glass in their cores rather than the conventional Fe–Si or permalloy.

25.8.3 Ferrimagnetic Materials

Ferrites are iron oxide compounds with a composition given by $MO(Fe_2O_3)$, where M is a divalent cation such as Zr^{2+}, Cu^{2+}, Ni^{2+}, Co^{2+}, Mn^{2+}, or Fe^{2+}. To improve the saturation magnetization, which is limited by the cancellation by opposing spins, it is possible to mix both normal and inverse spinel structures. For example, $(ZnOFe_2O_3)_x(MnOFe_2O_3)_{1-x}$ has a net magnetic moment of $7.5\mu_B$ when $x = (1/2)$ compared with $4\mu_B$ for $NiFeO_4$. Garnets have compositions $X_3Y_2(SiO_4)_3$, where X is a divalent cation (Ca^{2+}, Mg^{2+}, Fe^{2+}) and Y is a trivalent cation (Al^{3+}, Fe^{3+}, Cr^{3+}). The synthetic yttrium iron garnet (YIG) has a magnetic moment of $9\mu_B$.

There are both soft and hard ferrimagnetic materials. The soft magnets with low coercivities are used as cores for high frequency transformers to reduce the hysteresis losses. Ferrite inductors are often found on computer cables to suppress unwanted high frequency electromagnetic interference (EMI).

The hard magnetic ferrites are used as permanent magnets. They are the magnets usually found in loudspeakers. Ferrites are also used in magnetic audio and video tapes. Early computers used small ferrite doughnuts for their random access memories, each doughnut equals 1 bit of information. (32 kB was considered a large memory.) $BaFeO_4$ and $SrFeO_4$ form in plates with a perpendicular easy axis. These small platelets are embedded in plastic to form flexible magnets used in refrigerator doors and in other appliances. YIG is transparent in the infrared and is a Faraday rotator and is used in magneto-optic data storage applications.

25.8.4 Magnetostrictive Materials

Magnetic ordering causes dimensional changes in materials from the spontaneous alignment of the magnetic moment of the atoms in a domain and from the alignment of the domains in an applied magnetic field. The magnetism of a material can be slightly changed by lattice strains, an effect call piezomagnetism. Conversely, the lattice should respond to the magnetic state of the material, an effect call magnetostriction. Because of the anisotropy of the magnetization, the coupling between the magnetic state and the lattice strain is a

tensor of rank 4 for single crystals. However, for polycrystalline materials, the overall effect can be averaged and the strain written as

$$\frac{\partial \ell}{\ell} = \frac{3\lambda \left(\cos^2 \theta - 1/3\right)}{2}, \tag{25.37}$$

where

θ is the angle between the direction of M and the direction the strain is measured in

λ is the average of the saturated magnetostrictive constants λ_{100} and λ_{111} for the material

This average may be positive or negative, meaning the material may contract in one direction and expand in another, or the volume may contract or expand, depending on the material. The 60 Hz hum associated with power transformers is a result of the expansion and contraction of the laminations in the core.

The effect is small $O(10^{-5})$ for the d-electron magnetic materials, but can be larger for f-electron magnetic materials. The materials with the largest magnetostriction are the Fe_2RE alloys where $RE = Dy$ or Tb. However, it requires a large magnetic field to drive the magnetostriction in these materials. Terfenol-D ($Tb_xDy_{1-x}Fe_2$) can produce a strain of 1600 μm/m at 2000 A/m and is the most commonly used material for magnetostrictive transducers (see Clark, 1980).

Spontaneous magnetization below the Curie temperature is usually accompanied by an increase in volume. Therefore, the magnetostriction that results as the spontaneous magnetization is destroyed by heating can be used to counteract the normal thermal expansion and produce materials with a controlled coefficient of thermal expansion (CTE). Invar ($Fe_{64}Ni_{36}$) has virtually zero CTE and is used for clock pendulums and other applications where good dimensional stability is required. $Fe_{58}Ni_{42}$ matches the CTE of Si and FeNiCo alloys (Kovar) can match the CTE of glasses. These Kovar alloys are used for glass–metal seals in vacuum tubes.

25.9 Magnetic Information Storage Technology

25.9.1 Magneto-Optic Effects

25.9.1.1 Faraday Rotation

As seen in Chapter 24, photons can interact with electrons in a material, and as we saw in Section 25.3, electrons interact with a magnetic field. As a consequence, the speed of light passing through a material in the presence of a magnetic field becomes dependent on the angle between the **k**-vector and the **B**-vector with the maximum change occurring when the two are parallel. Thus radiation arriving at some angle relative to the applied field will have one component along the field and one normal to the field, traveling at different speeds, which produces a phase shift. This, of course, is a condition for birefringence and the polarization of the emerging radiation will be changed. This effect was first discovered by Michael Faraday in 1845 and is referred to as Faraday rotation. The magnitude or amount of rotation in a diamagnetic material depends on the Verdet coefficient (a material property), the magnetic field, and the path length. Materials such as terbium gallium garnet have a very large Verdet coefficient and are used in Faraday rotators, devices that can actively control the plane of polarization of light and are used as modulators and optical isolators.

Faraday rotation also occurs in a free electron gas where the amount of rotation depends on the product of B and the electron density integrated along the path. This is a tool used in astrophysics for estimating the magnetic field on the sun, stars, and in the interstellar medium. It also provides a method of determining the electron density in the ionosphere.

25.9.1.2 Magneto-Optic Kerr Effect

The magneto-optic Kerr effect (MOKE), discovered by John Kerr in 1877, is similar to the Faraday effect, except it rotates the plane of polarization of light reflected from a solid that contains a magnetic field. As such it is a valuable tool for visualizing magnetic domains.

The MOKE is also used in magneto-optical recording. A standard polycarbonate disc is coated with vertically polarized MnBi or similar ferromagnetic material with a T_C of 150°C–200°C. Information is written into small domains of the ferromagnetic material by an electromagnetic write head. A short laser pulse accompanies the write event to quickly heat the magnetic domain above its T_C in order to change its polarization state. The information is retrieved by reflecting a polarized beam of laser light off the domains and sensing their polarization state with an analyzer and a photodiode. Up to 4.6 GB can be stored on a 5.5 in. magneto-optic disc as compared to 0.62 GB on a standard CD-ROM.

25.9.2 Magnetoresistance

25.9.2.1 Ordinary and Anisotropic Magnetoresistance

A change on the resistance of a conductor in a magnetic field was first noticed by Lord Kelvin in 1856. The effect is caused by the Lorentz force which deflects the trajectory of the current carriers as in the Hall effect (see Section 18.5) and is maximized when the current flow is perpendicular to the magnetic field. The effect is called ordinary magnetoresistance (OMR) or ordinary magnetoresistance.

In a nonmagnetic material, the number of conduction electrons with spin up will on the average equal the number of spin-down electrons. However, in a ferromagnetic material, there is a small energy difference between the two spin states so that we will have a majority of spins that align themselves with the internal magnetic field. The scattering times τ_{up} and τ_{down} will be different and there will be a coupling of the resistivity with the magnetic field. Because of the anisotropy of the magnetism, the coupling will be in the form of a tensor and is called anisotropic magneto resistance (AMR).

Even though the effect is small, magnetoresistance is used for sensing magnetic fields and for read heads on magnetic hard drives.

25.9.2.2 Giant Magnetoresistance

In order to reduce the bit size and increase the storage density on magnetic hard drives, a never-ending search goes on for improving the sensitivity of the read heads. This search was aided considerably by the discovery of giant magnetoresistance (GMR) by Albert Fert and Peter Grünberg (2007 Nobel Laureates). If two ferromagnetic materials are separated by a thin (nanometer) nonmagnetic or antiferromagnetic conductor, the resistance is much lower if the two magnets are aligned in the same direction than if antiparallel. The reason is that the majority of the conduction electrons have their spin aligned with the magnetic field of the ferromagnetic material they are in. If they try to enter a material with an opposite field, their spin must flip, which costs energy. Thus the sandwich of very thin ferromagnetic-antiferromagnetic-ferromagnetic materials can act as a spin valve, transporting electrons of one spin and restricting those of opposite spin with a resistivity change of ~30%–50%, an order of magnitude larger than AMR materials.

To incorporate this technology into a read head, one of the magnets is a hard permanent magnet and the other is a soft magnetic material such as Supermalloy. The information is coded on the disc as domains that are N or S for logical 1 or 0. As the disc moves under the read head, the spin valve opens and closes depending on the field induced in the soft magnetic head.

25.9.2.3 Colossal Magnetoresistance

In 1993, von Helmolt and coworkers at the Siemens Research Laboratories discovered that layered oxygen perovskite structures could change their resistivity by as much as two orders of magnitude in the presence of a magnetic field. Because of this enormous change in resistivity, the effect became known as colossal magnetoresistance (CMR). The mechanism by which this effect occurs is of great interest, not only because of the technology and potential applications but also from the theoretical aspect because of the similarity between the CMR structures ($RE[1 - x]M[x]MnO_3$, where RE denotes rare earth, M denotes Ca, Sr, Ba, Pb) and the layered cuprite structures of the high temperature superconductors (see Section 26.10), neither of which are completely understood. One of the major problems in utilizing CMR in practical devices is that, in the materials discovered thus far, the effect occurs only at temperatures well below ambient.

25.9.3 Magnetoelectronics (Spintronics)

As we all know, electrons carry spin as well as charge and, since the spin is quantized up or down, the spin state is a natural way for coding binary 1's and 0's. Much research is now being devoted to managing and manipulating the spin of electrons to develop a new form of recording and processing information. The discovery of the GMR in 1988 is considered to be the birth of this new technology. We have already seen that it is possible to inject a spin-polarized stream of electrons from a ferromagnet and to control the passage of these electrons with a spin valve. Work is underway to develop spin-based field-effect transistors in which the flow of spin-polarized current is controlled by a voltage applied at the gate. However, there is much to be done in the generation and propagation of a pure spin current, especially at a conductor–semiconductor junction, and extending the relaxation time of the spin state if the goals of a spin-based quantum computer and a spin-based nonvolatile memory are to be realized.

25.10 Summary

All materials interact to some degree with a magnetic field. Some materials are weakly repelled by a magnetic field and are said to be diamagnetic. They have a negative susceptibility (a measure of the ability to be magnetized). Other materials are weakly attracted by a magnetic field and are said to be paramagnetic. Metals that produce a spontaneous magnetic field are said to be ferromagnetic, and ceramics that produce a spontaneous magnetic field are said to be ferrimagnetic. The magnetism in ferro- and ferrimagnetic materials decreases with temperature and is lost completely at their Curie temperature, which is a property of the material. Above the Curie temperature, ferro- and ferrimagnetic materials become paramagnetic. Some metals are antiferromagnetic, which means that they are not magnetic until they reach their Néel temperature where they become paramagnetic.

Diamagnetism occurs through the interaction of the core electrons with an applied magnetic field, which causes the electrons to produce an opposing magnetic field. Paramagnetism occurs when there are unpaired electrons, which the applied field tries to align, while thermal motion tends to randomize their alignment. As a result, paramagnetism decreases as the reciprocal of the absolute temperature. Classically, the free electrons in a metal should exhibit strong paramagnetism, but the only electrons that are free to participate are those near the Fermi level. Both diamagnetism and paramagnetism are weak effects. Diamagnetism occurs in all metals, but in some metals the paramagnetism is greater than the diamagnetism and prevails.

Ferromagnetism occurs in the first row of the transition metals with unpaired $3d$-electrons (Fe, Co, Ni) and in the first row of the rare earths with unpaired f-electrons (Dy, Gd) when the amount of overlap between the wavefunctions of the d- or f-electrons falls in a certain critical range. A quantum mechanical exchange force between neighboring atoms keeps their electronic spins aligned until the thermal energy becomes high enough at the Curie temperature to destroy the alignment. If the overlap is greater, the exchange forces cause the spins in the neighboring atoms to become antiparallel and the material is antiferromagnetic (Cr, Mn). Ferrites are compounds of divalent metal oxides and ferromagnetic Fe_2O_3. In order to form these compounds, some of the ferromagnetic atoms must go into the molecule with their spins antiparallel and the material is said to be ferrimagnetic. Ferrimagnetic materials behave like ferromagnetic materials, but are generally weaker magnetically because of the partial cancellation of the ferromagnetic spins. However, since they are dielectrics and do not conduct electricity, ferrites are suitable for tasks that ferromagnetics cannot do.

A ferromagnetic material is divided into a bunch of small regions called domains where the spins are in the same direction within the domain, but may be in a different direction in an adjacent domain. Domains are separated from each other by Bloch walls where the electrons gradually change the directions of their spins. If the spins of the domains are in random directions, there is no net magnetic moment. The material may be magnetized by applying a strong magnetic H-field. Domains that are aligned with the field will grow at the expense of the others by moving the domain walls. When all the domains are aligned with the field, the material is saturated. When the field is removed, the remaining magnetism B_{rem} is called the remanence. The H_{coer} field required to demagnetize the material is the coercivity. A plot of the magnetization–demagnetization cycle is the hysteresis. The strength of a magnet is characterized by its energy product $B_{rem}H_{coer}$ or by the area of the largest B–H rectangle that can be drawn under the demagnetization curve.

Hard magnetic materials (materials with high energy products) are used for permanent magnets. Materials can be made magnetically hard in much the same manner as metals are hardened (e.g., alloying, precipitation hardening, and grain refining). If the grains are small enough for only one domain to form, the demagnetizing field must be strong enough to flip the spins in the domain rather than just move the domain wall. Soft magnetic materials have low coercive strength and small BH product in order to minimize the energy loss in the hysteresis loop. Materials are made magnetically soft by texturing the grains to allow the Bloch walls to move easily in preferred directions or by using amorphous metals to eliminate the grain boundaries.

Bibliography

Ashcroft, N.W. and Mermin, N.D., *Solid State Physics*, Brooks Cole, Philadelphia, 1976.
Baibich, M.N., et al., Giant magnetoresistance of (001)Fe/(001)Cr magnetic superlattices, *Phys. Rev. Lett.*, 61/21, 1988, 2472–2475.

Binasch, G., et al., Enhanced magnetoresistance in layered magnetic structures with antiferromagnetic interlayer exchange, *Phys. Rev. B.*, 39/7, 1989, 4828–4830.

Christman, J.R., *Fundamentals of Solid State Physics*, John Wiley & Sons, 1988.

Clark, A.E., *Ferromagnetic Materials*, Vol. 1, ed. Wolfhart, E.P. North Holland, Amsterdam.

Das Sarma, S., Spintronics, *Am. Sci.*, 89, 2001, 516–523.

Earnshaw, W., On the nature of the molecular forces which regulate the constitution of the luminiferous ether, *Trans. Camb. Phil. Soc.*, 7, 1842, 97–112.

Gersten, J.I. and Smith, F.W., *The Physics and Chemistry of Materials*, John Wiley & Sons, 2001.

Hummel, R.E., *Electronic Properties of Materials*, 3rd edn., Springer-Verlag, New York, 2000.

Kittel, *Introduction to Solid State Physics*, John Wiley & Sons, New York, 1966.

Srivastava, C.M. and Srinivasan, C., *Science of Engineering Materials*, Wiley Eastern Ltd. New Delhi, India, 1987.

Wohlfart, E.P., Ed., Ferromagnetic materials, in *Handbook of Magnetic Materials*, Vol. 1, North-Holland, Amsterdam, 1980.

Problems

1. Show that $(ZnOFe_2O_3)_x(MnOFe_2O_3)_{1-x}$ has a magnetic moment of $7.5\mu_B$ for $x = (1/2)$.

2. Would it be possible to directly measure the angular momentum of an electron? Do a feasibility analysis for such an experiment. Assume you have a 1 g piece of Fe that is magnetized to saturation that you can levitate in an electrostatic levitator. While the Fe is levitated, heat the sample to above the Curie temperature. What will be the angular momentum of the sample?

3. What geometry would you use for the sample in the above problem to maximize the observed spin of the sample? What would be the expected angular rate? Would this be observable? Would the ability to levitate a larger sample help?

4. You have two 1 cm cubic magnets with a $B = 1$ T. What force would be required to pull them apart? (Hint: use the principle of virtual work.)

5. Construct a model for the local magnetic field in a magnetic material similar to the Clausius–Mossotti model for the local electric field in a dipolar field. Can such a model explain the ordering in a ferromagnetic material? (Compare the predicted Weiss constant against the observed Weiss constant for Fe.)

26

Superconductivity

Superconductivity is not just a phenomenon in which the resistance of a substance drops to zero at a certain temperature, but as we shall see in this chapter, it is a new state of matter in which the quantum nature is revealed macroscopically.

26.1 Historical Perspective

In 1908, at the University of Leiden, Kamerlingh Onnes succeeded in liquefying He at 4.2 K. Since this was the lowest temperature ever achieved at the time, he naturally started investigating physical properties of various materials at these low temperatures. For example, he measured the resistance of Pt down to this temperature and found that, like most good conductors, the resistivity decreased as T was lowered, and approached some finite value ρ_0 as $T \to 0$. However, in 1911 when he tried Hg, much to his surprise, the resistivity suddenly dropped to 0 at around 4 K. Subsequent work revealed that 27 elemental materials as well as many alloys exhibit this unusual property at very low temperatures. The elements that become superconductors are found in the left portion of the transition elements and in group II, III, and IV to the right of the transition elements. Also, some of the lanthanides and actinides become superconductors. Some of these are listed in Table 26.1 along with their transition temperature, T_C.

Other elements become superconducting under pressure (e.g., As, Ba, Ce, Ge). No alkali metals or noble metals are superconductors. Matthias pointed out that the transition temperature is generally higher if the electron per atom ratio is odd, as can be seen from the above tables (group IVA being the exception).

As can be seen, most of the elemental superconductors have T_C that are too close to 0 K to be useful. Also, the presence of a magnetic field or the passage of a current through a superconductor lowers its T_C. Therefore, extensive searches have been made for practical superconducting alloys or systems. The A-15 compounds, Nb_3Sn, Nb_3Ge, and Nb_3Si had the highest T_C before the recent discovery of the high T_C ceramic systems (T_C for $Nb_3Ge = 23\,K$). However, these are metastable systems and can only be produced by a nonequilibrium solidification process such as rapid quenching. They are also brittle and it is difficult to form wires with these systems. The most widely used superconducting alloy for commercial applications (primarily large superconducting magnets for MRI systems) is Nb46.5 W% Ti. This alloy has a $T_C = 9\,K$ at $B = 0$, and a critical field $H_{C2} = 11\,T$ at 4.2 K. Its popularity is primarily due to the fact that it is more economical to produce, can easily be drawn into flexible wires for winding magnets, and has fairly robust superconducting properties at liquid He temperatures.

TABLE 26.1

Elemental Superconductors

Group IVB	Group VB	Group VIB	Group VIIB	Group VIII	Group VIII
Ti (0.39)	V (5.39)				
Zr (0.55)	Nb (9.64)	Mo (0.92)	Tc (7.77)	Ru (0.51)	
Hf (0.12)	Ta (4.48)	W (0.01)	Re (1.4)	Re (0.66)	Ir (0.14)

Group IIB	Group IIIA				Group IVA
		Al (1.14)			
Zn (0.88)		Ga (1.09)			
Cd (0.56)		In (3.40)			Sn (3.72)
Hg (4.15)		Tl (2.39)			Pb (7.19)

Source: Data from Ashcroft, N.W. and Mermin, N.D., *Solid State Physics*, Brooks Cole, Philadelphia, 1976.

Since liquid He is expensive, difficult to keep, and is a nonrenewable resource, there is a strong desire to be able to go to a higher boiling point material. Liquid H_2 boils at 20 K. In principal, the A-15 alloys could be used with liquid H_2, but the critical current and critical magnetic field drops off rapidly as T approaches the T_C, so this really is not practical. Liquid N_2 boils at 77 K. This is why there was so much excitement when the high T_C ceramic systems were discovered with T_C well above 77 K. However, getting high critical currents and being able to fabricate these brittle materials into wires has proven more difficult than originally anticipated, hence systems based on these materials have still not made a significant impact on the more conventional systems based on Nb–Ti technology.

It was not until 1957 that the phenomena of superconductivity was explained by a theory developed by Bardeen, Cooper, and Schreiffer (BCS theory) in which it was postulated that two electrons became loosely bound by exchanging a virtual phonon (another way of saying a cooperative interaction with the lattice ions). By becoming paired with opposite spins, the so-called Cooper pairs act as bosons and are no longer controlled by the Pauli principle; therefore, many can exist in the same quantum state. This pairing creates an energy gap about the Fermi level in which no states are available for the Cooper pair to be scattered into; hence, they can move through the lattice unimpeded, very much like the superfluid state of liquid He which exhibits zero viscosity.

26.2 Basic Properties of Superconductors

In addition to the conductivity going to zero at some transition temperature >0, superconductors (at least the metallic ones) share the following properties.

26.2.1 Isotope Effect

For pure superconductors, it has been observed that by substituting a different isotope, the transition temperature depends on the average mass of the system according to $T_C \sim M^{-1/2}$. Since phonon frequency is inversely proportional to the square root of the mass of the vibrating atoms, this observation is crucial to the BCS theory because it implies an electron–phonon interaction.

26.2.2 Persistent Supercurrents and Perfect Diamagnetism

A B-field does not exist in the interior of superconductors. Since $B = \mu_0(H + M) = 0$, $\chi_m = M/H = -1$. which implies that superconductors are perfect diamagnetic materials. This diamagnetism results from surface currents that circulate to cancel external fields. These supercurrents are given by

$$\mathbf{J} = n_s e \mathbf{v}_{c/m} = \frac{n_s e \hbar}{m} \mathbf{k}_{c/m}, \tag{26.1}$$

where

n_s is the number of electrons in the superconducting state

$\mathbf{k}_{c/m}$ is the momentum of the center of mass of the Cooper pair

These currents, once set in motion, persist for as long as the temperature is maintained below the T_C. Measurement of the magnetic moment from these circulating currents is one of the ways of determining that the material is in a superconducting state. The persistent currents are not just an effect of zero resistivity, but constitute a macroscopic quantum state. Just as electrons, because they are in definite quantum states, can circulate in atoms without radiating energy and can only transition to other allowed quantum states, one can think of circulating supercurrents as being in some particular quantum state, in which they will remain until they transition to another allowed state.

26.2.3 Meissner Effect

The persistent currents not only prevent an applied magnetic field from entering the superconductor, but will also expel any magnetic field that may have been present when T is lowered below T_C. This expulsion of magnetic field, known as the Meissner effect, is a unique feature of the superconducting state. A normal conductor, when cooled to the point that its resistance approaches zero, will prevent an applied field from penetrating it, but will retain whatever magnetic field that was present before the conductor was cooled.

26.2.4 Electronic Heat Capacity

Recall that in normal metals the electronic heat capacity increases linearly with temperature near absolute zero. However, in a superconductor, the electronic heat capacity $\sim \exp(-2\Delta_0/kT)$ for $T < T_C$, and then drops to the normal linear value at $T \geq T_C$. This implies an energy gap of $2\Delta_0$ between the empty and filled states in the vicinity of the Fermi level, i.e., the Fermi level is centered in an energy gap of width $2\Delta_0$ between the filled states and empty states. The value $2\Delta_0/kT_C$ is seen in Table 26.2 to be more or less constant

TABLE 26.2

Bandgaps and Transition Temperatures for Elemental Superconductors

Element	$2\Delta_0$ (meV)	T_C	$2\Delta_0/kT_C$
Nb	3.05	9.5	3.8
Pb	2.90	7.19	4.3
Hg	1.65	4.15	4.6
Ta	1.40	4.48	3.6
Sn	1.15	3.72	3.5

Source: Data taken from Ashcroft, N.W. and Mermin, N.D., *Solid State Physics*, Brooks Cole, Philadelphia, 1976.

for different systems. (We will see later that the BCS theory predicts that the binding energy of the Cooper pair is related to the transition temperature by $2\Delta_0 = 3.53kT_C$.)

26.2.5 Optical Properties

A surprising experimental result is that the optical properties of a material in the superconducting state are no different than the normal state. This is surprising because from the discussions in Chapter 24, one would expect a perfect conductor to be a perfect reflector. As it turns out, the perfect conductivity occurs only for DC or for very low frequency AC. The dielectric function of the superconducting state is no different from the normal state at optical frequencies. There is an absorption edge in the far infrared (100–10,000 μm range) that corresponds to the energy gap of the superconducting state.

26.2.6 Thermal Conductivity

The bound Cooper pairs can neither transport energy, nor can they participate in phonon scattering. As a result, the electronic component of thermal conductivity is diminished in the superconducting state while the phonon contribution is enhanced because of the longer phonon mean free path.

26.3 BCS Theory

As stated previously, the basic concept behind the BCS theory is that two electrons with opposite spins and opposite momenta (in the center of mass system) can, through phonon interactions with the lattice atoms, experience an attractive force sufficient to overcome their Coulomb repulsion and form a loosely bound pair with zero net spin. Because they have zero net spin, they behave as Bosons and any number may occupy the ground state. The details of the BCS theory are too complicated to be presented here (the reader is referred to the original paper (Bardeen, Cooper, and Schrieffer, 1957) cited in the references). Instead, we shall just present some of the predictions of the theory that can be tested.

The essential result from the BCS theory is that the energy gap $2\Delta_0$ between the superconducting state and the normal state at 0 K is given by

$$2\Delta_0 = 4\hbar\omega_D e^{-1/UN(E)} = 4k\Theta_D e^{-1/UN(E)}, \tag{26.2}$$

where
  ω_D is the Debye frequency
  U is the attractive potential between the two electrons that make up the Cooper pair
  $N(E)$ is the density of states at the Fermi level

The transition temperature is found to be

$$T_C = 1.14\Theta_D e^{-1/UN(E)}. \tag{26.3}$$

Combining Equations 26.2 and 26.3

$$2\Delta_0 = \frac{4}{1.14}kT_C = 3.51kT_C. \tag{26.4}$$

The condensation energy between the superconducting and normal states at $T = 0$ is

$$G_N - G_{SC} = \frac{\mu_0 H_{C0}^2}{2} = \frac{2N(E)(\hbar\omega_D)^2}{e^{(2/UN(E))} - 1} \approx \frac{N(E)\Delta_0}{2}. \tag{26.5}$$

The spike in the electronic heat spike at T_C is given by

$$C_{SC} - C_N = 1.43\gamma T_C, \tag{26.6}$$

where γ is the linear coefficient of electronic heat capacity.

Finally, the corrections to Tuyn's approximation (see Equation 26.38) for the temperature dependence of the critical field are predicted to be

$$\frac{H_C(T)}{H_C(0)} = 1 - 1.06\frac{T^2}{T_C^2}; \quad T \approx 0 \quad \text{and} \quad \frac{H_C(T)}{H_C(0)} = 1.74\left(1 - \frac{T}{T_C}\right); \quad T \approx T_C. \tag{26.7}$$

From Equation 26.3 the T_C depends linearly on the Debye frequency, which predicts the isotope effect since the Debye frequency is proportional to the inverse square root of the mass of the vibrating atoms. We shall examine some of the other predictions in the following section.

26.4 Thermodynamics of Superconductivity

26.4.1 Free Energy of the Superconducting State

Let $G_N(H,T)$ and $G_{SC}(H,T)$ be the free energies of the normal and superconductive states, respectively, as a function of applied magnetic field and temperature. The work required to destroy the superconducting state in a type-1 superconductor is obtained from

$$W = -\int_0^{B_C} \mathbf{M} \cdot d\mathbf{B} = \mu_0 \int_0^{H_C} \mathbf{H} \cdot d\mathbf{H} = \frac{\mu_0}{2} H_C^2. \tag{26.8}$$

The free energy of the superconducting state as a function of applied field can be written as

$$G_{SC}(H,T) = G_{SC}(0,T) + \frac{\mu_0}{2} H(T)^2. \tag{26.9}$$

Since the free energy of the normal state is unaffected by the presence of a magnetic field

$$G_N(H,T) = G_N(0,T) = G_{SC}(0,T) + \frac{\mu_0}{2} H_{C0}^2. \tag{26.10}$$

The difference in free energy between the normal and superconductive states is given by subtracting Equation 26.9 from Equation 26.10

$$\Delta G(H,T) = \frac{\mu_0}{2} H_{C0}^2 - \frac{\mu_0}{2} H(T)^2 = \frac{\mu_0}{2} H_C(T)^2, \tag{26.11}$$

where $H_C(T)^2 = H_{C0}^2 - H(T)^2$ is the critical field required to destroy the superconducting state.

At the temperatures of interest, the heat capacity of the normal state is primarily due to the electronic contribution γT. The entropy is therefore given by γT and the free energy is

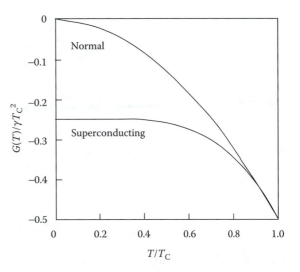

FIGURE 26.1
Free energies of the normal and the superconducting state as a function of T/T_C. The slopes of
the two free energies are the same at T_C, indicating a second-order transformation. The free
energy of the superconductivity state is found
by integrating the entropy Equation 26.17.

given by $-\gamma T^2/2$. The free energies of the normal and superconducting state are shown in
Figure 26.1. Note that the free energies of the two states merge at $T = T_C$ with identical
slopes, indicating a second-order phase transition.

Destroying the superconductivity by heating, i.e., raising the temperature above the
transition temperature, constitutes a second-order phase transition. There is no change in
the slope of the free energy, so the entropy is a continuous function as seen in Figure 26.3
and there is no latent heat involved. There will be a discontinuity in heat capacity, as will
be shown. Conversely, destroying the superconductivity by increasing the magnetic field
above the H_C has the effect of raising the superconducting free energy curve until it crosses
that of the normal state. Now there will be a change in slopes of the free energy and the
phase transition would be first order.

26.4.2 Heat Capacity from BCS Theory

An abrupt change in heat capacity is observed as the material is cooled into the superconducting state. Instead of following the γT behavior characteristic of the electronic component at low temperatures, the heat capacity in the superconducting state exhibits a
Boltzmann-like behavior,

$$C_{SC}(T) = \text{constant } e^{-(2\Delta_0/2kT)}, \quad T \leq T_C. \tag{26.12}$$

The $2\Delta_0$ represents the energy gap centered on the Fermi level and the $2kT$ is to account for
the two electrons forming the Cooper pair. The BCS theory (Equation 26.4) predicts that
this energy gap, $2\Delta_0 = 3.51kT_C$ and this relationship appears to be satisfied by a wide range
of superconductors. The constant can be determined by integrating $C_{SC}(T)/T$ from 0 to T_C
and equating this to the entropy at T_C which is γT_C.

26.4.3 Thermodynamically Consistent BCS Model

We now start with the exponential form of the heat capacity equation (Equation 26.12) and
work backward to find the free energy of the superconducting state and eventually obtain
the critical magnetic field as a function of temperature.

The entropy of the superconducting state must equal the electronic entropy of the
normal state at $T = T_C$, which requires

$$S_{SC}(T_C) = \int_0^{T_C} \frac{C_{SC}(T)}{T}\,dT = \text{constant} \int_0^1 \frac{e^{-1.76/x}}{x}\,dx = \gamma T_C. \tag{26.13}$$

The integral has the form of an exponential integral (Ei) and the constant thus becomes

$$\text{Constant} = \frac{\gamma T_C}{\text{Ei}(1.76)} = \frac{\gamma T_C}{0.067}. \tag{26.14}$$

The heat capacity may then be related to the critical temperature by

$$C_{SC}(T_C) = \frac{\gamma T_C}{0.067} e^{-1.76} = 2.511\gamma T_C \tag{26.15}$$

and the ΔC by

$$\Delta C(T_C) = 2.511\gamma T_C - \gamma T_C = 1.511\gamma T_C. \tag{26.16}$$

This result shown in Figure 26.2 is close to but not identical to the $\Delta C = 1.43\ \gamma T_C$ (Equation 26.6) predicted by the BCS theory.

The entropy of the superconducting state can now be calculated from

$$S_{SC}(T) = \int_0^T \frac{C_{SC}(T)}{T}\,dT = \frac{\gamma T_C^2}{\text{Ei}(1.76)} \int_0^{T/T_C} \frac{e^{-1.76/x}}{x}\,dx \tag{26.17}$$

and is shown in Figure 26.3.

Taking the free energy of the normal state to be zero at $T=0$, the free energy of the superconducting state at $T=0$ is obtained by integrating the entropy from T_C to 0 and adding this to the $G_{CS}(T_C)$ which is the same as $G_N(T_C)$.

$$G_{SC}(0) = G_N(T_C) - \int_{T_C}^0 S_{SC}(T)\,dT = -\frac{\gamma T_C^2}{2} + \frac{\gamma T_C^2}{\text{Ei}(1.76)} \int_0^1 du \int_0^u \frac{e^{-1.76/x}}{x}\,dx = -0.251\gamma T_C^2. \tag{26.18}$$

The result is plotted in Figure 26.1.

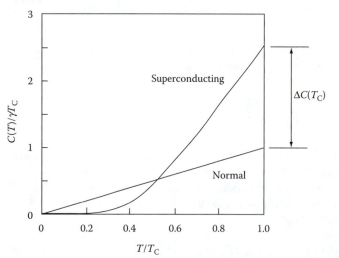

FIGURE 26.2
Heat capacity of the electrons in the superconducting state computed from the model compared with the electronic heat capacity in the normal state. The model predicts the difference in heat capacity $\Delta C(T_C) = 1.51\gamma T_C$.

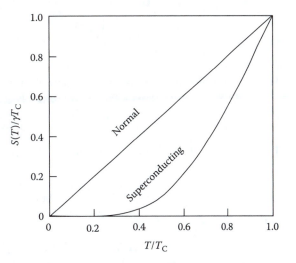

FIGURE 26.3
The entropy in the superconducting state computed from the model is compared to the entropy in the normal state. The two must be the same at T_C.

Equating this result to the work required to destroy the superconducting state at $T=0$ from Equation 26.8, we obtain a relation between the critical field and the transition temperature at 0 K.

$$\frac{\mu_0 H_{C0}^2}{2} = 0.251 \gamma T_C^2, \quad \gamma T_C^2 = 1.99 \mu_0 H_{C0}^2. \tag{26.19}$$

The free energy at T between 0 and T_C is obtained by integrating the entropy from 0 to T and subtracting it from $G_{SC}(0)$,

$$G_{SC}(T) = G_{SC}(0) - \int_0^T S_{SC}(T)\mathrm{d}T = -0.251 \gamma T_C^2 - \frac{\gamma T_C^2}{\mathrm{Ei}(1.76)} \int_0^{T/T_C} \mathrm{d}u \int_0^u \frac{e^{-1.76/x}}{x}\mathrm{d}x. \tag{26.20}$$

Since the free energy of the electron gas in the normal state is $-\gamma T^2/2$, the ΔG is

$$\Delta G_{SC}(T) = G_N(T) - G_{SC}(T) = -0.5 \gamma T^2 + 0.251 \gamma T_C^2 + \frac{\gamma T_C^2}{\mathrm{Ei}(1.76)} \int_0^{T/T_C} \mathrm{d}u \int_0^u \frac{e^{-1.76/x}}{x}\mathrm{d}x. \tag{26.21}$$

The ratio of the critical magnetic field required to destroy the superconductive state at temperature T to the critical field at $T=0$ is obtained from

$$\frac{H_C(T)}{H_C(0)} = \sqrt{\frac{\Delta G_{SC}(T)}{\Delta G_{SC}(0)}}. \tag{26.22}$$

The results are plotted in Figure 26.4 along with the Tuyn approximation, $1-T^2/T_C^2$, and two asymptotic solutions suggested by the BCS theory.

Other predictions can be obtained from this thermodynamically consistent model. We have already shown that $C_{SC}(T_C) = 2.511 \gamma T_C$ (Equation 26.15) and that $\gamma T_C^2 = 1.99\ \mu_0 H_{C0}^2$ (Equation 26.19). It therefore follows that

$$C(T_C) = 5.00 \mu_0 H_{C0}^2/T_C \tag{26.23}$$

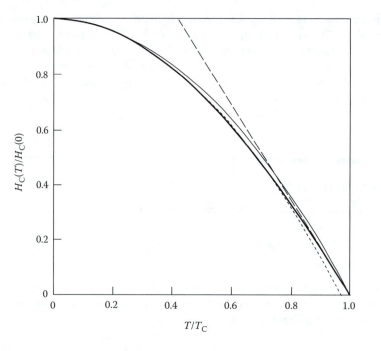

FIGURE 26.4

The ratio of the critical magnetic field at temperature T to the critical magnetic field at $T=0$ as a function of the ratio of T/T_C. The heavy line is the model prediction, the thin solid line slightly above it is the Tuyn approximation. The dashed line that falls below as T approaches T_C is the low temperature BCS solution ($1-1.06\ T^2/T_C^2$), and the straight dashed line above the curve is the BCS solution near T_C ($1.74\ (1-T/T_C)$).

or

$$\Delta C(T_C) = 3.00\mu_0 H_{C0}^2/T_C. \qquad (26.24)$$

We now compare our model with data from various type-1 superconductors in Table 26.3.

TABLE 26.3

Critical Parameters for Elemental Superconductors

Element	H_{C0} (Oe)	T_C (°K)	γ (mJ/mol K^2)	$\mu_0 H_{C0}^2$ (mJ/mol)	$\gamma T_C^2/\mu_0\, H_{C0}^2$	$\Delta C_{SC}(T_C)$ (mJ/mol K)	$\Delta C_{SC}(T_C)/\gamma T_C$
Al	105	1.175	1.350	0.87	2.13	2.23	1.41
Ga	59.2	1.083	0.596	0.33	2.12	0.91	1.41
In	282	3.408	1.690	9.93	1.98	8.74	1.52
Zn	54	0.850	0.640	0.33	1.42	1.15	2.11
Cd	28	0.517	0.688	0.08	2.27	0.47	1.32
Hg	411	4.154	1.790	19.89	1.55	14.37	1.93
Sn	305	3.720	1.780	12.25	2.01	9.88	1.49
Pb	803	7.200	2.980	93.63	1.65	39.01	1.82
Ti	56	0.400	3.350	0.26	2.02	1.99	1.48
Nd	2060	9.250	7.790	365.94	1.82	118.68	1.65
Ta	829	4.470	5.900	59.57	1.98	39.98	1.52
					1.91		1.61

Source: Data taken from Gersten, J.I. and Smith, F.W., *The Physics and Chemistry of Materials*, John Wiley & Sons, New York, 2001.

The magnetic data was given in Oersteds, which are converted to ampere per meter by multiplying by 79.6. The ratio $\gamma T_C^2 / \mu_0 H_{C0}^2$ ranges from 1.55 to 2.13 with an average value of 1.91, which is very close to the 1.99 predicted by Equation 26.19. Lacking data for $\Delta C_{SC}(T_C)$, this value was calculated from Equation 26.24 and then used to calculate $\Delta C_{SC}(T_C)/\gamma T_C$. We see this ratio ranges from 1.41 to 2.11 with an average value of 1.61. This is reasonably close to the model prediction of 1.51, which is closer to the experimental value than the BCS prediction of 1.43 (Equation 26.6).

26.5 London Equations

The equation of motion for an electron can be written

$$\dot{\nu} + \frac{\nu}{\tau} = -\frac{eE}{m},$$

(26.25)

but for a superconductor, we eliminate the damping term by setting $\tau = \infty$ which is equivalent to setting ρ to zero. The current density $j_s = -en_s\nu$ where n_s is the superconducting carrier density. Therefore we can write

$$\frac{d}{dt}\mathbf{J} = -en_s\dot{\mathbf{v}} = \frac{n_s e^2 \mathbf{E}}{m},$$

(26.26)

which is known as the first London equation.

Now put the Maxwell equation $\nabla \times \mathbf{E} = -\dot{\mathbf{B}}$ into Equation 26.26

$$\nabla \times \left(\frac{m}{n_s e^2} \frac{d\mathbf{J}}{dt} \right) = -\frac{d\mathbf{B}}{dt},$$

(26.27)

which can also be written as

$$\frac{d}{dt}\left[\nabla \times \mathbf{J}_s + \frac{n_s e^2}{m} \mathbf{B} \right] = 0$$

(26.28)

Equation 26.28 implies that either the term in the brackets is a constant or 0. For a superconductor above the transition temperature, $\mathbf{B}$ can penetrate the material even though $j_s = 0$ because $n_s = 0$. As the material is cooled below T_C, $j_s \neq 0$ and $n_s \neq 0$; therefore, $\mathbf{B} = 0$, which is required by the Meissner effect. We therefore assume the term is zero because it is consistent with the observed behavior of superconductors and the persistent supercurrents are given by

$$\nabla \times \mathbf{J}_s + \frac{n_s e^2}{m} \mathbf{B} = 0,$$

(26.29)

which is the second London equation.

26.5.1 Penetration Depth

We can use the London equations to describe the behavior of a magnetic field inside a superconductor. Taking the curl of the Maxwell equation $\nabla \times \mathbf{B} = \mu_0 j_s$ in which the $\dot{\mathbf{D}}$ term has been dropped since electric fields cannot exist in a superconductor,

$$\nabla \times \nabla \times \mathbf{B} = \mu_0 \nabla \times \mathbf{J}_s \tag{26.30}$$

and substituting the second London equation,

$$\nabla \times \nabla \times \mathbf{B} = \nabla(\nabla \cdot \mathbf{B}) - \nabla^2 \mathbf{B} = -\frac{\mu_0 n_s e^2}{m} \mathbf{B}. \tag{26.31}$$

Assume $\mathbf{B}$ is along the x-direction, parallel to the surface of the superconductor and z is perpendicular to the surface directed inward. Since $\nabla \cdot \mathbf{B} = 0$, the Equation 26.31 can be written

$$\frac{d^2 B_x}{dz^2} - \frac{\mu_0 n_s e^2}{m} B_x = 0, \tag{26.32}$$

which has the solution $B_x(z) = B_x(0) e^{-z/\lambda_L}$, where λ_L is the London penetration depth given by

$$\lambda_L = \left(\frac{m}{\mu_0 n_s e^2} \right)^{1/2}. \tag{26.33}$$

Typical values for λ_L are 200–300 Å.

26.6 Coherence Length

The coherence length is an important parameter that characterizes the behavior of super-conductors. This is the length scale over which the paired carriers interact, or more specifically the size of the Cooper pair. One can make a rough estimate of this length from the uncertainty principle. If the binding energy of the Cooper pair is ΔE, we know the energy to this precision. Since

$$\delta E = \delta \left(\frac{p^2}{2m} \right) = \delta \left(\frac{\hbar^2 k^2}{2m} \right) = \frac{\hbar^2 k_F \delta k}{m}, \tag{26.34}$$

then

$$\delta k = \frac{m \Delta E}{\hbar^2 k_F}. \tag{26.35}$$

The uncertainty principle states that $\Delta x \Delta p = h = \Delta x \hbar \Delta k$. Therefore

$$\Delta x = \frac{\hbar/h}{\Delta k} = \frac{2\pi \hbar^2 k_F}{m \Delta E} = \frac{2\pi \hbar v_F}{\Delta E} = \frac{\pi \hbar v_F}{\Delta_0}, \tag{26.36}$$

since $\Delta E = 2\Delta_0$. Pippard obtained a more precise result from BCS theory which is called the Pippard coherence length given by

$$\xi_0 = \frac{2\hbar v_F}{\pi \Delta_0} \sim 1 \ \mu\text{m} \tag{26.37}$$

for pure materials. The estimate from the uncertainty principle is remarkably close to Pippard's result.

26.7 Type-I and Type-II Superconductors

As stated above, increasing the applied H lowers the T_C. Furthermore, there is a critical field H_C that destroys the superconducting state. For pure systems (which are usually type-I superconductors), this critical field decreases with temperature according to Tuyn's approximation

$$\frac{H_C(T)}{H_C(0)} \approx 1 - \left(\frac{T}{T_C}\right)^2. \tag{26.38}$$

In other words, it takes very little field to destroy the superconducting state as T approaches T_C. This is one of the reasons that superconducting systems must be operated well below their transition temperature.

For alloy-type or impure systems (which are usually type-II superconductors), there are two critical fields, H_{C1} and H_{C2}. Like type-I superconductors, type-II superconductors exhibit perfect diamagnetism up to H_{C1}. The field starts to penetrate above H_{C1}, but superconductivity remains until H_{C2} is reached. Between H_{C1} and H_{C2}, the system is in a mixed state. Lines of magnetic flux start to penetrate in regions that have become normal while the rest of the material remains superconducting. The bulk resistivity is still zero because the current is carried by the superconducting regions.

For example, H_C for pure Pb is 600 Gauss at 4 K. For Pb2W%In, $H_{C1} = 400$ Gauss and H_{C2} 3600 Gauss. The area under the M versus H curve is the same for both systems as illustrated in Figure 26.5, but a type-II system can be operated at much higher fields. Therefore, all practical superconductors will be type-II.

The factor that determines whether a material will by a type-I or type-II superconductor is the ratio of the London penetration depth to the coherence length. If $\lambda/\xi < 0.71$, the material will be type-I; if $\lambda/\xi > 0.71$, it will be a type-II. (The reason for this will be discussed later.) Both penetration depth and coherence length are influenced by the addition of impurities which disrupt the regularity of the lattice. The penetration depth increases according to

$$\lambda = \lambda_L \left(\frac{\xi_0}{\ell} + 1\right)^{1/2} \tag{26.39}$$

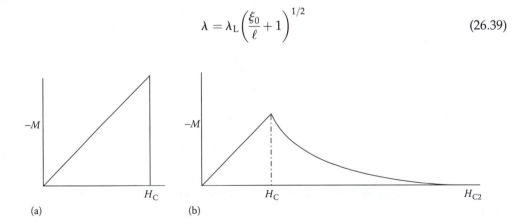

(a)   (b)

FIGURE 26.5
Schematic of the behavior of a type I superconductor (a) and a type-II superconductor (b). The type I exhibits perfect diamagnetic behavior ($-M = H$) until H_{C1} where the M goes abruptly to zero. A type II semiconductor is perfectly diamagnetic until H_{C1} at which point the magnetic field starts to penetrate. However the material remains superconductive until the second critical field H_{C2} is reached. The area under the two curves is the same.

and the coherence length decreases according to

$$\frac{1}{\xi} \approx \frac{1}{\xi_0} + \frac{1}{\ell}, \tag{26.40}$$

where ℓ is the mean free path between impurity atoms. Thus it is seen that pure materials tend to be type-I superconductors, but in alloys, $\ell \ll \xi_0$; hence they tend to be type-II superconductors.

26.8 Flux Quantization

Consider a superconducting ring that encloses a magnetic flux density **B**. The magnetic flux that is enclosed is $\Phi = \iint \mathbf{B} \cdot \mathbf{ds} = \iint \nabla \times \mathbf{A} \cdot \mathbf{ds} = \oint \mathbf{A} \cdot \mathbf{dL}$, where **A** is the vector potential. The supercurrent density from Equation 26.1 is $\mathbf{J} = -en_s\mathbf{v} = -(en_s/m)\langle \mathbf{p} \rangle$. However, in the presence of a magnetic field, the momentum operator is $\mathbf{p} = -i\hbar\nabla - e\mathbf{A}$. Therefore, to get the momentum, we use

$$\mathbf{J} = \frac{e}{m} \int \psi^*(-i\hbar\nabla - 2e\mathbf{A})\psi d\tau, \tag{26.41}$$

where $\psi = n_P^{1/2}e^{i\phi}$ and n_P is the number of Cooper pairs and ϕ is a phase factor. The factor 2 was added to the vector potential **A** because $2n_P = n_S$. Integrating Equation 26.41 to find **J** yields

$$\mathbf{J} = \frac{n_P e \hbar}{m} \nabla\phi - \frac{2n_P e^2}{m} \mathbf{A}. \tag{26.42}$$

(Note: If we take the curl of **J** in Equation 26.19, since $\nabla \times \nabla\phi = 0$, we get $\nabla \times \mathbf{J} = -(2n_P e^2/m)\nabla \times \mathbf{A} = -(n_S e^2/m)\mathbf{B}$, which is the second London equation (Equation 26.29).)

We integrate inside of a superconducting ring containing magnetic flux Φ at a sufficient depth where $J = 0$,

$$\oint \mathbf{J} \cdot \mathbf{dL} = \frac{n_P e}{m}\left[\oint \hbar \frac{\partial\phi(L)}{\partial L} - 2e\oint \mathbf{A} \cdot \mathbf{dL}\right] = 0. \tag{26.43}$$

For the wavefunction ψ to be single valued when integrated around a closed path, the phase change $\oint \frac{\partial\phi}{\partial L} dL = \Delta L = 2\pi n$ where n is an integer. Since $\oint \mathbf{A} \cdot \mathbf{dL} = \Phi$, we see that the flux inside the ring is quantized in units of

$$\Phi_0 = \frac{h}{2e} = 2.06 \times 10^{-15} Tm^2. \tag{26.44}$$

26.9 Critical Currents

26.9.1 Critical Current Related to the Binding Energy of a Cooper Pair

For a single electron, the change in energy with respect to k is $\delta E = \hbar^2 k_F \delta k/m$, so if the energy of the Cooper pair exceeds the binding energy 2Δ, the pair will break up and the supercurrent will collapse. Therefore the critical $\mathbf{k}_{c/m}$ can be found by equating

$$2\delta E = 2\frac{\hbar^2 k_F \delta k_{c/m}}{m} = 2\Delta. \tag{26.45}$$

So the critical current density at which the Cooper pair will break up is found from

$$J_C = \frac{n_s e\hbar}{m}\delta k_{c/m} = \frac{n_s e\hbar}{m}\frac{m\Delta}{\hbar^2 k_F} = \frac{n_s e\Delta}{\hbar k_F}, \tag{26.46}$$

where n_s is the number of superconducting electrons in the Cooper pairs. Notice that Δ was used here instead of Δ_0, the binding energy per electron at $T = 0$ K. The quantity Δ diminishes with temperature and with the magnetic field as will be shown, and therefore, so does J_C.

26.9.2 Critical Current Related to Magnetic Field

The critical current density in a superconductor can be found from $\nabla \times \mathbf{H} = \mathbf{J}$, the differential form of Ampere's law (or the Maxwell equation in the absence of E-fields). Integrating this quantity over a surface and applying Stokes theorem,

$$\iint \mathbf{J} \cdot d\mathbf{s} = \iint \nabla \times \mathbf{H} \cdot d\mathbf{s} = \oint \mathbf{H} \cdot d\ell. \tag{26.47}$$

Assume a wire with radius r. The field at the surface of the wire is

$$\oint \mathbf{H} \cdot d\ell = 2\pi r H. \tag{26.48}$$

Since the field only penetrates to depth λ, the current density will only exist to depth λ. Therefore, if $\lambda \ll r$, $\iint \mathbf{J} \cdot d\mathbf{s} = 2\pi r \lambda J_C$. Therefore from Equation 26.33, the critical field is given by

$$J_C = \frac{H_C}{\lambda} = H_C \left(\frac{\mu_0 n_s e^2}{m}\right)^{1/2} \quad \text{for } \lambda \ll r. \tag{26.49}$$

We see that the limiting current is not affected by the wire size provided $r \gg \lambda$. If the wire with radius $r < \lambda$ is carrying current J, the critical field at the surface will be reached when $J_C = 2H_C/r$. In this case, the critical current density increases inversely with r, but since the current $I = \pi r^2$, the critical current will be directly proportional to r.

The temperature dependence of the critical field, H_C, was given by Equation 26.38 as $H_C(T)/H_C(0) \approx 1 - (T/T_C)^2$. Since both Δ and J_C are directly proportional to H_C, they follow the same temperature dependence.

26.9.3 Critical Fields in Type-II Superconductors

For type-II superconductors, the onset of the mixed state occurs when the applied H field can form a flux vortex which contains a quantum of flux in the material. Such a flux vortex is sometimes called a fluxoid or a fluxon. This requires

$$2\pi \int B_0 e^{-r/\lambda} r \, dr = \Phi_0. \tag{26.50}$$

Integrating, we get $2\pi B_0 \lambda^2 = \Phi_0$, or

$$H_{C1} = \frac{B_0}{\mu_0} = \frac{\Phi_0}{2\pi\mu_0\lambda^2} = \frac{\hbar}{4\pi e \mu_0 \lambda^2}. \tag{26.51}$$

As the applied flux density increases, more flux vortices are created and their diameter shrinks as they crowd together. Eventually their diameter shrinks to approximately the coherence length and the supercurrents can no longer repel the field. The H_{C2} occurs when the flux vortices fill up all of the available space as illustrated in Figure 26.6, which occurs when the flux density equals $\Phi_0/\pi\xi^2$,

$$H_{C2} = \frac{\Phi_0}{\pi\mu_0\xi^2} = \frac{h}{2\pi e \mu_0 \xi^2}. \tag{26.52}$$

Given these results, it is easy to see why the transition from type-I to type-II takes place when $\lambda/\xi = 0.71$. In a type-I superconductor, the $H_{C1} = H_{C2}$ and this occurs when $\lambda^2 = \xi^2/2$.

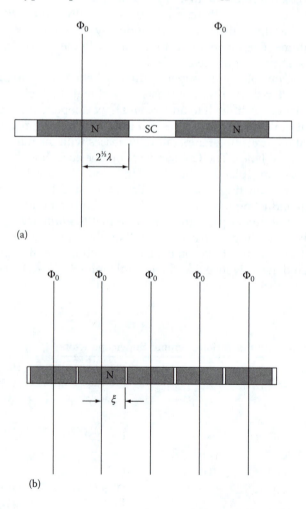

FIGURE 26.6

Fluxoids at H just above H_{C1} (a) and just below H_{C2} (b). When the flux density reaches the point where each quanta occupies a circle with radius $= \xi$, there is no superconducting regions left and the material becomes normal.

26.10 High Temperature Superconductors

The advent of the high temperature superconductors (HTSCs) came in 1986 from the work of Bednortz and Muller who showed that a ceramic, La_2CuO_4, became superconductive with a transition temperature $T_C = 38$ K, substantially higher than the best of the A-15 class of alloy superconductors. This result set off a flurry of activity to discover other ceramic compounds with higher transition temperature. A major breakthrough came when Ashburn and Wu, along with Chu synthesized the $YBa_2Cu_3O_{7-}$ and found a transition temperature of 92 K, well above the 77 K boiling point of LN_2. The existence of superconductors that can operate at LN_2 opens many new applications for superconductivity by eliminating the need for LHe which is expensive, difficult to store, and represents a nonrenewable resource. (He atoms find their way to the exosphere where they can escape earth's gravity.) The discovery of a ceramic superconductor further stimulated research to discover new ceramic materials with even higher transition temperatures. To date there are more than 40 such systems, known generically as high T_C materials. Presently, $HgBa_2Ca_2Cu_3O_8$ holds the record with a $T_C = 135$ K.

The structure of the ceramic superconductors is basically a sandwich in which CuO_2 planes, which carry the supercurrent, are surrounded by perovskite-like structures containing the rest of the atoms. This layered structure along with the compositions of $YBa_2Cu_3O_7$ and related HTSC systems is indicated in Table 26.4.

The mechanism responsible for the superconducting behavior of these materials remains an unsolved problem for theoretical physicists. Given that the high T_C materials have a positive Hall coefficient, it is difficult to understand how Cooper pairs could be responsible for the supercurrent, unless somehow holes can form a bound Cooper pair.

It can be seen that the transition temperature increases with the number of CuO_2 planes in which the carriers are holes. The CuO and BaO planes are thought to donate holes to the CuO_2 superconducting planes. The Y^{3+}, Hg^{2+}, and Ca^{2+} are insulating and donate electrons to the superconducting planes. The BaO, SrO, and LaO layers serve to isolate the groups of superconducting planes from each other. Since a number of ceramic systems become superconducting under pressure, it may be that the primary function of the other nonconducting layers is simply structural in that they provide the necessary lattice parameter for the CuO_2 planes to become superconducting. Another piece of the puzzle is the role of oxygen defects. Notice that the formula for $YBa_2Cu_3O_{7-}$ has a minus behind

TABLE 26.4

Layered Structure of the Various Ceramic Superconductors

			$Bi^{3+}O^{2-}$	Hg^{2+}
	$2O^{2-}La^{3+}$	$Cu^{2+}O^2$	$O^{2-}Sr^{2+}$	$O^{2-}Ba^{2+}$
	$La^{3+}O^{2+}$	$O^{2-}Ba^{2+}$	$Cu^{2+}2O^{2-}$	$Cu^{2+}2O^{2-}$
$Ti^{4+}2O^-$	$O^{2-}La^{3+}$	$Cu^{2+}2O^2$	Ca^{2+}	Ca^{2+}
$Ba^{2+}O^{2-}$	$Cu^{2+}2O^2$	Y^{3+}	$Cu^{2+}2O^{2-}$	$Cu^{2+}2O^{2-}$
$Ti^{4+}2O^-$	$O^{2-}La^{3+}$	$Cu^{2+}2O^2$	Ca^{2+}	Ca^{2+}
	$La^{3+}O^{2+}$	$O^{2-}Ba^{2+}$	$Cu^{2+}2O^{2-}$	$Cu^{2+}2O^{2-}$
	$2O^{2-}La^{3+}$	$Cu^{2+}O^2$	$O^{2-}Sr^{2+}$	$O^{2-}Ba^{2+}$
			$Bi^{3+}O^{2-}$	Hg^{2+}
$BaTiO_3$	La_2CuO_4	$YBa_2Cu_3O_{7-}$	$Bi_2Sr_2Ca_2Cu_3O_{10}$	$HgBa_2Ca_2Cu_3O_8$
Insulating	$T_C = 38$ K	$T_C = 92$ K	$T_C = 110$ K	$T_C = 135$ K

the O_7 meaning that on the average there are fewer than seven oxygen atoms in the unit cell. The properties of the material can vary from nonconducting, to semiconducting, and to superconducting, depending on the number of oxygen defects.

A major problem with the high T_C materials is the fact that they are superconducting only in the **a–b** planes. The initial preparations of $YBa_2Cu_3O_{7-}$ had very low critical currents because of the weak coupling between the CuO_2 planes. Techniques have since then been developed to texture the material so that the **c** axes are aligned, which dramatically improved the critical current density. Flux creep caused by the interaction of the magnetic field with the fluxons (tubes of magnetic flux that penetrate type-II superconductors) was also a problem in these materials. One solution to the flux creep problem has been to precipitate out a nonconductive phase such as the 211 phase in the case of $YBa_2Cu_3O_{7-}$. The nonconducting 211 phase acts as pinning sites to keep the fluxons from moving in the presence of a high magnetic field.

Permanent magnets using high T_C materials with flux densities of several tesla are currently in use serving as superconducting bearings and other functions such a maglev trains. However, high T_C materials have found little use in transmission lines and in large bore superconducting magnets such as those used in MRI systems and similar applications because of the difficulties in forming wires from these very hard and brittle ceramic materials.

26.11 Recent Advances in Superconductivity

Although research in the CuO_2 ceramic high-T_C superconductors seems to have stagnated in recent years, an exciting advance has occurred in the intermetallic systems previously dominated by the A-15 compounds. In 2001 Jan Akimitsu (University of Tokyo) announced at an American Physical Society meeting the discovery of a magnesium diboride (MgB_2) intermetallic superconductor with a T_C of approximately 40 K. Though not as exciting as the high T_C materials that can operate at LN_2 temperatures, a T_C of 40 K will allow operation with LH_2 (20 K) or possibly even with LNe, either of which is far less expensive than LHe. More good news was found when it was discovered that by doping the material with SiC, the upper critical field H_{C2} can be raised almost to that of Nb_3Sn. Even better news is that the material can be fabricated in long wires by diffusing Mg vapor into long boron filament under high pressure.

MgB_2 does exhibit the isotope effect indicating that it is most likely a normal BCS-type, although its 40 K transition temperature is about $10°$ higher that the highest estimated theoretical transition temperature for BCS-coupling.

26.12 Applications

By far, the largest use of superconductors remains in superconducting magnets using the Nb46.5W%Ti alloy although Nd_3Sn is finding its way into these applications because of its higher critical currents at LHe temperatures. Application for the high temperature superconducting ceramic systems have been slow to emerge because of their brittleness and difficulties in forming them into cables with high current carrying capabilities. There are, however, some rather exotic applications of superconductors based on their tunneling properties, which are described in the following sections.

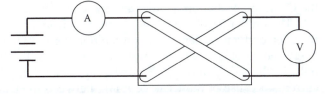

FIGURE 26.7
JJ formed by depositing a thin stripe of metal on an insulator, coating this stripe with a thin oxide insulating layer, and then adding a second metallic stripe.

26.12.1 Josephson Effect

If two strips of metal are separated by a very thin insulator as illustrated in Figure 26.7, the electrons can tunnel through the insulator and the behavior is ohmic, i.e., the tunneling current is directly proportional to the applied voltage. If one of the metal stripes is cooled below its superconducting transition temperature, an energy gap equal to $2\Delta_0$ is created with a Fermi level the same as in the normal metal as seen in Figure 26.8. Now if a small voltage is applied to the normal metal stripe, no current will flow until the Fermi level is raised above the top of the energy gap in the superconducting metal. Measuring this voltage provides an accurate method for measuring the energy gap in superconductors.

26.12.2 DC Josephson Effect

If both of the metal stripes are made superconducting (an s-i-s configuration), a super-current will be present in both sides of the junction that can tunnel through the insulating barrier even with no voltage applied to the terminals. Passing a current from an external source through the junction will show no voltage drop until a critical current J_0 is reached. For a narrow and weak barrier, J_0 can be approximated by

$$J_0 = \frac{e\hbar}{ma}n_s, \tag{26.53}$$

where a is the barrier half-width.

Since the metal stripes are also conductors, a normal current can also be present once the potential is above $q\Delta_0$. When the current supplied from an external source exceeds J_0, the current switches from a supercurrent to a normal current and a voltage now appears across the other end as shown in Figure 26.9. Thus the Josephson junction (JJ) acts in the same manner as a transistor switch, except that the switching time is on the order of picoseconds—orders of magnitude faster than the fastest transistors. For this reason JJs are attractive for very high speed logic circuits.

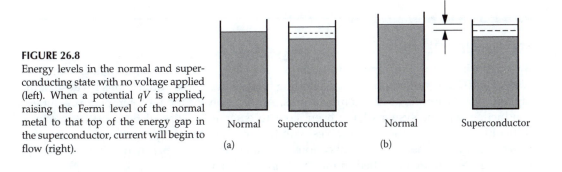

FIGURE 26.8
Energy levels in the normal and super-conducting state with no voltage applied (left). When a potential qV is applied, raising the Fermi level of the normal metal to that top of the energy gap in the superconductor, current will begin to flow (right).

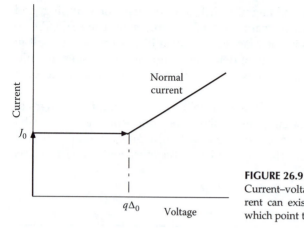

FIGURE 26.9
Current–voltage for the DC Josephson effect. The supercurrent can exist with no voltage drop until it reaches J_0, at which point the supercurrent switches to the normal current.

26.12.3 AC Josephson Effect

If the current through the left side of the junction is maintained at a value less than J_0 and a potential is applied across the junction from the right side, the supercurrent through the junction will oscillate with a frequency given by

$$\omega = \frac{2eV}{\hbar}. \tag{26.54}$$

For $V = 1$ μV and $\omega \sim 3 \times 10^9$ rad/s or $\sim$500 MHz. This equation says that a photon of energy $\hbar\omega = 2eV$ is emitted every time a Cooper pair crosses the barrier. Since voltage and frequency can be measured with a high degree of precision, this effect provides a very precise measurement of $e/\hbar$. Conversely, the NIST definition of a volt is now based on frequency measurements and Equation 26.54.

26.12.4 Superconducting Quantum Interference Detectors

Connecting two JJs in parallel forms a superconducting quantum interference detector (SQUID) as illustrated in Figure 26.10. SQUIDS are primarily used as magnetometers. The current in the loop oscillates as magnetic flux enters the loop, reaching a maximum each time the flux passing through equals a multiple of a magnetic flux quantum. Just as in laser interferometry where it is possible to perform wavefront analysis on fractional portions of interference fringes, it is possible for SQUIDs to measure the magnetic flux to a fraction of a magnetic flux quantum. The sensitivity of such a device is $O(10^{-14})$ T. This sensitivity is enough to measure the extremely small magnetic effects that occur in living organisms.

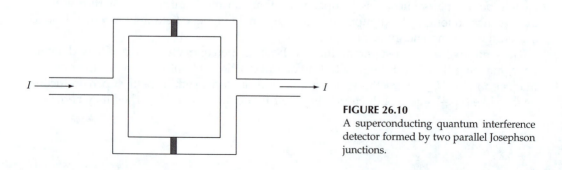

FIGURE 26.10
A superconducting quantum interference detector formed by two parallel Josephson junctions.

They are also used in magnetic resonance imaging (MRI), for geomagnetic mapping, and for mineral prospecting. John Lipa and his team at Stanford University used SQUIDs to measure temperature-induced changes in the magnetic moment of a paramagnetic salt in order to build a thermometer that was capable of detecting temperature changes of 1 nK. (This device was the heart of the Lambda point experiment that was carried out on a space shuttle mission in 1992.)

SQUIDs are also being used in the Gravity Probe-B satellite to detect minute changes of less than 0.5 milliarcseconds in the orientation of the superconducting gyroscope by measuring its London moment. The London moment is the magnetic moment that is generated by a rotating superconductor. The Gravity Probe-B experiment is a critical test of Einstein's general theory of relativity.

26.13 Summary

Superconductors are materials that conduct electricity with no resistance at temperatures below their critical temperatures. A number of metallic elements, alloys, intermetallic compounds, and ceramics exhibit this property, although few have transition temperatures high enough for practical applications. The maximum current that can be carried by a superconductor is limited by the magnetic field associated with the current that tends to destroy the superconducting state. The critical magnetic field and thus the critical current increase as the temperature is lowered, thus it is necessary to operate superconductors well below their transition temperatures in most applications. Until the discovery of high temperature ceramic superconductors in 1986, it was necessary to cool superconductors with LHe, which is expensive, hard to keep, and nonrenewable. For this reason much effort has gone into the search for materials with higher transition temperatures so they may be cooled with less expensive fluids such as LN_2.

No magnetic fields can exist in a superconductor. Any field that may have existed in the normal state will be ejected when the material is cooled into the superconducting state by supercurrents induced at the surface of the superconductor. The ejection of any magnetic field is known as the Meissner effect. The supercurrents at the surface of the superconductor are persistent and will flow indefinitely. The superconducting state can be likened to a macroscopic quantum state which is separated from the normal state by a small energy gap 2Δ that is Δ below the Fermi level. Just as electrons in a atom may circulate indefinitely in their ground state until they are kicked into a higher quantum state by absorbing energy, supercurrents may flow in a superconductor until they are kicked into the normal state by absorbing energy.

The theory of superconductivity was developed by Bardeen, Cooper, and Schrieffer in 1957 and is known as the BCS theory. Electrons interacting with the lattice ions form loosely bound pairs known as Cooper pairs that can move through the lattice without interference so long as the temperature and external magnetic field are low enough so as exceed the binding energy of the pairs.

There are two types of superconductors. Type-I superconductors are perfectly diamagnetic, meaning they reject the penetration of a magnetic field until they reach their critical magnetic field, H_C, where the superconductivity is destroyed. Type-II superconductors behave like type-I superconductors up to their first critical field, H_{C1}. Then they become a

mixed state in which there are holes of normal material that have been penetrated by magnetic flux lines called fluxoids or flux vortices within the superconducting material. As the applied magnetic field increases, the number of these flux vortices increases until eventually at H_{C2} there is no superconducting material left. Since type-II superconductors can operate at much higher fields than type-I superconductors, they are used in most practical applications.

Whether a material is type-I or type-II is determined by the ratio of the penetration depth λ (the distance into the surface a magnetic field can penetrate) to the coherence length ξ (essentially the diameter of the Cooper pair). Type-II behavior occurs when $\lambda > \sqrt{2}\xi$. Most pure materials are type-I, but adding impurities such as alloying with another metal increases λ and decreases ξ, so most practical superconductors are alloys or intermetallic compounds.

The major application of superconductors is their use in superconducting magnets used in medical imaging, nuclear magnetic resonance (NMR), particle accelerators, crystal growth, and other industrial processes. Despite their great promise, the ceramic HTSCs have been slow to find commercial applications, although they are finding limited use in bulk permanent magnets for frictionless rotating bearings used for flywheel energy storage systems and for maglev systems. Progress is being made in fabricating superconducting tapes and cables for power transfer and demonstration projects are underway for their use in the power grid. However, because of the high cost of fabricating these power cables, their wide-spread use is presently not commercially viable. Other, more exotic applications of superconductivity, lie in the ability to detect a single quantum of magnetic flux using superconducting quantum interference detectors (SQUIDs). Such highly sensitive devices open new possibilities in biomedicine, magnetic imaging, nondestructive testing, prospecting, and surveillance.

Bibliography

Ashcroft, N.W. and Mermin, N.D., *Solid State Physics*, Brooks Cole, Philadelphia, PA, 1976.

Bardeen, J., Cooper, L.N., and Schrieffer, J.R., Theory of superconductivity, *Phy. Rev.*, 108, 1175–1204, 1957.

Bednorz, J.G. and Müller, K.A., Possible high Tc superconductivity in the Ba–La–Cu–O system. *Z. Physik.*, B64, 189–193, 1986.

Christman, J.R., *Fundamentals of Solid State Physics*, John Wiley & Sons, New York, 1988.

Fujimoto, H., Developing a high-temperature superconducting bulk magnet for the Maglev Train of the future, *JOM*, 50/10, 16–18, 1998.

Gersten, J.I. and Smith, F.W., *The Physics and Chemistry of Materials*, John Wiley & Sons, New York, 2001.

Ibach, H. and Lüth, H., *Solid State Physics*, 3rd edn., Springer-Verlag, New York, 1990.

Kittel, C., *Introduction to Solid State Physics*, 7th edn., John Wiley & Sons, New York, 1966.

Nagamatsu, J., et al., Superconductivity at 39 K in MgB_2, *Nature*, 410, 63, 2001.

Fujimoto, H., Owens, F.J., and Poole, C.P. Jr., *Progress in Superconductivity, The New Superconductors*, Springer US, 2002.

Post, R.F., New look at an old idea, the electromechanical battery, *Sci. Technol. Rev.*, April 1996.

Tanaka, S., High-temperature superconductivity: Outlook and history, *JSAP Int.*, 4, 17–22, 2001.

Wu, M.K., Ashburn, J.R., and Torng, C.J., Superconductivity at 93 K in a new mixed-phase Y-Ba-Cu-O compound system at ambient pressure. *Phys. Rev. Lett.* 58(9), 908–910, 1987.

Problems

1. Equation 26.45 gives the critical current in terms of the binding energy of the Cooper pair

$$J_C = \frac{n_s e \Delta}{\hbar k_F}.$$

Equation 26.49 gives critical current in terms of magnetic field at the surface of a cylindrical conductor

$$J_c = H_c \left(\frac{\mu_0 n_s e^2}{m} \right)^{1/2}.$$

Show that these are the same.

2. I have a He-cooled superconducting solenoid magnet with a 30 cm diameter core and 1 m in length that produces a B of 1 T. Suddenly, I lose my He supply and the magnet goes normal. How much energy is released?

Index